D0065111

Noetherian Semigroup Algebras

Algebra and Applications

Volume 7

Algebra and Applications aims to publish well written and carefully refereed monographs with up-to-date information about progress in all fields of algebra, its classical impact on commutative and noncommutative algebraic and differential geometry, K-theory and algebraic topology, as well as applications in related domains, such as number theory, homotopy and (co)homology theory, physics and discrete mathematics.
Particular emphasis will be put on state-of-the-art topics such as rings of differential operators, Lie algebras and super-algebras, group rings and algebras, C*algebras, Kac-Moody theory, arithmetic algebraic geometry, Hopf algebras and quantum groups, as well as their applications. In addition, Algebra and Applications will also publish monographs dedicated to computational aspects of these topics as well as algebraic and geometric methods in computer science.

Noetherian Semigroup Algebras

by

Eric Jespers
Vrije Universiteit Brussel, Brussels, Belgium

and

Jan Okniński
Warsaw University, Warsaw, Poland

A C.I.P. Catalogue record for this book is available from the Library of Congress.

ISBN-10 1-4020-5809-8 (HB)
ISBN-13 978-1-4020-5809-7 (HB)
ISBN-10 1-4020-5810-1 (e-book)
ISBN-13 978-1-4020-5810-3 (e-book)

Published by Springer,
P.O. Box 17, 3300 AA Dordrecht, The Netherlands.

www.springer.com

Printed on acid-free paper

To Griet *Eric Jespers*

To Iwona *Jan Okniński*

Contents

Chapter 1

Introduction

The first aim of this work is to present the main results and methods of the theory of Noetherian semigroup algebras. These general results are then applied and illustrated in the context of certain interesting and important concrete classes of algebras that arise in a variety of areas and have been recently intensively studied.

One of the main motivations for this project has been the growing interest in the class of semigroup algebras (and their deformations) and in the application of semigroup theoretical methods. Two factors seem to be the cause for this. First, this field covers several important classes of algebras that recently arise in a variety of areas. Furthermore, it provides methods to construct a variety of examples and tools to control their structure and properties, that should be of interest to a broad audience in algebra and its applications. Namely, this is a rich resource of constructions not only for the noncommutative ring theorists (and not only restricted to Noetherian rings) but also to researchers in semigroup theory and certain aspects of group theory. Moreover, because of the role of new classes of Noetherian algebras in the algebraic approach in noncommutative geometry, algebras of low dimension (in terms of the homological or the Gelfand-Kirillov dimension) recently gained a lot of attention. Via the applications to the Yang-Baxter equation, the interest also widens into other fields, most notably into mathematical physics. A second factor for the growing interest in semigroup algebras is the natural relation to the extensively studied class of group algebras. However, one must realize that the scope of research on semigroup algebras must be in some sense more specialized. Namely, one has to study the "correct" classes of semigroups in order to get deep and applicable results. Our feeling is that this amounts to saying that again one should focus on important motivating constructions and examples that show up in other areas of algebra and its potential applications. It is also worth mentioning that, within the last decade, and at a rapidly increasing scale, semigroup theoretical methods have occurred naturally in many aspects of ring theory, algebraic combinatorics and representation theory. However, the knowledge of these methods is not so common among algebraists.

The selection of the material is based on the principle that only those topics are dealt with for which a comprehensive general theory has been developed and that lead to nontrivial convincing results and applications. From the semigroup theoretic point of view, it turned out that the common feature for most of these topics is the required focus on a variety of methods based on cancellative semigroups and linear semigroups. The techniques built seem to be of independent interest, they form a core for the presented results that also contributes to the integrity of selected material. Some "isolated" topics, that often require a lot of prerequisites that would not fit in the scope of the methods developed in this book, are not included. However, all such exceptions are mentioned in the comments following every chapter.

We denote by $K[S]$ the semigroup algebra of a given field K and a semigroup S. So $K[S]$ is a K-space with basis S and with multiplication extending, in a natural way, the operation on S. A typical nonzero element $\alpha \in K[S]$ is of the form $\alpha = \sum_{i=1}^{n} k_i s_i$ fore some nonzero $k_i \in K$, $s_i \in S$ and $n \geq 1$. Then the support of α is defined by $\operatorname{supp}(\alpha) = \{s_1, \ldots, s_n\}$. For a semigroup S with zero θ, a related object, the contracted semigroup algebra $K_0[S]$, is defined as the factor algebra $K[S]/K\theta$. So we identify the zero of S with the zero of the algebra. As, in this case, $K[S] \cong K_0[S] \oplus K$, the two algebras are very strongly related and sometimes it is more convenient to deal with $K_0[S]$. So every nonzero $\alpha \in K_0[S]$ can be considered as an element of the form $\alpha = \sum_{i=1}^{n} k_i s_i$ with $k_i \neq 0$ and $s_i \neq \theta$, and in this case we define $\operatorname{supp}(\alpha)$ as the set of all s_i.

The following result ([54, Corollary 5.4]) is one of the starting points for the study of Noetherian semigroup algebras, both from the abstract point of view and from the point of view of applications: the semigroup algebra $K[S]$ of an abelian monoid S is Noetherian if and only if S is finitely generated. It should be pointed out that the proof of the necessity is quite nontrivial. However, in the noncommutative case the situation is much more complicated and it is even hard to expect a complete characterization. Nevertheless, as in the group algebras case, one of the goals is to see whether the homogeneous information (that is, the information on S) reflects and determines the properties of $K[S]$.

A substantial part of the book is devoted to algebras $K[T]$ of submonoids T of a group (in most cases, a polycyclic-by-finite group). One of the obvious reasons to study such algebras $K[T]$ is that they provide a natural class of orders and of Noetherian algebras, related to the extensively studied, quite well understood, and rich in beautiful results class of group algebras of polycyclic-by-finite groups [129, 130]. In this context, one should also point out the important role of such groups and of their algebras in several aspects of algebra, topology and of their applications. Not only the class of algebras $K[T]$ for subsemigroups T of such groups seems of a general interest, but such algebras arise in a natural way in a number of concrete problems. For example, in the problem of determining set theoretical solutions of the Yang-Baxter equation, which originated from the theory of quantum groups, as a tool for tackling certain problems in physics [33]. Another source of motivation comes from powerful homological methods developed for the study of Sklyanin algebras, a class of low dimensional algebras playing an important role in the attempt to build foundations for noncommutative geometry [150]. Surprisingly, these two problems have a meeting point, culminating in a fascinating class of semigroup algebras, presented in Chapter 8. A wider class, providing also several interesting examples of low dimensional algebras, is then considered in Chapter 9. Because of the general tendency to investigate concrete classes of algebras, we describe in full detail several examples in Chapter 10. It is a real satisfaction to see that the theory developed in the preceding chapters allows us to show that several ingredients of the structural description of these combinatorially defined algebras can be effectively computed.

Looking from another perspective, the necessity for studying algebras $K[T]$ of cancellative semigroups T in the context of general Noetherian algebras is an obvious consequence of the structure theorem for linear semigroups [123], see Theorem 2.4.3. This results from the fact that a monoid S with $K[S]$ right Noetherian has important homomorphic images that are linear, that is, subsemigroups of the full multiplicative monoid $\mathrm{M}_n(D)$ of $n \times n$ matrices over a division algebra D. Unexpectedly, we are able to show that the structure of S itself shares, in many cases, the flavor of the structure of a linear semigroup.

As mentioned before, the main motivation comes from several important concrete classes of examples. Though, a general theory leads to surprising and very useful results of general character. For example, such are the main results of Chapter 4 and Chapter 5: Theorem 4.4.7 and its main consequences, Theorem 5.1.5, Theorem 5.1.6, as well as Theorem 5.2.3 and Theorem 5.3.7. All these results are heavily used in the next chapters of the book and they also illustrate the strategy of studying arbitrary Noetherian algebras $K[S]$ via certain related algebras $K[T]$ of cancellative semigroups T.

The material is organized as follows. Chapter 2 and Chapter 3 are included for the convenience of the reader. These chapters contain the main prerequisites of semigroup and ring theory used in the book. However, they are not intended as an overview of all relevant important classical results on semigroups and rings. We only state those results that are relevant for the main purpose of this monograph. In Chapter 4 we deal with algebras of submonoids of polycyclic-by-finite groups. Right Noetherian algebras of this type are completely characterized. It is shown that these algebras are finitely presented, they are Jacobson rings and information on their prime spectrum is obtained. Chapter 5 is devoted to the structure of arbitrary right Noetherian semigroup algebras. Certain necessary and sufficient conditions for $K[S]$ to be Noetherian are proved, in terms of certain ideal chains of special type in S. Relations with the PI (polynomial identity) property and general results on the Gelfand-Kirillov dimension are also proved. The next two chapters deal with two special classes of algebras defined on the ring level: principal ideal rings and Noetherian maximal orders. In Chapter 6 a full description of principal ideal rings of the form $K[S]$ is obtained. The structure of the underlying S is entirely determined, see Theorem 6.2.1, Theorem 6.3.6, Theorem 6.4.1 and Corollary 6.4.2. Noetherian PI domains $K[S]$ of cancellative monoids S that are maximal orders are described in Chapter 7. The main results show that, also in this case, the information can be expressed in terms of S only, Theorem 7.2.5 and Theorem 7.2.7. A class of intriguing finitely presented algebras defined by quadratic relations, called algebras of skew type, is considered in Chapter 9. It provides us with many examples of Noetherian PI algebras. See Theorem 9.3.7, Theorem 9.4.2, and also, Theorem 10.1.2 and Theorem 10.1.3 for the main general results on these algebras. Before this, in Chapter 8, we study the special subclass arising independently from the set theoretical solutions of the Yang-Baxter equation, homological methods of Sklyanin algebras and from some considerations concerning Gröbner bases techniques in quadratic algebras. The main accomplishments of the theory are stated in Theorem 8.1.4, Theorem 8.2.3, Theorem 8.3.10, Theorem 8.4.3, Theorem 8.5.1 and Theorem 8.5.6. These algebras are Noetherian PI domains, and also maximal orders of finite homological dimension. Their intriguing combinatorial properties lead to deep results of a structural nature. In particular, the underlying monoids can be characterized as certain submonoids of the semidirect product $\mathrm{Fa}_n \rtimes \mathrm{Sym}_n$ of the free abelian group Fa_n of rank n and the symmetric group Sym_n.

Several examples are studied in detail in Chapter 10. Each of the chapters, except for the chapters with prerequisites, contains bibliographic comments and a list of problems related to the main results presented.

Most of the presented material is due to the authors. Some results have been obtained in joint work with F.Cedo and T.Gateva-Ivanova. In Chapter 8, we also incorporate some of the latest developments on set theoretic solutions of the Yang-Baxter equation and on algebras of I-type, due to T.Gateva-Ivanova and M.Van den Bergh, P.Etingof, T.Schedler and A.Soloviev, T.Gateva-Ivanova, and W.Rump.

For the previous work on semigroup algebras we refer to the monographs [54] and [119]. Clearly, we strongly rely on results on group algebras [129] and, more generally, on graded rings [130]. Some of our results can be proved in the more general setting of certain quantized deformations of the considered algebras, such as for example twisted semigroup algebras. However, this is beyond the scope of the monograph. Some material related to the present work was presented in the survey paper [82].

This project was supported by the Flemish-Polish bilateral agreements BIL 01/31 and BIL2005/VUB/06, by the KBN research grant 2P03A 033 25 (Poland), Onderzoeksraad of Vrije Universiteit Brussel and by Fonds voor Wetenschappelijk Onderzoek (Belgium).

Chapter 2

Prerequisites on semigroup theory

This chapter is devoted to the necessary background on semigroups and especially on semigroups of matrices over a division algebra, referred to as linear semigroups. The material is a mixture of very classical results of semigroup theory, basic structural results of the theory of linear semigroups and certain rather special technical results that provide us with important tools used in the book. Unless explicit references are given, the results can be found in the monograph [123]. The reader should be warned however that the latter is mostly devoted to linear semigroups over a field. So, a few proofs are given in Section 2.5, as they cannot be found in the required form in the existing literature and also since they might put more light onto the topic for readers with little experience in semigroup theory.

2.1 Semigroups

In this section we recall some basic facts about semigroups and we introduce notation and terminology. Most of the material can be found in [27] and [66]. If not stated otherwise, multiplicative notation will be always used for the operation in a semigroup S.

If A is a subset of a semigroup S, then by $\prec A \succ$ we denote the subsemigroup of S generated by A. If S is a monoid, then $\langle A \rangle$ denotes the submonoid generated by A. If $A = \{a_1, \ldots, a_k\}$, then we also write $\prec a_1, \ldots, a_k \succ$ and $\langle a_1, \ldots, a_k \rangle$ for the corresponding subsemigroup or submonoid of S. In particular, $\prec a \succ$ is the cyclic semigroup generated by $a \in S$. Then a is called a periodic element if $\prec a \succ$ is finite. In this case a^n is an idempotent for some

$n \geq 1$ and $c \prec a \succ$ is a an ideal and a subgroup of $\prec a \succ$. If G is a group and $A \subseteq G$, then $\mathrm{gr}(A)$ stands for the subgroup of G generated by A.

If S is a semigroup that is not a monoid, then by S^1 we mean the monoid $S \cup \{1\}$ obtained by adjoining an identity element to S. Otherwise, we put $S^1 = S$. Moreover $\mathrm{U}(S)$ denotes the set of all invertible elements of S, called the unit group of S. If S has a zero element, we often write θ for the zero of S. By S^0 we mean the semigroup S with zero adjoined.

Let $e = e^2 \in S$. Then eSe is a monoid and hence $\mathrm{U}(eSe)$ is a subgroup of S with identity e. Subgroups of this type are exactly the maximal subgroups of S. For example, if $S = \mathrm{M}_n(K)$ the multiplicative semigroup of $n \times n$ matrices over a field K, then every $e\,\mathrm{M}_n(K)e$ is isomorphic to $\mathrm{M}_j(K)$, where j is the rank of e. Hence, the maximal subgroup of $\mathrm{M}_n(K)$ containing e is isomorphic to $\mathrm{GL}_j(K)$.

A nonempty subset I of S is a right (left) ideal of S if $IS \subseteq I$ ($SI \subseteq I$, respectively). Green's relations on S reflect the ideal structure. They are defined as follows:

- $a\mathcal{R}b$ if $aS^1 = bS^1$
- $a\mathcal{L}b$ if $S^1a = S^1b$
- $a\mathcal{H}b$ if $a\mathcal{R}b$ and $a\mathcal{L}b$
- $a\mathcal{J}b$ if $S^1aS^1 = S^1bS^1$.

An equivalence relation ρ on S is a right congruence if $s\rho t$ implies $su\rho tu$ for every $u \in S$. A left congruence is defined symmetrically. A left and right congruence is simply called a congruence. For convenience sake we sometimes write $s\rho t$ as $(s,t) \in \rho$. Clearly $\mathcal{R}$ is an example of a left congruence, and $\mathcal{L}$ is a right congruence. If ρ is a congruence on S then we denote by S/ρ the set of ρ-classes which has a natural semigroup structure inherited from S. If K is a field then the kernel of the natural epimorphism $K[S] \longrightarrow K[S/\rho]$ is denoted by $I(\rho)$ and it is the K-span of the set consisting of all elements $s - t$ with $s\rho t$. If $\rho = S \times S$ then $I(\rho)$ is called the augmentation ideal of $K[S]$ and is usually denoted by $\omega(K[S])$.

If I is a (two-sided) ideal of S, then one can form the Rees factor S/I, which is the semigroup with zero $(S \setminus I) \cup \{\theta\}$ with operation defined by $a \cdot b = ab$ if $ab \in S \setminus I$ and $ab = \theta$ otherwise. Clearly I corresponds to a congruence τ_I on S determined by $s\tau_I t$ if and only if either $s = t$ or $s, t \in I$. It is often convenient to define $S/I = S$ for $I = \emptyset$.

If $a \in S$, then the set I_a of non-generators of the ideal S^1aS^1 is an ideal of S, if nonempty. The Rees factor S^1aS^1/I_a is called the principal factor of a in S. It is clear that, if $a \neq \theta$, then the set of nonzero elements of S^1aS^1/I_a can be identified with the $\mathcal{J}$-class of a in S. Principal factors are used to study the structure of S 'locally'. This approach is especially useful in the class of finite semigroups.

A semigroup S is called 0-simple if $S = S^2$ and S has no ideals other than S and possibly $\{\theta\}$ if S has a zero element θ. If additionally S has a primitive idempotent, then S is completely 0-simple. (Notice that, for practical reasons, we include the class of completely simple semigroups into the class of completely 0-simple semigroups, which is not a standard convention.)

It is known that every principal factor of a semigroup S is either completely 0-simple or a null semigroup (that is, a semigroup with zero multiplication).

Let X, Y be nonempty sets, G a group and $P = (p_{yx})$ a $Y \times X$ matrix with entries in G^0. (Clearly, by this we mean a function $(y, x) \mapsto p_{yx}$ from $Y \times X$ to G^0.) By $S = \mathcal{M}(G, X, Y; P)$ we denote the set of all triples $(g, x, y) \in G^0 \times X \times Y$ with $(g, x, y,) \cdot (g'x', y') = (gp_{yx'}g', x, y')$, where all triples (θ, x, y) are identified with the zero of S. In particular, θ also denotes the zero of S. S is called a semigroup of matrix type over the group G with sandwich matrix P. Then the sets $S_{(x)} = \{(g, x, y) \in S \mid g \in G^0, y \in Y\}, x \in X$, are called the rows of S, while $S^{(y)} = \{(g, x, y) \in S \mid g \in G^0, x \in X\}, y \in Y$, are the columns of S. If $|X| = k < \infty$, then we also write $S = \mathcal{M}(G, k, Y; P)$, identifying X with the set $\{1, \ldots, k\}$. The same convention applies to the set Y of columns of S.

For any subsets $A \subseteq X, B \subseteq Y$ we put $S_{(A)} = \bigcup_{x \in A} S_{(x)}, S^{(B)} = \bigcup_{y \in B} S^{(y)}, S^{(y)}_{(x)} = S_{(x)} \cap S^{(y)}$ and

$$S^{(B)}_{(A)} = S_{(A)} \cap S^{(B)} = \bigcup_{x \in A, y \in B} S^{(y)}_{(x)}.$$

Clearly, $S^{(B)}_{(A)} = \mathcal{M}(G, A, B; P_{BA})$ is a semigroup of matrix type, where P_{BA} is the $B \times A$ submatrix of P.

S can be represented as the set of $X \times Y$ matrices over $G \cup \{0\}$ with at most one nonzero entry (under the identification of every (g, x, y) with the matrix with g in position (x, y) and zeros elsewhere) with $a \cdot b = aPb$, where the latter is the usual multiplication of matrices.

It is known that a semigroup S with zero is completely 0-simple if and only if it is isomorphic to a semigroup of matrix type $\mathcal{M}(G, X, Y; P)$ such that every row and every column of the matrix P has a nonzero entry. If S is a semigroup with no zero element, then it is completely 0-simple if and only if S^0 is of the latter type (in particular, the sandwich matrix P has no zero entries). The basic information on the properties of semigroups of the above type is given below.

Lemma 2.1.1. *Let $S = \mathcal{M}(G, X, Y; P)$ be a semigroup of matrix type. Then*

1. *for every nonempty subsets $A \subseteq X, B \subseteq Y$, $S_{(A)}$ is a right ideal of S and $S^{(B)}$ is a left ideal of S,*
2. *$E = \{(g, x, y) \mid p_{yx} \in G, g = p_{yx}^{-1}\}$ is the set of nonzero idempotents of S. If $e = (g, x, y) \in E$, then $S_{(x)}^{(y)} \cong G^0$ via the map $(h, x, y) \mapsto hp_{yx}$,*
3. *if $p_{yx} = \theta$, then $S_{(x)}^{(y)}$ is a semigroup with zero multiplication,*
4. *if $a, b, c \in S$ and $ab \neq \theta, bc \neq \theta$, then $abc \neq \theta$; if $a \neq \theta$, then any of the conditions $ab = a, ba = a$ implies that b is an idempotent,*
5. *if $ax = bx \neq \theta$ for some $a, b \in S^{(y)}$, then $a = b$,*
6. *if S is completely 0-simple, then for every nonzero $a = (g, x, y) \in S$ there exist $e, f \in E$ such that $a = ea = af$ and there exist $b, c \in S$ such that bac lies in a maximal subgroup of S. Moreover, $aS = S_{(x)}$ and $Sa = S^{(y)}$.*

Note that the last assertion identifies the $\mathcal{R}, \mathcal{L}$ and $\mathcal{H}$-classes of a completely 0-simple S. In particular, we get an egg-box pattern on $S \setminus \{\theta\}$ (the boxes being the nonzero $\mathcal{H}$-classes $S_{(x)}^{(y)} \setminus \{\theta\}$), which will be crucial for the general structural approach presented in Section 2.4.

Let $S = \mathcal{M}(G, X, Y; P)$ be a semigroup of matrix type over a group G. If H is a normal subgroup of G, then $\phi_H((g, x, y)) = (gH, x, y)$ determines a homomorphism $\phi_H : S \longrightarrow S_H$, where $S_H = \mathcal{M}(G/H, X, Y; P_H)$ is the semigroup of matrix type over G/H with sandwich matrix $P_H = (p_{yH,xH})$ defined by $p_{yH,xH} = p_{yx}H$.

It is well that $\mathcal{M}(G, X, Y; P) \cong \mathcal{M}(G', X', Y'; P')$ if and only if $|X| = |X'|, |Y| = |Y'|$ and there exists an isomorphism $\phi : G \longrightarrow G'$, such that $\phi(P) = Q_1 P' Q_2$ for some $Y \times Y'$ and $X' \times X$ matrices Q_1, Q_2 respectively, over $G' \cup \{\theta\}$ with exactly one nonzero entry in each row and each column. Here $\phi(P) = (\phi(p_{yx}))$.

Using this fact, it is often useful to normalize the given sandwich matrix P. Namely, if we fix $x \in X$ and $y \in Y$, then the above allows to change the presentation of S so that all entries of the row y and of the column x of P lie in the set $\{\theta, 1\}$.

If $S = \mathcal{M}(G, X, X; I)$, for the identity $X \times X$ matrix I, then S is called a Brandt semigroup over the group G. An equivalent characterization is that S is completely 0-simple and every $\mathcal{R}$-class, and every $\mathcal{L}$-class, of S contains only one maximal subgroup of S.

More generally, we shall sometimes consider a semigroup of matrix type $\mathcal{M}(T, X, Y; P)$ over a semigroup T (here p_{yx} are simply chosen from T^0).

If R is an algebra over a field then $R' = \mathcal{M}(R, X, Y; P)$ stands for the set of all $X \times Y$ matrices over R with finitely many nonzero entries, subject to the natural addition of matrices and to the sandwich multiplication $ab = a \circ P \circ b$ for $a, b \in R'$, with $\circ$ standing for the usual matrix multiplication. It becomes an algebra and it is called a Munn algebra over R. In particular, if $R = K[T]$ for a semigroup T then $\mathcal{M}(K[T], X, Y; P) \cong K_0[\mathcal{M}(T, X, Y; P)]$. The ideal structure of R', in terms of that of R, is pretty well understood. In particular, the following comes from Lemma 5.11 and Proposition 5.14 in [119]. A more detailed discussion of prime ideals and their role in the theory of Noetherian algebras, needed in this book, is given in the next chapter. A $Y \times Y$ matrix over R is said to lie over a subset $Z \subseteq R$ if all its entries are in Z.

Proposition 2.1.2. *Assume that $R' = \mathcal{M}(R, X, Y; P)$ is a Munn algebra over an algebra R with an identity. Assume that every row and every column of P contains a unit of R. Then the map $J \mapsto \{x \in R' \mid P \circ x \circ P \text{ lies over } J\}$ is a bijection between the set of prime (semiprime, respectively) ideals of R and R'. Moreover, if I is an ideal of R' and $(r, x, y) \notin I$ for some fixed $x \in X, y \in Y$ and every nonzero $r \in R$, then $R'IR' = 0$.*

We continue with a discussion of the semigroups of quotients of U in the sense studied in [41, 42, 43]. Let S be a (right and left) cancellative semigroup. This means that $ax = bx$ implies $a = b$ and $xa = xb$ implies $a = b$ for all $a, x, b \in S$. We say that S satisfies the right Ore condition if for every $a, b \in S$ we have $aS \cap bS \neq \emptyset$. It is well known that this happens if and only if S has a group of right quotients $G = SS^{-1} = \{ab^{-1} \mid a, b \in S\}$, which is unique up to isomorphism. If S also has a group of left quotients H (defined symmetrically), then G, H are isomorphic and are referred to as the group of (two-sided) quotients of S. One of the special important classes of such semigroups arises from the following simple observation, which is folklore for cancellative semigroups.

Lemma 2.1.3. *Assume that T is a left cancellative semigroup with no free nonabelian subsemigroups. Then, for every right ideals I, J of T we have $I \cap J \neq \emptyset$. In particular, if T is cancellative, then T has a group of quotients.*

Proof. Suppose $I \cap J = \emptyset$. Let $s \in I, t \in J$. Then $s \prec s,t \succ \cap t \prec s,t \succ = \emptyset$. Therefore, since $\prec s,t \succ$ is not a free nonabelian semigroup, cancellation on the left allows us to find $x \in \prec s,t \succ$ such that $z = zx$ for some $z \in \{s,t\}$. Then $zy = zxy$ for all $y \in T$, so that $y = xy$. This easily implies that $sT \cap tT \neq \emptyset$. Therefore $I \cap J \neq \emptyset$, a contradiction. In particular, S satisfies the right Ore condition. The left Ore condition is proved similarly. □

More generally, we say that a semigroup Q is a semigroup of right quotients of a semigroup S (or, equivalently, S is a right order in Q) if: (1) every square-cancellable element of S lies in a subgroup of Q and (2) every element of Q can be written in the form $q = ab^{\#}$ where $a, b \in S$, b lies in a subgroup of Q and $b^{\#}$ denotes the inverse of b in this group. Here $a \in S$ is square-cancellable if $a^2x = a^2y$ implies $ax = ay$ and $xa^2 = ya^2$ implies $xa = ya$, for all $x, y \in S^1$. Left quotients are defined in a symmetric way.

A completely 0-simple semigroup Q is a semigroup of right quotients of its subsemigroup S if every element of Q can be written in the form $q = ab^{\#}$ where $a, b \in S, b^2 \neq \theta$ and $b^{\#}$ denotes the inverse of b in the maximal subgroup of Q to which b belongs.

A semigroup S (with a zero element) is prime if $IJ = \{\theta\}$ for some ideals I, J of S implies that $I = \{\theta\}$ or $J = \{\theta\}$. S is called categorical at zero if $abc = \theta$ implies $ab = \theta$ or $bc = \theta$, for $a, b, c \in S$. Notice that every semigroup $S = \mathcal{M}(G, X, Y; P)$ of matrix type over a group G has this property by Lemma 2.1.1. Let ρ be the relation on S defined by $a\rho b$ if $a = b = \theta$ or $aS \cap bS \neq \{\theta\}$. The following result comes from [42].

Theorem 2.1.4. *A semigroup S with zero has a completely 0-simple semigroup Q of (left and right) quotients if and only if it is prime, categorical at zero and satisfies the following conditions and their duals*

1. *for $a, b, s, t \in S$, if $a\rho b$ and $as \neq \theta, bt \neq \theta$, then $as\rho bt$,*

2. *for $a, b, c \in S$, if $a\rho b$ and $ca = cb \neq \theta$, then $a = b$.*

Moreover, Q is determined uniquely up to isomorphism.

A congruence η on S is called a 0-matrix congruence if $\{\theta\}$ is a congruence class and S/η is a completely 0-simple semigroup with trivial maximal subgroups. Also S is called weakly 0-cancellative if $ax = bx \neq \theta$ and $xa = xb \neq \theta$ imply that $a = b$ and S is 0-cancellative if $S \setminus \{\theta\}$ is a cancellative semigroup. The next three results were proved in [43].

Theorem 2.1.5. *A semigroup S with zero has a completely 0-simple semigroup of (left and right) quotients if and only if S is weakly 0-cancellative* *and has a 0-matrix congruence whose non-null classes are cancellative and satisfy the left and right Ore conditions.*

Theorem 2.1.6. *A semigroup S with zero has a Brandt semigroup of (left and right) quotients if and only if S is weakly 0-cancellative and has a 0-matrix congruence η such that S/η is a Brandt semigroup and non-null η-classes are cancellative and satisfy the left and right Ore conditions.*

Theorem 2.1.7. *S has a Brandt semigroup $Q = \mathcal{M}(G, X, X; I)$ of right quotients if and only if $S = \{(g, i, j) \in Q \mid g \in Q_{ij}\} \cup \{\theta\}$ for a collection $Q_{ij} \subset G$ of subsets satisfying $Q_{ij}Q_{jk} \subseteq Q_{ik}$ and $G = Q_{ji}Q_{ii}^{-1}$ for all $i, j, k \in X$.*

It is known that a completely 0-simple semigroup of right quotients of a semigroup S is not unique (up to isomorphism) in general. However, if $M = \mathcal{M}(G, X, Y; P)$ is such a semigroup, then G and X are determined uniquely. Moreover, if $S \cap H$ is left Ore for every maximal subgroup H of M then M is a two-sided semigroup of quotients of S and in particular M is determined uniquely. For more information on completely 0-simple semigroups of one-sided quotients we refer to [6, 57, 58].

The following observation is clear.

Lemma 2.1.8. *Assume that S is a cancellative subsemigroup of a semigroup $M = \mathcal{M}(G, X, Y; P)$ of matrix type over a group G. If S satisfies the left and right Ore conditions then $S \subseteq H$ for a maximal subgroup H of M.*

For any semigroup S, the least right cancellative congruence on S is the intersection of all congruences τ on S such that S/τ is right cancellative. Similarly, one defines the least cancellative congruence on S, denoted by ρ_{can}.

2.2 Uniform semigroups

Our first aim is to introduce a class of subsemigroups of completely 0-simple semigroups that will be crucial for the approach presented in this chapter. We follow [123].

Proposition 2.2.1. *Let U be a subsemigroup of a completely 0-simple semigroup S with zero θ, which intersects nontrivially every nonzero $\mathcal{H}$-class of S. Fix a maximal subgroup H of S. Let G denote the subgroup of H generated by $U \cap H$. Then there exists a subsemigroup I of U and a Rees matrix presentation $S = \mathcal{M}(H, X, Y; Q)$ of S such that $I = \mathcal{M}(T, X, Y; Q)$ is a* *semigroup of matrix type over the subsemigroup* $T = U^2 \cap H$ *of* H *and* $\widehat{I} = \mathcal{M}(G, X, Y; Q)$ *is the smallest completely* 0*-simple subsemigroup of* S *containing* U*. Moreover, the construction of* $\widehat{I}$ *does not depend on the choice of* H*, in particular for any other maximal subgroup* H' *of* S *the subgroup of* H' *generated by* $U \cap H'$ *is equal to* $\widehat{I} \cap H'$*, so that it is isomorphic to* G*.*

Let $S = \mathcal{M}(H, X, Y; P)$ be a completely 0-simple semigroup. A subsemigroup U of S will be called a uniform subsemigroup of S if U intersects each nonzero $\mathcal{H}$-class of a completely 0-simple subsemigroup $S' = \mathcal{M}(H, X', Y'; P')$, where $X' \subseteq X, Y' \subseteq Y$ and P' is the corresponding $Y' \times X'$ submatrix of P. From the above result it follows that S has the smallest completely 0-simple subsemigroup $\widehat{U}$ containing U. This $\widehat{U}$ will be called the completely 0-simple closure of U in S. It can be viewed as a 'group approximation' of U. The common (up to isomorphism) group G generated by the intersections $U \cap F$ with maximal subgroups F of S' will be called the group associated to the uniform subsemigroup U of S. (Note that this definition does not imply that $\theta \in \widehat{U}$. In fact, this is not so if $\theta \notin U$. However, including θ in $\widehat{U}$ is sometimes convenient and produces no ambiguity.) All such $U \cap F$ are called the cancellative parts or the cancellative components of U. Uniform semigroups were introduced in [120].

Remark 2.2.2. We note that a semigroup U that is a uniform subsemigroup of two completely 0-simple semigroups S_1, S_2 with non-isomorphic completely 0-simple closures U_1, U_2 can be constructed. For example, every polycyclic-by-finite group G which is not nilpotent-by-finite, has a free nonabelian subsemigroup X, see Corollary 3.5.8. Therefore, the subgroup of G generated by X and the free group of rank equal to the rank of X both can be considered as completely 0-simple closures of X.

Another example of this type can be constructed as follows. Let $S_1 = \mathcal{M}(G, 1, 2; P)$, where G is the free group on generators x, y and the sandwich matrix P is given by $p_{11} = 1, p_{21} = 1$. Let $U = V_1 \cup V_2$ with $V_1 = \{(g, 1, 1) \mid g \in \langle x, y \rangle x\}$ and $V_2 = \{(g, 1, 2) \mid g \in \langle x, y \rangle y\}$. It is easy to see that U is a cancellative subsemigroup of S_1 and S_1 is its completely 0-simple closure. On the other hand, U is isomorphic to the free semigroup $U' = \prec x, y \succ \subseteq S_2 = G$ via the map $(g, 1, i) \mapsto g$. Clearly, S_2 is the completely 0-simple closure of U' in S_2.

However, such a situation cannot occur if every cancellative subsemigroup T of U satisfies the right and left Ore conditions. Namely, assume that $a, b \in U$ are nonzero and such that $a\mathcal{R}b$ in S_1. Since U is a uniform subsemigroup of S_1 and S_1 is completely 0-simple, there exist $x, y \in U$ such that $ax, by \in H$ for a maximal subgroup H of S_1. But $U \cap H$ is an Ore semigroup, so $axs = byt$ for some $s, t \in U \cap H$. It follows that

$$a\mathcal{R}b \text{ in } S_1 \text{ if and only if } au = bw \text{ for some } u, w \in U.$$

Since the same applies to S_2, the $\mathcal{R}$-class partition of U coming from S_1 is the same as that coming from S_2. A symmetric argument works for the $\mathcal{L}$-class partitions. Hence, if H_1 is a maximal subgroup of S_1 nontrivially intersected by U, then $U \cap H_1 = U \cap H_2$ for a maximal subgroup H_2 of S_2. The maximal subgroups G_1, G_2 of U_1, U_2 are the classical groups of quotients of the respective intersections, because they are generated by these intersections by Proposition 2.2.1. It follows that $G_1 \cong G_2$ for an isomorphism which is the identity on $U \cap G_1$. We may write $U_1 = \mathcal{M}(G_1, X, Y; P_1), U_2 = \mathcal{M}(G_2, X, Y; P_2)$.

It can be easily verified that, in the above case, the semigroup $\widehat{U} = U_1$ is the semigroup of quotients of U in the sense discussed in the previous section. Namely, every $s \in \widehat{U}$ can be written as $s = ux^{-1} = y^{-1}v$ for some $u, v \in U$ and some $x \in U \cap H, y \in U \cap H'$, where H, H' are maximal subgroups of $\widehat{U}$ and the inverses are taken in the respective groups. In particular, $U_1 \cong U_2$ by Theorem 2.1.4.

We continue with some examples of uniform semigroups.

Example 2.2.3. 1. A subsemigroup of a group; the multiplicative semigroup of a subring of a division algebra.
2. More generally, let S be the multiplicative semigroup of a prime right Goldie ring R, see Chapter 3. Then S has an ideal I that is a uniform subsemigroup of some $\mathcal{M}(\mathrm{GL}_j(F), X, Y; P)$, where F is a division algebra such that the classical quotient ring of R is isomorphic to $\mathrm{M}_n(F)$ for some $n \geq j$. The same can be said about the image of a semigroup S under a homomorphism of its semigroup algebra $K[S]$ onto a prime right Goldie ring R. This is a consequence of the structure theorem presented in the next section, see also Lemma 5.1.2.
3. Let $S = \mathcal{M}(G, 2, 2; I)$, where G is the free nonabelian group on x, y and I is the 2×2 identity matrix. Then S can be considered as a subsemigroup of $\mathrm{M}_2(K[G])$. Let U be the subsemigroup of S of the form

$$U = \begin{pmatrix} \langle x \rangle & \langle x \rangle y \\ y^{-1}\langle x \rangle & y^{-1}\langle x \rangle y \end{pmatrix}$$

(meaning that the only nonzero entry of every nonzero matrix of U lies in the corresponding indicated subset of G.) It is clear that U is a uniform subsemigroup of S and the associated group is the cyclic infinite group.

4. Let $S = \mathrm{M}_n(\mathbb{Z}) \subseteq \mathrm{M}_n(\mathbb{Q})$. Since for every $a \in \mathrm{M}_n(\mathbb{Q})$ there exists $z \in \mathbb{Z}$ such that $za \in S$, it follows that S intersects all $\mathcal{H}$-classes of $\mathrm{M}_n(\mathbb{Q})$. Therefore, using the notation of Theorem 2.3.4, we get that $(S \cap \mathrm{M}_j)/(S \cap \mathrm{M}_{j-1})$ is a uniform subsemigroup of $\mathrm{M}_j / \mathrm{M}_{j-1}$ for every $j = 1, \ldots, n$.

For certain important classes of semigroups 'uniform' implies 'completely 0-simple'. A semigroup S is called π-regular if a power of every element of S lies in a subgroup of S. Clearly, every periodic semigroup (meaning that a power of every element is an idempotent) is π-regular. This class also includes regular semigroups $S \subseteq \mathrm{M}_n(D)$ of matrices over a division algebra D, see Proposition 2.5.4. The latter are defined by the condition that every element $a \in S$ is regular, that is, there exists $x \in S$ such that $axa = a$. Other important examples of π-regular semigroups are given in Theorem 2.6.1.

Proposition 2.2.4. *Let U be a uniform subsemigroup of a completely 0-simple semigroup S. If a power of every $a \in U$ is a regular element of U, then U is completely 0-simple. In particular, this applies if U is π-regular.*

We introduce the notion of a triangular set of idempotents. This notion arises from a technical condition, which occurs in the proof of the structural theorem for arbitrary linear semigroups (Theorem 2.4.3). Apparently, it is also connected with nilpotency of nil semigroups. Recall that a semigroup with zero θ is a nil semigroup if every $s \in S$ is nilpotent, so there exists $k \geq 1$ with $s^k = \theta$. If $S^k = \{\theta\}$ for some k, then S is called nilpotent. The key features of triangular sets of idempotents show up already in the context of subsemigroups of completely 0-simple semigroups. We follow [139] (see also [123]) where it was introduced in full generality and used to study the more general case of linear semigroups.

A linearly ordered set $T, \prec$ of nonzero idempotents of a semigroup S with zero θ is called a triangular set of idempotents if $e \prec f$ implies that $ef = \theta$ for $e, f \in T$.

Lemma 2.2.5. *Let $M = \mathcal{M}(G, X, Y; P)$ be a semigroup of matrix type. Assume that every triangular set of idempotents in M has at most n elements. If S is a nil subsemigroup of M, then $S^{n+2} = \{\theta\}$, and $S^n = \{\theta\}$ whenever M is a completely 0-simple semigroup. Conversely, if there exists a triangular set of idempotents of cardinality n in M, then there exists a π-regular semigroup $S \subseteq M$ which has n $\mathcal{J}$-classes containing idempotents and a nil ideal N which is not nilpotent of index $< n$.*

2.3 Full linear semigroup

Throughout this section D denotes a division ring. If $n \geq 1$, then we write $\mathrm{M}_n(D)$ for the multiplicative monoid of $n \times n$ matrices over D. By a linear semigroup over D we mean a subsemigroup of some $\mathrm{M}_n(D)$. As usual, $\mathrm{GL}_n(D)$ is the group of invertible matrices. $\mathrm{M}_n(D)$ acts on the column (right D-) vector space D^n by left multiplication. By $\ker(a), \mathrm{im}(a) \subseteq D^n$ we mean the kernel and the image of $a \in \mathrm{M}_n(D)$ under this action. Moreover $\mathrm{rank}(a)$ denotes the rank of the matrix a, that is the dimension of aD^n as a D-subspace.

In this section we present the basic information on the structure of the full linear monoid $\mathrm{M}_n(D)$, see [123]. This includes a discussion of the ideal structure and the Rees presentations of the principal factors of $\mathrm{M}_n(D)$.

We start with a description of $\mathcal{J}, R$, and $\mathcal{L}$-classes of $\mathrm{M}_n(D)$.

Lemma 2.3.1. *Green's relations on* $\mathrm{M}_n(D)$ *are given by*

1. $a\mathcal{J}b$ *if and only if* $\mathrm{GL}_n(D)b\,\mathrm{GL}_n(D) = \mathrm{GL}_n(D)a\,\mathrm{GL}_n(D)$, *which is also equivalent to* $\mathrm{rank}(a) = \mathrm{rank}(b)$,
2. $a\mathcal{R}b$ *if and only if* $a\,\mathrm{GL}_n(D) = b\,\mathrm{GL}_n(D)$, *which is also equivalent to* $\mathrm{im}(a) = \mathrm{im}(b)$,
3. $a\mathcal{L}b$ *if and only if* $\mathrm{GL}_n(D)a = \mathrm{GL}_n(D)b$, *which is also equivalent to* $\ker(a) = \ker(b)$.

It is now clear that every $\mathcal{R}$-class R of $\mathrm{M}_n(D)$ contains an idempotent, namely any projection e onto the common image of the endomorphisms from R. Moreover, conjugating by an element $g \in \mathrm{GL}_n(D)$, we may assume that e is of the form $e = \begin{pmatrix} I & 0 \\ 0 & 0 \end{pmatrix}$. Thus, R consists of all matrices of rank equal to the rank of e which are of the form $\begin{pmatrix} a & b \\ 0 & 0 \end{pmatrix}$. Similarly, the transpose R^t of R is the $\mathcal{L}$-class of e.

Recall that a matrix $a \in \mathrm{M}_n(D)$ of rank $j \leq n$ is said to be in reduced row elementary form if the following conditions are satisfied:

1. the leading coefficient of every nonzero row of a is 1,
2. for every $i \leq j$, the $i+1$-th row of a has more leading zero coefficients than the i-th row,
3. the leading coefficient of a nonzero row is the only nonzero coefficient in its column.

Let Y_j be the set of all matrices of rank j which are in the reduced row elementary form. Put $X_j = Y_j^t$, the transpose of Y_j. Assume that $x, x' \in X_j$. If $\mathrm{im}(x) = \mathrm{im}(x')$ then $x = x'$.

It is well known that every matrix a of rank j is column equivalent to a matrix $x \in X_j$ (that is, a can be brought to x by a sequence of elementary column operations). Moreover, if two matrices a, b are column equivalent, then $a = bg$ for some $g \in \mathrm{GL}_n(D)$, so $\mathrm{im}(a) = \mathrm{im}(b)$. It follows that X_j is a set of representatives of the set of $\mathcal{R}$-classes of $\mathrm{M}_n(D)$ consisting of matrices of rank j. Similarly, Y_j can be interpreted as the set of $\mathcal{L}$-classes.

Lemma 2.3.2. *Let $0 < j < n$. If $e \in Y_j$ is the diagonal idempotent, then let G_j be the group of units of the monoid $e\,\mathrm{M}_n(D)e$. Then $G_j \cong \mathrm{GL}_j(D)$, each xgy, where $x \in X_j, y \in Y_j, g \in G_j$, has rank j and every matrix of rank j has a unique presentation in this form.*

For a given semigroup $S \subseteq \mathrm{M}_n(D)$ the nonempty intersections $S \cap H$ with maximal subgroups H of $\mathrm{M}_n(D)$ will play a crucial role in our approach.

Lemma 2.3.3. *The following conditions are equivalent for a matrix $a \in \mathrm{M}_n(D)$*

1. *$a \in H$ for a maximal subgroup H of $\mathrm{M}_n(D)$,*
2. *$\mathrm{rank}(a^2) = \mathrm{rank}(a)$,*
3. *D^n is the direct sum of the subspaces $\ker(a)$ and $\mathrm{im}(a)$,*
4. *a is conjugate to a matrix $\begin{pmatrix} g & 0 \\ 0 & 0 \end{pmatrix}$ for some $g \in \mathrm{GL}_j(D)$ and some j, $0 \le j \le n$.*

Let $\mathrm{M}_j = \{a \in \mathrm{M}_n(D) \mid \mathrm{rank}(a) \le j\}$ for $j = 0, 1, \ldots, n$. We are now able to describe the semigroup structure of $\mathrm{M}_n(D)$.

Theorem 2.3.4. *The sets $0 = \mathrm{M}_0 \subset \mathrm{M}_1 \subset \cdots \subset \mathrm{M}_n = \mathrm{M}_n(D)$ are the only ideals of the monoid $\mathrm{M}_n(D)$. Each Rees factor $\mathrm{M}_j \,/\, \mathrm{M}_{j-1}$ is isomorphic to the completely 0-simple semigroup $\mathcal{M}(\mathrm{GL}_j(D), X_j, Y_j; Q_j)$, where the matrix $Q_j = (q_{yx})$ is defined for $x \in X_j, y \in Y_j$ by $q_{yx} = yx$ if yx is of rank j and 0 otherwise.*

The rank of a matrix $a \in \mathrm{M}_n(D)$ can also be defined using the dual action of $\mathrm{M}_n(D)$ on the space D^n of row vectors by right multiplication. It is easy to verify that the two notions of rank coincide. For example, use the fact that $a \mathcal{J} e$ for some diagonal idempotent $e \in \mathrm{M}_n(D)$, so that rank(a) = rank(e). Since the dual ranks of a, e are also equal and rank(e) is obviously equal to its dual rank, the assertion follows.

Every principal factor $\mathrm{M}_j \,/\, \mathrm{M}_{j-1}$ will be approached via its egg-box pattern arising from Theorem 2.3.4. The following observation will be often used for M equal to one of these factors.

Proposition 2.3.5. *Let S be a completely 0-simple subsemigroup of a completely 0-simple semigroup M. If $\mathcal{S}, \mathcal{S}_S$ is one of the Green relations $\mathcal{R}, \mathcal{L}, \mathcal{H}$, on M, and on S, respectively, then we have $\mathcal{S}_{|S} = \mathcal{S}_S$.*

The Rees presentation of $\mathrm{M}_j \,/\, \mathrm{M}_{j-1}$ in Theorem 2.3.4 is given in terms of subspaces of D^n. However, there is also a group theoretic description, which is sometimes useful, because it reflects the representation theory of the monoid $\mathrm{M}_n(D)$ and allows to develop combinatorics of $\mathrm{M}_n(D)$. For this, we refer to [123].

Let $R = \{a = (a_{ij}) \in \mathrm{M}_n(D) \mid a_{ij} \in \{0,1\}$ and a has at most one nonzero entry in each row and in each column$\}$. It is clear that R is the semigroup of all one-to-one partial transformations of the set $\{1, \ldots, n\}$. R is called the symmetric inverse monoid. It is the semigroup analogue of the symmetric group, that in particular plays an important role in the theory of the full linear monoid $\mathrm{M}_n(D)$, see [123, 133, 134]. The structure of R is particularly nice.

Proposition 2.3.6. *The sets $R_0 \subset R_1 \subset \cdots \subset R_n = R$, $R_j = R \cap \mathrm{M}_j$, are the only ideals of R. The Rees factors R_j/R_{j-1} are completely 0-simple semigroups with presentations $\mathcal{M}(W_j, \binom{n}{j}, \binom{n}{j}; I_j)$, where W_j is the group of $j \times j$ permutation matrices and I_j is the identity matrix. Moreover, Green's relations on R are the restrictions of the corresponding relations on $\mathrm{M}_n(D)$ and $a\mathcal{R}b$ if and only if $aW_n = bW_n$, $a\mathcal{L}b$ if and only if $W_n a = W_n b$, $a\mathcal{J}b$ if and only if $W_n a W_n = W_n b W_n$ for $a, b \in R$.*

2.4 Structure of linear semigroups

In this section we present the general structure theorem for arbitrary subsemigroups S of $\mathrm{M}_n(D)$. One of the motivations for developing this theory comes from problems on associative algebras that have 'many' finite dimensional linear representations. Algebras satisfying polynomial identities and Noetherian algebras are the main classes of this type. The latter often lead to representations over a division algebra, which is not necessarily a field. The starting idea is to study the ideal chain determined on S by the ideals $\mathrm{M}_0 \subseteq \mathrm{M}_1 \subseteq \cdots \subseteq \mathrm{M}_n = \mathrm{M}_n(D)$ and to see how S fills the egg-box pattern at every level $\mathrm{M}_j \setminus \mathrm{M}_{j-1}$. The theorem associates with S a collection of linear groups G_α and corresponding sandwich matrices P_α over these groups, or more precisely a collection of uniform subsemigroups U_α of the principal factors $\mathrm{M}_j / \mathrm{M}_{j-1}$ of $\mathrm{M}_n(D)$. Up to conjugation, G_α are all groups generated by the nonempty intersections $S \cap H$ with maximal subgroups H of $\mathrm{M}_n(D)$. This description of S shows a strong analogy with Wedderburn's structure theorem for finite dimensional algebras. The philosophy is then to study S via the 'associated linear groups' G_α and the corresponding U_α. Throughout this section we follow [123].

First we look at the uniform subsemigroups of principal factors $\mathrm{M}_j / \mathrm{M}_{j-1}$ of the monoid $\mathrm{M}_n(D)$.

Lemma 2.4.1. *Let $S \subseteq \mathrm{M}_n(D)$ be a semigroup. Assume that for some maximal subgroups H_1, H_2 of $\mathrm{M}_n(D)$, consisting of matrices of the same rank, there exist $a, b \in S$ such that $a(S \cap H_1)b \subseteq S \cap H_2$. Then the groups G_1, G_2 generated by $S \cap H_1, S \cap H_2$ in H_1, H_2, respectively, are conjugate in $\mathrm{M}_n(D)$.*

The following is an immediate consequence of Lemma 2.4.1 and of the definition of a uniform subsemigroup.

Corollary 2.4.2. *Let U be a uniform subsemigroup of some $\mathrm{M}_j / \mathrm{M}_{j-1}$. Assume that U intersects maximal subgroups H_1, H_2 of $\mathrm{M}_j / \mathrm{M}_{j-1}$. If G_1, G_2 are the groups generated by $U \cap H_1, U \cap H_2$, then G_1, G_2 are conjugate in $\mathrm{M}_n(D)$.*

As before, all spaces will be right spaces over D. So the set of column vectors D^n is a right D-space and $\mathrm{M}_n(D)$ acts on D^n by left multiplication. The structure of $\mathrm{M}_n(D)$ is exactly as described in Lemma 2.3.1 and Theorem 2.3.4. Keeping the notation of Section 2.3, we identify the nonzero elements of the Rees factor $\mathrm{M}_j / \mathrm{M}_{j-1}$ with the corresponding elements of $\mathrm{M}_j \setminus \mathrm{M}_{j-1}$.

We say that a subset A of a semigroup T is a 0-disjoint union of certain $A_\alpha, \alpha \in \mathcal{A}$, if $A = \bigcup_\alpha A_\alpha$ and $\bigcap_\alpha A_\alpha$ is either empty or it is the zero of T. The main structural theorem for $S \subseteq \mathrm{M}_n(D)$ may be now stated in the following way.

Theorem 2.4.3. *Let $S \subseteq \mathrm{M}_n(D)$ be a semigroup. Define the sets*

$$S_j = \{a \in S \mid \mathrm{rank}(a) \leq j\}$$

$$T_j = \{a \in S_j \mid S^1 a S^1 \text{ does not intersect maximal subgroups of } \mathrm{M}_n(D) \text{ contained in } \mathrm{M}_j \setminus \mathrm{M}_{j-1}\}.$$

Then

$$S_0 \subseteq T_1 \subseteq S_1 \subseteq T_2 \subseteq \cdots \subseteq S_{n-1} = T_n \subseteq S_n = S$$

are ideals of S (if nonempty). Moreover for every j we have

1. *$N_j = T_j/S_{j-1}$ is a nil ideal of S/S_{j-1} and it is a union of nilpotent ideals of nilpotency index 2,*

2. *$(S_j \setminus T_j) \cup \{\theta\} \subseteq \mathrm{M}_j / \mathrm{M}_{j-1}$ is a 0-disjoint union of uniform subsemigroups $U_\alpha^{(j)}, \alpha \in \mathcal{A}_j$, of $\mathrm{M}_j / \mathrm{M}_{j-1}$ that intersect different $\mathcal{R}$- and different $\mathcal{L}$-classes of $\mathrm{M}_j / \mathrm{M}_{j-1}$; moreover N_j does not intersect $\mathcal{H}$-classes of $\mathrm{M}_j / \mathrm{M}_{j-1}$ intersected by $S_j \setminus T_j$,*

3. *$U_\alpha^{(j)} U_\beta^{(j)} \subseteq N_j$ for every $\alpha \neq \beta$; moreover $U_\alpha^{(j)} N_j, N_j U_\alpha^{(j)} \subseteq N_j$ and $U_\alpha^{(j)} N_j U_\alpha^{(j)} = \{\theta\}$ in $\mathrm{M}_j / \mathrm{M}_{j-1}$. In particular, $U_\alpha^{(j)}$ can be considered as ideals of S/T_j.*

4. *Define*

$$S'_j = \{a \in \mathrm{M}_j / \mathrm{M}_{j-1} \mid \text{there exist images } b, c \text{ of elements of } S \text{ in } \mathrm{M}_j / \mathrm{M}_{j-1} \text{ such that } a\mathcal{L}b \text{ and } a\mathcal{R}c\}.$$

If every triangular set of idempotents in S'_j has at most m_j elements, where $m_j < \infty$, then N_j is nilpotent of index $\leq m_j + 2$ and S_j/T_j is a 0-disjoint union of at most m_j uniform ideals, so $|\mathcal{A}_j| \leq m_j$. In particular, the latter applies if D is a field, since every triangular set of idempotents in $M = \mathrm{M}_j / \mathrm{M}_{j-1}$ has at most $\binom{n}{j}$ elements in this case.

The structure of S is therefore described in terms of the ideal chain $S = S_n \supseteq S_{n-1} \supseteq \cdots \supseteq S_0, S_j = S \cap \mathrm{M}_j$ (note that we may have $S_i = \emptyset$ for some i). The Rees factors $S_j/S_{j-1} \subseteq \mathrm{M}_j / \mathrm{M}_{j-1}$ represent the 'layers' of S and are approached via the egg-box pattern on the completely 0-simple semigroup $\mathrm{M}_j / \mathrm{M}_{j-1}$. Recall that we adopt the convention $S_j/S_{j-1} = S_j$ if $S_{j-1} = \emptyset$.

Each nonempty $S_{j-1}, j \geq 1$, is an ideal in S_j and the Rees factor $S_{(j)} = S_j/S_{j-1}$ is a 0-disjoint union $\bigcup_\alpha U_\alpha \cup N$ of the maximal nil ideal N of $S_{(j)}$ and of certain semigroups $U_\alpha, \alpha \in A$, intersecting different $\mathcal{R}$- and different $\mathcal{L}$-classes of $\mathrm{M}_j / \mathrm{M}_{j-1}$. In particular, $U_\alpha U_\beta \subseteq N$ for $\alpha \neq \beta$. If the minimal rank of matrices in S is $j \geq 1$, so that $S_{j-1} = \emptyset$, then we must have $S_j = U_\alpha, A = \{\alpha\}$ and $N = \emptyset$. By Proposition 2.2.1, every U_α is contained in a smallest completely 0-simple subsemigroup $\widehat{U}_\alpha$ of $\mathrm{M}_j / \mathrm{M}_{j-1}$. That is,

U_α intersects all nonzero $\mathcal{H}$-classes of $\widehat{U}_\alpha$ and every maximal subgroup H of $\widehat{U}_\alpha$ is generated by $U_\alpha \cap H$. These U_α are uniquely determined by the above conditions. All U_α coming from all possible layers S_j/S_{j-1} are called the uniform components of S. With a slight abuse of language, the inverse image of $U_\alpha \setminus \{\theta\}$ in the corresponding $S_j \setminus S_{j-1}$ (θ denoting the zero of S_j/S_{j-1} here) will sometimes also be called a uniform component of S, depending on whether we view the matrices in $\mathrm{M}_j \setminus \mathrm{M}_{j-1}$ as elements of $\mathrm{M}_j / \mathrm{M}_{j-1}$ or $\mathrm{M}_n(D)$. If a uniform component U_α of S, treated as a subset of S, is an ideal of S or $U_\alpha \cup \{0\}$ is an ideal of S, then we call U_α an ideal uniform component of S. In this case $\widehat{U}_\alpha$, originally defined as a subset of $\mathrm{M}_j / \mathrm{M}_{j-1}$, will be identified with $(\widehat{U}_\alpha \setminus \{\theta\}) \cup \{0\} \subseteq \mathrm{M}_n(D)$ (that is, the zero of $\widehat{U}_\alpha$ is replaced by the zero matrix). Also, if $0 \in S$, this allows to include the zero matrix into U_α, so that the latter is indeed an ideal of S.

By the nilpotent components of S we mean the maximal nil ideals N_j of $S_{(j)}$ (or, again, the inverse images in S of the sets of nonzero elements of N_j, if convenient), for all j.

The maximal subgroups of $\widehat{U}_\alpha$ are called the groups associated to S. In particular, the groups associated to S that come from the same U_α are conjugate by Corollary 2.4.2. As said in Section 2.2, $\widehat{U}_\alpha$ is called the completely 0-simple closure of U_α.

Thus, to every $S \subseteq \mathrm{M}_n(D)$ there corresponds a collection of associated linear groups (these are linear groups generated by the nonempty intersections $S \cap H$ with the maximal subgroups H of $\mathrm{M}_n(D)$) and sandwich matrices over these groups (in fact, over $S \cap H$) arising from Proposition 2.2.1. Notice that linear groups over division algebras that are not fields have been also studied, see [141].

Theorem 2.4.3 is a natural extension of the structural approach to finite semigroups - via completely 0-simple principal factors, whose role is now played by the uniform components of S. Namely, Proposition 2.3.5 and Theorem 2.4.3 can be used to prove the first part of the following result. Then it is also easy to see that any homomorphic image and any ideal of a π-regular semigroup is also π-regular. Therefore, if $S \subseteq \mathrm{M}_n(D)$ is π-regular, then from Proposition 2.2.4 it follows that every uniform component of S is completely 0-simple.

Corollary 2.4.4. *Let $S \subseteq \mathrm{M}_n(D)$. Then every completely 0-simple principal factor of S is a uniform component of S. Moreover, if S is π-regular, then the uniform components of S coincide with the principal factors of S. In particular, they do not depend on the used matrix embedding.*

From the point of view of the ideal structure of a semigroup S the knowledge of 0-simple semigroups that can arise as principal factors of S is crucial.

Theorem 2.4.5. *Let J be a 0-simple principal factor of a semigroup $S \subseteq \mathrm{M}_n(D)$. If S has no free nonabelian subsemigroups or J contains a nonzero idempotent, then J is a completely 0-simple semigroup.*

Remark 2.4.6. i) If $S \subseteq \mathrm{M}_n(D)$ has a zero element θ, then Proposition 2.4.9 allows us to identify θ with the zero matrix by passing to $S \cong eSe \subseteq e\,\mathrm{M}_n(D)e \cong \mathrm{M}_t(D)$ for $e = 1-\theta$ and $t = n - \mathrm{rank}(\theta)$. If we first conjugate θ to a diagonal idempotent, then from the description of the egg-box patterns on $\mathrm{M}_n(D), e\,\mathrm{M}_n(D)e$ it is clear that this identification does not affect the structure of S coming from Theorem 2.4.3.
ii) We note that uniform components can be described in terms of the rank of certain matrices. Namely, $a \in S$ is in a uniform component of S, if $\mathrm{rank}(a) = \mathrm{rank}(xay) = \mathrm{rank}((xay)^2)$ for some $x, y \in S$. It may he checked that the uniform components of S consisting of matrices of rank j are the equivalence classes of the relation: $a \leftrightarrow b$ if $a\mathcal{H}xby$ for some $x, y \in S_j \setminus S_{j-1}$ (defined on every nonempty $T = S_j \setminus T_j$). Moreover, this relation coincides with the one given by the following condition: there exist $x, y \in S_j \setminus S_{j-1}$ such that $x\mathcal{R}a, x\mathcal{L}b, y\mathcal{R}b, y\mathcal{L}a$.
iii) More generally, the proof of the theorem shows that any subsemigroup T of a completely 0-simple semigroup J is a 0-disjoint union $\bigcup_{\alpha\in A} U_\alpha \cup N$ for a nil of index two (but not necessarily nilpotent) ideal N of T and uniform subsemigroups U_α of T that satisfy the remaining assertions of the theorem. These U_α are called uniform components of T.
iv) There is a striking analogy of the assertion of Theorem 2.4.3 with that of Wedderburn's theorem on the structure of finite dimensional algebras. Namely, N_j plays the role of the nilpotent radical, while the uniform components $U_\alpha^{(j)}$ behave like the simple blocks of the algebra modulo the radical.

The structural approach resulting from Theorem 2.4.3 raises the question of the status of arbitrary cancellative subsemigroups of the given $S \subseteq \mathrm{M}_n(D)$ as well as of the independence of the structural description of S of the embedding of S into a full matrix monoid $\mathrm{M}_m(D')$ for a division ring D' and $m \geq 1$.

First, as seen in Remark 2.2.2, two isomorphic free nonabelian semigroups $X, X' \subseteq \mathrm{GL}_n(D)$ can have non-isomorphic completely 0-simple closures. If G is a free nonabelian subgroup of $\mathrm{GL}_j(D), j > 1$, then the semigroup $\mathrm{M}_j / \mathrm{M}_{j-1}$, defined for $\mathrm{M}_n(D)$, contains an isomorphic copy of the semigroup $S_1 = \mathcal{M}(G, 1, 2; P)$ described in this remark. Therefore, the free semigroup $\prec x, y \succ$ has two faithful representations into $\mathrm{M}_j / \mathrm{M}_{j-1}$ with non-isomorphic completely 0-simple closures. There is also a further reason for the dependence of the decomposition of S on the matrix representation. For example, if J is a completely prime ideal of $S \subseteq \mathrm{GL}_n(D)$ (that is, $S \setminus J$ is a subsemigroup of S), consider the embedding $\phi : S \longrightarrow \mathrm{M}_{2n}(D)$ given by $s \mapsto \begin{pmatrix} s & 0 \\ 0 & I \end{pmatrix}$ for $s \in S \setminus J$ and $s \mapsto \begin{pmatrix} s & 0 \\ 0 & 0 \end{pmatrix}$ for $s \in J$. Clearly, S and $\phi(S)$ intersect different number of layers $\mathrm{M}_j \setminus \mathrm{M}_{j-1}$ of $\mathrm{M}_n(D), \mathrm{M}_{2n}(D)$ respectively, so their decompositions coming from Theorem 2.4.3 are different.

So, speaking about uniform components of S we shall always have in mind the given embedding $S \subseteq \mathrm{M}_n(D)$.

If $m_j < \infty$ for all j, as in condition 4) of Theorem 2.4.3, then we get several important consequences. In particular, the ideal chain $S_0 \subseteq T_1 \subseteq S_1 \subseteq \cdots \subseteq S_n = S$ can be refined in a very useful way.

Corollary 2.4.7. *Assume that $S \subseteq \mathrm{M}_n(D)$ satisfies the assumption in statement 4) of Theorem 2.4.3.*

1. *There exists an ideal chain*

$$I_1 \subset I_2 \subset \cdots \subset I_r = S$$

such that $r = r(m_1, \ldots, m_{j-1})$ and I_1 and each Rees factor I_k/I_{k-1} is either nilpotent of index at most $m_j + 2$ for some $j \leq n$ or it is a uniform subsemigroup of some $\mathrm{M}_j / \mathrm{M}_{j-1}$.

2. *There exists $t = t(m_1, \ldots, m_{n-1})$ such that if $a_1 S^1 \supset a_2 S^1 \supset \cdots \supset a_t S^1 \neq 0$ for some $a_i \in S$ and $a_0 = 1$ then there exists a uniform component U of S such that $a_k = a_i u, a_m = a_i uw$ for some $i < k < m$ and $u, w \in U$ such that $u\mathcal{R}uw$ in $\mathrm{M}_n(D)$.*

3. *If S is π-regular, then the length of every chain of nonzero principal right (two-sided) ideals of S is less than t (t^2 respectively).*

4. *There exists $z = z(m_1, \ldots, m_{n-1})$ such that if J is an ideal of S and $S \setminus J$ does not intersect maximal subgroups of $\mathrm{M}_n(D)$, then S/J is nilpotent of index not exceeding z. In particular, this holds if S/J is a nil semigroup.*

A chain of the type as described in 1) above will be called a structural chain of S. (Note that it may be not unique because of the possible permutation of the order of uniform components coming from the same rank in $\mathrm{M}_n(D)$.) For $S \subseteq \mathrm{M}_n(K)$, where K is a field, the numerical bounds

m_j, r, t, z can be defined by $m_j = \binom{n}{j}, r = 2^n + n - 1, t = 2^n \prod_{i=1}^{n} \binom{n}{i}$ and $z = \prod_{k=1}^{n} \binom{n}{k}$. This is proved through an exterior power argument.

The assertions of Corollary 2.4.7 can be interpreted as results on repetitions of certain patterns in sufficiently long products of elements of S. In this context, the following useful result, obtained in [95], is also of interest. It is an application of an extension of van der Waerden's theorem on arithmetic progressions, see Problem 4.1.1 in [104].

Theorem 2.4.8. *Let $a_1, a_2, \dots$ be a sequence of elements of a finite semigroup S. Then there exist an idempotent $e \in S$ and $r \geq 1$ such that for every integer $q > 1$ there exist indices $i_1 < \cdots < i_q$ such that $i_{j+1} - i_j \leq r$ and $a_{i_j} \cdots a_{i_{j+1}-1} = e$ for $j = 1, \dots, q-1$.*

The following special instance of assertion 4) in Lemma 2.4.7 is classical.

Proposition 2.4.9. *Let $S \subseteq \mathrm{M}_n(D)$ be a semigroup. Then if S has a zero θ, then $S \subseteq (I - \theta)\mathrm{M}_n(D)(I - \theta) + \theta \cong \mathrm{M}_t(D)$, where $t = n - \mathrm{rank}(\theta)$. Moreover, if S is a nil semigroup, then $S^t = \{\theta\}$.*

It is worth mentioning that the following more general result establishes nilpotency of certain nil semigroups, see [38, 17.22].

Theorem 2.4.10. *Let S be a monoid with a zero element. If S satisfies the ascending chain condition on one-sided ideals then every nil ideal of S is nilpotent.*

Our basic approach, resulting from the structure theorem, is to study $S \subseteq \mathrm{M}_n(D)$ via its uniform components. The aim here is to reduce problems on S to questions on the associated linear groups and the properties of the corresponding sandwich matrices. This approach can be successful once we are able to transfer the properties from S, or from the cancellative subsemigroups of type $S \cap H$ where H is a maximal subgroup of $\mathrm{M}_n(D)$, to the corresponding properties of the associated groups $\mathrm{gr}(S \cap H)$. An example of such a transfer of properties reads as follows.

Proposition 2.4.11. *Let T be a cancellative subsemigroup of a semigroup S. Assume that T has a group of (left and right) quotients. Then there exists a strict order preserving mapping from the lattice of right congruences on G to the lattice of right congruences on S. Hence, if S satisfies the ascending chain condition on right congruences, then G satisfies the ascending chain condition on subgroups.*

A related assertion, namely that $K[TT^{-1}]$ is right Noetherian if $K[S]$ is right Noetherian, can be sometimes proved. This will appear in Proposition 5.1.3 and its applications.

In order to understand the global structure of a linear semigroup S one often needs also to handle the relationship between different uniform components. It seems that the action of the associated groups on the uniform components of S that lie below is crucial here. This approach may be described as follows.

Lemma 2.4.12. *Let H be a maximal subgroup of $\mathrm{M}_n(D)$ and $e = e^2 \in H$. Consider the monoid $M(e) = e\,\mathrm{M}_n(D)e \cong \mathrm{M}_{\mathrm{rank}(e)}(D)$. If U is a uniform component of $S \subseteq \mathrm{M}_n(D)$ intersecting $M(e)$, then $U \cap M(e)$ is a uniform component of $S \cap M(e)$. If additionally S is π-regular, then $S \cap H$ is a group and it acts by conjugation on $U \cap M(e)$.*

We conclude with some examples.

Example 2.4.13. 1. Let $S \subseteq \mathrm{M}_n(D)$ be the semigroup of all diagonal matrices. Then S is a disjoint union of groups (each of type $(D^*)^j, j \leq n$, where $D^* = D \setminus \{0\}$), S is regular, and at every rank j it has exactly $\binom{n}{j}$ groups which are the uniform components of S.
2. Let $S \subseteq \mathrm{M}_n(D)$ be the set of all monomial matrices. That is, $a \in S$ if each row (and each column) of a has at most one nonzero entry. It is easy to check that S is a regular semigroup whose only ideals are $S_j = S \cap \mathrm{M}_j$, $j = 0, 1, \ldots, n$. Moreover, S_j/S_{j-1} is isomorphic to a completely 0-simple semigroup $\mathcal{M}(G_j, \binom{n}{j}, \binom{n}{j}; I_j)$, where G_j is the group of all monomial matrices in $\mathrm{GL}_j(D)$ and I_j is the identity matrix.
3. Let $S \subseteq \mathrm{M}_n(K)$ be a Zariski closed connected monoid over a field K with group of units G. Then S is π-regular and the uniform components of S are of the form GaG, where $\mathrm{rank}(a) = \mathrm{rank}(a^2)$, see [133].
4. Assume that $\mathrm{M}_n(D)$ is such that there exists an infinite triangular set of idempotents T in some $\mathrm{M}_j / \mathrm{M}_{j-1}, j < n$ (division algebras D of this type were constructed in [139], see [123]). Define $U = \{s \in \mathrm{M}_j / \mathrm{M}_{j-1} \mid$ there exist $e, f \in T$ such that $e \preceq f, se = e$ and $fs = f\}$. Then $S = U \cup \mathrm{M}_{j-1}$ is a semigroup. Moreover, it can be checked that S has infinitely many uniform components consisting of matrices of rank j and the remaining matrices of rank j yield a nil ideal of $S/\mathrm{M}_{j-1} = S_j/S_{j-1}$ which is not nilpotent.

2.5 Closure

Our next aim is to introduce a closure operation that is an important technical tool when applying the uniform components of a linear semigroup. Let $J = \mathcal{M}(G, X, Y; P)$ be a semigroup of matrix type over a group G. If for every $a, b \in J$ the condition $ax = bx$ for all $x \in J$ implies that $a = b$, and the condition $xa = xb$ for all $x \in J$ implies that $a = b$, then we say that J is annihilator-free. This applies for example if $J = \mathcal{M}(G, n, n; I)$.

Lemma 2.5.1. *Let I be an ideal of a semigroup S. Assume that J is a completely 0-simple semigroup such that I is a uniform subsemigroup of J and J is a completely 0-simple closure of I. If either J is a semigroup of (right and left) quotients of I or J is annihilator-free, then there is a unique semigroup structure on the disjoint union $S' = (S \setminus I) \cup J$ that extends the operation on S.*

Proof. Let $J = \mathcal{M}(G, X, Y; P)$. Choose a maximal subgroup $H = J_{(x)}^{(y)} = \{(g, x, y) \mid g \in G\}$ of J, where $x \in X, y \in Y$. Since I is a uniform subsemigroup of J, we know that there exist $a_i \in I \cap J_{(i)}^{(y)}$ and $b_j \in I \cap J_{(x)}^{(j)}$ for all $i \in X, j \in Y$. Then $J = \bigcup_{i,j} a_i H b_j \cup \{\theta\}$ because the right side is a completely 0-simple semigroup intersecting every $\mathcal{H}$-class of J and containing H. Also every non-zero element of J has a unique presentation of the form $a_i h b_j$ where $i \in X, j \in Y, h \in H$. Let $s \in S \setminus I$ and $h \in H$. On the disjoint union $(S \setminus I) \cup J$ we define the products

$$s \cdot a_i h b_j = (s a_i)(h b_j),\ a_i h b_j \cdot s = (a_i h)(b_j s),$$

where the products $s a_i$ and $b_j s$ are considered in S, while the other products on the right sides of these equalities are considered in J. Also, define $s \cdot t = st$ (the latter product is considered in S) if $s, t \in S \setminus I$. Finally, let $s \cdot \theta = \theta \cdot s = \theta$ for $s \in S$. First we claim that this extends the operation on S. Clearly $s \cdot t = st$ if $s, t \in I$. So we have to show that $s \cdot t = st$ and $t \cdot s = ts$ if $s \in S, t \in I$. We may assume that $t \neq \theta$. Write $t = a_i h b_j$.

In order to prove that $t \cdot s = ts$, first notice that for every $u \in I$ we get

$$\begin{aligned}
(t \cdot s)u &= [(a_i h b_j) \cdot s]u = \text{(using the definition)} \qquad (2.1)\\
& [(a_i h)(b_j s)]u = \text{(using associativity in } J)\\
& (a_i h)[(b_j s)u] = \text{(using associativity in } S)\\
& (a_i h)[b_j(su)] = \text{(using associativity in } J)\\
& [(a_i h) b_j](su) = t(su) = \text{(using associativity in } S)\\
& (ts)u.
\end{aligned}$$

If $t \cdot s = \theta$ then $(ts)u = \theta$ for every $u \in I$. Since $IJ = J$, we get $tsJ = \theta$, and hence $ts = \theta$. The converse follows similarly. So we may assume that $t \cdot s \neq \theta \neq ts$. Choosing u such that $(ts)u \neq \theta$ we see that $ts\mathcal{R}(t \cdot s)$ in J.

If J is annihilator-free then (2.1) implies that $ts = t \cdot s$. We now will show that this equality also holds if J is a semigroup of (right and left) quotients of I then for every $\theta \neq x, y \in I$ such that $x\mathcal{L}y$ in J there exist $x', y' \in I$ such that $x'x = y'y \neq \theta$. Suppose that, for $s \in S$, one of the elements xs, ys is non-zero, say $xs \neq \theta$. Then $xsz \neq \theta$ for some $z \in I$ and we get $y'y(sz) = x'x(sz) \neq \theta$. This implies that $xs\mathcal{L}ys$ in J. So S acts by right multiplication on I in such a way that it permutes the sets $L \cap I$, where L are $\mathcal{L}$-classes of J. Since $b_j\mathcal{L}t$ in J, this implies that $b_js\mathcal{L}ts$. As we also have $b_js\mathcal{L}(t \cdot s)$, it follows that $ts\mathcal{L}(t \cdot s)$. Therefore (2.1) easily implies that $t \cdot s = ts$ also in this case.

A symmetric argument shows that $s \cdot t = st$. This completes the proof of the claim.

Furthermore, if we get a semigroup structure on S', that extends those on S and on J, it must be unique. So it remains to show that the defined operation is associative. For this consider any $s, t \in S$ and $a, b \in J$. We need to check associativity only for the products $sta, tas, ast, sab, asb, abs$ (we simplify notation, writing as, sa in place of $a \cdot s, s \cdot a$, respectively). Let $a = a_m f b_n$ and $b = a_p f' b_q$ for some $m, p \in X, n, q \in Y, f, f' \in H$. Put $z = f b_n a_p f'$. Then $z \in H \cup \{\theta\}$ and we get

$$\begin{aligned}(ab)s &= (a_m f b_n a_p f' b_q)s = (a_m z b_q)s = (a_m z)(b_q s) = \\ &\quad (a_m f b_n a_p f')(b_q s) = (a_m f b_n)(a_p f' b_q s) = a(bs)\end{aligned} \tag{2.2}$$

(the third equality by the definition above and the fifth because J is a semigroup). The case sab is symmetric. For the cases ast, sta, by symmetry we may consider ast only:

$$\begin{aligned}(as)t &= [(a_m f b_n)s]t = [(a_m f)(b_n s)]t = \ \text{(using (2.2))} \\ &\quad (a_m f)[(b_n s)t] = \ \text{(using associativity in } S) \\ &\quad (a_m f)[b_n(st)] = \ \text{(using (2.2))} \\ &\quad [(a_m f)b_n](st) = a(st).\end{aligned} \tag{2.3}$$

The cases asb and tas are dealt with as follows:

$$\begin{aligned}(as)b &= [(a_m f b_n)s]b = [(a_m f)(b_n s)]b = \ \text{(using associativity in } J) \\ &\quad (a_m f)[(b_n s)b] = \ \text{(using associativity in } S) \\ &\quad (a_m f)[b_n(sb)] = \ \text{(using associativity in } J) \\ &\quad [(a_m f)b_n](sb) = a(sb),\end{aligned}$$

$$
\begin{aligned}
t(as) &= t[(a_m f b_n)s] = t[(a_m f)(b_n s)] = \text{ (by the case symmetric to (2.2))} \\
&\quad [t(a_m f)](b_n s) = \text{ (using (2.2))} \\
&\quad [(t(a_m f))b_n]s = \text{ (by the case symmetric to (2.2))} \\
&\quad [t((a_m f)b_n)]s = (ta)s.
\end{aligned}
$$

This completes the proof. □

A related result on linear semigroups can be proved under slightly weaker assumptions. The following is an easy consequence of Lemma 2.5.3.

Lemma 2.5.2. *Let $S \subseteq \mathrm{M}_n(D)$ be a semigroup and I the ideal of elements of the least nonzero rank (with 0 adjoined if it is in S). If I is a uniform subsemigroup of a completely 0-simple subsemigroup of $\mathrm{M}_n(D)$ (so $I \setminus \{0\}$ is a uniform component of S) and $\widehat{I}$ is its completely 0-simple closure then $\widehat{S} = (S \setminus I) \cup \widehat{I}$ (a disjoint union) is a subsemigroup of $\mathrm{M}_n(D)$.*

When studying the structure of S it is often convenient to replace S by its linear homomorphic image T in such a way that the given uniform component U of S becomes an ideal of T. The following lemma, though rather technical, is essential for this approach.

Lemma 2.5.3. *Assume that U is a uniform component of a semigroup $S \subseteq \mathrm{M}_n(D)$. Let $\widehat{U}$ be the completely 0-simple closure of U in the corresponding $\mathrm{M}_j / \mathrm{M}_{j-1}$ and G a maximal subgroup of $\widehat{U}$.*

1. *S and the semigroup S_G generated by $S \cup G$ intersect the same $\mathcal{H}$-classes of $\mathrm{M}_n(D)$ contained in $\mathrm{M}_n(D) \setminus \mathrm{M}_{j-1}$.*

2. *$\widehat{U}$ is a uniform component of S_G. Therefore $\widehat{U}$ is an ideal of the Rees factor $\widehat{S} = S_G/J$, where*

$$
\begin{aligned}
J &= \{a \in S_G G S_G \mid \operatorname{rank}(a) < j \text{ or } \operatorname{rank}((bac)^2) < j \\
&\qquad \text{for all } b, c \in S_G\}.
\end{aligned}
$$

In particular, if U is an ideal uniform component of S, then $\widehat{S} = S \cup \widehat{U} = S_G$.

Proof. Since $G \cup U$ generates $\widehat{U}$, it is clear that $\widehat{U}$ is contained in a uniform component V of S_G. Let $a \in \widehat{U}$ be nonzero (the set of nonzero elements of $\widehat{U}$ is identified with a subset of S_G) and let $s \in S$. Then, since U is a uniform subsemigroup of $\widehat{U}$, there exists $a' \in \widehat{U}$ such that $a' \mathcal{H} a$ in $\mathrm{M}_n(D)$. If $\operatorname{rank}(as) = j$, then we have $a\,\mathrm{M}_n(D) = as\,\mathrm{M}_n(D)$ and $\mathrm{M}_n(D)a' = \mathrm{M}_n(D)a$.

Hence $\text{rank}(a's) = j$ and $as\mathcal{H}a's$. This means that the $\mathcal{H}$-class of as in $M_n(D)$ contains an element of S. The same holds for sa. A similar argument shows that, if $z \in S_G$ is of rank j and is in the $\mathcal{H}$-class of an element of S, then, for any $t \in S \cup G$, each of the elements tz, zt also is in the $\mathcal{H}$-class of $M_n(D)$ intersected by S, whenever it is of rank j. Therefore, an induction on the length of the elements of S_G as words in $S \cup G$ allows to show that S and S_G intersect the same $\mathcal{H}$-classes of $M_n(D)$ consisting of matrices of rank j. In view of the Theorem 2.4.3 this implies that $\widehat{U}$ intersects the same $\mathcal{H}$-classes of $M_n(D)$ as V.

Assume that as is in the $\mathcal{H}$-class of an element of $\widehat{U}$ for some $a \in \widehat{U}$ and $s \in S$. Choose $e \in \widehat{U}$ such that $as = ase$. Since $U\widehat{U} = \widehat{U}$, we can write $e = xy$ for some $x \in U$ and $y \in \widehat{U}$. Then $asx\mathcal{H}a'sx$ for an element $a' \in U$ such that $a'\mathcal{H}a$. The structure theorem applied to S implies that $a'sx \in U$ and, since sx and $a'sx$ are in the same $\mathcal{L}$-class of $M_n(D)$, we must also have $sx \in U$. Hence $as = asxy \in \widehat{U}U\widehat{U} \subseteq \widehat{U}$. This shows that $\widehat{U}S \cap V \subseteq \widehat{U}$ because V, U intersect the same $\mathcal{H}$-classes of $M_n(D)$. Similarly one gets $S\widehat{U} \cap V \subseteq \widehat{U}$. Again by an induction argument, this implies that $V = \widehat{U}$ is a uniform component of S_G. Since the matrices of rank j in $S_G G S_G$ are contained in $\widehat{U} \cup N$, where N is the nilpotent component of S_G consisting of matrices of rank j (by Theorem 2.4.3) $\widehat{U}$ is an ideal of $\widehat{S}$.

Finally, assume that U is an ideal uniform component of S. Suppose that $a \in \widehat{U}$ and $s \in S$ are such that $as \in J$. As above, there exist $v \in U, u \in \widehat{U}$ such that $uv \in \widehat{U}$ satisfies $uva = a$. Choose If $u' \in U$ with $u'\mathcal{H}u$. Then $uvas \in J$ implies that $u'vas \in J \cap U$. Therefore $u'vas = 0$. It follows that $as = 0$ because $u'va\mathcal{L}a$. Similarly, $sa \in J$ implies $as = 0$. This proves that $S \cup \widehat{U}$ is a semigroup and $\widehat{U}$ is its ideal. Consequently $S_G = S \cup \widehat{U}$. Moreover $J = \{0\}$ or $J = \emptyset$, so that $S_G = \widehat{S}$. This completes the proof. □

If $S_G \subseteq M_n(D)$ is the semigroup constructed above, then $\widehat{U}$ can be considered as a subsemigroup of $S_G/(S_G \cap M_{j-1}) \subseteq M_n(D)/M_{j-1}$. In particular, Lemma 2.6.7 often allows us to study $\widehat{U} \subseteq S_G/(S_G \cap M_{j-1})$ as linear semigroups.

The next step is to try to construct a 'group approximation' of S in the sense that every $a \in S$ such that $\text{rank}(a) = \text{rank}(a^2)$ lies in a subgroup of S. This is more complicated and is based on the following observation.

Proposition 2.5.4. *$M_n(D)$ is π-regular. Moreover, if $a \in M_n(D)$, then the set $A_a = \{a^k \mid k \geq n\}$ is contained in a subgroup of $M_n(D)$. More generally, if $S \subseteq M_n(D)$ is a regular semigroup, then, for every $a \in S$, A_a is contained in a subgroup of S.*

Theorem 2.5.5. *For every semigroup* $S \subseteq \mathrm{M}_n(D)$ *there exists the smallest* π*-regular semigroup* $\mathrm{cl}(S) \subseteq \mathrm{M}_n(D)$ *containing* S. *Moreover, for every maximal subgroup* H *of* $\mathrm{M}_n(D)$ *intersecting* $\mathrm{cl}(S)$, $H \cap \mathrm{cl}(S)$ *is a maximal subgroup of* $\mathrm{cl}(S)$ *and, for every* $a \in \mathrm{cl}(S)$, a^n *lies in a maximal subgroup of* $\mathrm{cl}(S)$.

The semigroup $\mathrm{cl}(S)$ will be called the π-regular closure of S. In the simplest case, where $S \subseteq \mathrm{GL}_n(D)$, $\mathrm{cl}(S)$ is the group generated by S. If $S \subseteq \mathrm{M}_n(K)$ for a field K then $\mathrm{cl}(S)$ usually is much closer to S than the Zariski closure $\overline{S}$, see Theorem 2.6.1, so it is more useful when studying the algebraic properties of S. More insight into the construction of $\mathrm{cl}(S)$ is given in [123]. Some important results in this direction are recalled in Section 2.6.

We finish this section with a preliminary analysis of the consequences of the situation described in Lemma 2.5.1 and Lemma 2.5.2 in the context of the corresponding semigroup algebras. For a group H, we denote by $\mathcal{M}_n(H) \subseteq \mathrm{M}_n(K[H])$ the monoid of all monomial $n \times n$ matrices over H. That is, $\mathcal{M}_n(H)$ consists of all matrices $s = (s_{ij})$ such that each row, and each column, of s contains at most one nonzero entry and all entries are in $H^0 = H \cup \{0\}$. Note that, if H is the trivial group, then $\mathcal{M}_n(H)$ is the full symmetric inverse monoid. The set of matrices with at most one nonzero entry is an ideal of $\mathcal{M}_m(H)$ and is isomorphic to the completely 0-simple semigroup $\mathcal{M}(H, m, m; I)$ (that is, a Brandt semigroup over H). More generally, if X is any set, then one can define the monoid $\mathcal{M}_X(H)$ of all monomial $X \times X$ matrices over H. Then $\mathcal{M}_X(H)$ can be considered as a submonoid of the algebra $\mathrm{M}_X^{fin}(K[H])$ of matrices over $K[H]$ with finitely many nonzero entries in each column.

The hypothesis of the following result is satisfied if J is an ideal uniform component of a subsemigroup $S \subseteq \mathrm{M}_n(D)$ or if the assumptions of Lemma 2.5.1 are met. We write $\mathrm{la}(R)$ for the the left annihilator of an algebra R.

Proposition 2.5.6. *Assume that* J *is an ideal of a semigroup* S *such that* J *has a completely* 0*-simple closure* $\widehat{J} = \mathcal{M}(G, X, Y; P)$ *and* $(S \setminus J) \cup \widehat{J}$ *has a semigroup structure extending that of* S. *Then there exists a homomorphism* $\phi : K_0[S] \longrightarrow \mathrm{M}_X^{fin}(K[G])$, *such that* $\ker(\phi) \cap K_0[J] = \mathrm{la}(K_0[J])$, $(\ker(\phi) \cap K_0[J])^2 = \{0\}$ *and* $\phi(S)$ *consists of column monomial matrices over* G. *Moreover, if* X *is finite and* S *is finitely generated, then* G *is finitely generated.*

Proof. Choose $e = e^2 \in D$ for a maximal subgroup D of $\widehat{J}$. So $D \cong G$. Put $J' = \widehat{J}e$. Then $D = e\widehat{J}e \backslash \{\theta\}$ acts on J' by right multiplication. Hence $K_0[J']$

is a right $K[D]$-module. Note that it is a free $K[D]$-module of rank X, by the hypothesis. Indeed, picking one element from every nonzero $\mathcal{H}$-class of $\widehat{J}$ intersected by J', we get a basis $e_x, x \in X$ of this module. We may assume that $e = e_x$ for some $x \in X$. Now, each $s \in S$ acts by left multiplication on $K_0[J']$, so it determines an endomorphism of $K_0[J']$. Hence, we get a homomorphism $\phi : K_0[S] \longrightarrow \mathrm{End}_{K[D]}(K_0[J']) \subseteq \mathrm{M}_X^{fin}(K[D])$. From the definition of ϕ it is clear that each column of s has at most one nonzero entry. Hence ϕ maps S into column monomial matrices $\mathcal{M}_X(D)$ over D.

Since $\widehat{J}e\widehat{J} = \widehat{J}$, it is clear that $\ker(\phi) \cap K_0[J]$ coincides with the left annihilator $\mathrm{la}(K_0[J])$ of $K_0[J]$. Therefore $(\ker(\phi) \cap K_0[J])^2 = 0$.

Now assume also that $|X| = n < \infty$. Notice that $\mathrm{M}_X^{fin}(K[D]) \cong \mathrm{M}_n(K[D])$. If S is finitely generated, then $\phi(S) = \prec\phi(s_1), \ldots, \phi(s_t)\succ$ for some $s_i \in S$. Let $H \subseteq D$ be the group generated by the nonzero entries of all matrices $\phi(s_i)$. By the definition of ϕ it is clear that $S \cap D \subseteq H$. Since D is generated by $S \cap D$, it follows that $D = H$ is a finitely generated group. □

Under some extra assumptions, much more can be said in the context of the last proposition. Namely, we have the following classical result, see Proposition 5.25 and Corollary 4.8 in [119].

Lemma 2.5.7. *Assume that J is an ideal of a semigroup S with zero such that $J \cong \mathcal{M}(G, X, Y; P)$ is a semigroup of matrix type over a semigroup G. Then $K_0[J]$ has an identity element if and only if X, Y are finite sets such that $n = |X| = |Y|$ and P is invertible as a matrix in $\mathrm{M}_n(K[G])$. In this case $K_0[J] \cong \mathrm{M}_n(K[G])$ and $K_0[S] \cong K_0[S/J] \oplus K_0[J]$; moreover, J is completely 0-simple if G is a group.*

2.6 Semigroups over a field

We recall two powerful special techniques that can be used in case of linear semigroups over a field and we state some of their consequences. Because of Theorem 3.5.2 and Theorem 5.1.5, this applies to right Noetherian PI-algebras $K[S]$, whence to several algebras considered in the main body of the book.

Assume that K is an algebraically closed field. A subset $X \subseteq K^n$ is closed if it is the zero set of a collection of polynomials in $K[x_1, \ldots, x_n]$. This defines the Zariski topology on K^n. Closed subsets of K^n are in one-to-one correspondence with radical ideals of $K[x_1, \ldots, x_n]$. In particular, since $K[x_1, \ldots, x_n]$ is Noetherian, we have the descending chain condition on closed subsets of K^n.

When the K-space $\mathrm{M}_n(K)$ is identified with K^{n^2}, the polynomial ring $K[x_{ij} \mid i, j = 1, 2, \ldots, n]$ can be used to introduce the Zariski topology on $\mathrm{M}_n(K)$. Here the matrix units e_{ij} form the standard basis of $\mathrm{M}_n(K)$ and they correspond to the variables x_{ij}. The closure of a subset A of $\mathrm{M}_n(K)$ is denoted by $\overline{A}$. It is well known that right (and left) multiplication by any $a \in \mathrm{M}_n(K)$ is a continuous map in Zariski topology. More generally, if $A \subseteq B$ are subsets of $\mathrm{M}_n(K)$, then A is closed in B if it is of the form $X \cap B$ for a closed subset X of $\mathrm{M}_n(K)$. If A is a subset of $\mathrm{GL}_n(K)$, then by the closure of A in $\mathrm{GL}_n(K)$ we mean $\overline{A} \cap \mathrm{GL}_n(K)$. This defines the Zariski topology on $\mathrm{GL}_n(K)$. For the theory of linear groups, linear algebraic groups and linear algebraic semigroups (that is, closed subgroups of $\mathrm{GL}_n(K)$ and closed subsemigroups of $\mathrm{M}_n(K)$) we refer to [10, 67, 113, 133, 134, 153].

The basic property of the closure of a semigroup $S \subseteq \mathrm{M}_n(K)$ reads as follows.

Theorem 2.6.1. *Let $S \subseteq \mathrm{M}_n(K)$ be a semigroup. Then $\overline{S}$ is a π-regular subsemigroup of $\mathrm{M}_n(K)$ containing S and $\mathrm{cl}(S) \subseteq \overline{S} \subseteq K\{S\}$, where the latter denotes the K-linear span of S.*

The Zariski closure is especially useful when dealing with identities expressible in terms of the coordinate ring $K[x_{ij} \mid i, j = 1, \ldots, n]$ of $\mathrm{M}_n(K)$. This applies in particular to semigroup identities satisfied on a linear semigroup S. Recall that S satisfies a semigroup identity if there are two elements $w = w(x_1, \ldots, x_n), u = u(x_1, \ldots, x_n)$ of a free semigroup $X = \langle x_1, \ldots, x_n \rangle$ such that $w \neq u$ but $w(a_1, \ldots, a_n) = u(a_1, \ldots, a_n)$ for all $a_1, \ldots, a_n \in S$. The main result in this direction reads as follows.

Theorem 2.6.2. *Assume that $S \subseteq \mathrm{M}_n(K)$ satisfies a semigroup identity. Then $\mathrm{cl}(S) \subseteq \overline{S}$ also satisfy this identity.*

More structural information on semigroups of this type is given in Theorem 3.5.7 in case $S \subseteq \mathrm{GL}_n(K)$.

We continue with a discussion of the connections between the structure of S and of $\mathrm{cl}(S)$ for $S \subseteq \mathrm{M}_n(K)$. The results come from [123]. The first result explains the basic relation between the components of S and the components of its closure.

Theorem 2.6.3. *Let U be a uniform component of $S \subseteq \mathrm{M}_n(K)$ and N a nilpotent component of S.*

1. *$N = N' \cap S = N'' \cap S$ for some nilpotent components N', N'' of the Zariski closure $\overline{S}$, and of* $\mathrm{cl}(S)$ *respectively,*

2. *$U = U' \cap S = U'' \cap S$ for some uniform components U', U'' of $\overline{S}, \mathrm{cl}(S)$ respectively.*

The following example shows that $\mathrm{cl}(S)$ may have more uniform components than S.

Example 2.6.4. Let $S = \prec x, e \succ \subseteq \mathrm{M}_3(\mathbb{Q})$, where

$$x = \begin{pmatrix} 1 & 1 & 0 \\ 1 & 0 & 0 \\ 0 & 0 & 1 \end{pmatrix}, \quad e = \begin{pmatrix} 1 & 0 & 0 \\ 0 & 0 & 0 \\ 0 & 0 & 1 \end{pmatrix}.$$

Then $ex^{-1}e = \begin{pmatrix} 0 & 0 & 0 \\ 0 & 0 & 0 \\ 0 & 0 & 1 \end{pmatrix} \in \mathrm{cl}(S)$. The $(1,1)$ entry of x^n is positive, so that ex^ne is a matrix of rank 2 for every $n \geq 1$. The same is true of every $a \in \prec ex^n, x^ne, ex^ne \mid n \geq 1 \succ$. This easily implies that the set of elements of rank 2 in S forms a uniform component U of S and $S = U \cup \prec x \succ$. Therefore $\mathrm{cl}(S)$ has more components than S because $ex^{-1}e$ lies in a uniform component of $\mathrm{cl}(S)$.

In general, a uniform component U of S can intersect less $\mathcal{R}$- (and also less $\mathcal{L}$-) classes of $\mathrm{M}_n(D)$ than the uniform component of $\mathrm{cl}(S)$ containing U. However, this does not happen in the following case that is of special interest in view of applications to right Noetherian semigroup algebras.

Theorem 2.6.5. *Let U be a uniform component of a semigroup $S \subseteq \mathrm{M}_n(K)$ such that U intersects finitely many $\mathcal{R}$-classes of $\mathrm{M}_n(K)$.*

1. *The uniform components U', U'' of the Zariski closure $\overline{S}$, and of $\mathrm{cl}(S)$ respectively, containing U intersect the same $\mathcal{R}$-classes of $\mathrm{M}_n(K)$ as U.*

2. *For every maximal subgroup G of $\mathrm{cl}(S)$ intersecting U we have $G = \mathrm{gr}(S \cap G)$.*

The second general method mentioned in this section is concerned with the exterior powers. Throughout the rest of this section K stands for any field. Let $\Lambda^j V$ be the j-th exterior power of the column vector space $V = K^n$. Recall that, if $e_1, \ldots, e_n$ is a basis of K^n, then $e_{i_1} \wedge \cdots \wedge e_{i_j}, 1 \leq i_1 <$

$\ldots < i_j \leq n$, is a basis of $\Lambda^j V$. If W is a vector space, then every linear map $\phi : V \longrightarrow W$ uniquely extends to a linear map $\Lambda^j \phi : \Lambda^j V \longrightarrow \Lambda^j W$. The exterior power map $\Lambda^j : \mathrm{End}(V) \longrightarrow \mathrm{End}(\Lambda^j V)$ is defined by $\Lambda^j(\phi) = \Lambda^j \phi$. Therefore, Λ^j can be viewed as a map $\mathrm{M}_n(K) \longrightarrow \mathrm{M}_{\binom{n}{j}}(K)$. More precisely, we have

$$\Lambda^j(a)(a_1 \wedge \cdots \wedge a_j) = a(a_1) \wedge \cdots \wedge a(a_j)$$

for every $a_1, \ldots, a_j \in K^n$.

The following well known observations will be very useful, see [11].

Lemma 2.6.6. $\Lambda^j : \mathrm{M}_n(K) \longrightarrow \mathrm{M}_{\binom{n}{j}}(K)$ *is a semigroup homomorphism and the following assertions hold.*

1. $\mathrm{rank}(\Lambda^j(a)) = \binom{r}{j}$ *if* $\mathrm{rank}(a) = r \geq j$, *and* $\mathrm{rank}(\Lambda^j(a)) = 0$ *otherwise.*
2. *For every vectors* $a_1, \ldots, a_j \in K^n$ *and every* $a_{ik} \in K, i, k = 1, \ldots, j$, $(\sum_{k=1}^j a_{1k} a_k) \wedge \cdots \wedge (\sum_{k=1}^j a_{jk} a_k) = \det(a_{ik})(a_1 \wedge \cdots \wedge a_j)$.
3. *If* $V \subseteq K^n$ *is a subspace with a basis* $f_1, \ldots, f_j$, *and* $a \in \mathrm{M}_n(K)$ *is a matrix of rank* j *which determines an isomorphism* $a_{|V} : V \longrightarrow V$, *then* $\Lambda^j(a)(f_1 \wedge \cdots \wedge f_j) = \det(a_{|V})(f_1 \wedge \cdots \wedge f_j)$.

Exterior power provides a convenient tool to approach linear semigroups inductively. This is a consequence of the following observation, see [123, Lemma 1.8].

Lemma 2.6.7. *Let* $1 \leq j \leq n$. *For every* $r > 1$ *there exists a homomorphism* $\phi : \mathrm{M}_n(K)/\mathrm{M}_{j-1} \longrightarrow \mathrm{M}_t(K), t = \binom{rn}{r(j-1)+1}$, *such that* $\phi(a) = \phi(b)$ *if and only if* $a = \lambda b$ *for an* $r(j-1)+1$*-th root of unity* $\lambda \in K$. *Moreover, if the field* K *is finitely generated, then* r *can be chosen so that* ϕ *is an embedding.*

Chapter 3

Prerequisites on ring theory

This chapter is devoted to the necessary background on ring theory. Our investigations force us to consider a rather wide variety of topics. Only results essential for later use are stated. For the convenience of the reader we mostly will refer to standard references. Topics on Noetherian rings, rings of quotients and rings satisfying a polynomial identity can be found mainly in [56, 112, 137], relevant results on finite dimensional algebras in [34] and on the Gelfand-Kirillov dimension in [100, 112], topics on principal ideal rings in [70, 112] and on group rings and graded rings in [129, 130]. Because of their relevance for the motivating class of Noetherian group algebras, results on polycyclic-by-finite groups also are included in this chapter (and not in Chapter 2). For earlier work on semigroup rings of arbitrary semigroups we refer to [119] and to [54] for semigroup rings of abelian semigroups.

3.1 Noetherian rings and rings satisfying a polynomial identity

In this section we recall some basic terminology and some well known fundamental structural results on Noetherian rings and algebras satisfying a polynomial identity.

A ring R will mean a nonzero associative ring, usually assumed to have an identity 1, except specified otherwise. Ring homomorphisms will be expected to preserve the identity and subrings to have the same identity. A right R-module M is assumed to be unital, that is, $m1 = m$ for all $m \in M$. A faithful right R-module M is one such that $Mr \neq \{0\}$ if $0 \neq r \in R$. A nonzero module M is said to be simple, or irreducible, if $\{0\}$ and M are its only submodules. A module that is the direct sum of simple submodules is called semisimple. A ring R is (right) semisimple if it is semisimple as a right R-module.

For a subset X of a ring R (or more general for a subset X of a semigroup with zero) we put $\text{ra}_R(X) = \{r \in R \mid Xr = \{0\}\}$, the right annihilator of X in R. If $X = \{r\}$ then we simply denote this set by $\text{ra}_R(x)$. One says that an element r is right regular if $\text{ra}_R(r) = \{0\}$. Similarly one defines the left annihilator $\text{la}_R(X)$ and left regular elements. An element that is both right and left regular is said to be regular.

A ring R is said to be prime if the product of two nonzero ideals is nonzero. Obviously, in this case, the center $\text{Z}(R)$ of R is a domain. A proper ideal P of R is prime if R/P is a prime ring. A minimal prime ideal in R is a prime ideal of R that does not properly contain an other prime ideal. Any prime ideal contains a minimal prime ideal.

The prime spectrum $\text{Spec}(R)$ of a ring R is the set of all prime ideals of R. The height $\text{height}(P)$ of a prime ideal P of R is the largest length of a chain of prime ideals contained in P. So $\text{height}(P) = 0$ means P is a minimal prime ideal. The set of minimal prime ideals is written as $\text{Spec}^0(R)$ and the set of height one prime ideals of R is $\text{Spec}^1(R)$. The classical Krull dimension $\text{clKdim}(R)$ of R is the supremum of the lengths of all finite chains of prime ideals.

An element r in a ring R is nilpotent if $r^n = 0$ for some positive integer n. If each element of a subset N of R is nilpotent then the subset N is said to be nil. If, furthermore, $N^n = \{0\}$ then N is called nilpotent. A ring R is said to be semiprime if it does not contain nonzero nilpotent ideals. An ideal I of R is semiprime if R/I is a semiprime ring. The prime radical $\mathcal{B}(R)$ of R is the intersection of all prime ideals of R. A ring R is semiprime if and only if $\mathcal{B}(R) = \{0\}$. Every nilpotent ideal is contained in $\mathcal{B}(R)$ and $\mathcal{B}(R) = \{0\}$ if and only if $\{0\}$ is the only nilpotent ideal of R.

Any prime ideal is contained in a maximal ideal and maximal ideals are prime ideals. A ring R is said to be simple if R has precisely two ideals, that is, $\{0\}$ is a maximal ideal. A ring R is right primitive if R has a faithful simple right R-module M. An ideal I is right primitive if R/I is a right primitive ring. Maximal ideals are right primitive. The Jacobson radical of R is denoted by $\mathcal{J}(R)$, it is the intersection of all right primitive ideals of R. If $\mathcal{J}(R) = \{0\}$ then R is said to be semiprimitive. Recall that $\mathcal{B}(R) \subseteq \mathcal{J}(R)$.

A ring R is said to be right Noetherian (respectively right Artinian) if it satisfies the ascending (respectively descending) chain condition on right ideals. Similarly one defines left Noetherian rings (respectively left Artinian rings). A Noetherian (respectively Artinian) ring is one that is both right and left Noetherian (respectively Artinian). A right Artinian ring is right Noetherian.

A well known result is the following ([56, Theorem 2.2.4]).

Theorem 3.1.1. *A right Noetherian ring contains only finitely many minimal prime ideals.*

An analogue of Theorem 2.4.10 for right Noetherian rings is the following result ([112, Theorem 2.3.7]). The requirement for the existence of an identity is of course not assumed for nil subrings.

Theorem 3.1.2. *Nil subrings of right Noetherian rings are nilpotent. In particular the prime radical of a right Noetherian ring is nilpotent.*

The well known Wedderburn-Artin theorem describes semisimple Artinian rings ([112, Theorem 0.1.10 and Theorem 0.1.11])

Theorem 3.1.3. *The following conditions are equivalent for a ring R.*

1. *R is a direct product of simple Artinian rings.*
2. *R is semiprime and right Artinian.*
3. *Every right R-module is semisimple.*

Furthermore, R is a simple Artinian ring if and only if $R \cong \mathrm{M}_n(D)$ for some (uniquely determined) n and a division ring D.

A multiplicatively closed subset X of a ring R is said to be a right Ore set if $xR \cap rX \neq \emptyset$ for any $x \in X$ and $r \in R$. If, moreover, X consists of regular elements then RX^{-1} denotes the ring of right quotients of R with respect to X. This also will be called the localization of R with respect to X. If, furthermore, X is the set consisting of all regular elements of R then RX^{-1} is called the (classical) ring of right quotients of R and it is also denoted by $\mathrm{Q}_{cl}(R)$, or simply $\mathrm{Q}(R)$. So $\mathrm{Q}_{cl}(R) = \{rx^{-1} \mid r, x \in R,\ x \text{ regular}\}$. In case the set of regular elements is a right Ore set one says that R has a classical ring of right quotients. The left versions of these notions are defined similarly.

Let R be a ring and let M be a right R-module. A nonzero submodule N of M is said to be uniform if N does not contain a direct sum of two nonzero submodules. Equivalently, $N_1 \cap N_2 \neq \{0\}$ for any two nonzero submodules N_1 and N_2 of N. A submodule E of M is said to be an essential submodule if $E \cap V \neq \{0\}$ for all nonzero submodules V of M.

A right R-module M is said to have finite uniform dimension if it does not contain infinite direct sums of nonzero submodules. In this case M contains an essential submodule which is a finite direct sum of uniform submodules. The (unique) number of direct summands is called the uniform dimension of R (or Goldie dimension).

A ring R is said to be a right Goldie ring if, as a right R-module, R has finite uniform dimension and R satisfies the ascending chain condition on the set of right annihilator ideals (the latter are right ideals of the form $\mathrm{ra}_R(X)$ with $X \subseteq R$). Similarly one defines a left Goldie ring. A ring that is both right and left Goldie is simply called a Goldie ring.

We can now state Goldie's theorem ([112, Theorem 2.3.6]).

Proposition 3.1.4. *The following conditions are equivalent for a ring R.*

1. *R is a semiprime (respectively prime) right Goldie ring.*
2. *R has a classical ring of right quotients which is semisimple (respectively simple) Artinian.*

Furthermore, when this occurs then the following properties hold.

1. *Every nonzero ideal I of R with $\mathrm{ra}_R(X) = \{0\}$ contains a regular element.*
2. *Every right regular element of R is regular.*
3. *If $r_1, r_2 \in R$ and $r_1 r_2 = 1$ then both r_1 and r_2 are invertible in R.*

So in a semiprime right Goldie ring a right (respectively left) invertible element is invertible. The set $\mathrm{U}(R)$ consisting of all invertible elements (also called units) of a ring R is a (multiplicative) group, called the unit group of R.

From results of Cohn and Bergman (see Theorem 5.32 and Theorem 4C in [29]) it follows that a semisimple Artinian algebra over a field can be embedded in a matrix ring over a division ring.

Theorem 3.1.5. *Let R be a semiprime right Goldie algebra over a field. Then there exists a division algebra D such that R embeds into the matrix algebra $\mathrm{M}_n(D)$ for some $n \geq 1$.*

An algebra R over a field K is said to satisfy a polynomial identity (abbreviated, R is a PI algebra) if there exists a nonzero polynomial $f(x_1, \ldots, x_n)$ in the free K-algebra on generators $x_1, \ldots, x_n$ so that $f(r_1, \ldots, r_n) = 0$ for all $r_1, \ldots, r_n \in R$. Examples of PI algebras are matrices $\mathrm{M}_n(A)$ over a commutative algebra A and algebras R that are a finite module over a commutative subalgebra.

The first part of the following theorem, due to Kaplansky (see [112, Theorem 13.3.8]), shows that primitive ideals in PI algebras are maximal ideals. Hence the Jacobson radical is the intersection of the maximal two-sided ideals. The second part can be found in [56, Theorem 8.12].

Theorem 3.1.6. *Let R be an algebra over a field K satisfying a polynomial identity.*

1. *If R is right primitive then R is a simple algebra that is finite dimensional over its center.*
2. *If R is right Noetherian then $\bigcap_{n\geq 1} \mathcal{J}(R)^n = \{0\}$.*

The following result due to Posner can be found in [112, Theorem 13.6.5].

Theorem 3.1.7. *Let R be an algebra over a field and with center $\mathrm{Z}(R)$. If R is prime and satisfies a polynomial identity then R is a (right and left) Goldie ring with classical ring of quotients $\mathrm{Q}_{cl}(R) = R(\mathrm{Z}(R) \setminus \{0\})^{-1}$ and with center $\mathrm{Z}(\mathrm{Q}_{cl}(R)) = \mathrm{Q}_{cl}(\mathrm{Z}(R))$.*

We mention another relevant result ([112, Lemma 13.9.10]).

Theorem 3.1.8. *Let R be a finitely generated algebra over a field. If R is a finite module over its center $\mathrm{Z}(R)$ then $\mathrm{Z}(R)$ is a finitely generated algebra.*

Theorem 5.3.7 and Theorem 5.3.9 in [129] give a characterization of group algebras that satisfy a polynomial identity. Corollary 19.13 in [119] extends this to semigroup algebras of finitely generated cancellative semigroups. Because of Lemma 4.1.9, standard techniques based on central localizations allow us to generalize this to semigroup algebras of arbitrary cancellative semigroups. In order to state the result we recall that if $\mathcal{P}$ and $\mathcal{Q}$ are group properties then a group G is said to be $\mathcal{P}$-by-$\mathcal{Q}$ if G has a normal subgroup N satisfying $\mathcal{P}$ so that G/N satisfies $\mathcal{Q}$. A group G is called poly-$\mathcal{P}$ if G has a subnormal series $\{1\} = G_0 \trianglelefteq G_1 \trianglelefteq \cdots \trianglelefteq G_n = G$ so that for each $1 \leq i \leq n$ the factor group G_i/G_{i-1} satisfies $\mathcal{P}$.

Theorem 3.1.9. *Let K be a field. If S is a cancellative semigroup then $K[S]$ satisfies a polynomial identity if and only if S has a group of quotients G so that $K[G]$ satisfies a polynomial identity. The latter holds if and only if G is abelian-by-finite and* $\mathrm{char}(K) = 0$ *or G is (finite p-group)-by-(abelian-by-finite) and* $\mathrm{char}(K) = p > 0$.

3.2 Prime ideals

In this section we state several results on semiprime rings and on the behaviour of prime ideals under various ring constructions. The first result states that within the class of semiprime algebras satisfying a polynomial identity the left and right Noetherian property are equivalent ([112, Theorem 13.6.15]).

Theorem 3.2.1. *If R is a semiprime PI algebra then the following conditions are equivalent.*

1. *R is right Noetherian.*
2. *R is left Noetherian.*
3. *R satisfies the ascending chain condition on two-sided ideals.*

The second result is on lifting matrix structures ([70, Proposition II.1.2, page 36] or [69, Theorem III.8.1]). Recall that a set $\{e_{ij} \mid 1 \leq i,j \leq n\}$ of a ring R is called a set of matrix units if $1 = \sum_{1\leq i\leq n} e_{ii}$ and $e_{ij}e_{kl} = e_{il}$ if $j = k$ and zero otherwise.

Proposition 3.2.2. *Let R be a ring. If $R/\mathcal{B}(R) \cong \mathrm{M}_n(\overline{A})$ for a ring $\overline{A}$ and some n then $R \cong \mathrm{M}_n(A)$ for some ring A with $A/\mathcal{B}(A) \cong \overline{A}$.*

Because of its use in the proof of Proposition 6.3.5 some comments on the proof of this result are needed. From the proof given in [69] one can recover a bit more information. If $e = e_1 + \cdots + e_k$ is a sum of orthogonal idempotents in R such that their natural images $\overline{e_i}$ form a part of a set of matrix units $\overline{E_{ij}}$ in $R/\mathcal{B}(R)$ (that is $\overline{e_l} = \overline{E_{ll}}$ for $1 \leq l \leq k$), then (as shown in [69, Proposition 5, Page 54]) there exist matrix units $e_{ij} \in R$, $1 \leq i,j \leq n$, such that $e_{ii} = e_i$ and $\overline{e_{ii}} = \overline{E_{ii}}$, for $1 \leq i \leq k$. It follows that R is a matrix ring $\mathrm{M}_n(A)$ where $A \cong e_1Re_1$.

The following result gives information on the decomposition of the identity into orthogonal idempotents ([34, Proposition III.4.1]). Recall that an idempotent e of a ring R is said to be minimal if 0 and e are the only idempotents in eRe.

Proposition 3.2.3. *Let $1 = e_1 + \cdots + e_n = f_1 + \cdots + f_m$ be two orthogonal decompositions of the identity of a finite dimensional algebra A with minimal idempotents e_i and f_j. Then $n = m$ and there is an invertible element $a \in A$ such that, up to a suitable reindexing, $f_i = ae_ia^{-1}$ for all i.*

The next result is the principal ideal theorem ([112, Theorem 4.1.11 and Corollary 4.1.12]) which involves the height of a prime ideal P of a ring R. Recall that an element n of a ring R is said to be a normal element if $Rn = nR$.

Theorem 3.2.4. *Let R be a right Noetherian ring and n a normal element of R which is not a unit. If P is a prime ideal of R minimal over Rn then* $\text{height}(P) \leq 1$. *More general, if Q is any prime ideal containing n then* $\text{height}\, Q \leq \text{height}(Q/Rn) + 1$.

Recall that a ring R is said to be catenary if for any two prime ideals P and Q of R with $P \subseteq Q$ any two chains of prime ideals between P and Q that cannot be refined have the same length. A fundamental result is the following ([112, Corollary 13.10.13]).

Theorem 3.2.5. *Every finitely generated algebra satisfying a polynomial identity is catenary.*

We now consider the behaviour of prime ideals under some ring extensions. We start with a property on localizations ([56, Theorem 9.22, Theorem 9.20 and Lemma 9.21]).

Theorem 3.2.6. *Let X be a right Ore set consisting of regular elements in a ring R. Assume that R is right Noetherian or that R satisfies a polynomial identity and RX^{-1} is right Noetherian. Then the following properties hold.*

1. *If I is an ideal of R then $I^e = IRX^{-1}$ is an ideal of RX^{-1}.*
2. *The maps $I \mapsto I^e$ and $J \mapsto J \cap R$ are inverse bijections between the set of prime ideals of RX^{-1} and the set of those prime ideals of R that intersect X trivially.*

Recall that a ring T is a finite normalizing extension of a subring R if there exist elements $t_1, \ldots, t_n \in T$ so that $T = \sum_{i=1}^{n} Rt_i$ and $Rt_i = t_iR$ for each $1 \leq i \leq n$. For the following result we refer to Section 10.2 in [112] and Theorem 16.9 in [130].

Proposition 3.2.7. *Let T be a finite normalizing extension of a subring R. The following properties hold.*

1. $\mathcal{B}(R) = \mathcal{B}(T) \cap R$.
2. $\mathcal{J}(R) \subseteq \mathcal{J}(T) \cap R$.

3. *If* $P \in \mathrm{Spec}(T)$ *then* $P \cap R$ *is a semiprime ideal in* R.

4. *If* $Q \in \mathrm{Spec}(R)$ *then there exists* $P \in \mathrm{Spec}(T)$ *so that* Q *is a minimal prime over* $P \cap R$*; one says that* P *lies over* Q.

5. *If* $Q_1, Q_2 \in \mathrm{Spec}(R)$, $P_1 \in \mathrm{Spec}(T)$ *with* $Q_1 \subset Q_2$ *and* P_1 *lies over* Q_1 *then there exists* $P_2 \in \mathrm{Spec}(T)$ *with* $P_1 \subset P_2$ *and* P_2 *lies over* Q_2.

6. *If* $P_1, P_2 \in \mathrm{Spec}(T)$ *with* $P_1 \subset P_2$ *then* $P_1 \cap R \subset P_2 \cap R$.

Furthermore, $\mathrm{clKdim}(R) = \mathrm{clKdim}(T)$.

We finish this section with a result on prime and semiprime semigroup algebras (see Section II.2 in [129] for group algebras and Lemma 7.15 and Theorem 7.19 in [119] for the extension to semigroup algebras).

For an arbitrary group G the set of torsion (we also say, periodic) elements is denoted by G^+. If $G^+ = \{1\}$ then G is said to be torsion free. If p is a prime number then G is said to be a p'-group if G does not contain nontrivial elements of order a power of p.

Recall that the finite conjugacy center (FC-center for short) $\Delta(G)$ of a group G is the set of those elements $g \in G$ that have finitely many conjugates, or equivalently the centralizer $\mathrm{C}_G(g)$ of g in G has finite index. The set $(\Delta(G))^+$ will be denoted as $\Delta^+(G)$. It is a characteristic subgroup of G and $\Delta(G)/\Delta^+(G)$ is a torsion free abelian group ([129, Lemma 4.1.6]). Clearly a finite normal subgroup is contained in $\Delta^+(G)$ and the latter group is trivial if and only if no such nontrivial subgroups exist. A group G so that $\Delta(G) = G$ is called a finite conjugacy group, or FC-group for short.

Theorem 3.2.8. *Let* K *be a field. Assume that* G *is a group generated by its subsemigroup* S. *Then the following properties hold.*

1. *If* $K[G]$ *is prime (semiprime) then* $K[S]$ *is prime (semiprime, respectively). The converse holds provided* G *is the group of right quotients of* S.

2. *The group algebra is prime if and only if* $\Delta(G)$ *is torsion free.*

3. *The group algebra* $K[G]$ *is semiprime if and only if* $\mathrm{char}(K) = 0$ *or* $\mathrm{char}(K) = p > 0$ *and* $\Delta^+(G)$ *is a* p'*-group.*

4. *If* G *is the group of right quotients of* S *and* $K[G]$ *is right Noetherian then* $\mathcal{B}(K[S]) = \mathcal{B}(K[G]) \cap K[S]$ *and* $\mathcal{B}(K[S])$ *is nilpotent.*

Later we shall need an extension of the above theorem to a more general context. For this recall that a semigroup S is a semilattice of semigroups S_α if $S = \bigcup_{\alpha \in \Gamma} S_\alpha$, a disjoint union, and $S_\alpha S_\beta \subseteq S_{\alpha\beta}$ for all $\alpha, \beta \in \Gamma$, where Γ is an abelian semigroup consisting of idempotents. Using standard techniques, as in Proposition 21.9 in [119], the following result can be obtained.

Corollary 3.2.9. *Let $S = \bigcup_{\alpha \in \Gamma} S_\alpha$ be a semilattice of semigroups. If K is a field and each $K[S_\alpha]$ is semiprime then so is $K[S]$. In particular, if each S_α is contained in a group and* $\operatorname{char}(K) = 0$ *then $K[S]$ is semiprime.*

3.3 Group algebras of polycyclic-by-finite groups

In this section we state the required background of Noetherian group algebras. For this recall that a group G is polycyclic-by-finite if and only if G has a normal subgroup N of finite index so that N has a subnormal series with cyclic factors. Equivalent properties are that G is poly-(finite or cyclic) or that G has a normal subgroup N of finite index that is either trivial or poly-(infinite cyclic) ([140, Proposition 1.2]). The number of infinite cyclic factors is independent of the choice of N. It is called the Hirsch rank of G and it is denoted by $\mathrm{h}(G)$. Basic references for polycyclic-by-finite groups are [96, 129, 140].

If K is a field and G is an arbitrary group then the group algebra $K[G]$ has an involution that K-linearly extends the classical involution on G which maps g onto g^{-1}. It follows that $K[G]$ is right Noetherian if and only if it is left Noetherian. Moreover, if $K[G]$ is right Noetherian, then G satisfies the ascending chain condition on subgroups, in particular, G is finitely generated. It remains an open problem to characterize when a group algebra is Noetherian. In case G is polycyclic-by-finite, an easy proof by induction on the Hirsch rank yields the following result ([129, Corollary 10.2.8]); it gives the only known class of groups for which the group algebra is Noetherian.

Theorem 3.3.1. *Let K be a field. If G is a polycyclic-by-finite group then the group algebra $K[G]$ is Noetherian.*

The following result provides natural classes of polycyclic-by-finite groups ([129, Lemma 3.4.2 and Lemma 4.1.5]).

Lemma 3.3.2. *Let G be a finitely generated group. If G is nilpotent or a finite conjugacy group then G is polycyclic-by-finite. In the latter case $[G : \mathrm{Z}(G)] < \infty$.*

Next we state some well known facts on a polycyclic-by-finite groups. Explanations for the consecutive statements can be found in Theorem 16.2.7 and Exercises 17.2.4 in [96], Theorem 16.2.8, Exercise 16.2.9 and 16.2.10 in [96], Theorem 2.4 and Theorem 5.5 in [140], Lemma 10.2.5 in [129] and Lemma 6.3.4 in [129].

Proposition 3.3.3. *The following properties hold for a finitely generated group G.*

1. *If G is nilpotent, then the torsion part G^+ of G is a finite normal subgroup.*
2. *Assume G is torsion free and nilpotent. Let $x, y \in G$. If, for some positive integer r, the elements x^r and y^r commute then x and y commute; and if $x^r = y^r$ then $x = y$. In particular, if G is abelian-by-finite, then it is abelian. Moreover $G/\mathrm{Z}(G)$ is also torsion free.*
3. *If G is polycyclic-by-finite then G embeds into $\mathrm{GL}_n(\mathbb{Q})$ for some n and it has a torsion free normal subgroup H of finite index which has nilpotent commutator subgroup $H' = [H, H]$. Moreover H is poly-(infinite cyclic).*
4. *If G is polycyclic-by-finite then G is (abelian-by-finite)-by-(abelian-by-finite) if and only if G is (abelian-by-abelian)-by-finite.*
5. *For any positive integer n the group G has only finitely many subgroups of index n.*

Because of Theorem 3.2.8 and Theorem 3.3.1 we know that the group algebra $K[G]$ of a torsion free polycyclic-by-finite group is a prime Noetherian algebra. Hence it has a simple Artinian classical ring of quotients. The following result, which is a solution to the zero divisor problem for group algebras of such groups, says that this ring of quotients actually is a division algebra ([130, Theorem 37.5]).

Theorem 3.3.4. *Let K be a field. If G is a torsion free polycyclic-by-finite group then the group algebra $K[G]$ is a domain.*

Through the work of Roseblade (see for example [129, 130]), group algebras $K[G]$ of polycyclic-by-finite groups G over a field K are rather well understood. In order to state the required properties some more notation and terminology is needed.

Suppose a group H acts on a set A. If $a \in A$ and $h \in H$ then we denote the action of h on a as a^h. An element $a \in A$ is said to be H-orbital (or simply orbital) if its H-orbit is finite. The set of H-orbital elements in A we denote by $\Delta_H(A)$. Thus $\Delta_H(A) = \{a \in A \mid [H : \mathrm{C}_H(a)] < \infty\}$ with $\mathrm{C}_H(a) = \{h \in H \mid a^h = a\}$.

Definition 3.3.5. Let H be a group. A finite dimensional $\mathbb{Q}[H]$-module V is said to be a rational plinth for H if V is an irreducible $\mathbb{Q}[H_1]$-module for all subgroups H_1 of finite index in H. Now let H act on a finitely generated abelian group A. Then A is a plinth for H if, in additive notation, $V = A \otimes_{\mathbb{Z}} \mathbb{Q}$ is a rational plinth.

Thus A is a plinth if and only if no proper pure subgroup of A is H-orbital. Recall that a subgroup B of A is pure if $nA \cap B = nB$ for all integers n.

A normal series $\{1\} = G_0 \subseteq G_1 \subseteq \cdots \subseteq G_n = G$ for the polycyclic-by-finite group G is called a plinth series for G if each quotient G_i/G_{i-1} is either finite or a plinth for G. It is not necessarily true that every G has a plinth series. However, any polycyclic-by-finite group G has a normal subgroup N of finite index with a plinth series, say $\{1\} = N_0 \subseteq N_1 \subseteq \cdots \subseteq N_n = N$. The number of infinite factors N_i/N_{i-1} is called the plinth length of G and is denoted by $\mathrm{pl}(G)$. This parameter is independent of the choice of N and of the particular series for N. If F is a normal subgroup of G so that G/F is abelian then $\mathrm{pl}(G) \leq \mathrm{pl}(F) + \mathrm{rk}(G/F)$, where $\mathrm{rk}(B)$ denotes the torsion free rank of an abelian group B.

We now can formulate a fundamental result on prime ideals of group algebras of polycyclic-by-finite groups ([130, Theorem 19.6]).

Theorem 3.3.6. *Let G be a polycyclic-by-finite group and K a field. The following properties hold.*

1. $\mathrm{clKdim}(K[G]) = \mathrm{pl}(G)$.
2. *If K is absolute (that is, an algebraic extension of a finite field) then every right primitive ideal M of $K[G]$ is maximal and $K[G]/M$ is finite dimensional.*

We finish with one more result ([103]).

Theorem 3.3.7. *A group algebra of a polycyclic-by-finite group is catenary.*

3.4 Graded rings

Let G be a group. A ring R is said to be G-graded if $R = \bigoplus_{g \in G} R_g$, the direct sum of additive subgroups R_g of R, such that $R_g R_h \subseteq R_{gh}$ for all

$g, h \in G$. If always $R_g R_h = R_{gh}$ then R is said to be strongly graded. The additive groups R_g are called the homogeneous components of R and R_e (also often denoted as R_1) is the identity component, where e is the identity of G. The elements of $\bigcup_{g\in G} R_g$ are called homogeneous. Obviously a group algebra $K[G]$ is an example of a strongly G-graded ring. This ring, however, has many other natural gradings. For example if N is a normal subgroup of G and T is a transversal of N in G then $K[G]$ is strongly G/N-graded with homogeneous components $K[N]t$ where $t \in T$. Actually $K[G] = K[N] * (G/N)$, a crossed product of G/N over $K[N]$ (see [130] for the definition).

Several of the proofs on semigroup algebras given in this monograph eventually rely on reductions to some graded rings. Hence the need to recall some structural results in this general framework. The nature of many results on G-graded rings R is to determine whether global information is determined by homogeneous information. In case G is finite one goes a step further and investigates which information is described by the properties of the identity component R_e.

One such result ([130, Theorem 1.3.7]) states that the ascending chain condition on right ideals for a ring graded by a polycyclic-by-finite group indeed is determined by homogeneous information. For an (left, right) ideal (respectively subring) I of a ring R graded by a group G put $I_G = \bigoplus_{g\in G}(I \cap R_g)$. If $I = I_G$ then I is said to be a homogeneous ideal (respectively subring) of R. The ring R is said to be graded right Noetherian if the ascending chain condition on homogeneous right ideals is satisfied. Similarly one defines graded left Noetherian. If both conditions are satisfied then R is said to be graded Noetherian.

Theorem 3.4.1. *A ring R graded by a polycyclic-by-finite group is right Noetherian if and only if R is graded right Noetherian.*

It is rather easy to verify that a minimal prime ideal of a $\mathbb{Z}$-graded ring is homogeneous. In [74] this was extended to rings graded by a unique product group. Recall ([129]) that a group is said to be a unique product group if for any two nonempty subsets X and Y of G there exists a uniquely presented element of the form xy with $x \in X$ and $y \in Y$. In [149] it is shown that such groups are two unique product groups, that is, given any two nonempty finite subsets X and Y of G with $|X|+|Y| > 2$ there exist at least two distinct elements g_1 and g_2 of G that have unique representations in the form $g_1 = x_1y_1$, $g_2 = x_2y_2$ with $x_1, x_2 \in X$ and $y_1, y_2 \in Y$. If a group G has a normal subgroup H so that both H and G/H are unique product groups then H is a unique product group as well. It follows that poly-(infinite cyclic) groups are unique product groups. More generally, any right ordered group is a unique product group (see Section 13.1 in [129]).

Theorem 3.4.2. *Minimal prime ideals of a ring R graded by a unique product group are homogeneous. In particular, the prime radical $\mathcal{B}(R)$ is homogeneous.*

Bergman showed the following result. A proof for $\mathbb{Z}$-graded rings can be found in [130, Theorem 22.6]; the general case easily can be reduced to this situation.

Theorem 3.4.3. *If R is a ring graded by a torsion free abelian group then the Jacobson radical $\mathcal{J}(R)$ is homogeneous.*

Recall that a ring is said to be a Jacobson ring if every prime factor ring has zero Jacobson radical. For example, the Hilbert Nullstellensatz implies that polynomial algebras $K[x_1, \ldots, x_n]$ are Jacobson rings. Actually every finitely generated commutative K-algebra is a Jacobson ring, as this property is inherited by homomorphic images. More generally, any finitely generated algebra K-algebra that satisfies a polynomial identity is a Jacobson ring ([112, Theorem 13.10.3]). The following property provides many more examples of Jacobson rings in the classes of rings graded by finite groups ([130, Theorem 2.3]), right Noetherian rings strongly graded by a polycyclic-by-finite group ([73, Corollary 5.2]) and certain natural graded subrings of rings strongly graded by an abelian group ([72, Theorem 3.6]).

Theorem 3.4.4. *Let R be a ring graded by a group G with identity e.*

1. *If G is a finite group then R is a Jacobson ring if and only if so is R_e.*
2. *Assume G is a polycyclic-by-finite group and R is strongly G-graded. If R_e is a right Noetherian Jacobson ring then so is R.*
3. *Assume G is an abelian group of finite torsion free rank and let S be a submonoid of G. If R is strongly G-graded and R_e is a right Noetherian Jacobson ring then so is $\bigoplus_{s\in S} R_s$.*

Also for rings graded by finite groups there are fundamental results on the prime radical and prime ideals. The following can be found in [130, Theorem 1.4.8]. Recall that a ring R graded by a finite group G is said to have no G-torsion if $|G|r \neq 0$ for any $0 \neq r \in R$.

Theorem 3.4.5. *Let R be a ring graded by a finite group G. If R has no nonzero homogeneous nilpotent ideals and R has no G-torsion then R is semiprime.*

A related result can be found in [28, Theorem 1.7].

Proposition 3.4.6. *Let R be a ring graded by a finite group G with identity e. Assume R has no nonzero homogeneous nilpotent ideals. If R_e is a semiprime Goldie ring then R is a Goldie ring with a classical ring of quotients equal to RT^{-1} where T consists of the regular elements of R_e.*

The next result is a generalized version of the Maschke theorem ([130, Proposition 1.4.10]).

Theorem 3.4.7. *Let R be a ring graded by a finite group G with identity e. If V is a completely reducible R-module then V is a completely reducible R_e-module.*

As for normalizing extensions in Proposition 3.2.7, there is a strong link between the primes of a finite group graded ring and those of its identity component ([130, Theorem 17.9]). Again, we only state the properties relevant for later use in the monograph.

Theorem 3.4.8. *Let $R = \bigoplus_{g \in G} R_g$ be a ring graded by a finite group G with identity e.*

1. *If $P \in \mathrm{Spec}(R)$ then $P \cap R_e = Q_1 \cap \cdots \cap Q_n$, where $n \leq |G|$ and $Q_1, \ldots, Q_n$ are all the prime ideals of R_e minimal over $P \cap R_e$. Furthermore, each prime Q_i has the same height as P and P is maximal among all ideals I of R such that $I \cap R_e \subseteq Q_i$.*

2. *If $Q \in \mathrm{Spec}(R_e)$ then there exits $P \in \mathrm{Spec}(R)$ (of the same height as Q) so that $P \cap R_e = Q_1 \cap \cdots \cap Q_n$ for some primes $Q_1 = Q, \ldots, Q_n$ of R_e minimal over $P \cap R_e$ (again all of the same height as Q). There are at most $|G|$ such prime ideals and these are precisely the primes P_i satisfying $(P_i)_G = P_G$).*

Furthermore $\mathrm{clKdim}(R) = \mathrm{clKdim}(R_e)$.

We finish this section with one more result that shows that some prime ideals necessarily contain a homogeneous element ([25] and [130, Corollary 23.3]).

Theorem 3.4.9. *Assume that R is a ring graded by a finitely generated nilpotent-by-finite group G. Let $P_0 \subset P_1 \subset \cdots \subset P_n$ be a chain of prime ideals of R. If either $n \geq 2^{\mathrm{h}(G)}$, or G is abelian-by-finite with $n > \mathrm{h}(G)$, then P_n contains a nonzero homogeneous element.*

3.5 Gelfand-Kirillov dimension

Let K be a field. The Gelfand-Kirillov dimension measures the rate of growth of K-algebras in terms of any generating set. For a semigroup algebra $K[S]$ it measures the rate of growth of the semigroup S. In this section we survey some fundamental results on this topic. Standard references are [100] and [112].

We begin with recalling the definition for finitely generated K-algebras R. Suppose R is as a K-algebra generated by a finite dimensional subspace V. Put $R_0 = V^0 = K$ and $R_n = \sum_{i=0}^{n} V^i$ for each positive integer n. Clearly each R_n is a finite dimensional K-space. The number

$$\limsup_{n\to\infty} \left(\frac{\log \dim_K R_n}{\log n} \right)$$

is independent of the choice of V. It is called the Gelfand-Kirillov dimension of R and it is denoted by $\mathrm{GK}(R)$. Now $\mathrm{GK}(R) = 0$ means that R is a finite dimensional K-algebra. Furthermore, $\mathrm{GK}(R)$ is finite if and only if there exists a positive integer m so that $\dim_K R_n \leq n^m$ for all sufficiently large positive integers n. For a not necessarily finitely generated K-algebra R one defines the Gelfand-Kirillov dimension as the supremum of all $\mathrm{GK}(R')$ where R' runs through all finitely generated K-subalgebras of R.

For any real number $r \geq 2$ there exist finitely generated K-algebras with Gelfand-Kirillov dimension r. There do not exist algebras R with $1 < \mathrm{GK}(R) < 2$ or with $0 < \mathrm{GK}(R) < 1$. For commutative algebras this dimension is an integer or is infinite. Furthermore, for finitely generated commutative algebras the Gelfand-Kirillov dimension corresponds with the classical Krull dimension ([100, Theorem 4.5]).

Theorem 3.5.1. *Let K be a field and R a commutative K-algebra. Then* $\mathrm{GK}(R)$ *is either infinite or a nonnegative integer. If, furthermore, R is finitely generated then* $\mathrm{GK}(R) = \mathrm{clKdim}(R) < \infty$.

This result has been generalized to finitely generated K-algebras satisfying a polynomial identity ([2] and [100, Corollary 10.7 and Section 12.10]).

Theorem 3.5.2. *Let K be a field. The following properties are satisfied for a finitely generated K-algebra R satisfying a polynomial identity.*

1. $\mathrm{GK}(R) < \infty$.
2. *If R also is right Noetherian then R embeds into a matrix algebra* $\mathrm{M}_n(L)$ *over a field extension L of K; moreover,* $\mathrm{GK}(R) = \mathrm{clKdim}(R) = \mathrm{GK}(R/\mathcal{B}(R))$ *and it is an integer.*

For arbitrary right Noetherian algebras R the Gelfand-Kirillov dimension of $R/\mathcal{B}(R)$ is determined by that of its prime images ([112, Corollary 8.3.3]). In general, to obtain upper bounds one can annihilate nilpotent ideals [100, Corollary 5.10]. These results, together with a statement on corners ([1]), are collected in the following theorem.

Theorem 3.5.3. *The following properties hold for an algebra R over a field K.*

1. *If R is right Noetherian then $\mathrm{GK}(R/\mathcal{B}(R)) = \mathrm{GK}(R/P)$ for a prime ideal P of R.*
2. *If a subalgebra T of R (not necessarily with identity) is a corner in R in the sense that $TRT \subseteq T$, then $\mathrm{GK}(RT^2R) = \mathrm{GK}(T)$.*
3. *If I is an ideal of R and $I^n = 0$ then $\mathrm{GK}(R) \leq n\,\mathrm{GK}(R/I)$.*

Finitely generated K-algebras of Gelfand-Kirillov dimension zero are finite dimensional and thus satisfy a polynomial identity. This has been extended to finitely generated algebras that are semiprime and of Gelfand-Kirillov dimension one ([143]).

Theorem 3.5.4. *Let R be finitely generated semiprime algebra over a field K. If $\mathrm{GK}(R) = 1$ then R is a finite module over its center.*

That the Gelfand-Kirillov dimension behaves well under polynomial extensions is stated in the following result ([100, Example 3.6]).

Proposition 3.5.5. *Let K be a field. If R is a K-algebra then*

$$\mathrm{GK}(R[x_1, \ldots, x_n]) = \mathrm{GK}(R) + n.$$

Gromov [60] proved that the Gelfand-Kirillov dimension of a finitely generated group algebra $K[G]$ is finite if and only if the group G is nilpotent-by-finite. Earlier Bass proved a formula for this Gelfand-Kirillov dimension ([8]). Later Grigorchuk extended this to semigroup algebras of finitely generated cancellative semigroups [59]. A proof for the latter, as well as the link with the classical Krull dimension in the PI case, can be found in [119, Theorem 8.3 and Theorem 23.4]. Recall that the rank $\mathrm{rk}(S)$ of a semigroup S is the supremum of the ranks of the free abelian subsemigroups of S. Also recall that the lower central series N_i of a group N is defined by $N_1 = N$ and $N_{i+1} = [N, N_i]$, for $i \geq 1$.

Theorem 3.5.6. *Let K be a field and let S be a cancellative semigroup.*

1. *If $K[S]$ satisfies a polynomial identity then* $\mathrm{GK}(K[S]) = \mathrm{clKdim}\, K[S] = \mathrm{rk}(S)$.

2. *The following conditions are equivalent if S is finitely generated.*

   (a) $\mathrm{GK}(K[S]) < \infty$.

   (b) *S has a group of quotients G and* $\mathrm{GK}(K[G]) < \infty$.

   (c) *S has a group of quotients G that is nilpotent-by-finite.*

   Moreover, if these equivalent conditions are satisfied then

$$\mathrm{GK}(K[S]) = \mathrm{GK}(K[G]) = \sum_{i=1}^{k} i r_i,$$

   where r_i is the rank of the i-th quotient of the lower central series $N = N_1 \supset N_2 \supset \cdots \supset N_k \supset N_{k+1} = \{1\}$ of a nilpotent subgroup N of finite index in G.

The well known Tits alternative [151] yields that finitely generated linear groups over fields are solvable-by-finite if and only if they do not contain a free nonabelian subgroup. This celebrated theorem has a semigroup analogue ([126], see also [123, Theorem 6.11]).

Theorem 3.5.7. *Let K be a field and let S be a finitely generated subsemigroup of* $\mathrm{GL}_n(K)$. *The following conditions are equivalent.*

1. *S satisfies a semigroup identity.*

2. *S has a group of quotients G that is nilpotent-by-finite.*

3. *S has no free nonabelian subsemigroups.*

In view of the third part of Proposition 3.3.3 one can now derive the following result.

Corollary 3.5.8. *If a polycyclic-by-finite group G is generated by a subsemigroup S that does not have free nonabelian subsemigroups then G is nilpotent-by-finite.*

We finish this section with recalling a result due to Malcev showing that nilpotent groups can be described by a semigroup identity [106] (see also [123, Theorem 7.2] and [117]).

Let x, y be elements of a semigroup S and let $w_1, w_2, \dots$ be elements of S. Consider the sequence of elements of S defined inductively as follows:

$$x_0 = x, \qquad y_0 = y,$$

and for $n \geq 0$

$$x_{n+1} = x_n w_{n+1} y_n, \qquad y_{n+1} = y_n w_{n+1} x_n.$$

We say the identity $X_n = Y_n$ is satisfied in S if $x_n = y_n$ for all $x, y \in S$ and $w_1, w_2, \dots$ in S. A semigroup S is said to be Malcev nilpotent of class n if S satisfies the identity $X_n = Y_n$ and n is the least positive integer with this property.

Theorem 3.5.9. *A cancellative semigroup S is Malcev nilpotent if and only if it has a nilpotent group of quotients.*

3.6 Maximal orders

It is well known that integrally closed Noetherian domains, or more generally, Krull domains ([40]), are of fundamental importance in several areas of mathematics. In this section we recall the definition and some properties of prime Noetherian maximal orders and Krull orders; these are generalizations to the class of noncommutative rings. In the vast literature one can find several types of such generalizations. Some of the relevant references for our purposes are [18, 19, 75, 107, 108, 109]. Within the class of algebras satisfying a polynomial identity all the considered generalizations agree ([18, 19, 93]).

We give a characterization of group algebras of polycyclic-by-finite groups that are Noetherian prime maximal orders. These results are due to Brown and can be found in [12, 13, 14]. It is also stated when such algebras are unique factorization rings or domains, in the sense as studied by Chatters and Jordan [24] and Chatters [21], respectively.

Next we state a characterization of commutative semigroup algebras that are Krull domains. This result is due to Chouinard [26] and was earlier proved by Anderson [4, 5] under some extra assumptions. For the Noetherian case we refer to Gilmer's book [54]. The main result reduces the problem to a semigroup problem and for this one needs the notion of a cancellative monoid that is a Krull order. Chouinard describes such monoids. Wauters in [152] considered generalizations to monoids that are not abelian.

Let R be a prime Goldie ring with classical ring of quotients $\mathrm{Q} = \mathrm{Q}_{cl}(R)$. One simply says that R is an order in Q. So, by Proposition 3.1.4, every regular element of R is invertible in Q and elements of Q can be written as $r_1 c_1^{-1} = c_2^{-1} r_2$ for some $r_1, r_2, c_1, c_2 \in R$ with c_1, c_2 regular.

If T is a subring of Q and a, b are invertible elements in Q so that $aTb \subseteq R$ then also T is an order in Q. This leads to an equivalence relation on the orders in Q. One says that two orders R_1 and R_2 in Q are equivalent if there are units $a_1, a_2, b_1, b_2 \in \mathrm{Q}$ so that $a_1 R_1 b_1 \subseteq R_2$ and $a_2 R_2 b_2 \subseteq R_1$. The order R is said to be maximal if it is maximal within its equivalence class.

A commutative integral domain D with field of fractions K is a maximal order if and only D is completely integrally closed, that is, if $x, c \in K$ with $c \neq 0$ and $cD[x] \subseteq D$ then $x \in R$. Clearly, a completely integrally closed domain is integrally closed, that is, if $x \in K$ is so that $D[x]$ is a finitely generated D-module then $x \in D$. If D is a Noetherian domain then both notions are equivalent. It is easily seen that if R is a maximal order then so are $\mathrm{M}_n(R)$ and $\mathrm{Z}(R)$ (in their respective classical rings of quotients).

The study of maximal orders is aided by the notion of a fractional ideal. So suppose R is an order in its classical ring of quotients Q. A fractional R-ideal (or simply, fractional ideal if the order R is clear from the context) is a two-sided R-submodule I of Q so that $aI \subseteq R$ and $Ib \subseteq R$ for some invertible elements $a, b \in \mathrm{Q}$. If, furthermore, $I \subseteq R$ then I is called an integral fractional ideal.

If A and B are subsets of Q then we put $(A :_r B) = \{q \in \mathrm{Q} \mid Bq \subseteq A\}$ and $(A :_l B) = \{q \in \mathrm{Q} \mid qB \subseteq A\}$. If I is a fractional R-ideal then $(I :_r I)$ and $(I :_l I)$ are orders in Q that are equivalent to R. Furthermore, R is a maximal order if and only if $R = (I :_r I) = (I :_l I)$ for every fractional R-ideal I (or equivalently, for every integral fractional R-ideal).

Assume now that R is a maximal order. It follows that $(R :_r I) = (R :_l I)$ for any fractional R-ideal I. One denotes this set simply as $(R : I)$ or also as I^{-1}. Put $I^* = (R : (R : I))$, the divisorial closure of I. If $I = I^*$ then I is said to be divisorial. Examples of such ideals are invertible fractional R-ideals J defined by the condition that $JJ' = J'J = R$ for some fractional R-ideal J' (in this case, clearly, $J' = (R : J)$). The following properties are readily verified for fractional R-ideals I and J: $I \subseteq I^*$, $I^{**} = I^*$, $(IJ)^* = (I^*J^*)^* = (I^*J)^* = (IJ^*)^*$, $(R : I^*) = (R : I) = (R : I)^*$, if $I \subseteq J$ then $I^* \subseteq J^*$. The divisorial product $I * J$ of two divisorial ideals I and J is defined as $(IJ)^*$. With this multiplication the set $\mathrm{D}(R)$ consisting of all divisorial ideals becomes a group with identity R and with I^{-1} the inverse of $I \in \mathrm{D}(R)$. Moreover, $\mathrm{D}(R)$ is a lattice (ordered by inclusion) so that every bounded set has a least upper bound and a greatest lower bound. It follows from a theorem of Iwasawa ([111, Proposition II.1.4]) that $\mathrm{D}(S)$ is an abelian group.

An order R is said to be a Krull order if R is a maximal order that satisfies the ascending chain condition on divisorial ideals that are integral. So commutative Noetherian Krull orders are precisely the Noetherian integrally closed domains. It turns out that the center $\mathrm{Z}(R)$ of a Krull order R is a Krull domain.

The divisor group $\mathrm{D}(R)$ of a Krull order R is a free abelian group with basis the set of prime divisorial ideals. The latter are primes of height one and for rings satisfying a polynomial identity the height one primes are precisely the prime divisorial ideals.

In the next result some fundamental properties are collected (see [18, 19, 71, 75, 93] and [111, Proposition II.2.6]).

Theorem 3.6.1. *Let R be a prime algebra satisfying a polynomial identity. Then, R is a Krull order if and only if R is a maximal order and $\mathrm{Z}(R)$ is a Krull domain. Moreover, in this case the following properties hold.*

1. *The divisorial ideals form a free abelian group with basis $\mathrm{Spec}^1(R)$, the height one primes of R.*

2. *If $P \in \mathrm{Spec}^1(R)$ then $P \cap \mathrm{Z}(R) \in \mathrm{Spec}^1(\mathrm{Z}(R))$, and furthermore, for any ideal I of R, $I \subseteq P$ if and only if $I \cap \mathrm{Z}(R) \subseteq P \cap \mathrm{Z}(R)$.*

3. *For a multiplicatively closed set of ideals $\mathcal{M}$ of R, the (localized) ring $R_{\mathcal{M}} = \{q \in Q_{cl}(R) \mid Iq \subseteq R, \text{ for some } I \in \mathcal{M}\}$ is a Krull order, and*
$$R_{\mathcal{M}} = \bigcap_{P} R(\mathrm{Z}(R) \setminus P)^{-1},$$
where the intersection is taken over those height one primes P of R for which $R_{\mathcal{M}} \subseteq R(\mathrm{Z}(R) \setminus P)^{-1}$.

4. *$R = \bigcap_P R(\mathrm{Z}(R) \setminus P)^{-1}$, where the intersection is taken over all height one primes P of R, and every regular element $r \in R$ is invertible in almost all (that is, except possibly finitely many) localizations $R(\mathrm{Z}(R) \setminus P)^{-1}$. Furthermore, for each height one prime P of R, the localization $R(\mathrm{Z}(R) \setminus P)^{-1}$ has $M = R(\mathrm{Z}(R) \setminus P)^{-1}P$ as a unique maximal ideal, M is invertible and every proper ideal can be written uniquely as M^n with n a positive integer.*

In order to state a characterization of group algebras $K[G]$ of polycyclic-by-finite groups G that are maximal orders more terminology is needed.

A normal element $\alpha \in K[G]$ is said to be G-normal if $K[G]\alpha = \alpha K[G] = \alpha^g K[G]$ for any $g \in G$. A group G is said to be dihedral free if the normalizer of any subgroup H isomorphic with the infinite dihedral group $D_\infty = \langle a, b \mid b^2 = 1,\ ba = a^{-1}b \rangle$ is of infinite index in G, that is, if $H \cong D_\infty$ then H has infinitely many conjugates in G. The following three results come from [12, 14].

Theorem 3.6.2. *Let G be a polycyclic-by-finite group and K a field. Then, K[G] is a Noetherian prime maximal order if and only if $\Delta^+(G) = \{1\}$ and G is dihedral free. Furthermore, in this case, a height one prime ideal P of $K[G]$ contains a nonzero normal element if and only if $P = K[G]\alpha = \alpha K[G]$ for some G-normal element $\alpha \in K[\Delta(G)]$.*

Brown also characterized when all height one primes satisfy the condition mentioned in the theorem.

Theorem 3.6.3. *Let G be a polycyclic-by-finite group and K a field. If $\Delta^+(G) = \{1\}$ then the following conditions are equivalent.*

1. *Every nonzero ideal of $K[G]$ contains an invertible ideal.*
2. *Every nonzero ideal of $K[G]$ contains a nonzero central element (because of Theorem 3.1.7, this holds for example if $K[G]$ is a PI algebra).*
3. *Every nonzero ideal of $K[G]$ contains a nonzero normal element.*

In the following result we collect the information obtained for group algebras that are PI Noetherian domains.

Theorem 3.6.4. *Let G be a finitely generated torsion free abelian-by-finite group and K a field. Then $K[G]$ is a Noetherian PI domain that is a maximal order and all its height one primes are principally generated by a normal element.*

A unique factorization ring (UFR for short) is a prime Noetherian ring R in which every nonzero prime ideal contains a prime ideal P generated by a normal element p, that is $P = Rp = pR$. Because of the principal ideal theorem P is of height one. If, moreover, R and R/P have no zero divisors for each height one prime ideal P then R is said to be a unique factorization domain (UFD for short). Brown also characterized when a group algebra of a polycyclic-by-finite group is a unique factorization ring or domain.

Also commutative semigroup algebras that are Krull domains have been described. In order to state the theorem we first need to establish the terminology of maximal orders for cancellative semigroups. This basically is as in the ring case, but for the convenience of the reader we also give the precise statements in the semigroup setting.

A cancellative monoid S which has a left and right group of quotients G is called an order. Such a monoid S is called a maximal order if there does not exist a submonoid S' of G properly containing S and such that $aS'b \subseteq S$ for some $a, b \in G$. For subsets A and B of G put $(A :_l B) = \{g \in G \mid gB \subseteq A\}$ and $(A :_r B) = \{g \in G \mid Bg \subseteq A\}$. A nonempty subset I of G is said to be a fractional ideal if $SIS \subseteq I$ and $cI, Id \subseteq S$ for some $c, d \in S$. If, furthermore, $I \subseteq S$ then I is called an integral fractional ideal. Note that S is a maximal order if and only if $(I :_l I) = (I :_r I) = S$ for every fractional ideal I of S (or equivalently, for every integral fractional ideal).

Suppose now that S is a maximal order. Then $(S :_l I) = (S :_r I)$ for any fractional ideal I; we simply denote this fractional ideal by $(S : I)$ or by I^{-1}. A fractional ideal I is said to be divisorial if $I = I^*$, where $I^* = (S : (S : I))$. The divisorial product $I * J$ of two divisorial ideals I and J is defined as $(IJ)^*$. With this operation, the set $\mathrm{D}(S)$ of divisorial fractional ideals is an abelian group.

A cancellative monoid S which is an order is said to be a Krull order if and only if S is a maximal order satisfying the ascending chain condition on integral divisorial ideals (the latter are by definition the divisorial fractional ideals contained in S). In this case the group $\mathrm{D}(S)$ is free abelian. If every ideal of S contains a central element (for example, if the group of quotients of S is finitely generated and abelian-by-finite, Lemma 4.1.9) then the minimal primes of S form a free basis for $\mathrm{D}(S)$.

If S is an abelian cancellative monoid then S is a maximal order in its group of quotients G if and only if S is completely integrally closed, that is, if $s, g \in G$ are such that $s\langle g \rangle \subseteq S$, then $g \in S$. If S satisfies the ascending chain condition on ideals then S is completely integrally closed if and only of S is integrally closed, that is, if $g^n \in S$ with $g \in G$ and some positive integer n then $g \in S$.

We now can state a characterization of commutative semigroup algebras that are Krull domains [26]. If F is a free abelian group with a free basis X then the positive cone of F (with respect to X) is defined by $F_+ = \langle X \rangle$.

Theorem 3.6.5. *Let S be an abelian cancellative monoid with torsion free group of quotients and let K be a field. The following conditions are equivalent.*

1. *$K[S]$ is a Krull domain.*

2. *S is a Krull order and its group of quotients satisfies the ascending chain condition on cyclic subgroups.*

3. *$S = \mathrm{U}(S) \times S_1$ where $S_1 = \mathrm{Fa}_+ \cap S_1 S_1^{-1}$ and Fa is a free abelian group containing S_1.*

In this case, all height one primes of $K[S]$ are principal if and only all minimal primes of S are principal.

Recall that for a commutative Krull domain R the height one prime ideals contain important information of the arithmetical properties of the domain ([40]). For example, if all height one primes are principal then R is a unique factorization domain. This occurs if R has only finitely many height one primes. In this case height one primes are maximal and thus R has classical Krull dimension at most one; actually all ideals of R are then principal. For several concrete classes, the last part of the statement of the previous result allows to simplify classical arguments in order to show that some Krull domains are unique factorization domains.

3.7 Principal ideal rings

In this section we give the necessary structural results on principal ideal rings. For general background the reader is referred to [70, 112].

Recall that a ring R is said to be a principal right ideal ring if every right ideal can be generated by a single element. Clearly such a ring is right Noetherian and epimorphic images again are principal right ideal rings. Similarly one defines principal left ideal rings. If a ring R is both a principal right and left ideal ring then we simply call it a principal ideal ring. Prime principal ideal rings clearly are Noetherian maximal orders.

We begin with giving several constructions of principal right ideal right ([112, Theorem 3.4.10], [70, Theorem II.3.1] and [39, Lemma 6]).

Theorem 3.7.1. 1. *If R is a principal right ideal ring then so are matrix rings $\mathrm{M}_n(R)$.*

2. *If D is a division ring and σ is an automorphism on $R = \mathrm{M}_n(D)$ then the skew polynomial ring $R[X, \sigma]$ is a prime principal ideal ring.*

3. *If σ is an automorphism on a semisimple Artinian ring R then the Laurent polynomial ring $R[X, X^{-1}, \sigma]$ is a semiprime principal ideal ring.*

The following example shows that, in general, part three does not hold for skew polynomial rings over semisimple rings. Let σ be the automorphism on $\mathbb{Q} \oplus \mathbb{Q}$ defined by $\sigma(q_1, q_2) = (q_2, q_1)$. Then it is easy to verify that the skew polynomial ring $(\mathbb{Q} \oplus \mathbb{Q})[X, \sigma]$ is neither a principal right ideal ring nor a principal left ideal ring.

Next we give a structure theorem for principal ideal rings ([70, Theorem II.8.1]). Recall that a ring R is said to be primary if for any two ideals I and J with $IJ = \{0\}$ one has $I = \{0\}$ or $J^n = \{0\}$ for some n.

Theorem 3.7.2. *The following are equivalent for a ring R.*

1. *R is a principal right ideal ring that satisfies the ascending chain condition on nilpotent left ideals.*
2. *R is a principal right ideal ring that satisfies the descending chain condition on nilpotent left ideals.*
3. *$R = R_1 \oplus \cdots \oplus R_m$ with each R_i a principal right ideal ring that is either prime or primary left and right Artinian.*

The description of principal ideal rings can be refined in case of prime rings or finite dimensional algebras over a field ([70, Lemma III.1.7], [112, Theorem 3.4.9] and [34, Theorem IX.4.1]).

Theorem 3.7.3. *The following properties hold for a ring R.*

1. *If R is a prime principal right ideal ring then every idempotent ideal of R is trivial.*
2. *If R is a prime principal right ideal ring then $R = \mathrm{M}_n(A)$ with A a right Noetherian domain.*
3. *If R is a finite dimensional algebra over a field then the following conditions are equivalent.*
   (a) *R is a principal right ideal ring.*
   (b) *R is a principal left ideal ring.*
   (c) *Every two-sided ideal of R is principal as a right ideal.*
   (d) *$R = R_1 \oplus \cdots \oplus R_n$ where $R_i \cong \mathrm{M}_{k_i}(A_i)$ with A_i a local algebra (that is, $A_i/\mathcal{B}(A_i)$ is a division ring) so that $\mathcal{B}(A_i)$ is principal as a right ideal.*

Because of Proposition 3.2.3 one obtains as an immediate consequence of the above theorem the following result.

Corollary 3.7.4. *Let R be a finite dimensional algebra over a field and let e be a nonzero idempotent of R. If R is a principal ideal ring then so is the ring eRe.*

Chapter 4

Algebras of submonoids of polycyclic-by-finite groups

As explained in Section 3.3, group algebras of polycyclic-by-finite groups are the only known examples of Noetherian group algebras. In the search for more classes of Noetherian rings, the first obvious step is to investigate semigroup algebras that are subalgebras of Noetherian group algebras. In view of a very rich theory of the latter class of algebras one might expect that many decisive results on the structure and properties of related Noetherian semigroup algebras can be obtained. There are however much deeper reasons justifying this approach. First, there are some natural important classes of such algebras arising from several different contexts. Chapter 8 is devoted to the study of these algebras. Secondly, the algebras $K[S]$ of cancellative semigroups S in some sense form the building blocks of arbitrary Noetherian semigroup algebras and, to a large extent, determine their properties. This will be discussed in Section 5.2 and Section 5.3.

The problem of describing when a semigroup algebra $K[S]$ is right Noetherian is very difficult. For an abelian monoid S, there is however a very satisfactory solution, see [54, Corollary 5.4].

Theorem 4.0.5. *Let K be a field and let S be an abelian monoid. Then the semigroup algebra $K[S]$ is Noetherian if and only if S is finitely generated.*

For nonabelian monoids the problem is much harder, even if S is contained in a group G. However, if G is polycyclic-by-finite, then the result of Chin and Quinn on graded rings, Theorem 3.4.1 implies that $K[S]$ is right Noetherian if and only if S satisfies the ascending chain condition on right ideals. Thus, one might hope for a structural characterization of $K[S]$ in terms of the underlying monoid S.

In this chapter we deal with the structural problems for submonoids S of a polycyclic-by-finite group that satisfy the ascending chain condition on right ideals (or equivalently, $K[S]$ is right Noetherian). The main result characterizes Noetherian semigroup algebras $K[S]$ of submonoids S of a polycyclic-by-finite group completely in terms of the structure of S: namely, the group G of right quotients group of S exists and it contains a normal subgroup H of finite index such that $[H, H] \subseteq S$ and $S \cap H$ is finitely generated. It follows that such algebras are also left Noetherian. As an application we show that these algebras are finitely presented and also that they are Jacobson rings. Finally, we study prime ideals P and the corresponding homomorphic images of $K[S]$. Those ideals P that do not intersect S are determined by prime ideals of the group algebra $K[SS^{-1}]$. More generally, we show that every prime ideal P is strongly related to a prime ideal of a group algebra of a subgroup of SS^{-1} via a generalized matrix ring structure on $K[S]/P$. If SS^{-1} is torsion free, P is of height one and $P \cap S \neq \emptyset$, then P is homogeneous, that is, $P = K[P \cap S]$. So, surprisingly, these height one primes behave as in the rings graded by torsion free abelian groups, Theorem 3.4.2. We also are able to determine the classical Krull dimension of $K[S]$.

As mentioned above, for algebras $K[S]$ of submonoids of polycyclic-by-finite groups, we obtain that S is necessarily finitely generated. In the next chapter we show that this is also the case for arbitrary right Noetherian semigroup algebras $K[S]$ provided that $K[S]$ has finite Gelfand-Kirillov dimension or satisfies a polynomial identity, or $K[S]$ is also left Noetherian.

4.1 Ascending chain condition

We begin with some general lemmas that relate the ascending chain condition of a semigroup S and of the algebra $K[S]$ to that of some subsemigroups and homomorphic images of S.

Recall that a subsemigroup T of a semigroup S is said to be left grouplike, if, for every $s \in S$ and $t \in T$, the condition $ts \in T$ implies that $s \in T$. Then T is called a grouplike subsemigroup if also $s \in S, t, st \in T$ imply that $s \in T$.

Lemma 4.1.1. *Let K be a field and let T be a left grouplike subsemigroup of a semigroup S.*

1. *If S satisfies the ascending chain condition on right ideals, then so does T.*

2. *If $K[S]$ is a right Noetherian K-algebra, then $K[T]$ is a right Noetherian K-algebra.*

Proof. Since T is left grouplike, it is easy to see that

$$IS^1 \cap T = I \text{ and } JK[S^1] \cap K[T] = J,$$

where I is any right ideal of T and J is any right ideal of $K[T]$ which is a K-subspace. The result is now clear. □

The simplest motivating examples of grouplike subsemigroups are obtained as follows. Let S be a subsemigroup of a group G. If H is a subgroup of G then $S \cap H$ is grouplike in S, provided that it is nonempty. In particular, if S satisfies the ascending chain condition on right ideals then also $S \cap H$ satisfies the ascending chain condition on right ideals.

Lemma 4.1.2. *Suppose G is the group generated by its subsemigroup S. If H is a subgroup of finite index in G, then $S \cap gH \neq \emptyset$ for every $g \in G$. Furthermore, if G is the group of right quotients of S, then H is the group of right quotients of $S \cap H$.*

Proof. Let $g \in G$. Let H' be a normal subgroup of finite index in G and such that $H' \subseteq H$. Since the image $\overline{S}$ of S in G/H' is a finite subsemigroup and it generates G/H' as a group, it follows that $\overline{S} = G/H'$. Thus we get $\emptyset \neq S \cap gH' \subseteq S \cap gH$.

Assume that $G = SS^{-1}$ and $g \in H$. Then there exists $s \in S$ so that $gs \in S$. We also have $s^n \in S \cap H$ for some $n \geq 1$. Thus $gs^n \in S \cap H$, so that H is the group of right quotients of $S \cap H$. □

Lemma 4.1.3. *Assume that H is a subgroup of a group G and let S be a submonoid of G that generates G as a group.*

1. *If $K[S]$ is right Noetherian then, for every $g \in G$, $K[S] \cap gK[H]$ is a finitely generated right $K[S \cap H]$-module.*

2. *Assume that $[G : H] < \infty$ and H is normal in G. Then $K[S]_g = K[S] \cap gK[G]$ defines a G/H-gradation on $K[S]$. Moreover, there is an embedding of right $K[S \cap H]$-modules $K[S] \longrightarrow K[S \cap H]^{|G/H|}$ (the latter denoting the direct sum of $|G/H|$ copies of $K[S \cap H]$) and $K[S]$ is right Noetherian if and only if $K[S \cap H]$ is right Noetherian, and then $K[S]$ is a right Noetherian $K[S \cap H]$-module.*

Proof. If $N_1 \subseteq N_2 \subseteq \cdots$ is a chain of right $K[S \cap H]$-submodules of $K[S] \cap gK[H]$ then $N_1 K[S] \subseteq N_2 K[S] \subseteq \cdots$ is a chain of right ideals of $K[S]$. Now $N_i K[S] \cap (K[S] \cap gK[H]) = N_i K[S \cap H] \cap (K[S] \cap gK[H]) = N_i$ for every i. Hence (1) follows.

Assume that $[G : H] < \infty$ and H is normal. Let X be a transversal for H in G. By Lemma 4.1.2, for every $g \in X$ there exists $s_g \in S \cap g^{-1}H$. Clearly $s_g(S \cap gH) \subseteq S \cap H$. Left multiplication by s_g induces an embedding of right $K[S \cap H]$-modules $K[S]_g \longrightarrow K[S \cap H]$. Consequently, $K[S] = \bigoplus_{g \in X} K[S]_g$ embeds into $K[S \cap H]^{|G/H|}$ as a right $K[S \cap H]$-module. Therefore $K[S]$ is a right Noetherian $K[S \cap H]$-module whenever $K[S \cap H]$ is right Noetherian. Hence the remaining assertions follow from (1). □

Lemma 4.1.4. *Assume that S is a submonoid of a group G. Let H be a normal subgroup of G and let $\overline{S}$ denote the image of S in G/H. Suppose that either $H \subseteq S$ or H is finite. Then S satisfies the ascending chain condition on right ideals if and only if $\overline{S}$ satisfies the ascending chain condition on right ideals.*

Proof. Clearly, one implication is obvious. For the converse, assume $\overline{S}$ satisfies the ascending chain condition on right ideals and suppose S has a strictly increasing chain of right ideals of the form

$$a_1 S \subset a_1 S \cup a_2 S \subset \cdots \tag{4.1}$$

Let $b_1 = a_i$ be such that $\overline{a_i}\overline{S}$ is maximal among the right ideals $\overline{a_k}\overline{S}$, $k \geq 1$, and such that $\overline{a_i}\overline{S}$ contains infinitely many right ideals $\overline{a_j}\overline{S}$ with $j \geq i$. Let I_1 denote the set of the corresponding indices j. Consider the right ideals $\overline{a_j}\overline{S}$ with $j \in I_1$ but $j \neq i$. Then, again, let $b_2 = a_m$, $m \in I_1$, be such that $\overline{a}_m\overline{S}$ contains infinitely many right ideals $\overline{a}_j\overline{S}$ with $j \in I_1, j > m$, and let $I_2 \subseteq I_1$ be the set of the corresponding indices. Repeatedly using the above argument we obtain elements $b_1, b_2, \ldots$ such that $b_n = a_{i_n}$ with $i_1 < i_2 < \cdots$ and such that

$$\overline{b_1 S} \supseteq \overline{b_2 S} \supseteq \cdots$$

So, for any $i \geq 2$,

$$\overline{b_i} = \overline{b_{i-1}}\overline{s_i}$$

for some $s_i \in S$. Hence

$$\overline{b_i} = \overline{b_1}\overline{s_2} \cdots \overline{s_i},$$

and therefore

$$b_i = h_i b_1 s_2 \cdots s_i = b_1 s_2 \cdots s_i h_i',$$

for some $h_i, h_i' \in H$. If $H \subseteq S$, the latter yields that $b_i \in b_1 S$, contradicting (4.1). On the other hand, if H is finite, then there exist $n > i \geq 1$ such that $h_n = h_i$. Therefore

$$b_n = h_i b_1 s_2 \cdots s_i s_{i+1} \cdots s_n = b_i s_{i+1} \cdots s_n.$$

So $b_n \in b_i S$, again a contradiction. □

The next lemma shows that a cancellative semigroup satisfying the ascending chain condition on right ideals contains many conjugates.

Lemma 4.1.5. *Let S be a left cancellative semigroup. If S satisfies the ascending chain condition on right ideals, then, for any $a, b \in S$, there exists a positive integer r such that $a^r b \in bS^1$. If S is also right cancellative, then S satisfies the right Ore condition, so it embeds into the group SS^{-1} of right quotients. If, moreover, $a \in \mathrm{U}(S)$ then $b^{-1}a^n b \in S$ and $b^{-1}(a^{-1})^n b \in S$, for some $n \geq 1$. Hence $b^{-1}a^n b \in \mathrm{U}(S)$.*

Proof. Let $a, b \in S$. Consider the following chain of right ideals of S

$$abS^1 \subseteq abS^1 \cup a^2bS^1 \subseteq abS^1 \cup a^2bS^1 \cup a^3bS^1 \subseteq \cdots$$

Since S satisfies the ascending chain condition on right ideals, there exist positive integers $k > i$ such that

$$a^k b \in a^i bS^1.$$

Because by assumption S is left cancellative, it follows that $a^{k-i}b \in bS^1$. Hence, all assertions follow easily. □

The following direct consequence of Theorem 3.4.1 reduces the problem of characterizing Noetherian algebras $K[S]$ considered in this chapter to a problem concerning the monoid S.

Theorem 4.1.6. *If K is a field and S is a submonoid of a polycyclic-by-finite group G then S satisfies the ascending chain condition on right ideals if and only if $K[S]$ is right Noetherian.*

Proof. The rule $K[S]_g = Kg \cap K[S]$, for $g \in G$, defines a natural G-gradation on $K[S]$. The class of graded right ideals of $K[S]$ coincides with the class of all $K[I]$ where I is a right ideal of S. Therefore the result is a direct consequence of Theorem 3.4.1. □

We continue with the following important property of the monoids considered in this chapter.

Theorem 4.1.7. *Let K be a field and let S be a submonoid of a polycyclic-by-finite group. If $K[S]$ is right Noetherian then S is finitely generated.*

Proof. By Lemma 4.1.5 S has a polycyclic-by-finite group G of right quotients. Suppose first that G is a poly-(infinite cyclic) group. We proceed by induction on the Hirsch rank of G. If $n = 0$ then $S = G = \{1\}$ and the assertion is trivial. So assume that $n \geq 1$. Let F be a normal subgroup of G such that G/F is infinite cyclic. Write $G = \bigcup_{j\in\mathbb{Z}} g^j F$ for some $g \in G$. Let $S_+ = \bigcup_{j\geq 0}(S \cap g^j F)$ and $S_+ = \bigcup_{j\leq 0}(S \cap g^j F)$. Define also $T = \bigcup_{j>0}(S \cap g^j F)$. Clearly, T is an ideal of S_+ and TS is a right ideal of S. By the hypothesis $TS = s_1 S \cup \cdots \cup s_m S$ for some $s_i \in T$ and $m \geq 1$. Let $s_i \in g^{j_i} F$. If $t \in S \cap g^l F$, where $l \geq j_i$ for $i = 1, \ldots, m$, then $t \in T \subseteq TS$, so that $t = s_{i_1} u$ for some $i_1 \in \{1, \ldots, m\}$ and $u \in S$. Thus $u \in S_+$, which shows that $S \cap g^l F \subseteq s_1 S_+ \cup \cdots \cup s_m S_+$. Moreover, if $u \notin \bigcup_{j\leq k}(S \cap g^j F)$ where $k = \max\{j_i \mid i = 1, \ldots, m\}$, then we can repeat the above procedure in order to find i_2 such that $u \in s_{i_2} S_+$. After a finite number of such steps we come to $t = s_{i_1} \cdots s_{i_r} z$, where $z \in \bigcup_{j\leq k}(S \cap g^j F), r \geq 1$. Therefore $S_+ = \langle s_1, \ldots, s_m, \bigcup_{j\leq k}(S \cap g^j F)\rangle$. On the other hand, every $K[S] \cap g^j K[F], j = 1, \ldots, k$, is a Noetherian right $K[S \cap F]$-module by Lemma 4.1.3. In particular, the induction hypothesis implies that $K[S \cap F]$ is a finitely generated semigroup, and so S_+ also is finitely generated. A similar argument shows that S_- also is finitely generated. Thus $S = S_+ \cup S_-$ is finitely generated.

Finally, let G be an arbitrary polycyclic-by-finite group. Then, by Proposition 3.3.3, and Lemma 4.1.3, $K[S \cap W]$ is right Noetherian for a normal subgroup W of finite index in G such that W is poly-(infinite cyclic). The foregoing implies that $S \cap W$ is finitely generated. Since Lemma 4.1.3 implies also that $K[S]$ is a finitely generated right $K[S \cap W]$-module, it follows that $S = \bigcup_{i=1}^{q} t_i(S \cap W)$ for some $t_i \in S$. Therefore S is finitely generated. □

We shall see later that the same can be proved for several other classes of right Noetherian semigroup algebras. However, it is not known whether all right Noetherian semigroup algebras are finitely generated.

For submonoids of FC-groups the converse can also be proved.

Theorem 4.1.8. *Let K be a field and let S be a submonoid of an FC-group. Then S is finitely generated if and only if $K[S]$ is right Noetherian. Moreover, in this case $K[S]$ is also left Noetherian.*

Proof. Notice that S has a group of quotients G by Lemma 2.1.3. Assume first that S is finitely generated, say $S = \langle s_1, \ldots, s_n \rangle$. Since G is a finitely generated FC-group, from Lemma 3.3.2 we know that $m = [G : \mathrm{Z}(G)] < \infty$. Let T denote the set of all elements of $S \cap \mathrm{Z}(G)$ that are products of at most m elements of the set $\{s_1, \ldots, s_n\}$. Consider an element $s = t_1 \cdots t_r \in S \cap \mathrm{Z}(G)$, where $r > m$ and $t_j \in \{s_1, \ldots, s_n\}$ for $j = 1, \ldots, r$. Then there exist k, l with $k < l$ such that the elements $t_1 \cdots t_k, t_1 \cdots t_l$ are in the same coset of $\mathrm{Z}(G)$ in G. Therefore $z = t_{k+1} \cdots t_l \in \mathrm{Z}(G)$. Now $s = t_1 \cdots t_r = z(t_1 \cdots t_k t_{l+1} \cdots t_r)$ and clearly $t_1 \cdots t_k t_{l+1} \cdots t_r \in \mathrm{Z}(G)$. Hence, proceeding this way we eventually come to a presentation $s = z_1 \cdots z_j t$ where t and all z_i are products of at most m elements of the set $\{s_1, \ldots, s_n\}$ and $z_1, \ldots, z_j \in \mathrm{Z}(G)$. Then also $t \in \mathrm{Z}(G)$. Therefore $s \in \langle T \rangle$ and, since $s \in S \cap \mathrm{Z}(G)$ is arbitrary, $S \cap \mathrm{Z}(G) = \langle T \rangle$ is a finitely generated monoid. It follows that $K[S \cap \mathrm{Z}(G)]$ is a Noetherian algebra. Therefore, Lemma 4.1.3 implies that $K[S]$ right Noetherian. Similarly it follows that $K[S]$ is left Noetherian.

Assume now that $K[S]$ is right Noetherian. Then $K[G]$ is right Noetherian as a right localization of $K[S]$. Therefore G is a finitely generated group. Thus, Lemma 3.3.2 implies that G is polycyclic-by-finite and S must be finitely generated by Theorem 4.1.7. □

Clearly, in view of Corollary 3.5.8, the first assertion of Theorem 4.1.8 cannot be extended to submonoids of all polycyclic-by-finite groups. We shall see later that it does not even hold for certain submonoids of finitely generated nilpotent groups.

We close this section with an observation on subsemigroups of abelian-by-finite groups. This is of special interest in view of Theorem 4.4.7, and it will be later used in the context of semigroup algebras satisfying polynomial identities.

Lemma 4.1.9. *Let S be a subsemigroup of a finitely generated abelian-by-finite group. Then S has a group G of (right and left) quotients, $\mathrm{Z}(S) \neq \emptyset$ and $G = S\,\mathrm{Z}(S)^{-1}$.*

Proof. The quotient group G exists by Lemma 2.1.3. G has the largest finite normal subgroup H since this is the same as the largest locally finite normal subgroup of G. Then $G_1 = G/H$ has no nontrivial finite normal subgroups. Define $S_1 \subseteq G_1$ as the image of S in G_1. Now $K[G_1]$ is a prime PI algebra, see Theorem 3.2.8, Theorem 3.1.9. We first claim that $G_1 = S_1\,\mathrm{Z}(S_1)^{-1}$. The classical quotient ring $Q = \mathrm{Q}_{cl}(K[S_1])$ of $K[S_1]$ is obtained by inverting the nonzero elements in the center of $K[S_1]$. Consequently, for every $g \in G_1$, there exists a central element $\alpha \in K[S_1]$ such that $g\alpha \in K[S_1]$. Hence we have $g\,\mathrm{supp}(\alpha) \subseteq S_1$. Now, for any $h \in G_1$, $h\alpha h^{-1} = \alpha$. Hence

$$h\,\mathrm{supp}(\alpha)h^{-1} = \mathrm{supp}(\alpha) \subseteq S_1.$$

Therefore

$$ghxh^{-1} \in S_1 \quad \text{for any} \quad h \in G_1 \quad \text{and} \quad x \in \mathrm{supp}(\alpha).$$

Since G_1 is finitely generated and abelian-by-finite, there exists an exponent n such that $x^n \in S_1 \cap \Delta(G_1) \cap A$, where $\Delta(G_1)$ is the FC-center of G_1 and A is an abelian normal subgroup of finite index in G_1. Since x^n has only finitely many conjugates in G_1, say $g_1x^ng_1^{-1}, \ldots, g_mx^ng_m^{-1}$, and all of them are in A, we get $z = (g_1xg_1^{-1})^n \cdots (g_mxg_m^{-1})^n \in \mathrm{Z}(S_1)$. Then $gz = g(g_1xg_1^{-1})^n \cdots (g_mxg_m^{-1})^n \in S_1$ and consequently $g \in S_1\,\mathrm{Z}(S_1)^{-1}$. Hence, indeed we have $G_1 = S_1\,\mathrm{Z}(S_1)^{-1}$.

Let Z be the inverse image of $\mathrm{Z}(S_1)$ in S. Take any $s \in Z$. Then $sxH = xHs$ for every $x \in S$. So s acts by conjugation on the coset xH, and it follows that s^n acts as the identity map, where $n = |H|!$. This means that $s^n \in \mathrm{Z}(S)$, since $x \in S$ is arbitrary. So $\mathrm{Z}(S) \neq \emptyset, \mathrm{Z}(S) \subseteq Z$ and Z is periodic modulo $\mathrm{Z}(S)$. Therefore $S\,\mathrm{Z}(S)^{-1} = SZ^{-1}$. But the image of SZ^{-1} in G_1 is equal to $S_1\,\mathrm{Z}(S_1)^{-1} = G_1$. Hence, for every $a \in S\,\mathrm{Z}(S)^{-1}$ there exists $b \in S\,\mathrm{Z}(S)^{-1}$ such that $ab \in H$. Then $(ab)^n = 1$, so $S\,\mathrm{Z}(S)^{-1}$ is a group. It follows that $G = S\,\mathrm{Z}(S)^{-1}$. □

An interesting consequence is that, if S is a submonoid of a finitely generated abelian-by-finite group and $\mathrm{Z}(S)$ is trivial, then S must be a group.

Remark 4.1.10. The same assertion as in Lemma 4.1.9 holds for any subsemigroup of a torsion free abelian-by-finite group. In this case, with the notation as in the proof, one gets $G = G_1$ and $S = S_1$, so indeed $S = S\,\mathrm{Z}(S)^{-1}$.

4.2 The unit group

In this section we study the unit group $\mathrm{U}(S)$ of S under the assumption that S satisfies the ascending chain condition on right ideals. In particular, we show that it contains a large subgroup that is closed under conjugation. By Lemma 4.1.5 we already know that conjugates of elements of S are periodic modulo S.

Lemma 4.2.1. *Assume that G is a torsion free group such that $[G,G]$ is a finitely generated nilpotent group. Let $S \subseteq SS^{-1} = G$ be a submonoid satisfying the ascending chain condition on right ideals. If S does not contain nontrivial normal subgroups of G then $\mathrm{U}(S)$ is trivial.*

Proof. We first show that $\mathrm{U}(S) \cap [G,G] = \{1\}$. Suppose the contrary, and let l be the maximal nonnegative integer so that $G^{(l)} \cap \mathrm{U}(S) \neq \{1\}$, where $G^{(l)}$ denotes the l-th term of the lower central series of $[G,G]$. Notice that $[G,G]$ satisfies the ascending chain condition on subgroups.

Write $W = G^{(l)} \cap \mathrm{U}(S)$. Let $\overline{W}, \overline{S}$ be the images of W, S respectively, in $\overline{G} = G/G^{(l+1)}$. Let H be the largest subgroup of the finitely generated abelian group $G^{(l)}/G^{(l+1)}$ that is a finite extension of $\overline{W}$. Since every subgroup $x^{-1}Hx, x \in \overline{S}$, is periodic modulo $\overline{W}$ by Lemma 4.1.5 and $x^{-1}Hx \subseteq G^{(l)}/G^{(l+1)}$, we must have $x^{-1}Hx \subseteq H$. Then

$$H \subseteq xHx^{-1} \subseteq x^2Hx^{-2} \subseteq \cdots$$

Therefore, the ascending chain condition on subgroups in G implies that $x^{-1}Hx = H$. Since $\overline{G}$ is the group of quotients of $\overline{S}$ and $x \in \overline{S}$ is arbitrary, it follows that H is a normal subgroup of $\overline{G}$. In particular, every $x^{-1}\overline{W}x, x \in \overline{S}$, is a subgroup of the same finite index in H, so there are only finitely many such conjugates of $\overline{W}$.

Suppose that for some $y \in S$ the images of the groups $W, y^{-1}Wy$ in $G/G^{(l+1)}$ are equal. If $w \in W$, then $y^{-1}wy = vg$ for some $v \in W, g \in G^{(l+1)}$. Hence, for $k \geq 1$, we have $y^{-1}w^ky = v^kh$ for some $h \in G^{(l+1)}$. Using Lemma 4.1.5, choose k so that $y^{-1}w^ky \in \mathrm{U}(S)$. Then $h = 1$ by the choice of l. Therefore $(y^{-1}wy)^k = v^k$ and since roots are unique in the nilpotent torsion free group $[G,G]$, see Proposition 3.3.3, we get $y^{-1}wy = v \in W$. This implies that $y^{-1}Wy \subseteq W$ and again the ascending chain condition on subgroups yields $y^{-1}Wy = W$.

The two preceding paragraphs imply that there are only finitely many conjugates $y^{-1}Wy$ for $y \in S$. So every $s \in S$ permutes these conjugates by $y^{-1}Wy \mapsto s^{-1}(y^{-1}Wy)s$. Since $G = SS^{-1}$, it follows easily that $V = \bigcap_{y \in S} y^{-1}Wy = \bigcap_{y \in S} yWy^{-1}$ is a normal subgroup of G. Let $w \in W, w \neq 1$. Lemma 4.1.5 implies that, for every y, yWy^{-1} contains some w^m, $m \geq 1$. Therefore V also contains a power of w, whence it is nontrivial. Since $V \subseteq S$, we come to a contradiction. We have shown that indeed $\mathrm{U}(S) \cap [G,G] = \{1\}$.

Finally, suppose that $1 \neq u \in \mathrm{U}(S)$. By Lemma 4.1.5, for any $x \in S$, we obtain that $u^{-k}x^{-1}u^kx \in \mathrm{U}(S) \cap [G,G]$ for some $k \geq 1$. Therefore $u^{-k}x^{-1}u^kx = 1$. Consequently, a fixed power of u commutes with every element of a finite subset of S that generates G as a group. Hence, it is central. Again this contradicts the fact that G is torsion free and $\mathrm{U}(S)$ does not contain nontrivial normal subgroups of G. The result follows. □

Proposition 4.2.2. *Let S be a submonoid of a polycyclic-by-finite group. If S satisfies the ascending chain condition on right ideals, then $\mathrm{U}(S)$ contains a subgroup of finite index which is normal in the group of right quotients G of S.*

Proof. We proceed by induction on the Hirsch rank $\mathrm{h}(G)$ of G. If $\mathrm{h}(G) = 0$ then $S = G$ is finite and the assertion is clear. Assume that $\mathrm{h}(G) \geq 1$. Let H be a maximal normal subgroup of G that is contained in S. Factoring out H we may assume that S does not contain nontrivial normal subgroups of G. Then it is enough to show that $\mathrm{U}(S)$ is finite.

By Proposition 3.3.3 there exists a normal subgroup F of finite index in G such that F is torsion free and $[F, F]$ is nilpotent. Suppose we know that $S \cap F$ contains a normal subgroup T of F such that $[\mathrm{U}(S \cap F) : T] < \infty$. By Lemma 4.1.2 we may write $G = \bigcup_{i=1}^{m} F g_i$ for some $g_i \in S$. Then $T_0 = \bigcap_{i=1}^{m} g_i T g_i^{-1}$ is normal in F, whence it is also normal in G. Since $T_0 \subseteq S$, we must have $T_0 = \{1\}$. Because of Lemma 4.1.5, for every $a \in g_i^{-1} T g_i$ there exists a positive integer n so that $a^n \in \mathrm{U}(S) \cap F = \mathrm{U}(S \cap F)$. Since $[\mathrm{U}(S \cap F) : T] < \infty$, we also get $a^k \in T$ for some $k \geq 1$. This implies that $g_i^{-1} T g_i$ is periodic modulo T. Therefore T/T_0 is a periodic group. Since F is torsion free, it follows that $T = \{1\}$. Therefore $\mathrm{U}(S \cap F)$ is finite and must have $\mathrm{U}(S \cap F) = \{1\}$. Since $[G : F] < \infty$, this implies that $\mathrm{U}(S)$ is finite, as desired.

It follows that it is enough to prove the assertion of the proposition for the monoid $S \cap F$, which inherits the hypotheses on S by Lemma 4.1.1. If F has a nontrivial normal subgroup R such that $R \subseteq \mathrm{U}(S \cap F)$, then $\mathrm{h}(F/R) < \mathrm{h}(F) = \mathrm{h}(G)$ because F is torsion free. Therefore, the inductive hypothesis yields this assertion, completing the proof. Otherwise, $S \cap F$ satisfies the hypotheses of Lemma 4.2.1. Thus we get that $\mathrm{U}(S)$ finite, so the assertion on $S \cap F$ also follows in this case. □

The following example shows that in the proposition we cannot expect, in general, the full unit group $\mathrm{U}(S)$ to be normal in G.

Example 4.2.3. Let $G = Z \times F$, where $Z = \mathrm{gr}(z)$ is an infinite cyclic group and F is a finite group with a subgroup H which is not normal. Let

$$S = \langle zF, H \rangle.$$

Clearly, $S \cap Z = \langle z \rangle$ is finitely generated. Hence, by Lemma 4.1.3, S satisfies the ascending chain condition on left and right ideals. On the other hand, $\mathrm{U}(S) = H$ is not a normal subgroup of the group $G = SS^{-1}$.

Corollary 4.2.4. *Let S be a submonoid of a polycyclic-by-finite group. Assume that S satisfies the ascending chain condition on right ideals. If $y \in S$, then the $\mathcal{J}$-class of y in S is equal to $\mathrm{U}(S)y\,\mathrm{U}(S)$.*

Proof. Assume that $SxS = SyS$ for some $x, y \in S$. Then $y = sxt = ss'yt't$ for some $s, s', t, t' \in S$. Write $a = ss', b = t't$. We show that $a, b \in \mathrm{U}(S)$. By Lemma 4.1.5 we may choose $n \geq 1$ such that $y^{-1}a^n y \in S$. Then $y = ayb = a^n yb^n = y(y^{-1}a^n yb^n)$ implies that $b \in \mathrm{U}(S)$. By Proposition 4.2.2 n can be chosen so that $b^n \in F$ for a normal subgroup F of the group $G = SS^{-1}$ contained in $\mathrm{U}(S)$. So $y = a^n dy$ for some $d \in F$. Hence we also get $a \in \mathrm{U}(S)$. The assertion follows. □

Lemma 4.2.5. *Let S be a submonoid of a group G. Assume that S satisfies the ascending chain condition on right ideals.*

1. *If $\mathrm{U}(S) = \{1\}$, then S satisfies the ascending chain condition on principal left ideals.*

2. *If S satisfies the ascending chain condition on principal left ideals, then $S = \langle s_1, \ldots, s_n, \mathrm{U}(S)\rangle$, where $s_1S, \ldots, s_nS$ are all the distinct maximal principal right ideals of S (note that the elements s_i are unique up to multiplication on the right by a unit of S). In case $\mathrm{U}(S) = \{1\}$, one has $\{s_1, \ldots, s_n\} = (S \setminus \{1\}) \setminus (S \setminus \{1\})^2$, and furthermore, $Ss_1, \ldots, Ss_n$ are all the maximal principal left ideals and $\{s_1, \ldots, s_n\}$ is the unique minimal set of generators of S.*

Proof. Since S satisfies the ascending chain condition on right ideals, S has finitely many maximal principal right ideals, say $s_1S, \ldots, s_nS$. It is easy to see that $s_1S = sS$ if and only if $s_1 = su$ for some $u \in \mathrm{U}(S)$. So these generators are unique up to right multiplication by units of S.

Now, let $s \in S$. Then, if $s \notin \mathrm{U}(S)$, there exists $1 \leq i_1 \leq n$ so that $s \in s_{i_1}S$ and thus $s = s_{i_1}t_1$, for some $t_1 \in S$. If $t_1 \notin \mathrm{U}(S)$, then $t_1 = s_{i_2}t_2$ for some $1 \leq i_2 \leq n$ and $t_2 \in S$. In general, if $s = s_{i_1}s_{i_2}\cdots s_{i_k}t_k$ for some $1 \leq i_1, \ldots, i_k \leq n$ and $t_k \in S$, then if $t_k \notin \mathrm{U}(S)$, there exists $t_{k+1} \in S$ such that $t_k = s_{i_{k+1}}t_{k+1}$ for some $t_{k+1} \in S$. Clearly

$$St_1 \subseteq St_2 \subseteq St_3 \subseteq \cdots$$

If we assume that S satisfies the ascending chain condition on principal left ideals, then there exists $m \geq 1$ such that $t_{m+1} = ut_m$ for some $u \in S$. Hence

$$t_m = s_{i_{m+1}} t_{m+1} = s_{i_{m+1}} u t_m$$

and thus $s_{i_{m+1}} u = 1$, so that $s_{i_{m+1}} \in \mathrm{U}(S)$, a contradiction. So there exists $k \geq 1$ such that $s = s_{i_1} s_{i_2} \cdots s_{i_k} t_k$ and $t_k \in \mathrm{U}(S)$. Therefore $S = \langle s_1, \ldots, s_n, \mathrm{U}(S) \rangle$.

To prove the remainder assume that $\mathrm{U}(S) = \{1\}$. Suppose

$$Sx_1 \subseteq Sx_2 \subseteq Sx_3 \subseteq \cdots$$

is an increasing chain of principal left ideals. Considering the right ideals $x_i S$ and using the ascending chain condition on right ideals, we see that there exists some i such that $x_i S \supseteq x_j S$ for infinitely many positive integers j. Repeating this argument, as in the proof of Lemma 4.1.4, we get a decreasing chain of right ideals

$$x_{i_1} S \supseteq x_{i_2} S \supseteq x_{i_3} S \supseteq \cdots$$

with $i_1 < i_2 < i_3 < \cdots$. Thus we have $Sx_{i_1} S \subseteq Sx_{i_2} S$ and also $Sx_{i_1} S \supseteq Sx_{i_2} S$, whence $Sx_{i_1} S = Sx_{i_2} S$. Since $\mathrm{U}(S) = \{1\}$, from Corollary 4.2.4 it follows that $x_{i_1} = x_{i_2}$. This proves that indeed S satisfies the ascending chain condition on principal left ideals. To complete the proof, note that since $\mathrm{U}(S) = \{1\}$, an element $s \in S$ generates a maximal principal right (or left) ideal if and only if s cannot be written as a product of two elements of $S \setminus \{1\}$. Then it is clear that the elements of the set

$$(S \setminus \{1\}) \setminus (S \setminus \{1\})^2$$

are the generators of the distinct maximal principal right ideals, and it is also the set of the generators of all the maximal principal left ideals. □

The automorphism group of a polycyclic-by-finite group is pretty well understood, see [140]. Hence, our next result provides an insight into the structure of $\mathrm{Aut}(S)$ for the considered class of monoids S.

Corollary 4.2.6. *Let S be a submonoid of a polycyclic-by-finite group. If S satisfies the ascending chain condition on right ideals, then there exists a finitely generated abelian normal subgroup A of* $\mathrm{Aut}(S)$ *such that* $\mathrm{Aut}(S)/A$ *is a finite extension of a subgroup of* $\mathrm{Aut}(\mathrm{U}(S))$.

Proof. Let $\sigma \in \mathrm{Aut}(S)$. Clearly σ permutes the maximal principal right ideals $s_1S, \ldots, s_nS$ of S. Hence σ permutes the corresponding $\mathcal{R}$-classes $s_1\,\mathrm{U}(S), \ldots, s_n\,\mathrm{U}(S)$. By Proposition 4.2.2 there exists a normal subgroup $H \subseteq S$ of the group SS^{-1} which has finite index in $\mathrm{U}(S)$. Then σ acts by permutation on the finite set of normal subgroups of $\mathrm{U}(S)$ of index equal to the index of H in $\mathrm{U}(S)$, see Proposition 3.3.3. Let $H_1 = \bigcap_{i\in\mathbb{Z}} \sigma^i(H)$. Then $\mathrm{U}(S)/H_1$ is a subdirect product of finite groups of equal order, so it is periodic. Therefore it is finite, as it is polycyclic-by-finite. Since $\sigma(H_1) = H_1$, it follows that σ permutes the cosets of H_1 in $\mathrm{U}(S)$. Write $\mathrm{U}(S) = \bigcup_{j=1}^{k} u_jH_1$ for some $u_j \in \mathrm{U}(S)$. We may assume that $u_1 = 1$. Then σ permutes the sets $s_iu_jH_1$. Therefore we have a homomorphism $\beta : \mathrm{Aut}(S) \longrightarrow \mathrm{Sym}_r, r = kn$. Let $\gamma : \mathrm{Aut}(S) \longrightarrow \mathrm{Aut}(\mathrm{U}(S))$ be the restriction map. Let $B = \ker(\beta)$ and $C = \ker(\gamma)$. Consider $A = B \cap C$. Let $\sigma \in A$. Then $\sigma(u) = u$ for $u \in \mathrm{U}(S)$ and for every $i = 1, \ldots, n$ we have $\sigma(s_i) = s_if_i$ for some $f_i \in H_1$. If $g \in H$, then $s_igs_i^{-1} \in H$ and therefore

$$s_igf_i = s_igs_i^{-1}s_if_i = \sigma(s_igs_i^{-1})\sigma(s_i) = \sigma(s_ig) = \sigma(s_i)\sigma(g) = s_if_ig.$$

Hence $f_ig = gf_i$ and so $f_i \in \mathrm{Z}(H)$. Consider the map $\phi : A \longrightarrow \mathrm{Z}(H)^n$ defined by $\phi(\sigma) = (f_1, \ldots, f_n)$. It is easy to see that ϕ is a homomorphism. Moreover, since $S = \langle s_1, \ldots, s_n, \mathrm{U}(S)\rangle$ by Lemma 4.2.5, it follows that $\ker(\phi)$ is trivial because $\sigma \in \ker(\phi)$ satisfies $\sigma(s_i) = s_i$ for $i = 1, \ldots, n$ and $\sigma(u) = u$ for $u \in \mathrm{U}(S)$. Hence A is abelian and finitely generated. Now $\mathrm{Aut}(S)/B$ is finite and $\mathrm{Aut}(S)/C$ embeds into $\mathrm{Aut}(\mathrm{U}(S))$. Since $\mathrm{Aut}(S)/A$ is a subdirect product of $\mathrm{Aut}(S)/B$ and $\mathrm{Aut}(S)/C$, the assertion follows. □

4.3 Almost nilpotent case

Our first step is to describe when a semigroup algebra $K[S]$ of a submonoid S of a finitely generated nilpotent-by-finite group G is right Noetherian. The following lemma on submonoids of finitely generated groups that are nilpotent of class 2 is crucial for the main theorem of this section.

Lemma 4.3.1. *Let S be a nonabelian submonoid of a group that is torsion free, finitely generated and nilpotent of class 2. Assume that S satisfies the ascending chain condition on right ideals. Then S contains a nontrivial central subgroup.*

Proof. Let x and y be elements of S such that $xy \neq yx$. Since the quotient group G of S is nilpotent class 2, the commutator element $z = y^{-1}x^{-1}yx$ is central in G. We claim that the group $\mathrm{gr}(x, z)$ is free abelian. Suppose the contrary. Then a power of x is central in G. Since G is torsion free, it follows that also x is central, see Proposition 3.3.3, a contradiction.

Consider the grouplike submonoid $T = \mathrm{gr}(x, z) \cap S$. By Lemma 4.1.1 T satisfies the ascending chain condition on right ideals. Hence, by Theorem 4.1.7, T is finitely generated. Since z is central and $y^{-1}xy = xz^{-1}$, we get $y^{-n}xy^n = xz^{-n}$ and $y^{-n}x^q y^n = x^q z^{-nq}$ for $n, q \geq 1$. Hence, by Lemma 4.1.5, for every n there exists $q \geq 1$ such that

$$x^q z^{-nq} = y^{-n}x^q y^n \in S. \tag{4.2}$$

Notice that $x^q z^{-nq} \in T \setminus \mathrm{gr}(z)$. Suppose that $T \subseteq \{x^i z^j \mid i > 0, \text{ or } i = 0 \text{ and } j \geq 0\}$. Since T is finitely generated, choose generators for T of the type

$$x^{i_1}z^{-j_1}, \ldots, x^{i_r}z^{-j_r}, z^{j_{r+1}}, \ldots, z^{j_l},$$

where all $i_k > 0$, $j_k \in \mathbb{Z}$ and all $j_{r+1}, \ldots, j_l \geq 0$. There exists $N \geq 1$ such that $j_k \leq N i_k$ for $k = 1, \ldots, r$. Consider any $w \in T \setminus \mathrm{gr}(z)$. Write

$$w = (x^{i_1}z^{-j_1})^{a_1} \cdots (x^{i_r}z^{-j_r})^{a_r}(z^{j_{r+1}})^{a_{r+1}} \cdots (z^{j_l})^{a_l},$$

with all $a_m \geq 0$. Then also

$$w = x^b z^{-c}$$

where

$$c = c_1 - c_2, c_1 = \sum_{k=1}^{r} j_k a_k, c_2 = \sum_{k=r+1}^{l} j_k a_k \geq 0$$

and

$$b = \sum_{k=1}^{r} i_k a_k > 0 \quad (\text{ at least one } a_k \neq 0).$$

Therefore

$$\frac{c}{b} = \frac{c_1}{b} - \frac{c_2}{b} \tag{4.3}$$

and $c/b \leq c_1/b \leq N$, because $c_2/b \geq 0$ and $j_k \leq N i_k$ for $i = 1, \ldots, r$. However, from (4.2) it follows that T contains elements of the type $w = x^q z^{-nq}$ with n arbitrarily large, and q positive. This contradiction shows that our supposition is false. Therefore T contains an element of the type $x^{-i}z^j$ with $i > 0$ and $j \in \mathbb{Z}$, or it contains an element of the type z^{-j} with $j > 0$.

We claim that there exists $k > 0$ such that $z^{-k} \in S$. To prove this, it is sufficient to assume that $x^{-i}z^j \in S$ with $i > 0$ and $j \in \mathbb{Z}$. Because of (4.2) we may assume that $j \neq 0$. Let $x^q z^{-nq} \in S$, $q > 0$. Then also $x^{-iq} z^{jq} = (x^{-i} z^j)^q \in S$ and $x^{iq} z^{-inq} = (x^q z^{-nq})^i \in S$. Hence $z^{(j-in)q} \in S$. Choosing n large enough, we get $j - in < 0$, and thus the claim follows.

Similarly as in (4.2), but working with the monoid $T' = \mathrm{gr}(y, z) \cap S$, we get, again from Lemma 4.1.5, that for any positive integer n there exists $q > 0$ with

$$y^q z^{nq} = x^{-n} y^q x^n \in S. \tag{4.4}$$

Applying an argument similar to that used before for the monoid $T = \mathrm{gr}(x, z) \cap S$, but replacing x, z by y, z^{-1}, one shows that

$$\mathrm{gr}(y, z) \cap S \not\subseteq \{y^i z^j \mid i > 0, \text{ or } i = 0 \text{ and } j \leq 0\}.$$

Therefore, $z^j \in S$ with $j > 0$, or $y^{-i} z^j \in S$ for some $i > 0$ and $j \in \mathbb{Z}$. Then, using (4.4) we get a positive power of z in S. So we obtain that $\mathrm{gr}(z^k) \subseteq S$ for some $k \geq 1$. This completes the proof. $\square$

Theorem 4.3.2. *Let S be a submonoid of a finitely generated nilpotent-by-finite group. Assume that S satisfies the ascending chain condition on right ideals. Then the quotient group G of S has a normal subgroup H of finite index such that $[H, H] \subseteq S$.*

Proof. We proceed by induction on the nilpotency class n of the maximal nilpotent normal subgroup N (of finite index) in G. The result is obvious if N is abelian, that is, $n = 1$. So assume $n > 1$.

Notice that G satisfies the ascending chain condition on subgroups. Hence, factoring out the maximal normal subgroup of G contained in S, we may assume that S does not contain nontrivial normal subgroups of G. Then from Proposition 4.2.2 it follows that $\mathrm{U}(S)$ is finite. It is now enough to show that G is abelian-by-finite.

Let $F \subseteq G$ be a subgroup of finite index. Then, by Lemma 4.1.2, F is the quotient group of $S \cap F$. Being a grouplike submonoid of S, it also satisfies the ascending chain condition on right ideals, see Lemma 4.1.1. It is enough to prove that F is abelian-by-finite. Hence, we may replace G by any of its subgroups F of finite index, replacing S, G by $S \cap F, F$. In particular, we may assume that G is a nilpotent group.

From Proposition 3.3.3, we know that G contains a torsion free subgroup W of finite index. Therefore, replacing S, G by $S \cap W$ and W again, we may assume that G is torsion free. Therefore $\mathrm{U}(S) = \{1\}$.

Consider the image $\overline{S}$ of S in $\overline{G} = G/G^{(n-1)}$, where $G^{(n-1)}$ is the $n-1$-th term of the lower central series of G. If $x \in G$ then we write $\overline{x}$ for the image of x in $\overline{G}$. From the inductive hypothesis it follows easily that there exists $m > 0$ with

$$\mathrm{gr}([\overline{x}^m, \overline{y}^m]) \subseteq \overline{S}, \text{ for all } x, y \in S.$$

Write $z = [x^m, y^m]$, then Then $\overline{z} = \overline{t}$ for some $t \in S$. Hence, $wz = t \in S \cap G^{(1)}$ for some $w \in G^{(n-1)} \subseteq \mathrm{Z}(G)$.

Let a be an arbitrary element of S. Clearly $\mathrm{gr}(a, wz)$ is nilpotent of class at most $n - 1$. So, by the induction hypothesis applied to the monoid $F = \mathrm{gr}(a, wz) \cap S$ (note that this monoid is grouplike in S and thus inherits the ascending chain condition on right ideals) there exists $r > 0$ such that

$$\mathrm{gr}([a^r, (wz)^r]) \subseteq S.$$

Since w is central, $[a^r, (wz)^r] = [a^r, z^r]$, and thus

$$\mathrm{gr}([a^r, z^r]) \subseteq S.$$

Since we know that $\mathrm{U}(S)$ is trivial, it follows that $[a^r, [x^m, y^m]^r] = 1$. As G is torsion free, a and $[x^m, y^m]$ commute by Proposition 3.3.3. So, $[x^m, y^m] \in \mathrm{Z}(G)$ for all $x, y \in S$, because $a \in S$ is arbitrary and S generates G. From Proposition 3.3.3 it follows that $G/\mathrm{Z}(G)$ is torsion free and $[x, y] \in \mathrm{Z}(G)$. Hence G is nilpotent of class 2. Since $\mathrm{U}(S) = \{1\}$, Lemma 4.3.1 shows that G must be abelian. This completes the proof. □

The main result of this section reads as follows.

Theorem 4.3.3. *Let K be a field and let S be a submonoid of a finitely generated nilpotent-by-finite group. Then the following conditions are equivalent.*

1. *S satisfies the ascending chain condition on right ideals.*
2. *$K[S]$ is right Noetherian.*
3. *The quotient group $G = SS^{-1}$ of S contains a normal subgroup H of finite index such that $S \cap H$ is a finitely generated monoid and $[H, H] \subseteq S$.*

Furthermore, if any of these equivalent conditions holds, then S is finitely generated.

Proof. We know from Theorem 4.1.6 that (1) and (2) are equivalent; and if any of these conditions holds, then S is finitely generated by Theorem 4.1.7.

If (3) holds, then by Lemma 4.1.4 S satisfies the ascending chain condition on right ideals if and only if $\overline{S} = S/[H,H] \subseteq G/[H,H]$ satisfies the ascending chain condition on right ideals. So, in order to prove that (2) is a consequence of (3), we may assume that G contains an abelian normal subgroup H of finite index and $S \cap H$ is finitely generated. Then $K[S \cap H]$ is Noetherian and thus, by Lemma 4.1.3, $K[S]$ is right Noetherian.

From Theorem 4.3.2 it follows that (3) is a consequence of (1). This completes the proof. □

Notice that, if G is a nilpotent group, then in condition (3) the assumption that $S \cap H$ is finitely generated may be replaced by the assumption that S is finitely generated. Indeed, as above, when proving that (2) follows from (3), we may assume that H is abelian. We know that the set G^+ of periodic elements of G forms a finite normal subgroup of G. Lemma 4.1.4 allows us to assume that G^+ is trivial. Then G is torsion free, abelian-by-finite and nilpotent, whence it must be abelian by Proposition 3.3.3. Then $K[S]$ is Noetherian because S is finitely generated.

The following example shows however that in general one cannot weaken the condition that $S \cap H$ is finitely generated to S being finitely generated.

Example 4.3.4. Let G be the group generated by a, b and c subject to the following relations: $ca = bc$, $cb = ac$, $ab = ba$ and $c^2 = 1$. Clearly, the subgroup A generated by a and b is free abelian, normal in G and of finite index. Let $S = \langle a, ac \rangle \subseteq G$. It is easily verified that $S \cap A = \{a^i \mid i \geq 0\} \cup \{a^k b^l \mid l \geq 1, k \geq 1\}$. Since $ab^n \in S \cap A$, for all $n \geq 1$, it follows that $S \cap A$ is not finitely generated. In particular, $K[S]$ is not right Noetherian, see Theorem 4.3.3 or Lemma 4.1.1.

Even if S is a finitely generated submonoid of a nilpotent group, one does not obtain, in general, that the semigroup algebra $K[S]$ is left or right Noetherian.

Example 4.3.5. Let G be the group generated by a, b and c such that c is central and $ab = bac$; that is, G is the Heisenberg group. Let $S = \langle a, b \rangle$. Using the principal right ideals $b^n aS$ one can construct easily a strictly increasing chain of right ideals of S. So $K[S]$ is not right Noetherian. It is also easy to see that $\mathrm{U}(S) = \{1\}$, whence the same conclusion follows from Theorem 4.3.3, because G is not abelian-by-finite. Similarly one shows $K[S]$ is not left Noetherian.

4.4 Structure theorem

Theorem 4.3.3 is an essential step in characterizing submonoids S of arbitrary polycyclic-by-finite groups that yield right Noetherian algebras $K[S]$. Before stating this main result we need a few technical lemmas. Recall that the Hirsch rank of a polycyclic-by-finite group G is denoted by $\mathrm{h}(G)$.

Lemma 4.4.1. *Let S be a monoid that S satisfies the ascending chain condition on right ideals and such that the group of right quotients $G = SS^{-1}$ is polycyclic-by-finite. Assume that G is not nilpotent-by-finite. Then there exists a subgroup $H = \mathrm{gr}(S \cap H)$ which is not nilpotent-by-finite and contains a nilpotent normal subgroup N such that H/N is infinite cyclic and $\mathrm{h}(H/[H,H]) = 1$. Moreover,*

1. *$\mathrm{h}(H_1/[H_1,H_1]) = 1$, for every subgroup H_1 of finite index in H; and*
2. *H contains a free nonabelian submonoid $\langle s,t\rangle$ of S such that $s \in tN$ and $H = \mathrm{gr}(s,t)$.*

Proof. Because of Proposition 3.3.3 and Lemma 4.1.2, we may assume that G is torsion free and $[G,G]$ is nilpotent. Since, by assumption, $G = SS^{-1}$ is not nilpotent-by-finite, we obtain from Corollary3.5.8 that S contains a free nonabelian submonoid. Let M be a subgroup of G of minimal Hirsch rank that contains such a free submonoid $\langle x,y\rangle$. Then S also contains the free submonoid $\langle s,t\rangle$, where $s = xy, t = yx$. Let $H = \mathrm{gr}(s,t)$. Clearly $H = \mathrm{gr}(S \cap H)$ and $H \subseteq M$. Moreover, for $N = [G,G] \cap H$, we have that H/N is cyclic and it is infinite because H is not nilpotent-by-finite. Also $s \in tN$.

By the minimality of the Hirsch rank, $\mathrm{h}(\mathrm{gr}(st,ts)) = \mathrm{h}(H) = \mathrm{h}(M)$. Since $\mathrm{gr}(st,ts)$ is cyclic modulo $[H,H]$, it follows that $\mathrm{h}(H/[H,H]) = 1$. So it remains to prove that (1) is satisfied. Assume H_1 is a subgroup of finite index in H. If $\mathrm{h}(H_1/[H_1,H_1]) > 1$, then the above proof yields a subgroup H_2 of H_1 with $\mathrm{h}(H_2) < \mathrm{h}(H_1) = \mathrm{h}(M)$ and containing a free nonabelian submonoid of S. Hence H_2 is not nilpotent-by-finite, a contradiction. □

Recall that the positive cone of a free abelian group F with a free basis X is defined by $F_+ = \langle X\rangle$. The following fact also will be needed.

Lemma 4.4.2. *Let B be a finitely generated submonoid of a free abelian group A. Assume that $B = \{a \in \mathrm{gr}(B) \mid a^n \in B \text{ for some } n \geq 1\}$. If $\mathrm{U}(B) = \{1\}$ then there exists a free abelian group F such that $B \subseteq F_+$ and the groups $\mathrm{gr}(B), F$ have equal (torsion free) ranks.*

Proof. From Theorem 3.6.5 we know that $K[B]$ is a Noetherian integrally closed domain and $\mathrm{gr}(B)$ is a subgroup of a free abelian group H, with generators $x_1, x_2, \ldots$, such that $B = \mathrm{gr}(B) \cap H_+$, where $H_+ = \langle x_1, x_2, \ldots \rangle$. We claim that $\mathrm{gr}(B)$ is isomorphic to its projection $\pi(\mathrm{gr}(B))$ onto the free abelian group $F = \mathrm{gr}(x_{i_1}, \ldots, x_{i_r})$ for some $i_1, \ldots, i_r$, where r is the rank of $\mathrm{gr}(B)$. Then $\mathrm{gr}(B) \cong \pi(\mathrm{gr}(B))$ can be considered as a subgroup of $F \cong \mathbb{Z}^r$. Therefore, replacing $\mathrm{gr}(B)$ by $\pi(\mathrm{gr}(B))$, we get that $B \subseteq F_+$. Hence the lemma follows.

To prove the claim let $D \cong \mathbb{Z}^r$ be a subgroup of H. Write $D = D_1 \times C$ where C is a cyclic group. Proceeding by induction on r we find $i_1, \ldots, i_{r-1}$ such that the projection $\pi_1(D_1)$ of D_1 onto $\mathrm{gr}(x_{i_1}, \ldots, x_{i_{r-1}})$ is of rank $r-1$. Then $\pi_1(D)$ is also of rank $r-1$, hence $\pi_1(d) = 1$ for some $1 \neq d \in D$. Choose i_r so that the i_r-th coordinate of d is not 1. Then it is easy to see that $\pi(D)$ has rank r. This implies that $D \cong \pi(D)$, and the claim follows with $D = \mathrm{gr}(B)$. □

The reduction step considered below is one of the keys for proving the main result of this section.

Proposition 4.4.3. *Let G be a group that is an extension of a finitely generated free abelian group A by an infinite cyclic group. Assume that $S \subseteq G$ is a submonoid that generates G as a group and S satisfies the ascending chain condition on right ideals. If $\mathrm{U}(S) = \{1\}$ then either $S \cap A = \{1\}$ or the FC-center of G intersects A nontrivially.*

Proof. Suppose that the FC-center of G intersects A trivially, $\mathrm{U}(S) = \{1\}$ and $S \cap A \neq \{1\}$. Notice first that, by Lemma 4.1.1 and Lemma 4.1.2, the hypotheses on S are inherited by every monoid $S \cap H$ for a subgroup H of finite index in G. Here $H = \mathrm{gr}(S \cap H)$ plays the role of G and $H \cap A$ plays the role of A. Moreover, the assumed conditions are also inherited, namely: $\mathrm{U}(S \cap H) = \{1\}, S \cap H \cap A \neq \{1\}$ because A is torsion free, and the FC-center of H intersects A trivially. This implies that in the proof we may always replace S by an appropriate $S \cap H$.

Let $G = \bigcup_{i \in \mathbb{Z}} t^i A$ for some $t \in G$. Clearly $S \not\subseteq A$. Then choose $t_1 \in t^i A \cap S$, for some $i \neq 0$, and consider the subgroup G_1 of G generated by A and t_1. Then $[G : G_1] < \infty$. By the first paragraph of the proof, $S_1 = S \cap G_1$ inherits the hypotheses on S and $G_1 = \mathrm{gr}(S_1)$, so replacing S, G by S_1 and G_1, respectively, and replacing t by t_1, we may assume that $t \in S$.

It is easy to see that the submonoid $S_+ = \bigcup_{i \geq 0} S \cap t^i A$ inherits the ascending chain condition on right ideals. Indeed, let $x_1, x_2, \ldots \in S_+$ with $x_i \in t^{n_i} A$. If $x_1 S_+ \subset x_1 S_+ \cup x_2 S_+ \subset \cdots$ then choose a sequence $m_1 <$

$m_2 < \cdots$ so that $n_{m_1} \leq n_{m_2} \leq \cdots$. Then $x_{m_1}S \subseteq x_{m_1}S \cup x_{m_2}S \subseteq \cdots$ stabilizes and thus $x_{m_k} \in x_{m_l}S$ for some $k > l$. Since $n_{m_k} \geq n_{m_l}$ it follows that $x_{m_k} \in x_{m_l}S_+$. This contradicts the choice of the sequence (x_i). So S_+ satisfies the desired chain condition.

Clearly $S \cap A = S_+ \cap A$. Moreover, if $s = t^{-i}a \in S$ for some $i > 0$ and $a \in A$ then $a = t^i s \in S \cap A$ and $s \in t^{-i}(S \cap A) \subseteq \mathrm{gr}(S_+)$. Hence it follows that $\mathrm{gr}(S) = \mathrm{gr}(S_+)$. Therefore it is enough to deal with the case $S = S_+$. So we may assume $S = \bigcup_{i \geq 0} t^i A_i$ for some subsets $A_i \subseteq A$. Notice that $A_i \subseteq A_{i+1}$ for every i. From Lemma 4.1.3 and Theorem 4.1.6 we know that every $\mathbb{Q}[A_i]$ is a finitely generated $\mathbb{Q}[A_0]$-submodule of the rational semigroup algebra $\mathbb{Q}[A]$.

Since $A_0 = S \cap A$ inherits the ascending chain condition on right ideals, Theorem 4.1.7 implies that it is a finitely generated monoid. Let $A_0^{(c)} = \{g \in A \mid g^k \in A_0$ for some positive integer $k\}$.

Since A is a finitely generated abelian group, the group of quotients $\mathrm{gr}(A_0^{(c)})$ is a finite extension of $\mathrm{gr}(A_0)$. So the field of quotients L of the group algebra $\mathbb{Q}[\mathrm{gr}(A_0^{(c)})]$ is a finite field extension of the field of quotients L_0 of the group algebra $\mathbb{Q}[\mathrm{gr}(A_0)]$. Hence, by a well known result (see for example [35, Theorem 4.14]), the integral closure of $\mathbb{Q}[A_0]$ in L is a finite module over the Noetherian semigroup algebra $\mathbb{Q}[A_0]$. Because $\mathbb{Q}[A_0^{(c)}]$ is contained in this integral closure, the semigroup algebra $\mathbb{Q}[A_0^{(c)}]$ is itself Noetherian. Therefore, again by Theorem 4.1.7, $A_0^{(c)}$ is a finitely generated monoid. In view of Lemma 4.1.5, it is readily verified that $t^{-1}A_0^{(c)}t \subseteq A_0^{(c)}$.

Let $H = \mathrm{gr}(A_0^{(c)}) \subseteq A$ and let $H_{(t)}$ be the group generated by t and H. Then $t^{-1}Ht \subseteq H$. Let $\sigma : H \longrightarrow H$ be defined by $\sigma(h) = t^{-1}ht$ for $h \in H$. Because of the ascending chain condition on subgroups in A, σ is an automorphism of H. Moreover $S \cap H_{(t)}$ inherits the hypotheses on S. Also $(S \cap H_{(t)}) \cap A = A_0 \cap H_{(t)} = A_0 = S \cap A$ and $\mathrm{U}(S \cap H_{(t)}) = \{1\}$. Let $\widehat{S} = S \cap H_{(t)}, \widehat{G} = \mathrm{gr}(S \cap H_{(t)}), \widehat{A} = A \cap \widehat{G}$ and $\widehat{A}_0 = S \cap \widehat{A}$. Then $\widehat{A}_0 = A_0$ and $\mathrm{gr}(A_0) \subseteq \widehat{A} \subseteq H$.

If the FC-center of $\mathrm{gr}(S \cap H_{(t)})$ has a nontrivial element in $\widehat{A}$ then this element has finitely many $\mathrm{gr}(t)$-conjugates, so it is also in the FC-center of G. Therefore, when proving the assertion, we may replace S, G, A by $\widehat{S}, \widehat{G}$ and $\widehat{A}$, respectively. Since H is finitely generated and periodic modulo $\mathrm{gr}(A_0)$, we get $[H : \mathrm{gr}(A_0)] < \infty$. Hence, the above replacement allows us to assume that $[A : \mathrm{gr}(A_0)] < \infty$.

Now, if $H = \mathrm{gr}(A_0^{(c)}) \neq A$ then we repeat the above reduction step, replacing again S, G, A by the corresponding $\widehat{S}, \widehat{G}$ and $\widehat{A}$. Then $\widehat{A} \subseteq H$

implies that $[\widehat{A} : \mathrm{gr}(\widehat{A}_0)] < [A : \mathrm{gr}(A_0)]$. Therefore, after finitely many of such reduction steps, we come to the situation where $H = A$. So we may also assume that $A = \mathrm{gr}(A_0^{(c)})$.

It now follows that Lemma 4.4.2 applies to $B = A_0^{(c)}$. Therefore we may assume that $A_0^{(c)} \subseteq F_+$ for a free abelian group F with a free basis $X = \{x_1, \ldots, x_r\}$ such that $[F : A] = j < \infty$ and $F_+ = \langle X \rangle$. Consider the group $F^j = \mathrm{gr}(x_1^j, \ldots, x_r^j)$. Then $F^j \subseteq A$ and $q = [A : F^j] < \infty$. Since A has finitely many subgroups of index q, it follows that $\sigma^i(F^j) = F^j$ for some $i \geq 1$. Then it is easy to see that $E = \mathrm{gr}(F^j, t^i)$ is of finite index in G, whence $\mathrm{gr}(S \cap E) = E$ by Lemma 4.1.2, and $F^j = A \cap E$. Notice that the positive cone of F^j is $(F^j)_+ = \langle x_1^j, \ldots, x_r^j \rangle \subseteq A$. So we have $E \cap A_0^{(c)} = F^j \cap A_0^{(c)} \subseteq (F^j)_+$. Moreover it is clear that $A_0^{(c)} \cap F^j = \{x \in F^j \mid x^k \in S \cap F^j$ for some positive integer $k\}$. Further, because $[A : F^j] < \infty$ and $A = \mathrm{gr}(A_0^{(c)})$, applying Lemma 4.1.2 we get that $\mathrm{gr}(A_0^{(c)} \cap F^j) = F^j$. Replacing S by $S \cap E$ (and consequently G by E, A by F^j, and $A_0^{(c)}$ by $A_0^{(c)} \cap F^j$) we therefore may assume that $A = F$. In this way we still have the property $A = \mathrm{gr}(A_0^{(c)})$ but we may additionally assume that $A_0^{(c)} \subseteq F_+$.

Let $a \in A_0$. For any nonnegative integer l we define $I_l = aS \cup taS \cup \cdots \cup t^l aS$. Because of the ascending chain condition we get that $I_m = I_{m+1} = \cdots$ for some $m \geq 1$. It is easy to see that this leads to an infinite sequence $m \leq m_1 < m_2 < \cdots$ so that $t^{m_i} a \in t^p aS$ for some $0 \leq p \leq m$. Therefore $t^{m_i} a = t^p a t^{m_i - p} b(i)$ for some $b(i) \in A_{m_i - p}$. So $a = t^{p - m_i} a t^{m_i - p} b(i) \in \sigma^{m_i - p}(a) A_{m_i - p}$.

Hence, for any given $a \in A_0$, there exist infinitely many positive integers n such that $a \in \sigma^n(a) A_n$. So, for such n, $a = \sigma^n(a) c$ for some $c = c_n \in A_n$.

Because G is polycyclic-by-finite and S satisfies the ascending chain condition on right ideals, we know by Theorem 4.1.7 that S is finitely generated. Let $S = \langle s_1, \ldots, s_k \rangle$ where $s_i = t^{n_i} a_i \in t^{n_i} A_{n_i}$. If $s \in t^n A_n$ then we may write $s = s_{j_1} \cdots s_{j_m}$ for some $j_1, \ldots, j_m \in \{1, \ldots, k\}$.

Taking into account the fact that $A_i A_0 \subseteq A_i$, this presentation of s can be rewritten as

$$s = b_0 t^{n_{i_1}} b_1 t^{n_{i_2}} b_2 \cdots t^{n_{i_q}} b_q \qquad (4.5)$$

with $n_{i_1} + \cdots + n_{i_q} = n$, all $n_{i_j} \geq 1$, $b_0 \in A_0$ and $b_j \in A_{n_{i_j}}$ for $j > 0$. In particular, all b_j belong to the finitely generated $\mathbb{Q}[A_0]$-submodule $\mathbb{Q}[A_p]$, where $p = \max\{n_1, \ldots, n_k\}$. Let $D \subseteq A_p$ be a finite generating set for this module.

We apply (4.5) to any of the elements $s = t^n c \in t^n A_n$, with the infinitely many pairs $n, c = c_n$ found above for the given $a \in A_0$. Hence we get

$$c = b_q \sigma^{n_{i_q}}(b_{q-1}) \cdots \sigma^{n_{i_1}+\cdots+n_{i_q}}(b_0) \tag{4.6}$$

(note that b_j and n_{i_j} depend on c and n).

From now on, we switch to the additive notation for $A = F$, identifying F with $\mathbb{Z}^r$ for some $r \geq 1$. So we view σ as an automorphism of $\mathbb{Z}^r \subseteq \mathbb{Q}^r$. In the additive notation (4.6) means that

$$c = b_q + \sigma^{n_{i_q}}(b_{q-1}) + \cdots + \sigma^{n_{i_1}+\cdots+n_{i_q}}(b_0).$$

On the other hand, $a = \sigma^n(a) + c$. So we get

$$a = \sigma^n(a) + b_q + \sigma^{n_{i_q}}(b_{q-1}) + \cdots + \sigma^{n_{i_1}+\cdots+n_{i_q}}(b_0). \tag{4.7}$$

Consider any b_j as in (4.7). Notice that $b_j \in A_p$ and $A_0 \subseteq F_+$. Therefore $b_j = c_j + d_j$ where $c_j \in A_0$ and $d_j \in D$. We know that $A = \mathrm{gr}(A_0^{(c)})$, whence A is a localization of $A_0^{(c)}$ with respect to a cyclic submonoid $\langle z \rangle$ for some nontrivial element $z \in A_0^{(c)}$. Since $A_0^{(c)}$ is periodic modulo A, we may assume that $z \in A_0$. Then we may write $d_j = e_j - Rz$ for some positive integer R (dependent on D only) and some $e_j \in A_0^{(c)}$. Put $f_j = c_j + e_j$. Let $M \in \mathrm{M}_r(\mathbb{Z})$ be the matrix corresponding to σ in the standard basis of F. Then (4.7) can be written as

$$\begin{aligned} a - M^n a &= b_q + M^{n_{i_q}} b_{q-1} + \cdots + M^{n_{i_1}+\cdots+n_{i_q}} b_0 = \\ &= (f_q - Rz) + M^{n_{i_q}}(f_{q-1} - Rz) + \cdots + M^{n_{i_1}+\cdots+n_{i_q}}(f_0 - Rz). \end{aligned}$$

Therefore, if $N \geq 1$ and we apply the above to the element $a = Nz$, then

$$N(1 - M^n)z = w - v \tag{4.8}$$

(for infinitely many positive integers n depending on N and for elements f_j dependent on n) where

$$\begin{aligned} v &= R(z + M^{n_{i_q}} z + \cdots + M^{n_{i_1}+\cdots+n_{i_q}} z) \\ w &= f_q + M^{n_{i_q}} f_{q-1} + \cdots + M^{n_{i_1}+\cdots+n_{i_q}} f_0. \end{aligned}$$

All coordinates of the vectors w and $M^i z, i \geq 1$, are nonnegative, because $f_j, z \in A_0^{(c)}$ and $M A_0^{(c)} \subseteq A_0^{(c)} \subseteq F_+$.

Suppose first that 1 is not an eigenvalue of M. Then $M - 1$ is nonsingular as a matrix in $\mathrm{M}_r(\mathbb{Q})$ and

$$\begin{aligned} v = R(z + M^{n_{i_q}} z + \cdots + M^{n_{i_1}+\cdots+n_{i_q}} z) &\leq R(z + Mz + \cdots + M^n z) \\ &= R(1 + M + \cdots + M^n)z \\ &= R((M-1)^{-1}(M^n - 1) + M^n)z. \end{aligned}$$

Here we write $u \leq u'$ for $u = (u_1, \ldots, u_r), u' = (u'_1, \ldots, u'_r) \in \mathbb{Z}^r$ if $u_i \leq u'_i$ for every i. Then, in view of (4.8), we get

$$\begin{aligned} N(1 - M^n)z = w - v &\geq w - R((M-1)^{-1}(M^n - 1) + M^n)z \\ &\geq -R((M-1)^{-1}(M^n - 1) + M^n)z. \end{aligned}$$

So

$$N(M^n - 1)z \leq R((M-1)^{-1} + 1)(M^n - 1)z + Rz. \tag{4.9}$$

Notice that we may assume that $M^n z - z$ has a positive coordinate for sufficiently large $n \geq 1$. (Otherwise $M^n z \leq z$ for infinitely many n. Since $M^n z \geq 0$ for every n, it follows that $M^n z = M^m z$ for some $m \neq n$, whence, because σ is an automorphism of $\mathbb{Z}^r$, we get that $M^l z = z$ for some positive integer l. Then $z \in A_0$ is central in the subgroup $\mathrm{gr}(A, t^l)$ of finite index in G, and hence it is in the FC-center of G, a contradiction.) Let $N' = N/R$, $u_n = (M^n - 1)z$ and $X = (M-1)^{-1} + 1$. Then (4.9) implies

$$N' u_n \leq X u_n + z. \tag{4.10}$$

Also write $u_n = (u_{n1}, \ldots, u_{nr})$, $z = (z_1, \ldots, z_r)$ and $X = (x_{ij})$ with $z_i, u_{ni} \in \mathbb{Z}$, $z_i \geq 0$ and $x_{ij} \in \mathbb{Q}$. By the above we know that, for sufficiently large n, there exists j such that $u_{nj} = \max\{u_{ni} \mid 1 \leq i \leq r\} > 0$. Comparing the j-th coordinates in (4.10) we obtain for such n

$$0 < N' u_{nj} \leq x_{j1} u_{n1} + \cdots + x_{jr} u_{nr} + z_j$$

and thus

$$\begin{aligned} 0 < N' u_{nj} &\leq |x_{j1} u_{n1} + \cdots + x_{jr} u_{nr} + z_j| \\ &\leq \gamma \max\{|u_{ni}| \mid 1 \leq i \leq r\} + |z_j|, \end{aligned}$$

where $\gamma = \max\{|x_{ik}| \mid 1 \leq i, k \leq r\}$. Let $\omega = \max\{|z_k| \mid 1 \leq k \leq r\}$. Clearly we have $u_n = M^n z - z \geq -z$. Hence we get $-z_i \leq u_{ni} \leq u_{nj}$, for every $i = 1, \ldots, r$, and consequently

$$0 < N' u_{nj} \leq \gamma \max\{\omega, u_{nj}\} + \omega.$$

As u_{nj} and ω are positive integers, this implies that $0 < N' \leq \gamma\omega + \omega$. Since γ and ω are fixed and N' can be taken arbitrarily large (because so can be chosen N), this yields a contradiction.

The second case to consider is when 1 is an eigenvalue of M. Then there exists $0 \neq u \in \mathbb{Q}^r$ such that $Mu = u$, whence there exists such $u \in \mathbb{Z}^r$. Thus u is a nontrivial element of $\mathrm{Z}(G)$, which contradicts the assumption on the FC-center of G, completing the proof. □

The following two simple observations also will be used.

Lemma 4.4.4. *Let V be a finite dimensional vector space over a field K and let $\sigma \in \mathrm{End}(V)$ be a unipotent automorphism. Let $H = V \rtimes C$ be the semidirect product of the abelian group V and the cyclic group C with the action of C on V determined by σ. Then H is a nilpotent group.*

Proof. We proceed by induction on $r = \dim_K V$. If $r = 0$ then the assertion is clear. Let $r \geq 1$. Then $V_1 = \{v \in V \mid \sigma(v) = v\}$ is a nontrivial subgroup of H and $V_1 \subseteq \mathrm{Z}(H) \cap V$. Since σ determines a unipotent automorphism σ_1 on the space V/V_1, the corresponding semidirect product of $V/V_1 \rtimes C$ is nilpotent by the induction hypothesis. The assertion follows. □

Lemma 4.4.5. *Let $k \geq 2$. Assume that $(d_n)_{n \geq 0}$ is a sequence of real numbers such that $d_n = d_{n-1} + d_{n-2} + \cdots + d_{n-k}$ for $n \geq k$. Then $|d_n| < 2^n$ for almost all natural numbers n.*

Proof. Write $f(x) = x^k - x^{k-1} - \cdots - x - 1$. Let $\lambda_1, \ldots, \lambda_k \in \mathbb{C}$ be the roots of the polynomial $f(x)$. Let $\lambda = \lambda_1$ be the biggest positive root of $f(x)$. We claim that $\lambda < 2$, and $|\lambda_i| \leq \lambda$ for every i. Indeed, $(x-1)f(x) = x^{k+1} - 2x^k + 1$ implies that $0 = (\lambda - 1)f(\lambda) = \lambda^{k+1} - 2\lambda^k + 1$. Since $\alpha^{k+1} - 2\alpha^k + 1 \geq 1$ for every real $\alpha \geq 2$, we get $\lambda < 2$. Notice that

$$|\lambda_i|^k = |\lambda_i^{k-1} + \cdots + \lambda_i + 1| \leq |\lambda_i|^{k-1} + \cdots + |\lambda_i| + 1.$$

Hence $f(|\lambda_i|) \leq 0$, and therefore $|\lambda_i| \leq \lambda$ by the choice of λ, which proves the claim.

By the theory of recurrent sequences, see [61, Theorem 3.1.1], we know that there exist polynomials $p_1(x), \ldots, p_r(x) \in \mathbb{C}[x]$ such that $d_n = \lambda_{i_1}^n p_1(n) + \cdots + \lambda_{i_r}^n p_r(n)$ for every n, where $\lambda_{i_1}, \ldots, \lambda_{i_r}$ are all the distinct roots of $f(x)$. Hence $d_n \leq \lambda^n(|p_1(n)| + \cdots + |p_r(n)|)$ and the assertion follows easily because $\lambda < 2$. □

Theorem 4.4.6. *Let S be a submonoid of a polycyclic-by-finite group. If S satisfies the ascending chain condition on right ideals, then S is group-by-abelian-by-finite, that is, the group G of right quotients of S contains a normal subgroup H of finite index such that $[H, H] \subseteq S$.*

Proof. We prove the result by induction on the Hirsch rank of G. The case $\mathrm{h}(G) = 0$ is obvious. Thus assume that $\mathrm{h}(G) \geq 1$. If G is nilpotent-by-finite, then the statement has been shown in Theorem 4.3.2.

Suppose that G is not nilpotent-by-finite. Performing reduction steps as in the beginning of the proof of Theorem 4.3.2, in view of Proposition 3.3.3, we may assume that G is torsion free, $[G, G]$ is nilpotent and $\mathrm{U}(S) = \{1\}$. It is now sufficient to show that all these conditions lead to a contradiction.

For this, because of Lemma 4.4.1, we may furthermore assume that $G = \mathrm{gr}(t, A)$, where $t \in S$ and A is a nilpotent normal subgroup of G. Moreover, G/A is infinite cyclic and $\mathrm{h}(G/[G, G]) = 1$. By Lemma 4.4.1 these conditions are also inherited by subgroups of finite index. Furthermore, there exist $t, tg \in S$, with $g \in A$ so that $\langle t, tg\rangle$ is a free nonabelian monoid and $G = \mathrm{gr}(t, tg)$.

First we deal with the case $S \cap A = \{1\}$. Since $\mathrm{U}(S) = \{1\}$, we get that $S \subseteq \bigcup_{i\geq 0} t^i A$. Suppose that $S \cap t^i A$ is infinite for some $i \geq 0$. Then choose an infinite sequence of different elements $x_1, x_2, \ldots \in S \cap t^i A$. The ascending chain condition on right ideals implies that $x_k^{-1} x_n \in S$ for some $k < n$. But then $x_k^{-1} x_n \in S \cap A$, a contradiction. Therefore every $S \cap t^i A$ is finite. Since $\langle t, tg\rangle \subseteq S$ is a free monoid, it follows easily that $|S \cap t^i A| \geq 2^i$ for every $i \geq 1$. By induction we will construct a sequence of integers $n_1 < n_2 < \cdots$ so that for every k we have:

$$|A_{n_k}| \;>\; |A_{n_k - n_1}| + \cdots + |A_{n_k - n_{k-1}}|, \tag{4.11}$$

where the sets $A_i \subseteq A$ are defined by $S \cap t^i A = t^i A_i$. Let $d(n) = |A_n|$. We have to find a sequence $n_1 < n_2 < \cdots$ of positive integers with $d(n_k) > d(n_k - n_1) + \cdots + d(n_k - n_{k-1})$. Assume $n_1 = 1$ and $n_1, \ldots, n_{k-1}$ have been chosen. Suppose that n_k with the desired property does not exist. Then

$$d(n) \leq d(n - n_1) + \cdots + d(n - n_{k-1})$$

for all $n > n_{k-1}$. Since $A_i \subseteq A_{i+1}$, the sequence $d(i)$ is non-decreasing, whence we have

$$d(n) \leq d(n-1) + \cdots + d(n-k+1)$$

for $n > n_{k-1}$. Let $d_i = d(i)$ for $i \leq n_{k-1}$, and

$$d_i = d_{i-1} + \cdots + d_{i-k+1}$$

for $i > n_{k-1}$. It follows that $d(n) \leq d_n$, for every $n \geq 1$. Since we also have $2^n \leq d(n)$, Lemma 4.4.5 leads to a contradiction. This proves that a sequence $n_1, n_2, \ldots$ with property (4.11) can be found.

This allows us to construct an infinite sequence $s_1, s_2, \ldots$ of elements of S with the properties $s_k \in S \cap t^{n_k} A$ and $s_k \notin s_1 S \cup \cdots \cup s_{k-1} S$ for

$k > 1$. Indeed, write $\sigma(x) = t^{-1}xt$ for $x \in A$. Assume that $s_i = t^{n_i}a_i$ for $i = 1, 2, \ldots, k-1$ have been found. For every $s_i \in S \cap t^{n_i}A$ we have

$$s_i S = \bigcup_{j \geq 0} t^{j+n_i} \sigma^j(a_i) A_j,$$

because $S = \bigcup_{j \geq 0} t^j A_j$. Hence, for every j,

$$|s_i S \cap t^{j+n_i} A_{j+n_i}| = |A_j|$$

and therefore, substituting $j = n_k - n_i$ and applying (4.11), we obtain

$$|(s_1 S \cup \cdots \cup s_{k-1} S) \cap t^{n_k} A_{n_k}| \leq |A_{n_k - n_1}| + \cdots + |A_{n_k - n_{k-1}}| < |A_{n_k}|.$$

So $S \cap t^{n_k} A \nsubseteq s_1 S \cup \cdots \cup s_{k-1} S$ and may indeed find an element $s_k \in S \cap t^{n_k} A$ such that $s_k \notin s_1 S \cup \cdots \cup s_{k-1} S$. This contradicts the ascending chain condition hypothesis, finishing the proof in case $S \cap A$ is trivial.

So now assume $S \cap A$ is not trivial. First consider the case where A is abelian. Then G is an extension of a finitely generated free abelian group A by an infinite cyclic group $C = \mathrm{gr}(t)$, where $t \in S$. We know also that $\mathrm{U}(S) = \{1\}$, G is not nilpotent-by-finite and $\mathrm{h}(G/[G,G]) = 1$. Let σ denote the automorphism of A determined by the action of C on A.

As before, using also Lemma 4.4.1, we may replace S by the monoid $S \cap E$ for any subgroup E of finite index in G. From Proposition 4.4.3 it follows that the FC-center of G contains a nontrivial element $a \in A$. Thus there exists $m \geq 1$ such that t^m is in the centralizer of a, whence replacing S by $S \cap E$ with $E = \mathrm{gr}(A, t^m)$, we may assume that $\mathrm{Z}(G) \cap A$ is nontrivial.

In the proof we identify A with $\mathbb{Z}^r$, for some r, and we extend σ to the space $Y = \mathbb{C}^r$. Then $Y_1 = \{y \in Y \mid (\sigma - 1)^k(y) = 0 \text{ for some } k\}$ is a σ-invariant subspace and $Y = Y_1 \oplus Y_2$ for some σ-invariant subspace Y_2. Moreover Y_1, Y_2 can be considered as normal subgroups of $\overline{G} = (Y_1 \oplus Y_2) \rtimes C$. Hence $G = A \rtimes C$ embeds into $G/U_1 \times G/U_2$ where $U_i = G \cap Y_i$. Since $\mathrm{Z}(G) \cap A$ is nontrivial, we know that $U_1 \neq \{0\}$. Moreover, $U_2 \neq \{0\}$ since otherwise $G \cong G/U_2 \subseteq \overline{G}/Y_2 \cong Y_1 \rtimes C$ is nilpotent, because of Lemma 4.4.4.

Therefore $U_1, U_2 \neq \{0\}$ and $\mathrm{h}(G/U_1), \mathrm{h}(G/U_1) < \mathrm{h}(G)$, so by the induction hypothesis we know that the images S_1, S_2 of S in G/U_1 and G/U_2 are group-by-abelian-by-finite. Replacing G by a subgroup F of finite index and S by $S \cap F$ we may assume that they are group-by-abelian. In other words, if G_i' denotes the image of $[G,G]$ in the i-th factor of $G/U_1 \times G/U_2$, we know that G_i' is a subgroup of S_i.

Suppose first that G_2' is nontrivial. Then we may choose a nontrivial element $g_2 \in G_2'$ and $g_1 \in G/U_1$ such that $(g_1, g_2) \in S$. Then for some

$(h_1, h_2) \in S$ the second component of $(g_1, g_2)(h_1, h_2)$ is trivial. Thus $g_1 h_1 \in S_1$ is nontrivial because $\mathrm{U}(S)$ is trivial. Hence $(g_1, g_2)(h_1, h_2)$ is a nontrivial element in $S \cap U_2$. Since the monoid $S \cap (Y_2 \rtimes C)$ inherits the ascending chain condition on right ideals (by Lemma 4.1.1), the induction hypothesis and $\mathrm{U}(S) = \{1\}$ imply that $\mathrm{gr}(S \cap (Y_2 \rtimes C))$ is abelian-by-finite. Since $T = \langle S \cap U_2, t \rangle \subseteq S \cap (Y_2 \rtimes C)$ it follows that $\mathrm{gr}(T)$ also is abelian-by-finite.

Then choose a nontrivial element $b \in S \cap U_2$ that has finitely many $\mathrm{gr}(T)$-conjugates, say $b_1, \ldots, b_q$. The transitive action of $\mathrm{gr}(T)$ on the set $\{b_1, \ldots, b_q\}$ determines a homomorphism $\phi : \mathrm{gr}(T) \longrightarrow \mathrm{Sym}_q$ and $\phi(T) = \phi(\mathrm{gr}(T))$ because $\phi(T)$ is a finite semigroup that generates $\phi(\mathrm{gr}(T))$ as a group. Therefore each b_j is of the form $s^{-1}bs$ for some $s \in T$. From Lemma 4.1.5 we know also that there exists $m \geq 1$ such that $b_j^m \in T$ for every j. So $b' = b_1^m \cdots b_q^m \in \mathrm{Z}(T)$. Note that all b_j^m are nontrivial as $\mathrm{gr}(T)$ is torsion free. Since, by assumptions, $\mathrm{U}(T)$ is trivial, it follows that b' is nontrivial. Hence $\mathrm{Z}(T) \cap U_2$ is nontrivial. However $\mathrm{Z}(T) \cap U_2 \subseteq \mathrm{Z}(S) \cap A$. Since $\mathrm{Z}(S) \cap A \subseteq U_1$ by the definition of Y_1, this leads to a contradiction.

So, consider the case where G_2' is trivial. This means that $G_2 = G/U_2$ is abelian. Thus $[G, G] \subseteq U_2$ and therefore $U_1 \cap [G, G]$ is trivial. We obtain that $\mathrm{h}(G/[G, G]) \geq 2$, again a contradiction. This completes the proof in case A is abelian.

Finally, we deal with the case where $S \cap A$ is not trivial and A is non-abelian. Since $\overline{G} = G/\mathrm{Z}(A)$ has smaller Hirsch rank than G, the induction hypothesis implies that the natural image $\overline{S}$ of S in $\overline{G}$ contains a normal subgroup $\overline{F}$ of $\overline{G}$ so that $\overline{G}/\overline{F}$ is abelian-by-finite. If $\overline{F}$ is trivial, we get that G is abelian-by-abelian-by-finite. Hence, again replacing S, G by their intersections with a subgroup of finite index in G, we may assume that G is abelian-by-abelian. However, since $\mathrm{h}(G/[G, G]) = 1$, this means that G is abelian-by-cyclic-by-finite. Therefore, this leads to the abelian-by-cyclic case, handled before. Thus, this case yields a contradiction. So, we may assume that $\overline{F}$ is nontrivial. Let F be the inverse image in S of $\overline{F}$, and let $1 \neq f \in F$. Then since $\overline{F} \subseteq \mathrm{U}(\overline{S})$, there exists $f' \in F$ with $ff' \in \mathrm{Z}(A)$. Because $\mathrm{U}(S) = \{1\}$, we get $S \cap \mathrm{Z}(A) \neq \{1\}$. By Lemma 4.1.3 and Theorem 4.1.7 we may write

$$S \cap \mathrm{Z}(A) = \langle s_1, \ldots, s_n \rangle$$

for some $s_i \in S$ and let

$$H = \mathrm{gr}(S \cap \mathrm{Z}(A)).$$

Note that $\mathrm{gr}(\mathrm{Z}(A), t)$ is abelian-by-cyclic and thus so is $\mathrm{gr}(H, t)$. Since $\mathrm{gr}(H, t)$ is as a group generated by its intersection with S, it follows again from the abelian-by-cyclic case that this group is actually abelian-by-finite. Hence, for some $m \geq 1$, the element t^m commutes with all s_i^m, where $1 \leq i \leq n$. Thus, $1 \neq z = s_1^m \in \mathrm{Z}(\mathrm{gr}(A, t^m))$. Hence, replacing G by the subgroup $\mathrm{gr}(A, t^m)$ of finite index and replacing S by $S \cap \mathrm{gr}(A, t^m)$, we may assume that $1 \neq z \in \mathrm{Z}(S) \cap A$.

Since $G_1 = G/\mathrm{gr}(z)$ has smaller Hirsch rank than G, the induction hypothesis implies that the image $\overline{S}$ of S in G_1 contains a normal subgroup $\overline{F}_1$ of G_1, so that $G_1/\overline{F}_1$ is abelian-by-finite. Let F_1 be the inverse image of $\overline{F}_1$ in G. Then $\overline{F}_1 \subseteq \overline{S}$ implies that $F_1 \subseteq S\,\mathrm{gr}(z)$. Since $z \in F_1$, we get

$$F_1 = (S \cap F_1)\,\mathrm{gr}(z) = \mathrm{gr}(S \cap F_1).$$

If $[G : F_1] = \infty$ then $\mathrm{h}(F_1) < \mathrm{h}(G)$ and by the induction hypothesis applied to $S \cap F_1$ it follows that F_1 is abelian-by-finite. So G is (abelian-by-finite)-by-(abelian-by-finite). Hence, in view of Proposition 3.3.3, G is abelian-by-abelian-by finite. As before, this leads to a contradiction.

So it remains to consider the case where $[G : F_1] < \infty$. Then, replacing S, G by $S \cap F_1$ and $F_1 = (S \cap F_1)\,\mathrm{gr}(z)$, we may assume that $G = S\,\mathrm{gr}(z)$, where $z \in A$. Then $A = (S \cap A)\,\mathrm{gr}(z) = \mathrm{gr}(S \cap A)$. Since $\mathrm{h}(A) < \mathrm{h}(G)$, the induction hypothesis therefore yields that A is abelian-by-finite, so G is abelian-by-cyclic-by-finite. As before, this leads to a contradiction. This completes the proof of the theorem. □

We are now ready to state the main result of this section, characterizing when a semigroup algebra $K[S]$ of a submonoid S of a polycyclic-by-finite group is right Noetherian. It extends Theorem 4.3.3.

Theorem 4.4.7. *Let S be a submonoid of a polycyclic-by-finite group and let K be a field. Then the following conditions are equivalent.*

1. *S satisfies the ascending chain condition on right ideals.*
2. *$K[S]$ is right Noetherian.*
3. *S has a group of right quotients G with a normal subgroup H of finite index such that $[H, H] \subseteq S$ and $S \cap H$ is finitely generated.*

Moreover, conditions (1) and (2) are equivalent to their left-right symmetric analogs.

Proof. Recall that (1) and (2) are equivalent by Theorem 4.1.6. Assume that S satisfies the ascending chain condition on right ideals. Then, by Theorem 4.4.6, there exists a normal subgroup H of G such that H is of finite index in G and $[H,H] \subseteq S$. Moreover, by Theorem 4.1.6 and Lemma 4.1.3, $K[S \cap H]$ is right Noetherian and by Theorem 4.1.7 $S \cap H$ is finitely generated.

Notice that the proof of the implication (3)⇒(1) in Theorem 4.3.3 remains valid in this case also, so (1) is a consequence of (3). Since condition (3) is left-right symmetric, the result follows. □

Let S, G, H be such as in Theorem 4.4.7. As in the case of group algebras, it follows that the semigroup algebra $K[S]$ is isomorphic with a crossed product

$$K[H](S/H, \sigma, \gamma),$$

where $\sigma : S/H \longrightarrow \mathrm{Aut}(H)$ is a homomorphism and γ is a two-cocycle with values in H.

The following example shows that one cannot hope for a generalization of the assertions of Theorem 4.4.7 to the case of submonoids of finitely generated solvable groups that are not polycyclic-by-finite.

Example 4.4.8. Consider the submonoid $S = \langle y, y^{-1}, x\rangle$ of the group $G = \mathrm{gr}(x, y \mid xyx^{-1} = y^2)$. Then $S = \{y^i x^j \mid i \in \mathbb{Z}, j \geq 0\}$. So S may be identified with the following submonoid of the full linear group $\mathrm{GL}_2(\mathbb{Q})$

$$\left\{\begin{pmatrix} 1 & 0 \\ i & 2^j \end{pmatrix} \mid i, j \in \mathbb{Z}, j \geq 0\right\} = \langle \begin{pmatrix} 1 & 0 \\ 1 & 1 \end{pmatrix}, \begin{pmatrix} 1 & 0 \\ -1 & 1 \end{pmatrix}, \begin{pmatrix} 1 & 0 \\ 0 & 2 \end{pmatrix}\rangle.$$

Consider $u = \begin{pmatrix} 1 & 0 \\ 1 & 2 \end{pmatrix}$ and $v = \begin{pmatrix} 1 & 0 \\ 0 & 2 \end{pmatrix}$. For any integers a, b, a', b', the equality

$$\begin{pmatrix} 1 & 0 \\ 1 & 2 \end{pmatrix}\begin{pmatrix} 1 & 0 \\ a & b \end{pmatrix} = \begin{pmatrix} 1 & 0 \\ 0 & 2 \end{pmatrix}\begin{pmatrix} 1 & 0 \\ a' & b' \end{pmatrix}$$

implies

$$1 + 2a = 2a',$$

which is impossible. Therefore $uS \cap vS = \emptyset$, and thus S is not a right Ore semigroup. In particular, $K[S]$ is not right Noetherian by Lemma 4.1.5. It is easy to verify that the left ideal of S generated by $\begin{pmatrix} 1 & 0 \\ a & 2^k \end{pmatrix} \in S$ consists of all matrices of the form $\begin{pmatrix} 1 & 0 \\ z & 2^{k+m} \end{pmatrix}$, where $z \in \mathbb{Z}$ and $m \geq 0$. Therefore every proper left ideal of S is also of this form and S satisfies the ascending chain condition on left ideals. In particular, S satisfies the left Ore condition, so $G = S^{-1}S$. Moreover, with the obvious identifications, using the notation of skew polynomial rings we have $K[S] = K[\mathrm{gr}(y)][x, \sigma] \subseteq$

$K[T][x,\sigma] \subseteq K[T][x,x^{-1},\sigma] = K[G]$, where the abelian torsion free group T is the unipotent radical of G and σ is an automorphism of $K[T]$. It is well known that $K[\mathrm{gr}(y)][x,\sigma]$ is a left Ore domain, see [101, Theorem 10.28].

So S is an example of a finitely generated submonoid S of a solvable group (which is not polycyclic-by-finite) such that S satisfies the ascending chain condition on left ideals, but S is not a right Ore semigroup; in particular, S does not satisfy the ascending chain condition on right ideals. Also, this example shows that the ascending chain condition on left ideals does not imply that $K[S]$ is left Noetherian. Indeed, otherwise $K[G]$ is also left Noetherian as a left localization of $K[S]$, and thus G satisfies the ascending chain condition on subgroups and therefore is polycyclic-by-finite, a contradiction.

We continue with two interesting applications of the characterization of Noetherian submonoids of polycyclic-by-finite groups. They are concerned with classical properties that are known to hold for group algebras of such groups. First, we need the following easy modification of the proof of Lemma 8.10 in [140], which shows that the finite presentation property assumed on a normal subgroup N of a group G and on the group G/N implies that G itself is finitely presented. Recall that a monoid S (an algebra R, respectively) is finitely presented if $S \cong F/I$ for a finitely generated ideal I of a finitely generated free monoid F ($R \cong K[F]/J$ for a finitely generated ideal of the free algebra $K[F]$).

Lemma 4.4.9. *Let N be a normal subgroup of a group G. Assume that S is a submonoid of G such that $N \subseteq S$. If N is finitely presented and the image S/N of S in G/N is finitely presented, then S is finitely presented.*

Proof. Let y_0 be the identity of S. We may write $N = \langle y_0, y_1, \ldots, y_r\rangle$ with a monoid presentation given by the finitely many relations of the form $u_i(y) = u_i'(y), i = 1, \ldots, m$ and $y_0 y_k = y_k y_0$ for $k = 0, \ldots, r$,, where y stands for the r-tuple $(y_1, \ldots, y_r)$ and u_i, u_i' are words in $y_1, \ldots, y_r$. Also, let $S/N = \langle \overline{x}_1, \ldots, \overline{x}_t\rangle$, subject to the finitely many defining relations $v_i(\overline{x}) = v_i'(\overline{x}), i = 1, \ldots, n$, where $\overline{x}$ is the t-tuple $(\overline{x}_1, \ldots, \overline{x}_t)$, and $v_i(\overline{x}), v_i'(\overline{x})$ are words in $\overline{x}_1, \ldots, \overline{x}_t$. Let $x_i \in S$ be an inverse image of $\overline{x}_i$. It is clear that $S = \langle x_1, \ldots, x_t, y_0, y_1, \ldots, y_r\rangle$. Moreover, because N is a normal subgroup, we have relations in S of the form

$$
\begin{aligned}
u_i(y) &= u'_i(y) \quad \text{for } i = 1, \ldots m, \\
y_0 y_k &= y_k y_0 \quad \text{for } k = 0, \ldots r, \\
v_i(x) &= \lambda_i(y) v'_i(x) \quad \text{for } i = 1, \ldots n, \\
y_i x_j &= x_j \mu_{ij}(y) \quad \text{for } i = 1, \ldots r, j = 1, \ldots, t, \\
x_j y_i &= \nu_{ij}(y) x_j \quad \text{for } i = 1, \ldots r, j = 1, \ldots, t, \\
y_0 x_l &= x_l = x_l y_0 \quad \text{for } l = 1, \ldots t,
\end{aligned}
\tag{4.12}
$$

where $\lambda_i(y), \mu_{ij}(y), \nu_{ij}(y)$ are some words in $y_1, \ldots, y_r$. We claim that this set of relations is a finite presentation for S.

So, let $\Gamma = \langle \xi_1, \ldots, \xi_t, \eta_0, \ldots, \eta_r \rangle$, where ξ_i corresponds to x_i and η_j corresponds to y_j, be the monoid defined by the relations corresponding to relations (4.12). Then we have a natural homomorphism $\pi : \Gamma \longrightarrow S$ defined by $\pi(\xi_j) = x_j, \pi(\eta_i) = y_i$. We will verify that π is an isomorphism. Let $\Delta = \langle \eta_1, \ldots, \eta_r \rangle$. Since the above set of relations contains the defining relations for N, it is clear that $\pi_{|\Delta} : \Delta \longrightarrow N$ is an isomorphism. Moreover, η_0 is the identity of Γ, Δ is a subgroup of Γ with identity η_0 and $c\Delta = \Delta c$ for all $c \in \Gamma$. Therefore we may consider the monoid Γ/Δ of cosets $c\Delta, c \in \Gamma$. It is then clear that Γ/Δ is defined by the relations $v_i(\xi) = v'_i(\xi), i = 1, \ldots, n$, where $\xi = (\xi_1, \ldots, \xi_t)$. Hence, the original presentation for S/N implies that π induces an isomorphism $\Gamma/\Delta \longrightarrow S/N$. Suppose that $\pi(a) = \pi(b)$ for some $a, b \in \Gamma$. Then $\pi(a\Delta) = \pi(a)N = \pi(b)N = \pi(b\Delta)$ implies that $a\Delta = b\Delta$. So $a = \delta b$ for some $\delta \in \Delta$ and $\pi(b) = \pi(\delta b) = \pi(\delta)\pi(b)$. Since S is cancellative, it follows that $\pi(\delta) = y_0$, the identity of S. Thus $\delta = \eta_0$ because $\pi_{|\Delta}$ is an isomorphism. Therefore $a = b$. This implies that π is an isomorphism and hence S is finitely presented. □

An induction based on Lemma 4.4.9 allows to show that polycyclic-by-finite groups are finitely presented, which is well known, see [140, Theorem 8.4].

Corollary 4.4.10. *Let S be a submonoid of a polycyclic-by-finite group. If S satisfies the ascending chain condition on right ideals, then S is a finitely presented monoid. In particular, the semigroup algebra $K[S]$ over any field K is finitely presented.*

Proof. By Theorem 4.4.6, S contains a normal subgroup H of $G = SS^{-1}$ so that G/H is abelian-by-finite. Since H is polycyclic-by-finite, it is finitely presented. By Lemma 4.4.9, it is enough to verify that the natural image of S in $G/[H, H]$ is finitely presented. In other words, we may assume that

G has an abelian normal subgroup A of finite index. Since $K[S \cap A]$ is Noetherian, we also know that $S \cap A$ is finitely generated. So let

$$S \cap A = \langle a_1, \ldots, a_k \mid p_1 = q_1, \ldots, p_l = q_l \rangle \tag{4.13}$$

be a finite presentation for $S \cap A$. Here p_i, q_j are words in $a_1, \ldots, a_k$. By Lemma 4.1.3 we may write

$$S = \bigcup_{i=1}^{m} s_i(S \cap A),$$

for some $s_i \in S$. Hence, S satisfies relations of the following types:

$$s_i s_j = s_{i,j} a_{i,j} \tag{4.14}$$

$$a_r s_i = t_{i,r} b_{i,r} \tag{4.15}$$

for $1 \leq i, j \leq m$, $1 \leq r \leq k$, for some $s_{i,j}, t_{i,r} \in \{s_1, \ldots, s_m\}$ and $a_{i,j}, b_{i,r} \in S \cap A$. To introduce some other relations we need more notation. For each $1 \leq i, j \leq m$, let

$$A_{i,j} = \{a \in S \cap A \mid s_i a \in s_j(S \cap A)\}.$$

Clearly $A_{i,j}$ is an ideal in $S \cap A$, if nonempty, and therefore it is a finitely generated ideal, say generated by $c_{i,j,q}$, where $1 \leq q \leq n_{i,j}$ for some $n_{i,j}$. For every $i \neq j$ with $A_{i,j} \neq \emptyset$, we have relations.

$$s_i c_{i,j,q} = s_j d_{i,j,q} \tag{4.16}$$

for some $d_{i,j,q} \in S \cap A$. In all the above we treat the elements $a_{i,j}, b_{i,r}, c_{i,j,q}$, $d_{i,j,q}$ as certain words in $a_1, \ldots, a_k$. We claim that $\langle a_1, \ldots, a_k, s_1, \ldots, s_m \rangle$ subject to the relations (4.13),(4.14),(4.15) and (4.16) is a presentation for S.

For this, define Z to be the monoid generated by $x_1, \ldots, x_k, y_1, \ldots, y_m$ subject to relations as in (4.13),(4.14),(4.15) and (4.16) with a_i replaced by x_i and s_j by y_j, and also $c_{i,j,q}, d_{i,j,q}$ replaced by the corresponding words $u_{i,j,q}, v_{i,j,q}$ in the generators of Z. Note that from relations (4.14) and (4.15) we get that

$$Z = \bigcup_{i=1}^{m} y_i X,$$

where $X = \langle x_1, \ldots, x_k \rangle \subseteq Z$. It is clear that the natural map $\varphi : Z \longrightarrow S$ is onto and restricted to X is an isomorphism onto $S \cap A$. We will show that φ is one-to-one. Suppose $\varphi(y_i w) = \varphi(y_j v)$ for some $v, w \in X$. If $i = j$, then the cancellativity in S implies that $\varphi(w) = \varphi(v)$, and hence $w = v$ and $y_i w = y_j v$, as required. So suppose $i \neq j$. Then $\varphi(w) \in A_{i,j}$. Write $\varphi(w) = c_{i,j,q} a$ for some $a \in S \cap A$ and some $c_{i,j,q} \in A_{i,j}$. Because φ yields an isomorphism of X onto $S \cap A$, there exists a unique $x \in X$ so that $\varphi(x) = a$. Then $\varphi(u_{i,j,q}) = c_{i,j,q}$ implies that $w = u_{i,j,q} x$. Consequently, $y_i w = y_i u_{i,j,q} x$. Using one of the relations defining Z, corresponding to (4.16), we therefore get that $y_i w = y_j v_{i,j,q} x$. Since $s_j \varphi(v) = \varphi(y_j v) = \varphi(y_i w) = \varphi(y_j v_{i,j,q} x) = s_j d_{i,j,q} a$, again by cancellativity we get $\varphi(v) = d_{i,j,q} a = \varphi(v_{i,j,q} x)$. Since φ is one-to-one on X, we therefore have $v = v_{i,j,q} x$. So, $y_i w = y_j v_{i,j,q} x = y_j v$, which shows that ϕ indeed is an isomorphism. Therefore, S is finitely presented. □

Our second application of Theorem 4.4.7 is concerned with an extension of Theorem 3.4.4.

Corollary 4.4.11. *Let S be a submonoid of a polycyclic-by-finite group and let K be a field. If S satisfies the ascending chain condition on right ideals, then $K[S]$ is a Jacobson ring.*

Proof. By Theorem 4.4.6, $G = SS^{-1}$ contains a normal subgroup H of finite index such that $[H, H] \subseteq S \cap H$ and $S \cap H$ is finitely generated. Consider $K[S]$ as a G/H-graded ring in a natural way, namely $K[S]_{gH} = \sum_{s \in S \cap gH} Ks$ for $gH \in G/H$. Notice that its identity graded component is $K[S \cap H]$. Since G/H is finite, by Theorem 3.4.4 it is enough to show that $K[S \cap H]$ is a Jacobson ring. For $F = [H, H]$ we consider the natural H/F-gradation on the latter ring; its identity component is $K[F]$. Since F is polycyclic-by-finite and H/F is a finitely generated abelian group, Theorem 3.4.4 yields that $K[F]$ and thus also $K[S \cap H]$ is a Jacobson ring. The result follows. □

As a consequence of Theorem 4.4.6 we obtain that every ideal of S contains a normal element, that is, an element x such that $Sx = xS$. The set $\mathrm{N}(S)$ of all such elements forms a submonoid. Elements of this type play an important role in the study of the structure of $K[S]$, as will be seen in Chapter 8. If $S = \mathrm{N}(S)$ then we say that S is a normalizing semigroup.

Corollary 4.4.12. *Let S be a submonoid of a polycyclic-by-finite group. If S satisfies the ascending chain condition on right ideals, then $\mathrm{N}(S)$ is finitely generated and satisfies the ascending chain condition on ideals. Furthermore, for every right ideal I of S, the intersection $I \cap \mathrm{N}(S)$ is nonempty.*

Proof. If $ns = n'$ for $n, n' \in \mathrm{N}(S)$ and $s \in S$, then $nsS = n'S = Sn' = Sns = nSs$, so cancellativity implies that $s \in \mathrm{N}(S)$. Similarly, $sn \in \mathrm{N}(S)$ implies $s \in \mathrm{N}(S)$. Hence $\mathrm{N}(S)$ is a grouplike submonoid of S. Therefore, by Lemma 4.1.1, $\mathrm{N}(S)$ satisfies the ascending chain condition on right ideals, so by Theorem 4.1.6 and Theorem 4.1.7 $K[\mathrm{N}(S)]$ is Noetherian and finitely generated.

Let H, G be as in Theorem 4.4.6. Assume that I is a right ideal of S. Since $[G : H] < \infty$, there exists $a \in I \cap (S \cap H)$. Let $\tilde{s}$ denote the image of $s \in S$ in $S/[H, H]$. By Lemma 4.1.9 we know that

$$\mathrm{Z}(S/[H,H])^{-1}\tilde{a}(S/[H,H]) = \tilde{a}(S/[H,H])\,\mathrm{Z}(S/[H,H])^{-1} = G/[H,H].$$

Therefore there exist $s, z \in S$ such that $\tilde{z} \in \mathrm{Z}(S/[H,H])$ and $z^{-1}as \in [H,H]$. Then $ash^{-1} = z$ for some $h \in [H,H]$ and $zx[H,H] = xz[H,H] = [H,H]xz$ for every $x \in S$. It follows that $zS = Sz$, so that $z \in \mathrm{N}(S) \cap I$. □

4.5 Prime ideals of $K[S]$

Let S be a cancellative monoid with a group of right quotients G such that either $K[S]$ is right Noetherian or G is a finitely generated and abelian-by-finite (so $K[G]$ is right Noetherian and it satisfies a polynomial identity in the latter case). Then by Theorem 3.2.6, for any two-sided ideal I of $K[S]$, the extension $I^e = IK[G]$ is a two-sided ideal of $K[G]$. Furthermore, the map $Q \mapsto Q^e$ determines a one-to-one correspondence between the set of prime ideals Q of $K[S]$ that do not intersect S and the prime ideals of $K[G]$. The latter is a consequence of the fact that $Q^e \cap K[S] = Q$. It follows also that, for such a prime Q, the ring $K[G]/Q^e$ is a ring of right quotients of $K[S]/Q$ and thus we get the equality of the classical rings of quotients

$$\mathrm{Q}_{cl}(K[S]/Q) = \mathrm{Q}_{cl}(K[G]/Q^e).$$

Prime ideals of the algebra $K[G]$, for a polycyclic-by-finite group G, have been extensively studied and several deep properties of the prime spectrum $\mathrm{Spec}(K[G])$ have been established, see Chapter3 and [12, 103, 130, 136]. On the other hand, we shall see later that primes Q of $K[S]$ intersecting S play a crucial role in the study of certain special classes of algebras (Chapter 7, Chapter 8 and Chapter 10).

One of the purposes of this section is to show that these properties can be extended to arbitrary primes of $K[S]$, where S is a cancellative monoid which satisfies the ascending chain condition on right ideals and $G = SS^{-1}$

is a polycyclic-by-finite group. We show that every prime ideal Q of $K[S]$ is strongly related to a prime ideal of the group algebra of a subgroup of G via a generalized matrix ring structure on $K[S]/Q$. Applications to the prime spectrum, classical Krull dimension and irreducible $K[S]$-modules are given. Since $S \cap Q$, if nonempty, clearly is a prime ideal of S, the primary focus is on prime ideals of the semigroup S. We start with a natural important class of prime ideals.

Proposition 4.5.1. *Let S be a subsemigroup of a group G. Let F be a subgroup of G such that $F \cap S \neq \emptyset$. Assume that P is maximal (hence, largest) among all ideals of S with respect to the property $F \cap P = \emptyset$. Then*

$$S \setminus P = \bigcup_{x,z \in A} x^{-1}Fz \cap S$$

for a subset A of S such that every $x^{-1}Fz \cap S$ is nonempty.

Proof. Let $u \in S \setminus P$. By the assumptions there exist $x, y \in S$ such that $xuy \in F \cap S$. So $u \in x^{-1}Fy^{-1} \cap S$. Now $xu \in Fy^{-1}$ and $uy \in x^{-1}F$. Therefore $S \setminus P \subseteq \bigcup_{x \in A, y \in B} x^{-1}Fy^{-1}$ for subsets $A = \{x \in S \mid x^{-1}F \cap S \neq \emptyset\}$ and $B = \{y \in S \mid Fy^{-1} \cap S \neq \emptyset\}$.

If $y \in B$ then $Fy^{-1} = Fs$ for some $s \in S$. Thus $y \in s^{-1}F$ and $s \in A$. Hence $x^{-1}Fy^{-1} = x^{-1}Fs$ and $S \setminus P \subseteq \bigcup_{x,s \in A} x^{-1}Fs$.

Assume that $x, z \in A$. Then there exists $t \in S$ such that $zt \in F$. Also $x^{-1}Fz \cap S \neq \emptyset$, as it contains $(x^{-1}F \cap S)z$. Moreover $x(x^{-1}Fz)t \subseteq F$ implies that $x^{-1}Fz \cap P = \emptyset$. It follows that $S \setminus P = \bigcup_{x,z \in A} x^{-1}Fz \cap S$, a union of nonempty sets. □

Clearly, P considered above is a prime ideal of S. We write $S_F = S/P$. It is easy to see that, if $F \cap S$ satisfies the right Ore condition, then in the above description of $S \setminus P$ the group F can be replaced by the group $F' = (F \cap S)(F \cap S)^{-1}$; in other words $S_{F'} = S_F$.

The collection of all subsets $gFh, g, h \in G$ (with obvious repetitions) can be visualized as an array

$$\begin{array}{cccc} F & Fa & Fb & \cdots \\ a^{-1}F & a^{-1}Fa & a^{-1}Fb & \cdots \\ b^{-1}F & b^{-1}Fa & b^{-1}Fb & \cdots \\ \cdots & \cdots & \cdots & \cdots \end{array}$$

where $a, b, \ldots$ come from a set of all coset representatives $Fc, c \in G$. We have seen that there exists a 'square subarray' that contains all elements of $S \setminus P$ and with all respective $b^{-1}Fa \cap S$ nonempty and with $a, b \in S$.

Recall that, for a group H and a set X, we denote by $\mathrm{M}_{\Gamma}^{fin}(K[H])$ the algebra of $\Gamma \times \Gamma$ matrices over $K[H]$ with finitely many nonzero entries in each column. Let $\mathcal{M}_X(H) \subseteq \mathrm{M}_X^{fin}(K[H])$ the monoid of all monomial $X \times X$ matrices over H. That is, $\mathcal{M}_X(H)$ consists of all matrices $s = (s_{ij})$ such that each row, and each column, of s contains at most one nonzero entry and all entries are in $H^0 = H \cup \{0\}$. The set of matrices with at most one nonzero entry is an ideal of $\mathcal{M}_X(H)$ and is isomorphic to the completely 0-simple semigroup $\mathcal{M}(H, X, X; I)$. We say that S is a monomial semigroup if $S \subseteq \mathcal{M}_X(H)$ for a group H and a set X. In case $|X| = n$ then we simply denote the monoid $\mathcal{M}_X(H)$ by $\mathcal{M}_n(H)$.

The square pattern on the Rees factor S/P arising in Proposition 4.5.1 is often of a much stronger nature. We shall see that it comes from a monomial structure on S/P. The latter comes from an ideal of any prime factor S/P if certain natural finiteness conditions are assumed. The considered conditions are motivated by applications to semigroup algebras, discussed later.

For a semigroup T with zero θ and an element $a \in T$ we define $\mathrm{ra}_T(a) = \{x \in T \mid ax = \theta\}$. Similarly, $\mathrm{la}_T(a)$ stands for the left annihilator of a in T. The main theorem reads as follows.

Theorem 4.5.2. *Let S be a cancellative semigroup. Let P be a prime ideal of S. Then S/P is monomial if either of the following holds.*

1. *S satisfies the ascending chain condition on right ideals.*
2. *S/P satisfies the ascending chain condition on right ideals of the form $\mathrm{ra}_{S/P}(x), x \in S/P$, and S has no free nonabelian subsemigroups.*

Moreover, in this case, there exists a subgroup H of the right quotient group $G = SS^{-1}$ of S such that the ideal P is maximal with respect to $P \cap H = \emptyset$ and we have $H = (S \cap H)(S \cap H)^{-1}$.

Proof. Let

$$J = \{a \in S \mid xay \in P, x, y \in S, \text{ implies } xa \in P \text{ or } ay \in P\}.$$

Clearly $P \subseteq J$. Let $a \in J, z \in S$ and assume that $xzay \in P$. Then $xza \in P$ or $ay \in P$, because $a \in J$. Hence $xza \in P$ or $zay \in P$. So $za \in J$. This shows that J is a left ideal in S. A symmetric argument allows us to show that J is an ideal of S.

We shall often view elements of $S \setminus P$ as nonzero elements of S/P. If $S \setminus P$ is a subsemigroup of S, then $J = S$ and in both cases $S \setminus P$ has a group of right quotients, by Lemma 4.1.5 and Lemma 2.1.3, whence the first assertion trivially follows. So assume that S/P has nonzero zero divisors. Now $J \neq P$ because elements $a \in S \setminus P$ with maximal $\mathrm{ra}(a) = \mathrm{ra}_{S/P}(a)$ are in J. It is clear that $J \setminus P$ coincides with the set of all elements $a \in S \setminus P$ such that $\mathrm{ra}(a) = \mathrm{ra}(xa)$ if $xa \notin P$. Similarly, if we define $\mathrm{la}(b) = \mathrm{la}_{S/P}(b)$ for $b \in S \setminus P$, then $J \setminus P$ consists of all $b \in S \setminus P$ such that $\mathrm{la}(b) = \mathrm{la}(by)$ whenever $by \notin P$.

Let $R_\gamma, \gamma \in \Gamma$, be the equivalence classes on $J \setminus P$ of the relation σ defined by $(x, y) \in \sigma$ if $\mathrm{la}(x) = \mathrm{la}(y)$. Clearly $R_\gamma \cup \{\theta\}$ is a right ideal of S/P and $J \setminus P$ is a disjoint union of $R_\gamma, \gamma \in \Gamma$. Similarly, if $L_\alpha, \alpha \in \Omega$, are the equivalence classes on $J \setminus P$ defined by the right annihilators in S/P, then $L_\alpha \cup \{\theta\}$ are 0-disjoint left ideals of S/J and

$$J \setminus P = \bigcup_{\gamma \in \Gamma} R_\gamma = \bigcup_{\alpha \in \Omega} L_\alpha.$$

If $a \in R_\gamma$ and $b \in L_\alpha$ then $a(S/P)b \neq \theta$ because S/P is prime. Hence $a(S/P)b \subseteq (L_\alpha \cap R_\gamma) \cup \{\theta\}$ and each $L_\alpha \cap R_\gamma$ is nonempty.

For every $\alpha \in \Omega$ define J_α as the set of all elements in $L_\alpha \subseteq S/P$ that are not nilpotent. First notice that $J_\alpha \neq \emptyset$. Indeed, otherwise $L_\alpha \cup \{\theta\}$ is a left nil ideal of S/P. Then $a^2 = \theta$ in S/P for every $a \in L_\alpha$ because $a \in J$. So if $b \in L_\alpha$ then $\mathrm{ra}(a) = \mathrm{ra}(b)$ implies that $ba = \theta$. Then the square of L_α is zero, contradicting the primeness of S/P.

Then J_α is a subsemigroup of $S \setminus P$. Indeed, from $a, b \in J_\alpha$ it follows that $a^2, b^2 \in J_\alpha$, which easily implies that $ab \in J_\alpha$ since $\mathrm{ra}(a) = \mathrm{ra}(b)$ and $a, b \in J$. We claim that J_α is a right Ore semigroup. This is clear in case (2) because of Lemma 2.1.3. If (1) is satisfied and $a, b \in J_\alpha$ then $b^n a = az$ for some $n \geq 1$ and $z \in S$ by Lemma 4.1.5. So $b^n aa = aza \in J_\alpha$, with $aza, aa \in J_\alpha$. Then za is not nilpotent as an element of S/P because $a \in J$ and $az, za \neq \theta$. Since $za \in L_\alpha$, we get $za \in J_\alpha$ and $aJ_\alpha \cap bJ_\alpha \neq \emptyset$, as desired.

Next we show that every J_α intersects only one R_γ. Indeed, suppose $a \in J_\alpha \cap R_\gamma, b \in J_\alpha \cap R_\delta$ for some $\gamma \neq \delta$. Then $xa \neq \theta$ and $xb = \theta$ for some $x \in S$. Now J_α is right Ore, whence there exist $u, v \in J_\alpha \subseteq S$ such that $au = bv \in J_\alpha$. Then $xau \neq \theta$ (because $a \in J$) and $xbv = \theta$, a contradiction since these elements are equal.

Hence $J_\alpha \subseteq L_\alpha \cap R_\gamma$ for a unique $\gamma \in \Gamma$ and it is now easy to see that $J_\alpha = L_\alpha \cap R_\gamma$. Indeed, if $a \in J_\alpha$ and $x \in L_\alpha \cap R_\gamma$, then $\mathrm{ra}(x) = \mathrm{ra}(a)$ and $\mathrm{la}(x) = \mathrm{ra}(a)$. If $x^2 = \theta$, this implies that $ax = \theta$ and hence $a^2 = \theta$, a contradiction. Therefore $x \in J_\alpha$, as desired.

Suppose now that $J_\alpha, J_\beta \subseteq R_\gamma$ for some $\alpha, \beta \in \Omega$ and $\gamma \in \Gamma$. Clearly $\mathrm{ra}(J_\alpha) \cap (J \setminus P)$ is a union of some $R_\delta, \delta \in \Gamma$. These are R_δ such that

$R_\delta \subseteq \mathrm{ra}(J_\alpha) = \mathrm{ra}(L_\alpha) = \mathrm{ra}(L_\alpha \cap R_\delta)$, so $L_\alpha \cap R_\delta \subseteq \mathrm{la}(R_\delta) = \mathrm{la}(L_\alpha \cap R_\delta)$. Since $L_\alpha \cap R_\delta$ consists of nilpotents if $\delta \neq \gamma$ and it is equal to J_α if $\delta = \gamma$, we get that $\mathrm{ra}(J_\alpha) \cap (J \setminus P)$ is the union off all R_δ with $\delta \neq \gamma$. Computing $\mathrm{ra}(J_\beta)$ in the same way we get $\mathrm{ra}(J_\alpha) = \mathrm{ra}(J_\beta)$, which means that $\alpha = \beta$. It follows that each R_γ contains at most one J_α. In particular, we see that J_α can also be defined as the set of all elements of R_γ that are not nilpotent.

Then R_γ contains exactly one J_α since otherwise $R_\gamma \cup \{\theta\}$ would be a nil right ideal of S/J, which also leads to a contradiction. Therefore there is a bijection between the sets Ω and Γ and we may assume they are equal. Moreover, with this identification, we get

$$J \setminus P = \bigcup_{\alpha,\gamma\in\Gamma} (L_\alpha \cap R_\gamma) \text{ and } J_\alpha = R_\alpha \cap L_\alpha$$

and the previous paragraph implies that $L_\alpha R_\beta \neq \theta$ only if $\alpha = \beta$. Then J/P can be considered as a subsemigroup of the Brandt semigroup $M = \mathcal{M}(G, \Gamma, \Gamma; I)$. Fix some $\alpha \in \Gamma$ and define $H = J_\alpha J_\alpha^{-1}$. Then H can be viewed as a subgroup of an $\mathcal{H}$-class of M but also as a subgroup of G. Since J/P intersects every $\mathcal{H}$-class of M, it is contained in a unique Brandt subsemigroup $\widehat{J}$ of M isomorphic to $\mathcal{M}(H, \Gamma, \Gamma; I)$, see Proposition 2.2.1. Then H, considered as an $\mathcal{H}$-class of $\widehat{J}$, contains an idempotent e.

By Lemma 2.5.1 $S' = (S \setminus J) \cup \widehat{J}$, a disjoint union, has a unique semigroup structure extending that of S/P. So the action of S/P on the left ideal $K_0[\widehat{J}e]$ of $K_0[\widehat{J}]$ by left multiplication, used in Proposition 2.5.6, yields a homomorphism

$$\varphi : K_0[S/P] \longrightarrow \mathrm{End}_{K[H]}(K_0[\widehat{J}e]) \subseteq \mathrm{M}_\Gamma^{fin}(K[H]).$$

Also $\varphi(S)$ consists of column monomial matrices over G.

If $(s-t)\widehat{J}e = 0$ for some $s, t \in S/P$ then, since $\widehat{J}e\widehat{J} = \widehat{J}$, we have $(s-t)J = 0$. So, either $sx = tx \neq 0$ for some $x \in J$ or $sJ = 0$. The former yields $s = t$ because S is cancellative. The latter is not possible because S/P is prime. So φ is one-to-one on S/P and hence S/P embeds into $\mathrm{M}_\Gamma^{fin}(K[H])$. This argument shows even more, namely that $K_0[S/P]$ embeds into $\mathrm{M}_\Gamma^{fin}(K[H])$.

We know that each column of $\varphi(s)$ has at most one nonzero entry. Also, if $a \in S \setminus P$ and $ax \neq \theta, ay \neq \theta$ for some $x \in R_\beta, y \in R_\gamma$, then $\beta = \gamma$ because $\widehat{J}$ is a Brandt semigroup. Therefore, every matrix in $\varphi(J)$ has at most one nonzero entry. Suppose that an element $\varphi(s), s \in S/P$, has at least two nonzero entries in some row. Then there exists $a \in J$ such that

$\varphi(as) \in \varphi(J)$ has at least two nonzero entries in some row, a contradiction. Therefore $S/P \cong \varphi(S/P) \subseteq \mathcal{M}_\Gamma(H)$ and $\varphi(J) \subseteq \mathcal{M}(H, \Gamma, \Gamma; I) \subseteq \mathcal{M}_\Gamma(H)$.

To prove the remaining assertions notice that, since S/P is prime, for every nonzero ideal T of S/P we have $\{\theta\} \neq R_\alpha T L_\alpha \subseteq T \cap J_\alpha$. (The case where $S \setminus P$ is a subsemigroup of S is interpreted as the case where $\Gamma = \{\alpha\}$ and $J_\alpha = S \setminus P$.) Since $P \cap J_\alpha = \emptyset$ in G, we also get $P \cap H = \emptyset$. So P is maximal among all ideals of S not intersecting H. The equality $H = (S \cap H)(S \cap H)^{-1}$ is clear because $H = J_\alpha J_\alpha^{-1}$ with $J_\alpha \subseteq S$. □

Remark 4.5.3. If Q is a prime ideal of a semigroup algebra $K[S]$ such that $K[S]/Q$ is a right Goldie ring, then the image $\overline{S}$ of S in $K[S]/Q$ inherits the ascending chain condition on right annihilator ideals. If $P = Q \cap S \neq \emptyset$ then $\phi^{-1}(0) = \{\theta\}$ for the natural map $\phi : S/P \longrightarrow \overline{S}$. Suppose that $x, y \in S/P$ are such that $\mathrm{ra}_{S/P}(x) \subseteq \mathrm{ra}_{S/P}(y)$ and $\mathrm{ra}_{\overline{S}}(\phi(x)) = \mathrm{ra}_{\overline{S}}(\phi(y))$. If $ys = \theta$ in S/P for some $s \in S/P$ then $\phi(ys) = 0$ implies that $\phi(xs) = 0$. Hence $xs = \theta$. It follows that $\mathrm{ra}_{S/P}(x) = \mathrm{ra}_{S/P}(y)$. Therefore S/P also satisfies the ascending chain condition on right annihilator ideals. So Theorem 4.5.2 applies if S has no free nonabelian subsemigroups. In particular, by Theorem 3.1.9, this is the case for every prime ideal Q of $K[S]$ provided that $S \subseteq G$ for an abelian-by-finite group G. Furthermore, under the homomorphism $\mathrm{M}_\Gamma(H) \longrightarrow \mathrm{M}_\Gamma(\{1\})$ induced by the trivial map $H \longrightarrow \{1\}$, the ideal J used in the proof maps onto the Brandt semigroup $\mathcal{M}(\{1\}, \Gamma, \Gamma, I)$ over the trivial group. Therefore, if S satisfies the ascending chain condition on right ideals or G is abelian-by-finite (so $K[S]$ is a PI algebra by Theorem 3.1.9), then Γ must be a finite set.

We can give more information on the structure of S/P in group-theoretic terms. This also shows that the square pattern on S/P arising from Proposition 4.5.1 agrees with the matrix type pattern of Theorem 4.5.2. By $\mathrm{N}_G(H)$ we denote the normalizer $\{g \in G \mid g^{-1}Hg \subseteq H\}$ of a subgroup H in a group G.

Proposition 4.5.4. *Let P be a prime ideal of a cancellative semigroup S such that the hypothesis of Theorem 4.5.2 is satisfied. Let J be the ideal of S and H the subgroup of the group of right quotients G of S used in the proof. Then*

1. *there exist sets $F_1, F_2 \subseteq S$ and a bijection $x \mapsto x^*, F_1 \longrightarrow F_2$ such that $S \setminus P \subseteq \bigcup_{x \in F_1, y \in F_2} xHy \cap S$ with $x^*x \in H$, while $J \setminus P = \bigcup_{x \in F_1, y \in F_2} xHy \cap J$ is a union of disjoint nonempty sets,*

2. *for* $x, y \in F_1$, *any of the conditions* $x\,\mathrm{N}_G(H) = y\,\mathrm{N}_G(H), \mathrm{N}_G(H)x^* = \mathrm{N}_G(H)y^*$ *implies that* $x = y$,

3. *the groups of quotients of the 'diagonal' subsemigroups* $xHx^* \cap J$ *of* $J \setminus P$ *are conjugate in* G *to the group* $(J \cap H)(J \cap H)^{-1} = (S \cap H)(S \cap H)^{-1}$.

Proof. We use the notation of the proof of Theorem 4.5.2. Let $J_{\alpha\beta} = R_\alpha \cap L_\beta$ for $\alpha, \beta \in \Gamma$. All $J_{\alpha\beta}$ are nonempty disjoint subsets of $J \setminus P$. Let $H_\alpha = J_{\alpha\alpha} J_{\alpha\alpha}^{-1}$. From the proof it follows that $H_\alpha \cap J = J_{\alpha\alpha}$ and $H_\alpha = (S \cap H_\alpha)(S \cap H_\alpha)^{-1}$ for every $\alpha \in \Gamma$.

Choose any $a_{\alpha\beta} \in J_{\alpha\beta}$. Then $J_{\alpha\beta} a_{\beta\gamma} \subseteq J_{\alpha\gamma}$ and $a_{\gamma\alpha} J_{\alpha\beta} \subseteq J_{\gamma\beta}$ for every $\alpha, \beta, \gamma \in \Gamma$ but $a_{\alpha\beta} a_{\gamma\delta} \in P$ for $\beta \neq \gamma$. Also $J_{\alpha\beta} J_{\beta\alpha} \subseteq J_{\alpha\alpha} \subseteq H_\alpha$. It follows that $J_{\alpha\beta} \subseteq H_\alpha a_{\beta\alpha}^{-1} = H_\alpha a_{\alpha\beta}$ and $J_{\alpha\beta} \subseteq a_{\beta\alpha}^{-1} H_\beta = a_{\alpha\beta} H_\beta$. So $a_{\beta\alpha} H_\alpha a_{\alpha\beta} = a_{\beta\alpha} H_\alpha a_{\beta\alpha}^{-1} = a_{\alpha\beta}^{-1} H_\alpha a_{\alpha\beta}$ is a group. Also $(a_{\beta\alpha} J_{\alpha\alpha} a_{\alpha\beta}^*)(a_{\beta\alpha} J_{\alpha\alpha} a_{\alpha\beta}^*)^{-1} = a_{\beta\alpha} J_{\alpha\alpha} J_{\alpha\alpha}^{-1} a_{\beta\alpha}^{-1} = a_{\beta\alpha} H_\alpha a_{\beta\alpha}^{-1}$. Moreover $J_{\gamma\beta} \subseteq a_{\alpha\gamma}^{-1} H_\alpha a_{\beta\alpha}^{-1} = a_{\gamma\alpha} H_\alpha a_{\alpha\beta}$. Since $a_{\alpha\gamma}((J \setminus P) \cap a_{\gamma\alpha} H_\alpha a_{\alpha\beta}) a_{\beta\alpha} \subseteq J_{\alpha\alpha}$, we also get $(J \setminus P) \cap a_{\gamma\alpha} H_\alpha a_{\alpha\beta} \subseteq J_{\gamma\beta}$, so the equality follows, for all β, γ. We fix some $\alpha \in \Gamma$ and we define $H = H_\alpha, F_1 = \{a_{\beta\alpha} \mid \beta \in \Gamma\}$ and $a_{\beta\alpha}^* = a_{\alpha\beta}$. If $z \in S \setminus P$, then $JzJ \not\subseteq P$ because P is prime. So there exist $\beta, \gamma \in \Gamma$ and $a \in J_{\alpha\beta}, b \in J_{\gamma\alpha}$ such that $azb \notin P$. Then $azb \in J_{\alpha\alpha}$, whence $z \in a^{-1} H b^{-1} = a_{\beta\alpha} H a_{\alpha\gamma}$. Therefore $z \in a_{\beta\alpha} H a_{\alpha\gamma}$ and assertions (1) and (3) follow.

Notice that $x\,\mathrm{N}_G(H) = y\,\mathrm{N}_G(H)$ if and only if $x^{-1} y H = H x^{-1} y$, or equivalently $xHx^{-1} = yHy^{-1}$. Hence, for a fixed $\alpha \in \Gamma$, (1) implies that $a_{\beta\alpha}, \beta \in \Gamma$, yield different cosets $a_{\beta\alpha}\,\mathrm{N}_G(H)$. Similarly, the cosets $Ha_{\alpha\beta}, \beta \in \Gamma$, are all different. □

In order to make the connection with Theorem 4.5.2 more explicit, and to see how Proposition 2.2.1 works there, let us note that from the above proof it follows that $H_\beta = J_{\beta\beta} J_{\beta\beta}^{-1} \subseteq a_{\beta\alpha} H_\alpha a_{\alpha\beta}$, since the latter is a group containing $J_{\beta\beta}$. Similarly $H_\alpha \subseteq a_{\alpha\beta} H_\beta a_{\beta\alpha}$, and $H_\beta \subseteq a_{\beta\alpha} H_\alpha a_{\alpha\beta} \subseteq a_{\beta\alpha} a_{\alpha\beta} H_\beta a_{\beta\alpha} a_{\alpha\beta} = H_\beta$, which implies that $H_\beta = a_{\beta\alpha} H_\alpha a_{\alpha\beta}$. In other words, we can visualize $J \setminus P$ as follows

$$J \setminus P \subseteq M = \begin{pmatrix} H & Ha_{\alpha\beta} & \cdots & Ha_{\alpha\gamma} & \cdots \\ a_{\beta\alpha} H & a_{\beta\alpha} H a_{\alpha\beta} & \cdots & a_{\beta\alpha} H a_{\alpha\gamma} & \cdots \\ \cdots & \cdots & \cdots & \cdots & \cdots \\ a_{\gamma\alpha} H & a_{\gamma\alpha} H a_{\alpha\beta} & \cdots & a_{\gamma\alpha} H a_{\alpha\gamma} & \cdots \\ \cdots & \cdots & \cdots & \cdots & \cdots \end{pmatrix} \cup \{\theta\},$$

where M is a semigroup of matrix type, with diagonal components being groups conjugate to H. So M is a Brandt semigroup isomorphic to

$\mathcal{M}(H, \Gamma, \Gamma; I)$. Notice that the semigroup M from the above proof is the semigroup of right quotients of J/P in the sense of Section 2.1.

If S satisfies the ascending chain condition on right ideals, then for any $a, b \in H \cap J$ there exists $n \geq 1$ such that $a^n b \in bS$, by Lemma 4.1.5. Hence $a^n b^2 \in bSb \cap J$, which easily implies that $a^n b \in b(H \cap J)$. So the property: $a\langle b\rangle \cap bS \neq \emptyset$ for every $a, b \in S$ is inherited by the semigroup $H \cap J$ from S.

The following special case is of interest in the context of Noetherian algebras $K[S]$. Our argument depends on the structural description of the underlying monoid S.

Corollary 4.5.5. *Let S be a submonoid of a finitely generated nilpotent group. Assume that S satisfies the ascending chain condition on right ideals. If P is a prime ideal of S then $S \setminus P$ is a subsemigroup of S.*

Proof. Let G be the group of quotients of S. From Theorem 4.4.7 we know that there exists a normal subgroup F of G such that $F \subseteq S$ and G/F is abelian-by-finite. Let $G' = G/F$ and let $P' \subseteq S' \subseteq G'$ be the natural homomorphic images of P and S. Every ideal of S' is the image J' of an ideal J of S. If $J_1' J_2' \subseteq P'$ then $J_1 J_2 \subseteq PF = P$. This shows that P' is a prime ideal of S'. Moreover, $S \setminus P$ is a subsemigroup of S if and only if $S' \setminus P'$ is a subsemigroup of S'. So, replacing P, S by P', S', we may assume that G is abelian-by-finite. Suppose $S \setminus P$ is not a semigroup. Then by Theorem 4.5.2 there exists a subgroup H of G and $m > 1$ such that $S/P \subseteq \mathcal{M}_m(H)$. Let J be the ideal of S described in the proof of this theorem. Since $m > 1$, we may choose $a \in J \cap H$ and $b \in J$ such that $ab \notin P$ but $ba \in P$. Since H is finitely generated and nilpotent, the torsion part H^+ of H is a finite normal subgroup and H/H^+ is torsion free, abelian-by-finite and nilpotent, whence abelian, see Proposition 3.3.3. So the commutator subgroup of H is finite. It follows that H is an FC-group and $a^n b = ba^n$ for some $n \geq 1$. However $a^n b \notin P$ by the choice of a, b and the description of J/P, while $ba \in P$, a contradiction. This proves the assertion. □

Next we discuss some applications to semigroup algebras.

Lemma 4.5.6. *Assume that a semigroup $\widehat{I} = \mathcal{M}(H, r, r; P)$ is the completely 0-simple closure of its uniform subsemigroup I. Assume also that the sandwich matrix P is invertible in $\mathrm{M}_r(K[H])$ and $I \cap G$ satisfies the right Ore condition for every maximal subgroup G of $\widehat{I}$. Then $K_0[\widehat{I}]$ is a right localization of $K_0[I]$.*

Proof. Recall from Section 2.1 that the element of $\widehat{I}$ with $g \in H$ in the (i, j) position, and zeroes elsewhere, is denoted by (g, i, j). Let $\widehat{I}_{ij}$ be the set of nonzero elements of $\widehat{I}$ that are in row i and column j. Put $I_{ij} = \widehat{I}_{ij} \cap I$. We use the fact that

$$\widehat{I}_{j1} = I_{j1} I_{i1}^{-1}$$

where $\widehat{I}_{i1}$ is a group from the first column and the inverse operation applies to this group. Indeed, $p_{1i} \neq \theta$, and $I_{i1} I_{i1}^{-1} = \widehat{I}_{i1}$ by the hypothesis, so that

$$\widehat{I}_{j1} \supseteq I_{j1} I_{i1}^{-1} \supseteq I_{j1} I_{i1} I_{i1}^{-1} \supseteq I_{j1} \widehat{I}_{i1} = \widehat{I}_{j1}. \tag{4.17}$$

In particular, for every element $y \in \widehat{I}_{j1}$ there exists $u \in I_{i1}$ such that $yu \in I_{j1}$.

View $K_0[I]$ as a subset of $\mathrm{M}_r(K[H])$, see Chapter 2 (but, as algebras, they have different multiplication of course). Define the set $C = \{P^{-1} \circ s \mid s \in E\} \subseteq \mathrm{M}_r(K[H])$, where E is the set of diagonal matrices with entries in H and $\circ$ is the ordinary matrix multiplication. Clearly, C consists of elements that are invertible in $K_0[\widehat{I}]$. Let C' consist of those elements of C that lie in $K_0[I]$. It is enough to show that for every matrix $z \in K_0[\widehat{I}]$ there exists $c = P^{-1} \circ s \in C'$ such that $zc \in K_0[I]$. But $zc = z \circ P \circ c = z \circ s$. So we need to find s such that $z \circ s$ and $P^{-1} \circ s$ are both in $K_0[I]$. So for each $1 \leq q \leq r$ and for the finitely many elements from the support of the q-th columns of P^{-1} and z (we treat them as elements of $K_0[\widehat{I}]$) it is enough to find $t = s_q = (g, q, q) \in I$ such that t multiplies (on the right) these matrices into I.

For simplicity assume that $q = 1$. So, given a finite set of nonzero elements $x_k = (h_k, i_k, 1)$ in $\widehat{I}$ we need an element $t = (g, 1, 1) \in I_{11}$ such that $x_k \circ t = (h_k g, i_k, 1) \in I$ for every k. But $I_{11} \circ P \circ I_{i1} = I_{11} I_{i1} \subseteq I_{11}$ implies that, defining $x'_k = x_k \circ a$ for any fixed $a \in I_{11}$, it is enough to find an element $u \in I_{i1}$ such that $x'_k u \in I_{i_k 1}$ for all k.

Existence of such u follows from (4.17). Namely, (4.17) allows us first to find u for the first of the elements x'_k, and then adjust it step by step by right multiplication by elements of I_{i1} so that it works for all x'_k, just like in the process of finding a common denominator for finitely many fractions. This proves the lemma. $\square$

Corollary 4.5.7. *Assume that the hypothesis and the notation of Theorem 4.5.2 hold. Then we have natural embeddings* $K[S \cap H] \subseteq K_0[S/P] \subseteq \mathrm{M}_\Gamma^{fin}(K[H])$ *such that* $K[S \cap H]$ *intersects nontrivially every nonzero ideal of* $K_0[S/P]$*. If* Γ *is finite (in particular, if* S *satisfies the ascending chain condition on right ideals or* G *is abelian-by-finite), then* $\mathrm{M}_\Gamma^{fin}(K[H]) = \mathrm{M}_n(K[H])$*, for* $n = |\Gamma|$*, is a right localization of* $K_0[S/P]$.

Proof. We have seen in the proof of Theorem 4.5.2 that $K_0[S/P]$ embeds into $\mathrm{M}_\Gamma^{fin}(K[H])$ with J/P mapping to a uniform subsemigroup of $\widehat{J} \cong \mathcal{M}(H,\Gamma,\Gamma,I) \subseteq \mathcal{M}_\Gamma(H)$. Moreover, if $e = e^2 \in \widehat{J}$ is as in the proof, then for every nonzero ideal T of $K_0[S/P]$, we we know that $T\widehat{J}e \neq 0$. Since J intersects every $\mathcal{H}$-class of $\widehat{J}$, it follows that $(\widehat{J}e \cap J)\widehat{J} = \widehat{J}$ and $0 \neq (e\widehat{J} \cap J)T(\widehat{J}e \cap J) \subseteq K[H]$. Consequently, $JTJ \cap K[J \cap H] \neq 0$. If Γ is finite then Lemma 4.5.6 implies that $\mathrm{M}_\Gamma^{fin}(K[H]) \cong K_0[\widehat{J}]$ is a right localization of $K_0[S/P]$. By Remark 4.5.3, this applies to the two listed cases. □

Recall that by $\mathrm{Spec}^1(K[S])$ we denote the set of height one prime ideals of $K[S]$, while $\mathrm{Spec}^1_h(K[S])$ stands for the set of all $Q \in \mathrm{Spec}^1(K[S])$ such that $Q \cap S \neq \emptyset$. The set of minimal prime ideals of S is denoted $\mathrm{Spec}^0(S)$.

Proposition 4.5.8. *Let S be a submonoid of a torsion free group G and let K be a field. Assume that S satisfies the ascending chain condition on right ideals, or G is abelian-by-finite.*

1. *If P is a prime ideal in S, then $K[P]$ is a prime ideal in $K[S]$.*
2. *If Q is a prime ideal in $K[S]$ with $Q \cap S \neq \emptyset$, then $K[Q \cap S]$ is a prime ideal in $K[S]$.*
3. *The height one prime ideals of $K[S]$ intersecting S are of the form $K[P]$, where P is a minimal prime ideal of S.*
4. *Suppose that G is polycyclic-by-finite and S satisfies the ascending chain condition on right ideals, or suppose that G is abelian-by-finite and $K[S]$ is a Krull domain. Then every ideal of the form $K[P]$, with P a minimal prime ideal of S, is a height one prime ideal of $K[S]$.*

Proof. Let P be a prime ideal of S. Notice that, by Remark 4.5.3, the hypotheses of Theorem 4.5.2 are satisfied. By Corollary 4.5.7, there exists a subgroup H of G such that $K[S \cap H]$ embeds into $K_0[S/P]$ and every nonzero ideal of $K_0[S/P]$ intersects $K[S \cap H]$ nontrivially. Since G is torsion free, $K[S \cap H]$ is prime by Theorem 3.2.8. It then follows that $K[P]$ is a prime ideal of $K[S]$.

Assume that Q is a prime ideal of the algebra $K[S]$ and $Q \cap S \neq \emptyset$. Clearly $Q \cap S$ is a prime ideal of S. Hence, by (1), $K[Q \cap S]$ is a prime ideal of $K[S]$.

Let $Q \in \mathrm{Spec}^1(K[S])$ with $Q \cap S \neq \emptyset$. Then, by the previous, $Q = K[Q \cap S]$ and $Q \cap S \in \mathrm{Spec}^0(S)$, because $K[S]$ is prime.

Finally, assume that S satisfies the hypothesis of statement (4) of the proposition. Let $P \in \mathrm{Spec}^0(S)$. Then (1) implies that $K[P]$ is a prime ideal of $K[S]$. By Corollary 4.4.12 and Remark 4.1.10, there exists $q \in P \cap \mathrm{N}(S)$. If G is polycyclic-by-finite and S satisfies the ascending chain condition on right ideals then $K[S]$ is right Noetherian by Theorem 4.1.6. From the principal ideal theorem, Theorem 3.2.4, it follows that $q \in Q \subseteq K[P]$ for some $Q \in \mathrm{Spec}^1(K[S])$. As $q \in Q \cap S \subseteq P$, from (3) we thus obtain that $Q = K[Q \cap S]$, $Q \cap S \in \mathrm{Spec}^0(S)$, and hence $Q \cap S = P$. Consequently, $K[P] = Q \in \mathrm{Spec}^1(K[S])$, as desired. So, assume that G is abelian-by-finite and $K[S]$ is a Krull domain. Then $K[Sq]$ is an invertible ideal in the Krull algebra $K[S]$. Therefore, in the divisor group $\mathrm{D}(K[S])$ we have $K[Sq] = I_1 * \cdots * I_k$, a product of height one prime ideals I_i of $K[S]$ and $q \in I_i$ for each i (see Section 3.6). So, in particular, $I_1 \cdots I_k \subseteq K[Sq] \subseteq K[P]$, and thus $I_j \subseteq K[P]$ for some j. Because of (2) we also know that $K[I_j \cap S]$ is a nonzero prime ideal of $K[S]$. Since I_j is of height one, it thus follows that $I_j = K[I_j \cap S]$. Hence $I_j \cap S \subseteq P$. As P is a minimal prime ideal of S one obtains that $I_j \cap S = P$ and thus $K[P] = I_j$ is a height one prime of $K[S]$. This completes the proof. □

Let S be a submonoid of a polycyclic-by-finite group G. Assume Q is a minimal prime ideal of $K[S]$. If H is a torsion free subgroup of finite index in G, then $K[H]$ is a domain by Theorem 3.3.4 and thus $Q \cap K[H] = \{0\}$ by Theorem 3.4.8. Hence $Q \cap S = \emptyset$. This shows that prime ideals of $K[S]$ intersecting S must be of height at least one.

Notice that rings graded by unique product groups have the property that minimal prime ideals are homogeneous, Theorem 3.4.2. Although not every torsion free polycyclic-by-finite group G is a unique product group, the previous proposition, surprisingly, shows that also height one primes of a Noetherian algebra $K[S]$ that intersect S are homogeneous, for any submonoid S of G.

A ring R is said to be a generalized matrix ring if, for some $n \geq 1$, there exist additive subgroups R_{ij} of R, $i, j = 1, \ldots, n$, such that

$$R = \bigoplus_{1 \leq i,j \leq n} R_{ij}$$

(as additive groups) and $R_{ij} R_{kl} \subseteq \delta_{jk} R_{il}$. If the identity of a ring R is the sum of orthogonal idempotents, say $e_1, \ldots, e_n$, then clearly $R = \bigoplus_{ij} e_i R e_j$ is a generalized matrix ring. If S is a multiplicative subsemigroup of R such that every element $s \in S$, represented as a generalized matrix in R, has at most one nonzero entry in each row and column then we say that S is in block monomial form with respect to the given set of idempotents. If X is a subsemigroup of an algebra A then we write $K\{X\}$ for the subalgebra spanned by X. We are now ready to interpret the results obtained above in terms of the image of S in the algebra $K[S]/Q$, for a prime ideal Q of $K[S]$ intersecting S. In particular, the following proposition shows that this leads to a prime ideal of the algebra $K[H]$ for a subgroup H of G.

Proposition 4.5.9. *Let S be a cancellative monoid with a group $G = SS^{-1}$ of right quotients. Assume that G is polycyclic-by-finite and S satisfies the ascending chain condition on right ideals, or G is abelian-by-finite. Let K be a field and let Q be a prime ideal in $K[S]$ such that $Q \cap S \neq \emptyset$.*

1. *There exists an ideal I of S and a cancellative right Ore subsemigroup T of I such that the image $\overline{I}$ of I in $K[S]/Q$ is a uniform subsemigroup in a Brandt subsemigroup J of the ring $\mathrm{Q}_{cl}(K[S]/Q)$ such that $1 = e_1 + \cdots + e_n$ for the nonzero idempotents e_i of J.*
2. $\mathrm{h}(TT^{-1}) < \mathrm{h}(G)$.
3. *The image $R = K\{\overline{I}\}$ of $K[I]$ in $K[S]/Q$ is a generalized matrix ring $R = \bigoplus_{i,j=1}^{n} R_{ij}$ with $R_{ij} = e_i \mathrm{Q}_{cl}(K[S]/Q)e_j \cap R$ and $R_{11} = K\{\overline{T}\}$.*
4. *The image $\overline{S}$ of S in $K[S]/Q$ is in block monomial form with respect to $e_1, \ldots, e_n$.*
5. *$R \subseteq R' \subseteq \mathrm{Q}_{cl}(K[S]/Q)$, where $R' \cong \mathrm{M}_n(K[TT^{-1}]/Q_0)$ is a right localization of R and Q_0 is a prime ideal in $K[TT^{-1}]$. Therefore $\mathrm{Q}_{cl}(K[S]/Q)$ is also the right quotient ring of $\mathrm{M}_n(K[TT^{-1}]/Q_0)$.*

Proof. Let $\overline{S}$ be the image of S in $K[S]/Q$. As noticed in Remark 4.5.3, $K[S]/Q$ is a right Goldie ring. We have the natural homomorphisms

$$K[S] \longrightarrow K_0[S/(Q \cap S)] \longrightarrow K[\overline{S}] \longrightarrow K[S]/Q \subseteq \mathrm{Q}_{cl}(K[S]/Q).$$

Clearly, $P = Q \cap S$ is a prime ideal of S. Therefore, by Proposition 4.5.2 and Corollary 4.5.7, $K_0[S/P] \subseteq \mathrm{M}_n(K[H])$ for a subgroup H of G and S/P has a structure of a monomial semigroup over H. Moreover, $\mathrm{M}_n(K[H])$ is the right localization of $K_0[S/P]$ with respect to the set C' as in the proof of Lemma 4.5.6. Let $I \subseteq S$ denote the ideal that is the inverse image of the ideal of rank one matrices in $S/P \subseteq \mathrm{M}_n(K[H])$. Then C' is identified with the set of all elements $s_1 + \cdots + s_n \in K_0[I/P]$ with each s_i in a different (cancellative) diagonal component of I/P. So $s_i = f_i s_i f_i$ for the rank one idempotents $f_1, \ldots, f_n$ in $\mathrm{M}_n(K[H])$.

Suppose that $s_1 + \cdots + s_n \in \tilde{Q}$, where $\tilde{Q}$ is the image of Q in $K_0[S/P]$. Then $s_1(s_1 + \cdots + s_n) \in \tilde{Q}$. However $s_1 s_i = 0$ for $i \neq 1$ and $s_1^2 \in (I/P) \cap F$, where $F = f_1 \mathcal{M}_n(H) f_1 \setminus \{0\} \cong H$. This contradiction shows that $\tilde{Q} \cap C' = \emptyset$. From Theorem 3.2.6 it then follows that there exists a prime ideal Q' in $\mathrm{M}_n(K[H])$ such that $\tilde{Q} = Q' \cap K_0[S/P]$. Clearly $Q' = \mathrm{M}_n(Q_0)$ for a prime ideal Q_0 in $K[H]$. Since F is mapped into a subgroup $\overline{F}$ of $\mathrm{M}_n(K[H]/Q_0)$, it follows that $(I/P) \cap F$ is mapped onto a subsemigroup of $\overline{F}$. We define T as the inverse image of $(I/P) \cap F$ in S. Then we may identify $H = TT^{-1}$.

The extension $K\{\overline{S}\} = K_0[S/P]/\tilde{Q} \subseteq \mathrm{M}_n(K[H])/Q'$ is a localization of a right Goldie ring. Thus we have an embedding $K\{\overline{S}\} \subseteq \mathrm{M}_n(K[H])/Q' \subseteq \mathrm{Q}_{cl}(K[S]/Q)$.

Now, if $A = K_0[I/P]$ then we have a decomposition $A = \bigoplus_{i,j=1}^n A_{ij}$, where $A_{ij} = f_i \, \mathrm{M}_n(K[H]) f_j \cap A$. It is clear that this decomposition leads to a decomposition $R = K\{\overline{I}\} = \bigoplus_{i,j} e_i \, \mathrm{Q}_{cl}(K[S]/Q) e_j \cap K\{\overline{I}\}$, where the images $e_1, \ldots, e_n$ of $f_1, \ldots, f_n$ form a set of orthogonal idempotents in $\mathrm{Q}_{cl}(K[S]/Q)$.

Next we show that $\mathrm{h}(TT^{-1}) < \mathrm{h}(G)$. Suppose the contrary. Then TT^{-1} is of finite index in G. So, for any $s \in P$ there exists $n > 0$ so that $s^n \in TT^{-1}$. Then $s^n = ab^{-1}$ for some $a, b \in T$. Consequently $s^n b = a \in T \cap P$, a contradiction because this intersection is empty. So (2) holds.

The remaining assertions of the proposition now follow readily. $\square$

Remark 4.5.10. Let S be as in Proposition 4.5.9. If Q is a prime ideal of $K[S]$ intersecting S and $P = Q \cap S$, then we exploit the block monomial pattern of $\overline{S}$, keeping in mind that it corresponds to the monomial pattern on S/P, as seen in the above proof. The proof associates to S a collection $T_1 = T, \ldots, T_n$ of subsemigroups of S. These are the inverse images in S of the diagonal parts of the ideal $\overline{I}$ of $\overline{S}$. Notice that, by Proposition 4.5.4, all groups $T_i T_i^{-1}$ are conjugate in G.

If G is polycyclic-by-finite, then G has a torsion free normal subgroup W of finite index. By Theorem 4.4.7, if S satisfies the ascending chain condition on right ideals, we may choose W in such a way that there exists a normal subgroup F of G with $F \subseteq S \cap W$ and W/F abelian. On the other hand, if G is abelian-by-finite, then let W be an abelian normal subgroup of finite index in G. In this case, let F be the trivial subgroup. First note that $S \cap W = \mathrm{N}(S \cap W)$, because for any $s, t \in S \cap W$, we have $st \in tsF = Fts$. Next we show the natural image $\overline{S \cap W}$ of $S \cap W$ in $K[S]/Q$ is diagonal (with respect to the block monomial pattern of $\overline{S}$). Indeed, suppose $x \in \overline{S \cap W}$ is such that $e_i x e_j \neq 0$ for some $i \neq j$. Pick $y \in \overline{T}_i$. Then, since x is block monomial, $0 \neq yx = ye_i x e_j \in e_i \overline{S} e_j$. But $\overline{S \cap W} y \subseteq \overline{S} e_i$. So $yx \notin$

$\overline{S \cap W}y$, contradicting the fact that $\overline{S \cap W} = \mathrm{N}(\overline{S \cap W})$. Now $Q \cap K[W]$ is a semiprime ideal of $K[S \cap W]$ and it is of the form $Q \cap K[W] = Q_1 \cap \cdots \cap Q_r$, where the intersection is taken over all the prime ideals Q_i of $K[S \cap W]$ that are minimal over $Q \cap K[W]$, see Theorem 3.4.8. Moreover $r \leq G/W$. Hence $Q \cap W = P \cap W = P_1 \cap \cdots \cap P_r$ for $P_i = Q_i \cap W$ and P_i are prime ideals of $S \cap W$. Since $S \cap W = \mathrm{N}(S \cap W)$, it is easy to see that each $(S \cap W) \setminus P_i$ is a subsemigroup of S. We have an embedding $S/P \subseteq \mathcal{M}_n(H)$ for some n and a group $H \cong TT^{-1}$. As above, $(S \cap W) \setminus P$ is diagonal. Also, let I be the ideal of S constructed in Proposition 4.5.9. Consider any diagonal element $s \in I/P \subseteq \mathcal{M}_n(H)$. Since S is periodic modulo W, it follows that $s^k \in W \cap I$ and $s^k \notin P$ for some $k \geq 1$. Hence every diagonal component of I/P intersects W and the number n of diagonal components of I/P is the same as the number of diagonal components intersected by W. Now $(I \cap W) \setminus P = \bigcup_{i=1}^{r} (I \cap W) \setminus P_i$, so it is a union of subsemigroups of $S \setminus P$, and it is contained in the diagonal part of I/P. Moreover each $I_i = (I \cap W) \setminus P_i$ must be contained in a single diagonal component of I/P. By the definition of Q_i we get that the diagonal components of the ideal $(I \cap W)/(P \cap W)$ of $(S \cap W)/(P \cap W)$ are exactly $(I \cap W) \setminus P_i$ (each minimal prime Q_i must contain the union of all diagonal components of $(I \cap W) \setminus P$ except for one) and we must have $r = n$. Since W is torsion free and $Q_i \cap (S \cap W) \neq \emptyset$, for every i, we know from Proposition 4.5.8 that $Q_i = K[P_i]$.

We continue with a simple example that can be used to illustrate some of the above features.

Example 4.5.11. Let M be the monoid defined by the presentation $M = \langle x, y \mid x^2 = y^2 \rangle$. We shall see in Example 8.1.3, Theorem 8.5.1 and Theorem 8.5.6, that $K[M]$ is a right and left Noetherian PI domain and it is a maximal order. The abelian group generated by xy, yx is of finite index in the torsion free group G of quotients of M. Every ideal of M contains a power of x^2. Since x^2 is central, every prime ideal of M contains x^2. Using matrix notation, we have

$$M \setminus Mx^2 = \{1\} \cup \begin{bmatrix} xy\langle xy \rangle & \langle xy \rangle x \\ y\langle xy \rangle & yx\langle yx \rangle \end{bmatrix}.$$

Because

$$K[M \setminus Mx^2] = K + \begin{bmatrix} K[xy\langle xy \rangle] & K[\langle xy \rangle x] \\ K[y\langle xy \rangle] & K[yx\langle yx \rangle] \end{bmatrix},$$

the generalized matrix ring techniques easily imply that $K_0[M/Mx^2] = K[M]/K[Mx^2]$ is a prime algebra. Therefore $K[Mx^2]$ is the only height one prime ideal of $K[M]$ intersecting M nontrivially. Also $H = \mathrm{gr}(xy)$ is a maximal subgroup of G not intersecting $P = Mx^2$.

Recall that height one prime ideals of $K[S]$ not intersecting S come from prime ideals in the group algebra $K[G]$. The algebra $K[G]$ is a maximal order. So the results of Brown provides us with a deep insight into the prime spectrum of $K[G]$, see Section 3.6. In particular, every height one prime is generated by a normal element.

We continue with some observations on the classical Krull dimension $\mathrm{clKdim}(K[S])$ of Noetherian algebras $K[S]$ of submonoids of a polycyclic-by-finite group G. Notice that $K[S]$ is a G-graded ring in a natural way. So some information on the classical Krull dimension of $K[S]$ follows from Theorem 3.4.9, in the special case where G is nilpotent-by-finite. Namely, given a chain $P_0 \subset P_1 \subset \cdots \subset P_n$ of prime ideals of $K[S]$ such that $n = 2^{\mathrm{h}(G)}$ ($n = \mathrm{h}(G)+1$ if G is abelian-by-finite), the ideal P_n must contain an element of S. By $\dim(S)$ we denote the supremum of the lengths of chains of prime ideals of S. Recall that the rank $\mathrm{rk}(S)$ of S as the supremum of the ranks of the free abelian subsemigroups of S. Clearly $\mathrm{rk}(S) \leq \mathrm{h}(G) < \infty$. Moreover, from Theorem 3.3.6 we know that $\mathrm{clKdim}\, K[G] = \mathrm{pl}(G) \leq \mathrm{h}(G)$. First we discuss the commutative case.

Lemma 4.5.12. *Let A be a finitely generated abelian monoid with a torsion free group of quotients and such that $\mathrm{U}(A) = \{1\}$. If K is a field then*

$$\mathrm{rk}(A) = \dim(A) = \mathrm{clKdim}(K[A]) \leq |\operatorname{Spec}^0(A)| \leq |\operatorname{Spec}(A)| < \infty.$$

Proof. Let $A = \langle a_1, \ldots, a_r\rangle$. If $P \in \operatorname{Spec}(A)$ then $A \setminus P$ is a submonoid of A and $a_i \in P$ for some i. It follows easily that $A \setminus P$ is generated by a proper subset of $\{a_1, \ldots, a_r\}$, whence $|\operatorname{Spec}(A)| < \infty$.

Let $q \in P$. By Proposition 4.5.8, there exists $J \in \operatorname{Spec}^1(K[A])$ with $q \in J \subseteq K[P]$. Moreover, $J = K[A \cap J]$ and $A \cap J \in \operatorname{Spec}^0(A)$. Hence $P = \bigcup Q$, where the union runs over all $Q \in \operatorname{Spec}^0(A)$ such that $Q \subseteq P$. So every chain of prime ideals of A has length at most $|\operatorname{Spec}^0(A)|$ and we get $\dim(A) \leq |\operatorname{Spec}^0(A)|$.

Next we prove by induction on $n = \dim(A)$ that $\dim(A) = \mathrm{clKdim}(K[A])$. If $n = 0$, then $A = \{1\}$, so this is obvious. Assume that $n > 1$. Then $\operatorname{Spec}^0(A)$ is nonempty. Choose $I \in \operatorname{Spec}^0(A)$ such that $\dim(A) = \dim(A/I) + 1$. Then $K[I] \in \operatorname{Spec}^1(K[A])$ by Proposition 4.5.8. The catenary property of $K[A]$, see Theorem 3.2.5, implies that $\mathrm{clKdim}(K[A]/K[I]) = \mathrm{clKdim}(K[A]) - 1$. Since $K[A]/K[I] \cong K[A \setminus I]$, by the induction hypothesis we obtain $\dim(A \setminus I) = \mathrm{clKdim}(K[A]/K[I])$. This implies that $\mathrm{clKdim}(K[A]) = \dim(A)$.

Since from Theorem 3.5.6 we know that $\mathrm{clKdim}(K[A]) = \mathrm{rk}(A)$, the assertion follows. □

We note that there are many examples of finitely generated abelian cancellative monoids A with $|\operatorname{Spec}^0(A)| > \mathrm{rk}(A)$.

Example 4.5.13. Let G be the free abelian group with free generators x, y, u, w. Consider $B = \langle xz, xw, yz, yw \rangle \subseteq G$. Clearly, $G/\operatorname{gr}(B)$ is an infinite cyclic group. So $\mathrm{rk}(B) = 3$. Moreover, since every prime ideal of B contains one of the generators and $(xz)(yw) = (xw)(yz)$, it is easy to see that $\operatorname{Spec}^0(B) = \{Q_1, Q_2, Q_3, Q_4\}$, where Q_1, Q_2, Q_3, Q_4 are the ideals generated by the sets $\{xz, xw\}, \{xz, yz\}, \{yw, xw\}, \{yw, yz\}$, respectively. It may also be verified that the abelian monoid D given by the presentation $D = \langle a_1, a_2, b_1, b_2 \mid a_1a_2 = b_1b_2 \rangle$ is isomorphic to B. This example was used in [5, page 474], and will be later applied in Section 10.5.

Now we are ready to deal with the general case.

Corollary 4.5.14. *Let S be a cancellative monoid satisfying the ascending chain condition on right ideals such that $G = SS^{-1}$ is polycyclic-by-finite. Let K be a field. Then the following properties hold.*

1. *Every prime ideal Q of $K[S]$ intersecting S contains a prime ideal in the finite set $\operatorname{Spec}^1_h(K[S])$.*

2. $\mathrm{clKdim}(K[S]) = \mathrm{clKdim}(K[F]) + \mathrm{rk}(G/F) = \mathrm{clKdim}(K[\mathrm{U}(S)]) + \mathrm{rk}(G/F) = \mathrm{pl}(\mathrm{U}(S)) + \mathrm{rk}(G/F) \leq \mathrm{h}(G)$, *where F is a normal subgroup of G such that $F \subseteq S$, F is of finite index in $\mathrm{U}(S)$ and G/F is abelian-by-finite.*

3. *If $Q \in \operatorname{Spec}^1_h(K[S])$, then* $\mathrm{clKdim}(K[S]/Q) = \mathrm{clKdim}(K[S]) - 1$ *and* $\mathrm{h}(TT^{-1}) = \mathrm{h}(G) - 1$, *where T is the Ore subsemigroup of S associated to Q in Proposition 4.5.9.*

Proof. Because of Theorem 4.4.7 and in view of Proposition 3.3.3, there exists a torsion free normal subgroup H of finite index in $G = SS^{-1}$ and a normal subgroup F in G such that $F \subseteq S \cap H$ and H/F is torsion free abelian. Moreover, F can be chosen of finite index in $\mathrm{U}(S)$ by Proposition 4.2.2. Hence $(S \cap H)/F$ has trivial units. We know also that $(S \cap H)/F$ inherits the ascending chain condition on right ideals, whence it is a finitely generated monoid.

If Q is a prime ideal of $K[S]$ which intersects S, then by Corollary 4.4.12 the ideal Q intersects the finitely generated monoid $\mathrm{N}(S)$, and hence it contains one of the generators of $\mathrm{N}(S)$, say f. By the Noetherian property, $K[S]$

contains only finitely many prime ideals minimal above $K[Sf]$. Moreover, because of the principal ideal theorem, Theorem 3.2.4, each of these primes is of height at most one and at least one of them is contained in Q. By the comment following Proposition 4.5.8 they are actually of height one. The first statement of the corollary now follows.

Let $A = (S \cap H)/F$. As H is of finite index in G, we have $H/F = \mathrm{gr}(A)$ by Lemma 4.1.2 and it follows that $\mathrm{rk}(A) = \mathrm{rk}(H/F) = \mathrm{rk}(G/F)$. Applying the natural G/H-gradation on $K[S]$, in view of Theorem 3.4.8 we get

$$\mathrm{clKdim}(K[S \cap H]) = \mathrm{clKdim}(K[S]). \tag{4.18}$$

Similarly, because of Theorem 3.3.6, $[\mathrm{U}(S) : F] < \infty$ leads to

$$\mathrm{clKdim}(K[\mathrm{U}(S)]) = \mathrm{clKdim}(K[F]) = \mathrm{pl}(\mathrm{U}(S)). \tag{4.19}$$

From Lemma 4.5.12 we know that $r = \mathrm{rk}(A) = \dim(A)$. So there exists a prime ideal chain $P_1 \subset P_2 \subset \cdots \subset P_r = A \setminus \{1\}$. Let X_i be the inverse image of P_i in $S \cap H$. Clearly each X_i is a prime ideal in $S \cap H$. Since H is torsion free we obtain from Proposition 4.5.8 that each $K[X_i]$ is a prime ideal in $K[S \cap H]$. Since $K[S \cap H]/K[X_r] = K[F]$ and $K[S \cap H]$ is a domain, Theorem 3.3.4, it follows that

$$\mathrm{clKdim}(K[S \cap H]) \geq \mathrm{clKdim}(K[F]) + \mathrm{rk}(A). \tag{4.20}$$

We now prove the converse inequality by induction on $\mathrm{rk}(A)$. If $\mathrm{rk}(A) = 0$, then $A = \{1\}$ and $S \cap H = F$, whence the inequality is obvious. So assume that $\mathrm{rk}(A) > 0$. Let

$$Q_0 \subset Q_1 \subset \cdots \subset Q_n$$

be a chain of prime ideals of $K[S \cap H]$. We consider two cases. First suppose that $Q_n \cap S = \emptyset$. Then, by the comments in the beginning of this section, the ideals $Q_i K[H]$ form a strictly increasing chain of prime ideals in $K[H]$. Consequently $n \leq \mathrm{clKdim}(K[H])$. On the other hand, since H/F is abelian, it follows from the definition of the plinth length $\mathrm{pl}(H)$ of H that $\mathrm{pl}(H) \leq \mathrm{pl}(F) + \mathrm{rk}(H/F)$. As $\mathrm{pl}(H) = \mathrm{clKdim}(K[H])$ and $\mathrm{pl}(F) = \mathrm{clKdim}(K[F])$, by Theorem 3.3.6, we therefore obtain that $n \leq \mathrm{clKdim}(K[F]) + \mathrm{rk}(A)$, as needed. Hence, suppose $Q_n \cap S \neq \emptyset$. We have $S \cap H = \mathrm{N}(S \cap H)$ by Remark 4.5.10. The principal ideal theorem implies that, choosing $a \in Q_n \cap S$, we get

$$n \leq \mathrm{height}(Q_n) \leq \mathrm{height}(Q_n/aK[S \cap H]) + 1.$$

Let $Q \in \operatorname{Spec}(K[S \cap H]$ be such that $a \in Q$ and $\operatorname{clKdim}(K[S \cap H]/Q) = \operatorname{clKdim}(K[S \cap H]/aK[S \cap H])$. (Note that Q is a minimal prime containing the normal element a of $S \cap H$.) Since $K[S \cap H]$ is a domain, we also know from the principal ideal theorem that Q is of height one. But, as $Q \cap S \cap H \neq \emptyset$, we therefore obtain from Proposition 4.5.8 that $Q = K[S \cap H \cap Q]$ and $S \cap H \cap Q \in \operatorname{Spec}^0(S \cap H)$. So

$$n \leq \operatorname{clKdim}(K[S \cap H]/Q) + 1 = \operatorname{clKdim}(K[V]) + 1, \tag{4.21}$$

where $V = (S \cap H) \setminus Q$ is a subsemigroup, because $S \cap H = \mathrm{N}(S \cap H)$. Since $S \cap H \cap Q \in \operatorname{Spec}^0(S \cap H)$, and the ideals of $S \cap H$ and of A are in one-to-one correspondence, we get $V/F \cong A \setminus J$ for the minimal prime ideal J of A which corresponds to $S \cap H \cap Q$. By Lemma 4.5.12 $\operatorname{rk}(A \setminus J) = \operatorname{rk}(A) - 1$, hence $\mathrm{h}(VV^{-1}) = \mathrm{h}(F) + \operatorname{rk}(A) - 1$. The induction hypothesis implies now that $\operatorname{clKdim}(K[V]) \leq \operatorname{clKdim}(K[F]) + \operatorname{rk}(A \setminus J)$. Therefore, by (4.21)

$$n \leq \operatorname{clKdim}(K[F]) + \operatorname{rk}(A \setminus J) + 1 = \operatorname{clKdim}(K[F]) + \operatorname{rk}(A).$$

This proves the required converse inequality. Hence, in view of (4.18) and (4.20), we get

$$\operatorname{clKdim}(K[S]) = \operatorname{clKdim}(K[S \cap H]) = \operatorname{clKdim}(K[F]) + \operatorname{rk}(A).$$

Since $\operatorname{clKdim}(K[F]) + \operatorname{rk}(A) \leq \mathrm{h}(F) + \operatorname{rk}(A) = \mathrm{h}(G)$ and $\operatorname{rk}(A) = \operatorname{rk}(G/F)$, the second statement of the corollary follows from (4.19).

Let $Q \in \operatorname{Spec}_h^1(K[S])$. As explained in Remark 4.5.10, and again using Theorem 3.4.8, there exists a height one prime ideal $Q_1 = K[P_1]$ of $K[S \cap H]$, with $P_1 \in \operatorname{Spec}^0(S \cap H)$, so that $\operatorname{clKdim}(K[S]/Q) = \operatorname{clKdim}(K[S \cap H]/Q_1])$. Since P_1 is a minimal prime containing a normal element, the argument of the preceding paragraph applied to the semigroup $V = (S \cap H) \setminus Q_1$ and using the appropriate ideal J of A, implies that

$$\begin{aligned} \operatorname{clKdim}(K[S \cap H]/Q_1]) &= \operatorname{clKdim}(K[V]) = \operatorname{clKdim}(K[F]) + \operatorname{rk}(A \setminus J) \\ &= \operatorname{clKdim}(K[F]) + \operatorname{rk}(A) - 1 \\ &= \operatorname{clKdim}(K[S]) - 1. \end{aligned}$$

Therefore $\operatorname{clKdim}(K[S]/Q) = \operatorname{clKdim}(K[S]) - 1$.

Finally we prove that $\mathrm{h}(TT^{-1}) = \mathrm{h}(G) - 1$. Again using the notation of Remark 4.5.10, note that $T = T_1$ is an ideal of $V = (S \cap H) \setminus P_1$. Hence T and V have the same group of right quotients. Therefore $\mathrm{h}(TT^{-1}) = \mathrm{h}(VV^{-1})$. We have seen that $\mathrm{h}(VV^{-1}) = \mathrm{h}(F) + \operatorname{rk}(A) - 1$. This implies that $\mathrm{h}(TT^{-1}) = \mathrm{h}(G) - 1$, as desired. □

The following example shows that the strict inequality $\mathrm{clKdim}(K[S]) > \mathrm{clKdim}(K[G])$ is possible in the case described in Corollary 4.5.14.

Example 4.5.15. Let F be a free abelian group of rank 2. Consider the group G which is the semidirect product of F and of the infinite cyclic group $C = \mathrm{gr}(z)$, where the automorphism of $\mathbb{Q}^2$ determined by the action of z on F has eigenvalues α, β such that $\alpha^n, \beta^n \notin \mathbb{Q}$ for every $n \geq 1$. Then $\mathbb{Q}^2$ is an irreducible $\mathbb{Q}[T]$-module for every subgroup T of finite index in G. Therefore, F is a plinth for G. It follows that $\{1\} = G_0 \subseteq F = G_1 \subseteq G = G_2$ is a plinth series for G and we get $\mathrm{clKdim}(K[G]) = \mathrm{pl}(G) = 2$, see Section 3.3. On the other hand, if $S = \langle F, z \rangle$, then $F = \mathrm{U}(S)$ and $\mathrm{clKdim}(K[S]) = \mathrm{clKdim}(K[F]) + \mathrm{rk}(G/F) = 3$, by Corollary 4.5.14.

Proposition 4.5.9 relates the prime images of a Noetherian algebra $K[S]$ to prime images of group algebras of polycyclic-by-finite groups. As an example of an application we give the following extension of a classical group algebra result.

Corollary 4.5.16. *Let S be a cancellative monoid satisfying the ascending chain condition on right ideals such that the group of right quotients $G = SS^{-1}$ is polycyclic-by-finite. If K is an absolute field, then all irreducible right $K[S]$-modules are finite dimensional over K.*

Proof. As in Remark 4.5.10 let W be a normal subgroup of finite index in G such that $\mathrm{N}(S \cap W) = S \cap W$. Let Q be a right primitive ideal of $K[S]$. Consider $K[S]$ as graded by the finite group G/W. Theorem 3.4.8 implies that $Q \cap K[S \cap W] = Q_1 \cap \cdots \cap Q_r$, where the intersection runs through all the prime ideals of $K[S \cap W]$ that are minimal over $Q \cap K[S \cap W]$. By Lemma 4.1.3, $K[S]$ is a finitely generated right module over its identity component $K[S \cap W]$. Let V be an irreducible $K[S]$-module such that $Q = \mathrm{ra}_{K[S]}(V)$. Then $K[S \cap W]/(Q \cap K[S \cap W])$ acts faithfully on V and by Theorem 3.4.7 V is a completely reducible $K[S \cap W]/(Q \cap K[S \cap W])$-module. Since V is cyclic as a $K[S]$-module, we obtain that V is finitely generated as a $K[S \cap W]/(Q \cap K[S \cap W])$-module and hence has finite length as a $K[S \cap W]$-module. By taking the annihilators of the composition factors of this module, we get that there exist right primitive ideals $L_1, \ldots, L_s$ of $K[S \cap W]$ such that $L_1 \cap \cdots \cap L_s = \{0\}$. Since every Q_i is a prime of $K[S \cap W]$ minimal over $Q \cap K[S \cap W]$, it follows that $Q_i = L_j$ for some j. Hence we obtain that $R = K[S \cap W]/Q_i$ is a right primitive ring, for every $1 \leq i \leq r$. Since $T = (S \cap W) \setminus Q_i$ is a normalizing semigroup, see Remark 4.5.10, we know that $R = K[T]/P$ for some right primitive ideal in the semigroup algebra $K[T]$, and $P \cap T = \emptyset$.

We claim that the algebra R is finite dimensional. Since R is a right primitive ring, let L be a maximal right ideal of $K[T]$ so that P is maximal among all ideals of $K[T]$ contained in L. Since T is a normalizing semigroup and $P \cap T = \emptyset$, it follows that $L \cap T = \emptyset$. Hence, in view of Theorem 3.2.6 we obtain that the group algebra $K[TT^{-1}]$ contains the maximal right ideal $LK[TT^{-1}]$ such that $PK[TT^{-1}]$ is a maximal among all ideals of $K[TT^{-1}]$ contained in $LK[TT^{-1}]$. So $K[TT^{-1}]/PK[TT^{-1}]$ is a right primitive algebra, and thus by Theorem 3.3.6 this algebra is finite dimensional. Since $K[T]/P = K[S \cap W]/Q_i$ embeds into this algebra, the claim follows.

The above implies that $K[S \cap W]/(P \cap K[S \cap W])$ is finite dimensional as subdirect product of finitely many finite dimensional algebras. Since $K[S]$ is a finite module over $K[S \cap W]$, the algebra $K[S]/P$ is also finite dimensional, as required. □

4.6 Comments and problems

Important partial steps in the proof of the main structural result, Theorem 4.4.7, were obtained in [79] and [83]. Then the missing case for the general statement was proved in [88]. The present proof is a simplified modification of these original steps. General results on prime ideals of cancellative semigroups come from [125]. Earlier, several results on the prime spectrum of $K[S]$ and of S were proved and applied in the context of semigroup algebras in [77, 78, 79, 83, 84, 92]. They will be extensively used in the next chapters. Much of the remaining material in this chapter comes from [83]. Some other results on the prime ideals of Malcev nilpotent semigroups were also obtained in [76, 80]. A result on $\mathbb{N}$-graded right Noetherian algebras proved in [148] is stronger that what we use in the proof of the main theorem, applying Lemma 4.4.5. Namely, every $\mathbb{N}$-graded right Noetherian algebra $R = \bigoplus_{n \geq 0} R_n$, with $R_0 = K$ and all R_n finite dimensional, has subexponential growth. The symmetry of Theorem 4.4.7 is similar in flavor to that occurring in the class of semiprime right Noetherian PI algebras, see Theorem 3.2.1.

Problems

1. Is the algebra $K[S]$ catenary whenever S is a submonoid of a polycyclic-by-finite group satisfying the ascending chain condition on right ideals? See Theorem 3.3.7 and Theorem 3.2.5.

2. Determine the primitive ideals of Noetherian algebras $K[S]$ for submonoids S of a polycyclic-by-finite group.

Chapter 5

General Noetherian semigroup algebras

In this chapter we prove certain fundamental general results on right Noetherian semigroup algebras $K[S]$. First, we show that in many important cases such algebras are finitely generated. In particular, this extends the observation made in Theorem 4.1.7 for submonoids of polycyclic-by-finite groups. The second main aim is to look for necessary and sufficient conditions on the monoid S for the algebra $K[S]$ to be Noetherian. In this direction we prove two main theorems, with the approach based on finite ideal chains of S of certain special type. Such ideal chains show that cancellative subsemigroups, and uniform semigroups containing them, can be considered, in some sense, as building blocks for the entire S, in a way similar to that occurring in the case of linear semigroups, described in Section 2.4. This feature surprisingly shows up in the study of concrete important motivating classes, considered in Chapter 8 and Chapter 9. As a tool in the proof, it is also shown that monomial monoids over polycyclic-by-finite groups that satisfy the ascending chain condition on right ideals yield Noetherian algebras, again extending the case of submonoids of polycyclic-by-finite groups, Theorem 4.1.6. In the final section, basic information on the Gelfand-Kirillov dimension of Noetherian algebras $K[S]$ is obtained. The main idea is to show that in some sense the dimension of $K[S]$ is determined by the dimensions of subalgebras $K[T]$ for cancellative subsemigroups T of S. The latter dimensions can be computed by applying the results of Gromov and Grigorchuk, together with the formula of Bass, Theorem 3.5.6. As an application of the description of the class of Noetherian semigroup algebras considered in Chapter 4, we prove that right Noetherian algebras of finite

Gelfand-Kirillov dimension that are defined by a homogeneous monoid presentation must satisfy a polynomial identity. This surprising result is another indication of the role of cancellative subsemigroups of S for the properties of any right Noetherian algebra $K[S]$.

5.1 Finite generation of the monoid

From Theorem 4.1.7 we know that $K[S]$ is finitely generated whenever S is a submonoid of a polycyclic-by-finite group such that $K[S]$ is right Noetherian. In view of Theorem 3.5.6, a consequence of this result can be stated in the following form. If S is a cancellative monoid such that $K[S]$ is right Noetherian and the Gelfand-Kirillov dimension $\mathrm{GK}(K[S])$ is finite, then S is finitely generated. In this section we extend this result to arbitrary monoids. More generally, our main aim is to show that S is finitely generated whenever the algebra $K[S]$ is right Noetherian in many other important cases. For example, if $K[S]$ satisfies a polynomial identity or if $K[S]$ also is left Noetherian. However, the general case still remains open.

Proposition 5.1.1. *Assume that S is a submonoid of the full linear monoid $\mathrm{M}_n(D)$ over a division ring D. If S satisfies the ascending chain condition on right ideals, then S intersects finitely many $\mathcal{R}$-classes of $\mathrm{M}_n(D)$ (so S has finitely many uniform components and thus finitely many $\mathcal{J}$-classes containing an idempotent) and every 0-simple ideal of S is completely 0-simple.*

Proof. Suppose that S intersects infinitely many $\mathcal{R}$-classes of the monoid $\mathrm{M}_n(D)$. Then it intersects infinitely many such classes, say $R_1, R_2, \ldots$, contained in certain $\mathrm{M}_j \setminus \mathrm{M}_{j-1}, j \geq 1$, where M_j is the ideal of the monoid $\mathrm{M}_n(D)$ consisting of all matrices of rank at most j. Now $\bigcup_{i=1}^{k}(S \cap R_i) \cup (S \cap \mathrm{M}_{j-1}), k = 1, 2, \ldots$, forms an increasing chain of right ideals of S. This contradiction, in view of Theorem 2.4.3 and Corollary 2.4.4, proves the first assertion.

Let I be a 0-simple ideal of S. If S has a zero element θ (notice that this does not imply $\theta = 0$), then define $N_I = \{s \in I \mid s \neq \theta\}$. If S has no zero element then let $I = N_I$. From Theorem 2.3.4 it follows that $N_I \subseteq \mathrm{M}_j \setminus \mathrm{M}_{j-1}$ for some j. Choose any $s \in N_I$. Write $s_1 = s$. Then $s = s_2 s t_1$ for some $s_2, t_1 \in I$. In this way we may construct a sequence $s_1, s_2, \ldots$ of elements of I such that $s_m = s_{m+1} s_m t_m$ for some $t_m \in I$. Clearly $s_m S \subseteq s_{m+1} S$ for every $m \geq 1$, so that $s_m \in N_I$. The ascending chain condition on right ideals implies that $s_q S = s_{q+1} S$ for some $q \geq 1$. Then there exists $z \in S$ such that $s_{q+1} = s_q z$ and therefore $s_q = s_{q+1} s_q t_q =$

$s_q z s_q t_q$. Thus $u = z s_q t_q \in N_I$ and the equality $s_q = s_q u$ can be considered in the completely 0-simple semigroup $\mathrm{M}_j / \mathrm{M}_{j-1}$. It then follows that u is a nonzero idempotent in $\mathrm{M}_j / \mathrm{M}_{j-1}$, see Lemma 2.1.1. Therefore, u treated as an element of I is a nonzero idempotent. As I is 0-simple and cannot contain infinite chains of idempotents, it must contain a primitive idempotent, and hence it must be be completely 0-simple. □

Notice that it may be shown that, if S is any monoid such that $K[S]$ is right Noetherian, then every 0-simple principal factor of S is completely 0-simple, [119, Proposition 12.1].

Let R be a semiprime right Goldie K-algebra generated by a multiplicative subsemigroup S. By Theorem 3.1.5, the classical ring of right quotients $\mathrm{Q}_{cl}(R)$ embeds into $\mathrm{M}_n(D)$, for some $n \geq 1$ and a division algebra D; and we may thus consider S as a subsemigroup of $\mathrm{M}_n(D)$. If S has a zero element θ, then we may assume that $\theta = 0$ in $\mathrm{Q}_{cl}(R)$, see Proposition 2.4.9. Moreover, since R is semiprime and nil subsemigroups of $\mathrm{Q}_{cl}(R)$ are nilpotent by Proposition 2.4.9, S has no nonzero nil ideals. It then follows from the structure theorem of linear semigroups, Theorem 2.4.3, that S has an ideal I contained in a completely 0-simple subsemigroup J of $\mathrm{M}_n(D)$ in such a way that I is uniform in J and $J = \widehat{I}$. Furthermore, if j denotes the minimal rank of nonzero matrices in S, then the set $\{a \in S \mid \mathrm{rank}(a) \leq j\}$ is a 0-disjoint union of some ideals I of this type. Every such ideal I is an ideal uniform component of S. Moreover, if R is prime then we may assume that $\mathrm{Q}_{cl}(R) = \mathrm{M}_n(D)$ and the nonzero elements of I are the elements of minimal nonzero rank (as matrices) of S in $\mathrm{M}_n(D)$. In particular, the following lemma follows from Lemma 2.5.2.

Lemma 5.1.2. *Let Q be a prime ideal of a semigroup algebra $K[S]$ such that $K[S]/Q$ is a right Goldie ring. Then the image S' of S in $K[S]/Q$ has a unique ideal uniform component I. Furthermore, if $\widehat{I}$ is the completely 0-simple closure of I, then $\widehat{S'} = S' \cup \widehat{I}$ is a semigroup in which $\widehat{I}$ is an ideal.*

We now prove a technical observation which plays an important role in the proofs of the main results of this section. It has a flavor similar to that of Proposition 4.5.9. Notice that $K[S] \cong K_0[S^0]$ for every semigroup S, so the following applies also to semiprime ideals of ordinary semigroup algebras $K[S]$.

Proposition 5.1.3. *Assume that P is a semiprime ideal of a right Noetherian algebra $K_0[S]$ of a semigroup S with a zero element. Let $\overline{S}$ be the image of S under the natural natural homomorphism $K_0[S] \longrightarrow K_0[S]/P$. Assume* *that* $\overline{I}$ *is an ideal uniform component of* $\overline{S}$ *(viewed as a linear semigroup) and let* $\overline{T} = G \cap \overline{S} = G \cap \overline{I}$*, where* G *is a maximal subgroup of the completely* 0*-simple closure* $\widehat{I}$ *of* $\overline{I}$*. Define* I, T *as the inverse images of* $\overline{I}, \overline{T}$ *in* S*.*

1. *If* $R = \{a \in S \mid Ta \subseteq T\}$*, then* R *is a submonoid of* S *such that* $K[R]$ *is a direct summand of the left* $K[R]$*-module* $K[S]$*, whence* $K[R]$ *is right Noetherian. Moreover,* T *is a right ideal of* R *and it is an ideal of* R *if* $\widehat{I}$ *is a Brandt semigroup.*

2. *If* $\overline{Z} = e\overline{R}$ *for the idempotent* $e \in G$*, then* $\overline{T} \subseteq \overline{Z} \subseteq G$*,* G *is the group of right quotients of* $\overline{T}$*, and* $K[\overline{Z}]$ *is a right Noetherian homomorphic image of* $K[R]$*; in particular,* G *is finitely generated.*

3. *If* $K[\overline{T}]$ *is a PI algebra, then* $K_0[\widehat{I}]$ *is a PI algebra, and if* P *is a prime ideal of* $K_0[S]$*, then* $K_0[S]/P$ *also is a PI algebra.*

Proof. It is clear that R is a submonoid of S. Assume that $ax \in R$ and $a \in R$ for some $x \in S$. Then $Tax \subseteq T$ and $Ta \subseteq T$. Hence $bx \in T$ for some $b \in T$. Let $\overline{d}$ be the inverse of the image $\overline{b} \in \overline{T}$ of b in the maximal subgroup G of $\widehat{I}$ containing $\overline{T}$. Then $\overline{T}\overline{x} = \overline{T}\overline{d}\overline{b}\overline{x} \subseteq G \cap \overline{S} = \overline{T}$. It follows that $Tx \subseteq T$, whence $x \in R$. Therefore $K[R]$ is a direct summand of the left $K[R]$-module $K[S]$. Since $K[S]$ is right Noetherian, also $K[R]$ is right Noetherian, see Lemma 4.1.1. It is clear that T is a right ideal of R. On the other hand, if $t \in T$ and $a \in R$ then $at \in I$ and $Tat \subseteq T$. Thus $\overline{T}\overline{at} \subseteq \overline{T}$. The latter condition implies that $\overline{at}\mathcal{L}\overline{t}$ in $\widehat{I}$. Moreover, if $\widehat{I}$ is a Brandt semigroup, we must have $\overline{at} \in G$, so that $\overline{at} \in G \cap \overline{S} = \overline{T}$. Hence $at \in T$ and it follows that T is an ideal of R, which completes the proof of assertion (1).

If $y \in \overline{R}$ then $\overline{T}ey = \overline{T}y \subseteq \overline{T}$. Since $ey \in \widehat{I}$, this implies that $ey\mathcal{L}e$ in $\widehat{I}$. As we also have $ey\mathcal{R}e$ in $\widehat{I}$, it follows that $ey \in G$. Hence the map $\overline{R} \longrightarrow G$ defined by $y \mapsto ey$ is a semigroup homomorphism. In view of (1), it follows that $K[\overline{Z}]$ is right Noetherian. Clearly, we have $\overline{T} = e\overline{T} \subseteq \overline{Z} \subseteq G$ and Proposition 2.2.1 implies that $\mathrm{gr}(\overline{T}) = G$. If $a, b, c \in \overline{T}$, then $c\overline{Z} = ce\overline{R} = c\overline{R} \subseteq \overline{T}$. By Lemma 4.1.5 we get $a^2\overline{Z} \cap b^2\overline{Z} \neq \emptyset$ and hence $\emptyset \neq aa\overline{Z} \cap bb\overline{Z} \subseteq a\overline{T} \cap b\overline{T}$. It follows that $\overline{T}$ is a right Ore semigroup. Hence G is the group of right quotients of $\overline{T}$, so also of $\overline{Z}$, and thus $K[G]$ is right Noetherian, as a localization of the right Noetherian algebra $K[\overline{Z}]$. In particular, G is finitely generated. Hence we have proved assertion (2).

If $K[\overline{T}]$ satisfies a polynomial identity, then so does $K[G]$ by Theorem 3.1.9. From Proposition 2.5.6 it then follows in view of Proposition 5.1.1 that $K_0[\widehat{I}]$ is a PI algebra. Since $K_0[I]/(P \cap K_0[I])$ is isomorphic to the image

$K\{\overline{I}\}$ of $K_0[I]$ in $K_0[S]/P$, it is a homomorphic image of $K_0[\overline{I}]$. Therefore it is also a PI algebra. If moreover $K_0[S]/P$ is prime, then it also must be a PI algebra because so is its nonzero ideal $K\{\overline{I}\}$. This completes the proof. □

Lemma 5.1.4. *Let A be a right Noetherian K-algebra with unity and let J be a nilpotent ideal of A. If A/J is finitely generated then A is finitely generated.*

Proof. We proceed by induction on the nilpotency index of J. Clearly, it is enough to consider the case where $J^2 = \{0\}$. By the hypothesis $J = \sum_{i=1}^{n} a_i A$ for some $a_i \in A, n \geq 1$. Let $\{b_1, \ldots, b_m\}$ be a subset of A mapping onto a finite set of generators of the algebra A/J under the natural homomorphism $A \longrightarrow A/J$. We claim that A is generated by the set $\{a_1, \ldots, a_n, b_1, \ldots, b_m\}$. Let B be the subalgebra generated by $\{b_1, \ldots, b_m\}$. If $a \in A$ then $a = b + j$ for some $b \in B$ and some $j \in J$. Hence $a = b + \sum_{i=1}^{k} a_i c_i$, for some $k \geq 1$ and $c_i \in A$. Every c_i may be written in the form $c_i = d_i + j_i$, where $d_i \in B$ and $j_i \in J$. Hence $a = b + \sum_i a_i(d_i + j_i) = b + \sum_i a_i d_i$ lies in the subalgebra of A generated by $\{a_1, \ldots, a_n, b_1, \ldots, b_m\}$, as desired. □

We are now ready to prove the first main result of this section.

Theorem 5.1.5. *Let K be a field and let S be a monoid such that $K[S]$ is right Noetherian. Then S is finitely generated in any of the following cases:*

1. *S satisfies the ascending chain condition on principal left ideals,*
2. *every cancellative subsemigroup of every homomorphic image of S has a locally (nilpotent-by-finite) group of quotients,*
3. *S is Malcev nilpotent,*
4. *the Gelfand-Kirillov dimension of $K[S]$ is finite,*
5. *$K[S]$ satisfies a polynomial identity.*

Proof. Since $K[S^0]$ is right Noetherian if and only if $K[S]$ is right Noetherian, we may assume that S has a zero element θ. Suppose S is not finitely generated. We will show that this leads to a contradiction. Since $K[S]$ is right Noetherian, S satisfies the ascending chain condition on congruences. Hence there exists a congruence ρ in S that is maximal with respect to the property that S/ρ is not finitely generated. Notice that every homomorphic image of S inherits the hypotheses on S. Hence, replacing S by S/ρ, we may assume that every proper homomorphic image of S is finitely generated. Because the prime radical $\mathcal{B}(K_0[S])$ of $K_0[S]$ is nilpotent and because an algebra is finitely generated if and only if it is finitely generated modulo the prime radical, by Lemma 5.1.4, it is enough to show that the image S' of S in $K_0[S]/\mathcal{B}(K_0[S])$ is finitely generated. Hence, by the comment preceding Lemma 5.1.2, we may assume that $S = S'$, and thus S is a subsemigroup of a simple Artinian ring $Q = \mathrm{M}_l(D)$, for a division ring D. Moreover $\theta = 0$ in $\mathrm{M}_l(D)$.

First consider the case where S has a minimal nonzero ideal M. Then M is not nilpotent, because otherwise $K_0[S]$ has a nonzero nilpotent ideal and Lemma 5.1.4 leads to a contradiction, since every proper homomorphic image of S is finitely generated. Hence M must be 0-simple and Proposition 5.1.1 implies that M must be completely 0-simple. We identify M with a Rees matrix semigroup $\mathcal{M}(H, X, Y; P)$. Let μ be the relation on S defined by: $(s,t) \in \mu$ if $s = t$ or $s,t \in M$ and s,t are in the same $\mathcal{H}$-class of M. It is easy to see that μ is a congruence on S. Suppose first that μ is trivial. Then H is a trivial group. Let L be the left annihilator of the algebra $K_0[M]$ and let ρ_L be the congruence on S defined by $(s,t) \in \rho_L$ if $s = t$ or $s,t \in M$ and $s - t \in L$. Then L is an ideal in $K_0[S]$ and $L^2 = \{0\}$, whence ρ_L must be trivial by Lemma 5.1.4. Thus, Proposition 2.5.6 yields an embedding $\phi : M \longrightarrow \mathrm{M}_{|X|}(\{0,1\})$. Since X is a finite set by the ascending condition on right ideals in S, it follows that M is finite. As S/M is finitely generated, also S is finitely generated.

Thus, we may assume that μ is not trivial. Then S/μ is finitely generated. Let $\{s_1, \ldots, s_p\} \subseteq S$ be a subset that maps onto a generating set of S/μ. If $0 \neq e = e^2 \in M$ then $eSe = eMe \cong H \cup \{0\}$. Moreover, H is a finitely generated group by Proposition 5.1.3. Let $t_1, \ldots, t_r \in eSe$ be such that $eSe = \langle t_1, \ldots, t_r \rangle$. We claim that $S = \langle s_1, \ldots, s_p, t_1, \ldots, t_r \rangle$. Clearly, since μ is trivial on $S \setminus M$, we get $S \setminus M \subseteq \langle s_1, \ldots, s_p \rangle$. Take $0 \neq u \in M$. Let $u = (g, j, m)$ where $g \in H, j \in X, m \in Y$. Then there exist $a, b \in M$ such that $aeb = u$ because M is 0-simple. Moreover, by the definition of μ, there exist $a', b' \in M$ such that $(a, a') \in \mu$ and $(b, b') \in \mu$ and $a', b' \in \langle s_1, \ldots, s_p \rangle$. Let $a' = (h, j', n), b' = (f, l, m')$ with $h, f \in H, j', l \in X, n, m' \in Y$. From $0 \neq u = aeb$ it follows that $j' = j$ and $m' = m$ because $(a, a') \in \mu$ and $(b, b') \in \mu$. If $e = (k, i, q)$, then the fact that $0 \neq u = aeb$ implies also that $p_{ni}, p_{ql} \neq 0$. So we may define $z = (p_{ni}^{-1} h^{-1} g f^{-1} p_{ql}^{-1}, i, q) \in M$. Then

$$a'zb' = (hp_{ni}p_{ni}^{-1}h^{-1}gf^{-1}p_{ql}^{-1}p_{ql}f, j, m) = (g, j, m) = u$$

and $z \in eSe$ because $eSe = \{(v, i, q) \mid v \in H\} \cup \{0\}$. Now $a'zb' \in \langle s_1, \ldots, s_p, t_1, \ldots, t_r \rangle$ and we get $M \subseteq \langle s_1, \ldots, s_p, t_1, \ldots, t_r \rangle$, since $u \in M$ is arbitrary. This proves the claim, contradicting the supposition that S is not finitely generated, and finishing the proof in this case.

So we may suppose that S has no minimal nonzero ideal. Let I be an ideal uniform component of S. Since $K[S]$ is right Noetherian, from Proposition 5.1.1 we know that S has finitely many $\mathcal{J}$-classes that contain a nonzero idempotent. If I contains a nonzero idempotent, then let $w \in I$ be such that SwS is minimal among the ideals SxS generated by a nonzero idempotent $x \in I$. Otherwise, choose any nonzero $w \in I$. Since, by assumption, S has no nonzero minimal ideals, the set $J = \{y \in SwS \mid SyS \neq SwS\}$ is a nonzero ideal of S. Clearly $J^2 \neq \{0\}$, because S has no nonzero nilpotent ideals. Notice that J is an ideal of I, and it follows easily that J is also a uniform subsemigroup of the completely 0-simple closure $\widehat{I}$ (defined in Q) of I, that is, J intersects all $\mathcal{H}$-classes of $\widehat{I}$. Moreover, if G is a maximal subgroup of $\widehat{I}$, then $J \cap G$ is an ideal of $I \cap G$. Hence, $\mathrm{gr}(J \cap G) = \mathrm{gr}(I \cap G) = G$, so that $\widehat{I}$ is also the completely 0-simple closure of J, see Proposition 2.2.1. By the choice of w, the ideal J has no nonzero idempotents. We will now show that $S = \langle S \setminus J^2 \rangle$. Then, as every proper homomorphic image of S is finitely generated, this implies that the monoid S/J^2 is finitely generated, and thus also S is finitely generated by Lemma 5.1.4; a contradiction completing the proof of the theorem.

Since $K[S]$ is right Noetherian, $J^2 = \bigcup_{i=1}^{m} w_i J$ for some $m \geq 1$ and $w_i \in J, w_i \neq 0$. Clearly, we may assume that all $w_i J$ are different and they are maximal among the right ideals of S of the form xJ, with $x \in J$. If $w_i \in J^2$ for some i, then there exists j such that $w_i \subseteq w_j J$. Then $w_i J = w_j J$ by the maximality of $w_i J$, and so $i = j$. Hence, $w_i = w_i s$ for some $s \in J \subseteq \widehat{I}$. Lemma 2.1.1 implies that s is an idempotent. Since J does not contain nonzero idempotents, we obtain that $w_i \in J \setminus J^2$ for every i. Put $W = \{w_1, \ldots, w_m\}$. To prove that S is generated by $S \setminus J^2$, it is sufficient to show that every element $u_1 \in J^2$ belongs to $\langle S \setminus J^2 \rangle$. So let $u_1 \in J^2$. Write $u_1 = t_2 u_2$, where $t_2 \in W$ and $u_2 \in J$. If $u_2 \in J^2$, then, similarly, $u_2 = t_3 u_3$, where $t_3 \in W$ and $u_3 \in J$. This procedure may be continued, so that either there exists $k \geq 1$ such that $u_1 = t_2 t_3 \cdots t_k u_k$ for some $t_i \in W$ and $u_k \in J \setminus J^2$, or for every $n \geq 1$ we have $u_1 = t_2 t_3 \cdots t_n u_n$, where $t_i \in W$ and $u_n \in J^2$. In the former case $u_1 \in \langle S \setminus J^2 \rangle$, as required. In the remainder of the proof we show that the latter case leads to a contradiction.

Notice that $Su_1 \subseteq Su_2 \subseteq \cdots$ is a chain of principal left ideals of S. Suppose that $Su_n = Su_{n+1}$ for some n. Then there exists $s \in S$ such that $su_n = u_{n+1}$, which implies that $u_n = t_{n+1} u_{n+1} = t_{n+1} s u_n$. Since $t_{n+1} s \in J \subseteq \widehat{I}$, it must be an idempotent, again by Lemma 2.1.1. This contradicts our assumption on J. In particular, $Su_1 \subset Su_2 \subset \cdots$ is a strictly increasing sequence. This completes the proof in case condition (1) is assumed.

Next, assume that condition (2) is satisfied. As before, we deal with the elements $u_i, t_{i+1} \in J, i = 1, 2, \ldots$, satisfying

$$u_i = t_{i+1}u_{i+1} \tag{5.1}$$

and such that the set $W' = \{t_i \mid i \geq 1\}$ is finite. Notice that every nonzero $\mathcal{R}$-class of $\widehat{I}$ intersects I nontrivially, so that the ascending chain condition on right ideals in S implies that $\widehat{I}$ has finitely many $\mathcal{R}$-classes. So, replacing the sequence (5.1) by some subsequence, we may assume that each u_i belongs to the same $\mathcal{R}$-class of $\widehat{I}$ and, if u_i and u_j are two consecutive terms in this sequence, then

$$u_i = t_{i+1}t_{i+2} \cdots t_j u_j.$$

It then follows that these elements u_i must be in the same $\mathcal{H}$-class of $\widehat{I}$. Notice also that the corresponding element $t_{i+1} \cdots t_j$ belongs to the same $\mathcal{R}$-class of $\widehat{I}$ as the elements t_{i+1} and u_i. Since all the elements t_j belong to the finite set W', we can refine the sequence further so that all the involved t_j are equal. So all products $t_{i+1}t_{i+2} \cdots t_j$ are also in a fixed $\mathcal{H}$-class of $\widehat{I}$. Because $\widehat{I}$ is the completely 0-simple closure of J, there exists $v \in J$ such that the refined sequence u_i multiplied on the right by v yields us a situation where our subsequence and all products $t'_{i+1} = t_{i+1} \cdots t_j$ (coming from the members of this subsequence) are in the same $\mathcal{H}$-class. Then this class must be a group, which we denote by G. Let $T = G \cap I$. We have constructed elements $u'_i = t'_{i+1}u'_{i+1}, i = 1, 2, \ldots$, with $u'_i, t'_i \in T$. For simplicity we denote these elements again by u_i and t_i, respectively.

Consider the submonoid Z of G defined as in Proposition 5.1.3, with respect to the ideal $\mathcal{B}(K_0[S])$, which is semiprime. So $Z = eR$, where $R = \{a \in S \mid Ta \subseteq T\}$ and e is the identity of G. From Proposition 5.1.3 it follows that G is finitely generated and Z satisfies the ascending chain condition on right ideals. Hence, if condition (2) is assumed, then G is nilpotent-by-finite. Therefore, Theorem 4.4.7 implies that Z satisfies the ascending chain condition on left ideals. Since $t_i \in T \subseteq Z$, we have $Zu_1 \subseteq Zu_2 \subseteq \cdots$ and this chain must stabilize. Thus, there exists $q \geq 1$ such that $Zu_q = Zu_{q+1}$. Hence $u_{q+1} = vu_q$ for some $v \in Z$. Then $u_q = t_{q+1}u_{q+1} = t_{q+1}vu_q$ and it follows that $t_{q+1}v = e$ because all considered elements are in G. Since $t_{q+1}v \in TZ \subseteq T$, this again contradicts the fact that J has no nonzero idempotents. This completes the proof of the theorem in case condition (2) is assumed.

If S is Malcev nilpotent, then G is a nilpotent group by Theorem 3.5.9. So case (2) yields the result if condition (3) holds.

Notice that, by Theorem 3.5.6, condition (2) is a consequence of (4).

Finally, assume that (5) holds. If T and G are as in the proof, then G is a group of quotients of T and G is finitely generated. Consequently, Theorem 3.1.9 implies that G is abelian-by-finite. So the assertion also follows from case (2). This proves the theorem. □

The assertion of the theorem also can be proved if S is a weakly periodic monoid, that is, for every $a \in S$ there exists $n \geq 1$ such that Sa^nS is an idempotent ideal of S, [64] or [119, Theorem 12.6]. In particular, this class includes periodic monoids and regular monoids.

Let $T \subseteq Z \subseteq G$ be as in the proof of Theorem 5.1.5. Then T satisfies the ascending chain condition on right ideals of the form AT, where A is a subset of T. This is because each AT is a right ideal in Z and $K[Z]$ is right Noetherian. Now, suppose that condition (2) of Theorem 5.1.5 is satisfied. If T is a monoid, so that $Z = T$ and $K[T]$ is right Noetherian, then we know its structure from Theorem 4.4.7. Otherwise, we claim that T is the semigroup generated by $T \setminus T^2$. To prove the latter, one proceeds as in the proof of Theorem 5.1.5 by decomposing any given $u_1 \in T$ in the form $u_1 = t_2 \cdots t_j u_j$ for $j = 2, 3, \ldots$, with $u_j, t_j \in T$ and t_jT maximal among all right ideals xT with $x \in T$. These t_j can be chosen from a finite subset of $T \setminus T^2$. If the chain of these decompositions does not terminate, then as in the proof of the theorem we get a contradiction. Therefore, $T \setminus T^2$ indeed generates T.

By Theorem 3.5.2, the Gelfand-Kirillov dimension of a finitely generated PI algebra is finite. We now prove a partial converse for finitely generated monoids defined by homogeneous relations. These are monoids with a presentation $S = \langle X; R \rangle$, where R is a set of defining relations in the generating set X, such that there exists a function $\ell : \mathrm{FM}_X \longrightarrow \mathbb{N}$ defined on the free monoid FM_X on X, such that

1. $\ell(w) = 0$ if and only if $w = 1$, and $\ell(uw) = \ell(u) + \ell(w)$ for all $u, w \in \mathrm{FM}_X$,
2. every relation in R is of the form $u = w$, with $u, w \in \mathrm{FM}_X$ satisfying $\ell(u) = \ell(w)$.

Clearly, in such a monoid we have a natural degree function given by $s \mapsto |s|$, where $|s| = \ell(w)$ for any inverse image $w \in \mathrm{FM}_X$ of $s \in S$. Important classes of such algebras, with $|s|$ defined as the length of $s \in S$ as a word in the corresponding generators of S, will be studied in Chapter 8 and Chapter 9.

Our second main result in this section reads as follows.

Theorem 5.1.6. *Let K be a field and let S be a monoid such that the algebra $K[S]$ is right Noetherian and $\mathrm{GK}(K[S]) < \infty$. Then S is finitely generated and if S has a monoid presentation of the form*

$$S = \langle x_1, \ldots, x_n; R\rangle$$

with R a set of homogeneous relations, then $K[S]$ satisfies a polynomial identity.

Proof. From Theorem 5.1.5 we know that S is finitely generated. So assume S has a homogeneous monoid presentation $S = \langle x_1, \ldots, x_n; R\rangle$. It is clear that the unit group $\mathrm{U}(S)$ is trivial. Let $W = S^0$, the monoid S with zero θ adjoined. We define a congruence ρ on W to be homogeneous if $(s,t) \in \rho$ and $(s,\theta) \notin \rho$ imply that $|s| = |t|$.

The contracted semigroup algebra $K_0[W]$ may be identified with $K[S]$. Suppose that $K_0[W]$ is not a PI algebra. Then, by the Noetherian condition, there exists a maximal homogeneous congruence η on W such that $K_0[W/\eta]$ does not satisfy a polynomial identity. So, replacing W by W/η, we may assume that every proper homogeneous homomorphic image of W yields a PI algebra.

Since there are only finitely many minimal prime ideals of $K_0[W]$ and the prime radical $\mathcal{B}(K_0[W])$ is nilpotent, there exists a minimal prime P such that $K_0[W]/P$ is not a PI algebra. Notice that $K_0[W]$ can be considered in a natural way as a $\mathbb{Z}$-graded algebra. Namely, the degree of a nonzero element $w \in W$ is defined as the value of the length function on W inherited from S. Hence, from Theorem 3.4.2 we know that P is a homogeneous ideal of $K_0[W]$, with respect to this gradation. Let ρ_P be the congruence on W determined by P. In other words, $(s,t) \in \rho_P$ if $s - t \in P$, for $s,t \in W$. Then ρ_P is a homogeneous congruence. Since $K_0[W]/P$ is a homomorphic image of $K_0[W/\rho_P]$, and because of the preceding paragraph of the proof, we get that $W = W/\rho_P$. As $K_0[W]$ is right Noetherian, we thus get natural embeddings

$$W \subseteq K_0[W]/P \subseteq \mathrm{M}_t(D)$$

for some division algebra D and some $t \geq 1$, where $\mathrm{M}_t(D) = \mathrm{Q}_{cl}(K_0[W]/P)$, the classical ring of right quotients of $K_0[W]/P$. Let I be an ideal uniform component of W viewed as a linear semigroup, see Lemma 5.1.2. Let G be a maximal subgroup of the completely 0-simple closure $\widehat{I}$ of I in $\mathrm{M}_t(D)$ and let $T = G \cap I \subseteq Z \subseteq G$ be as in Proposition 5.1.3. Recall that

$$Z = \{ex \mid x \in W, Tx \subseteq T\},$$

where e is the identity of G. From Proposition 5.1.3 it then follows that $K[T]$ is not a PI algebra and G is a finitely generated group. In particular, since G is generated by T, there exists a finitely generated subsemigroup $T_0 \subseteq T$ such that $G = \mathrm{gr}(T_0)$. Since $\mathrm{GK}(K_0[W]) < \infty$ by the hypothesis on $K[S]$, we also have $\mathrm{GK}(K[T_0]) < \infty$. From Theorem 3.5.6 it follows that $G = TT^{-1}$ is nilpotent-by-finite.

Next we claim that the group of units $\mathrm{U}(Z)$ of Z is a periodic group. For this, suppose $g, g^{-1} \in Z$. Then $Tg \subseteq T$ and $Tg^{-1} \subseteq T$ by the definition of Z. So $Tg = T$. Since $G = TT^{-1}$, we may write $g = ab^{-1}$ with $a, b \in T$. Then $Ta = Tb$. Let $M \subseteq T$ denote the subset consisting of the elements of minimal length in T, with respect to the length function inherited from S. Then a and b have equal length and $Ma = Mb$. Clearly $Mg = M$. As M is finite, it follows that there exists $k \geq 1$ such that the permutation of M determined by the right multiplication by g^k is trivial. This means that $g^k = e$, which proves the claim.

So $\mathrm{U}(Z)$ is a periodic subgroup of the finitely generated nilpotent-by-finite group G. Hence $\mathrm{U}(Z)$ is finite by Proposition 3.3.3. Also Z satisfies the ascending chain condition on right ideals by Proposition 5.1.3 and $ZZ^{-1} = G$. Therefore, from Theorem 4.4.7 it follows that G is finite-by-abelian-by-finite. As G is finitely generated, it must be abelian-by-finite and thus Theorem 3.1.9 implies that $K[G]$ is a PI algebra. Then $K[T]$ is a PI algebra, a contradiction. This completes the proof of the theorem. □

Notice that, in view of Theorem 3.5.2, algebras of the type considered in Theorem 5.1.6 are embeddable in matrix rings over a field.

The above proof shows also that the theorem remains valid in the more general situation of an algebra generated by a set X subject to a system R of relations of the form $u = w$, where $u, w \in \mathrm{FM}_X$ are such that $\ell(u) = \ell(w)$ or of the form $u = 0$, where $u \in \mathrm{FM}_X$, for a function $\ell : \mathrm{FM}_X \longrightarrow \mathbb{N}$ satisfying: $\ell(w) = 0$ if and only if $w = 1$, and $\ell(uw) = \ell(u) + \ell(w)$ for all $u, w \in \mathrm{FM}_X$.

5.2 Necessary conditions

A natural question, which might seem hopeless in full generality, is whether one can characterize the Noetherian property of a semigroup algebra $K[S]$ in terms of the underlying semigroup S. In cases it is possible, this usually provides us with new tools in the study of the structure of $K[S]$. In particular, if $K[S]$ is commutative, this is equivalent to $K[S]$ being finitely generated, Theorem 4.0.5. A very satisfactory solution was obtained for any submonoid S of a polycyclic-by-finite group in Theorem 4.4.7. Such algebras also turn out to be finitely presented. Some other special important classes of Noetherian algebras will be described in Chapter 6, Chapter 8 and Chapter 9.

By Theorem 3.5.2, every finitely generated right Noetherian algebra R satisfying a polynomial identity embeds into the matrix ring $\mathrm{M}_n(F)$ over a field F. In view of Theorem 5.1.5 this applies to every right Noetherian PI algebra $R = K[S]$. Therefore, in this case S embeds into the multiplicative semigroup of $\mathrm{M}_n(F)$ and its structure can be approached through the theory of linear semigroups, see Theorem 2.4.3. In particular, S has a finite ideal chain of a very special nature, namely, a chain with factors that are nilpotent or uniform, Corollary 2.4.7. One of our main aims in this section is to show that the latter is often the case if we assume that $K[S]$ is Noetherian.

Recall that, if X is a subset of an algebra R, then by $\mathrm{ra}_X(a), \mathrm{la}_X(a)$ we denote the right, respectively left, annihilator of an element $a \in R$ in X.

Lemma 5.2.1. *Assume that a semigroup algebra $K[S]$ satisfies the ascending chain condition on right annihilators. Then, for every subsemigroup T of S, there exists $u \in T$ such that for every $a, b \in uT$ and $x \in T$ the equality $xa = xb$ implies that $a = b$.*

Proof. Let $u \in T$ be such that $\mathrm{ra}_{K[S]}(u)$ is maximal in the family of all right ideals of $K[S]$ of the form $\mathrm{ra}_{K[S]}(x)$ for $x \in T$. Suppose that $xu(s - t) = 0$ for some $s, t \in S$ and $x \in T$. Since $xu \in T$ and we always have $\mathrm{ra}_{K[S]}(xu) \supseteq \mathrm{ra}_{K[S]}(u)$, it follows that $\mathrm{ra}_{K[S]}(xu) = \mathrm{ra}_{K[S]}(u)$. This implies that $u(s - t) = 0$, proving the assertion. □

The first indication of the role of cancellative subsemigroups in monoids S with a right Noetherian algebra $K[S]$ comes from the following result.

Lemma 5.2.2. *Let S be a semigroup with no free nonabelian subsemigroups. Assume that a semigroup algebra $K[S]$ satisfies the ascending chain condition on right annihilator ideals. Let T' be a cancellative subsemigroup of the image S' of S in the algebra $K[S]/N$ for a nilpotent ideal N of $K[S]$ and let T be the inverse image of T' in S.*

1. *There exists $y \in T$ such that yTy is a cancellative semigroup.*

2. *If every two right ideals of T intersect nontrivially, then the relation ρ defined on T by the rule: $(s, t) \in \rho$ if $sx = tx$ for some $x \in T$ is the least right cancellative congruence on T.*

Proof. From Lemma 5.2.1 we know that there exists $u \in T$ such that uT is left cancellative. Hence, by Lemma 2.1.3, every two right ideals of uT intersect nontrivially. Write $U = uT$. Let τ be the relation on U defined by the rule: $(s,t) \in \tau$ if $\mathrm{ra}_U(s-t) \neq \emptyset$. It is clear that τ is reflexive and symmetric. If $\mathrm{ra}_U(s-t) \neq \emptyset$ and $\mathrm{ra}_U(t-u) \neq \emptyset$ for some $s,t,u \in U$ then $\mathrm{ra}_U(s-u) \supseteq \mathrm{ra}_U(s-t) \cap \mathrm{ra}_U(t-u)$, the latter intersection being nonempty. Hence τ is transitive. Let $s,t \in U$ be such that $\mathrm{ra}_U(s-t) \neq \emptyset$. If $x \in U$ then also $\mathrm{ra}_U(xs-xt) \neq \emptyset$. Moreover, $\mathrm{ra}_U(s-t) \cap xU \neq \emptyset$ as an intersection of right ideals of U. Therefore $\mathrm{ra}_U(sx-tx) \neq \emptyset$. It follows that τ is a congruence on U. If $(sx,tx) \in \tau$ then $\mathrm{ra}_U(sx-tx) \neq \emptyset$ so that $\mathrm{ra}_U(s-t) \neq \emptyset$. Therefore τ is the least right cancellative congruence on U. The same argument allows us to prove assertion (2).

Let $I(\tau)$ be the ideal of $K[U]$ determined by τ, that is, the linear span of the set $\{s-t \mid (s,t) \in \tau, s,t \in U\}$. Let U' be the natural image of U in $K[S]/N$. Since $U' \subseteq T'$ is cancellative and τ_U is the least right cancellative congruence on U, we get that $I(\tau) \subseteq N$, whence it is nilpotent.

Let n be the nilpotency index of $I(\tau)$. If $n=1$ then τ is trivial and $uTu \subseteq uT = U \cong U/\tau$ is cancellative. So suppose that $n>1$. Choose $0 \neq \alpha \in I(\tau)^{n-1}$ and write $\alpha = \alpha_1 + \cdots + \alpha_r$, where each $\mathrm{supp}(\alpha_i)$ is contained in a different τ-class. Then

$$(s-t)\alpha_1 + \cdots + (s-t)\alpha_r = (s-t)\alpha = 0$$

for every $s,t \in U$ such that $(s,t) \in \tau$. But for $i \neq j$ the sets $\mathrm{supp}(s\alpha_i)$ and $\mathrm{supp}(s\alpha_j)$ are in different τ-classes because τ is cancellative, while $\mathrm{supp}(s\alpha_i)$ and $\mathrm{supp}(t\alpha_i)$ are in the same τ-class. Therefore we get $(s-t)\alpha_i = 0$ for every $i = 1,\ldots,r$. Let $\alpha_1 = \sum_{i=1}^{k} \lambda_i x_i$ for some $\lambda_i \in K$ and $x_i \in U$. Then $\mathrm{ra}_U(x_i - x_j) \neq \emptyset$ because $(x_i,x_j) \in \tau$. Hence $R = \bigcap_{i \neq j} \mathrm{ra}_U(x_i - x_j) \neq \emptyset$ by Lemma 2.1.3.

Let $w \in R$. Suppose that $(gx_1w - hx_1w)fx_1w = 0$ for some $g,h,f \in U$. Then $(g,h) \in \tau$. Consequently $(g-h)\alpha_1 = 0$. Since U is left cancellative this implies that there exists $j \in \{1,\ldots,k\}$ such that $gx_1 = hx_j$. Now $x_1w = x_jw$ since $w \in R$. It follows that $hx_1w = hx_jw = gx_1w$. This shows that Ux_1w is a right cancellative semigroup. But $Ux_1w \subseteq U$ also is left cancellative. Since $Ux_1w = uTx_1w$ and $x_1w = uq$ for some $q \in T$, we have $uqTuq \subseteq Ux_1w$. It is then enough to put $y = uq$. □

To prove the main result of this section, on one hand we have to refine the above argument in order to construct uniform ideals in certain homomorphic images of monoids S with Noetherian algebras $K[S]$. On the other hand, the following result can also be viewed as an analogue of Theorem 4.5.2, however much more effort is needed in the proof in order to create such uniform ideals.

Theorem 5.2.3. *Assume that S is a monoid such that the algebra $K[S]$ over a field K has finite Gelfand-Kirillov dimension and is right and left Noetherian. Assume also that for every $a, b \in S$ we have $a\langle a,b\rangle \cap b\langle a,b\rangle \neq \emptyset$ and $\langle a,b\rangle a \cap \langle a,b\rangle b \neq \emptyset$. Then S has an ideal chain $S_1 \subseteq S_2 \subseteq \cdots \subseteq S_n = S$ such that, if Z denotes any of $S_1, S_j/S_{j-1}, j = 2,\ldots,n$, then Z is either nilpotent or a uniform subsemigroup of a Brandt semigroup.*

Proof. We my assume that S has a zero element. We will show that every homomorphic image of S either has a nonzero nilpotent ideal or an ideal that is a uniform subsemigroup of a Brandt semigroup. Since, by the hypothesis, S satisfies the ascending chain condition on ideals, it will follow that a finite chain of the desired type exists.

So, suppose the assertion does not hold. Notice that every homomorphic image of S inherits the hypotheses on S. Thus, replacing S by a homomorphic image, we may assume that S has no nonzero ideal that is nilpotent or a uniform subsemigroup of a Brandt semigroup. Let $\overline{S}$ be the image of S in the algebra $K_0[S]/\mathcal{B}(K_0[S])$, where $\mathcal{B}(K_0[S])$ denotes the prime radical of $K_0[S]$. The latter algebra is right and left Noetherian and semiprime, so $\overline{S}$ embeds into a simple Artinian algebra Q, see the comment before Lemma 5.1.2. Moreover $\overline{S}$ has a uniform ideal $\overline{I}$ with a completely 0-simple closure $\widehat{I} \subseteq Q$. In view of Proposition 5.1.1, and because of the ascending chain condition on one-sided ideals in S, $\widehat{I}$ has finitely many $\mathcal{R}$-classes and finitely many $\mathcal{L}$-classes. Let $I \subseteq S$ be the inverse image of $\overline{I}$ under the natural homomorphism $\phi : S \longrightarrow \overline{S}$. Then $\phi^{-1}(0)$ is an ideal of S and for every $a \in \phi^{-1}(0)$ the element $a - a^2$ is nilpotent. Therefore a must be a periodic element of S, whence $\phi^{-1}(0)$ is a periodic ideal of S. Suppose $\phi^{-1}(0) \neq \{0\}$. Every principal factor of S contained in $\phi^{-1}(0)$ is completely 0-simple, see Corollary 2.2.4. From Proposition 5.1.1 we know that S has finitely many completely 0-simple principal factors. Then there exists a nonzero element $s \in \phi^{-1}(0)$ such that SsS is minimal among all ideals $SxS, x \in S$, such that the principal factor of x in S is completely 0-simple. Hence the set $\{t \in SsS \mid StS \neq SsS\}$ is a nil ideal of S. Since S has no nonzero nil ideals by Proposition 2.4.9, it follows that SsS is a completely 0-simple ideal of S. In particular, SsS is a uniform ideal of S.

We shall now assume that $\phi^{-1}(0) = \{0\}$. Notice that I has a matrix pattern induced from that in $\overline{I} \subseteq \widehat{I}$. Namely, let $\widehat{I} = \mathcal{M}(H, X, Y; P)$ be a Rees matrix presentation of the completely 0-simple semigroup $\widehat{I}$. We know that all $\mathcal{H}$-classes of $\widehat{I}$ are intersected by $\overline{I}$. Let $\overline{I} = \bigcup_{i,j} \overline{I}_{ij}$, where i, j

run through the finite sets X, Y, respectively, be the decomposition coming from the $\mathcal{H}$-classes (with zero adjoined) of $\widehat{I}$. Then $I = \bigcup_{i,j} I_{ij}$ where $I_{ij} = \phi^{-1}(\overline{I_{ij}})$ is a 0-disjoint union. Clearly $I_{ij}I_{kl} \subseteq I_{il}$, as the same is true of the image $\overline{I}$ of I.

If I_{ip}, I_{iq} have nonzero multiplication for some i and some p, q then the assumption on the 2-generated subsemigroups of S easily implies that $p = q$. Similarly one shows that every 'column' of I has at most one component with nonzero multiplication. Therefore, the number $|X|$ of rows of I is equal to the number $|Y|$ of columns and permuting the rows and columns of I we may assume that only the diagonal components have nonzero multiplication. In particular, $\widehat{I}$ is a Brandt semigroup. Moreover, $\widehat{S} = \widehat{I} \cup (\overline{S} \setminus \overline{I})$ (a disjoint union) forms a subsemigroup of Q', see Lemma 2.5.1.

Notice that the same argument shows that, in the case $\phi^{-1}(0) \neq \{0\}$, the completely 0-simple ideal SsS constructed above is a Brandt semigroup. Hence, we get a contradiction in this case.

Coming back to the case $\phi^{-1}(0) = 0$, first notice that I satisfies the condition: $abc = 0$ implies $ab = 0$ or $bc = 0$, because $\overline{I}$ has this property.

Let $\overline{T}$ be a cancellative part of $\overline{I}$, that is, $\overline{T} = \overline{I}_{ij} \setminus \{0\}$ for some $\overline{I}_{ij}$ with nonzero multiplication. Denote by T the inverse image of $\overline{T}$ in S. Notice that $T \cup \{0\} = I_{ij}$ is a subsemigroup of I. Let $R = \{a \in S \mid Ta \subseteq T\}$. By Proposition 5.1.3, since $K_0[S]$ is right Noetherian, also $K[R]$ is right Noetherian and T is an ideal of R. Clearly, $K[R]$ embeds into $K_0[S]$.

Let ρ' be the relation defined on T by the rule: $a\rho' b$ if $xa = xb$ for some $x \in T$. By the dual of Lemma 5.2.2, in view of the assumption on 2-generated subsemigroups of S, ρ' is the least left cancellative congruence on T. Since T is an ideal in R, the linear span $I(\rho')$ of all $a - b$, where $a, b \in T$ and $(a, b) \in \rho$, is a right ideal in $K[R]$ and an ideal of $K[T]$.

Now $I(\rho') = \sum_{i=1}^{r}(a_i - b_i)K[R]$ for some $a_i, b_i \in T$ since $K[R]$ is right Noetherian. Choose $c_i \in T$ such that $c_i(a_i - b_i) = 0$. Then $Tc_1 \cap \cdots \cap Tc_r \neq \emptyset$ because every two left ideals of T intersect nontrivially by the assumption on S. So the left annihilator C' of $I(\rho')$ in T is nonempty, hence it is an ideal of T. We have $C'I(\rho') = \{0\}$. Similarly, by a symmetric argument, one finds an ideal C of T such that $I(\rho)C = \{0\}$ where ρ is the least right cancellative congruence on T. We put $E = C \cap C'$.

Let $a, b \in E$ be such that $(a, b) \in \rho$ but $a \neq b$. Then $(a - b)E = \{0\}$ and so $(a - b)\langle a, b\rangle = \{0\}$. Also, by the assumption on S, we know that there exist $x, y \in \langle a, b\rangle$ such that $xa = yb$. It follows that $xaa = yba$ and $aa = ba$. In particular, $(x, y) \in \rho$. Then $(x - y)E = \{0\}$, and consequently $xa = ya$. So we have $yb = xa = ya$, which implies that $(a, b) \in \rho'$. It follows that $\rho \subseteq \rho'$. By a symmetric argument we show that $\rho = \rho'$. Hence $I(\rho) = I(\rho')$ and we also get $EI(\rho) = \{0\} = I(\rho)E$.

We claim that E contains a cancellative ideal of T. The cancellative semigroup $T' = T/\rho$ has a group of quotients G'. Since S satisfies the ascending chain condition on right congruences by the assumption on $K[S]$, from Proposition 2.4.11 it follows that G' satisfies the ascending chain condition on subgroups, whence it is finitely generated. Theorem 3.5.6 implies that G' is nilpotent-by-finite. The natural map $T \longrightarrow \overline{T}$ factors through a map $T' \longrightarrow \overline{T}$ and the latter extends to a group homomorphism $f : G' \longrightarrow G$, where $G \subseteq \widehat{I}$ is the group of quotients of $\overline{T}$. We know that G' has a maximal finite normal subgroup F. If T'_F denotes the image of T' in G'/F, then $K[T'_F]$ is prime by Theorem 3.2.8. However, the kernel of the natural homomorphism $K[T] \longrightarrow K\{\overline{T}\} \subseteq K_0[S]/\mathcal{B}(K_0[S])$ is nilpotent. Therefore the kernel of the map $\psi : K[T'] \longrightarrow K[\overline{T}]$ also is nilpotent. Hence $\ker(\psi)$ is contained in the kernel of the homomorphism $K[T'] \longrightarrow K[T'_F]$ and we have natural maps $K[T'] \longrightarrow K[\overline{T}] \longrightarrow K[T'_F]$. As $e\overline{R} \subseteq G$, where $e = e^2 \in G$, we may define $R' = f^{-1}(e\overline{R})$. So $T' \subseteq R'$. By Proposition 5.1.3 $K[e\overline{R}]$ is right Noetherian. If $(e\overline{R})_F$ denotes the image of $e\overline{R}$ in G'/F, then $K[(e\overline{R})_F]$ is right Noetherian and, as F is finite, from Lemma 4.1.4 it follows that $K[R']$ is right Noetherian. Therefore, by Theorem 4.3.2 we know that G' contains normal subgroups $U_1 \subseteq A_1$ such that A_1 has finite index in G', A_1/U_1 is abelian and $U_1 \subseteq R'$.

A symmetric argument, based on the application of the monoids $L = \{a \in S \mid aT \subseteq T\}$ and $L' = f^{-1}(\overline{L}e)$ in place of R and R', allows us to find normal subgroups $U_2 \subseteq A_2$ in G' such that $[G' : A_2] < \infty$, A_2/U_2 is abelian and $U_2 \subseteq L'$. Then $U = U_1 \cap U_2 \subseteq R' \cap L'$ is a normal subgroup in G' such that G'/U is abelian-by-finite.

Let E' be the image of E in T'. If T' is a group, then every $y \in E'$ satisfies $yT' = T'y$. On the other hand, if T' is not a group, then Lemma 4.1.9 allows us to find an element $z' \in E'$ so that $Uz's' = Us'z'$ for every $s' \in T'$. Then $T'z's' = T'Uz's' = T'Us'z' = T's'z'$ because $\overline{T}(e\overline{R}) = \overline{T}$ implies that $T'U = T'R' = T'$. So $T'z'T' = (T')^2 z'$ and by symmetry $(T')^2 z' = z'(T')^2$. So we always have an element $z' \in E'$ of the latter type. Now, since $EI(\rho) = \{0\} = I(\rho)E$, the equality $(T')^2 z' = z'(T')^2$ means that $ET^2 z = EzT^2$ and $T^2 zE = zT^2 E$, where $z \in E$ is an inverse image of z'. So $ET^2 z$ and $zT^2 E$ are ideals of T.

Consider any $a, b \in ET^2 z$ with $(a, b) \in \rho$. Since E is an ideal of T, we can write $a = e_1 z, b = e_2 z$ for some $e_1, e_2 \in E$. Hence $(e_1, e_2) \in \rho$, because ρ is a cancellative congruence on T. Therefore $(e_1 - e_2)z = 0$ (since $z \in E$) and thus $a = b$. This proves that $ET^2 z$ is right cancellative.

A symmetric argument allows us to show that zT^2E is left cancellative. Therefore $D = ET^2z \cap zT^2E$ is a cancellative ideal of T, which proves our claim.

Notice also that $IDI \cap T \subseteq D$. Indeed, if $adb \in T$ for some $a, b \in I$ and $d \in D$, then it follows from the matrix structure on I that $\overline{a}\mathcal{L}\overline{d}$ and $\overline{a}\mathcal{R}\overline{T}$ in $\widehat{I}$, whence $a \in T$. Similarly we get $b \in T$. Therefore $adb \in D$.

We can repeat the above argument for every inverse image T_k in I of a cancellative part of $\overline{I}$. In this way, we find a cancellative ideal D_k in T_k. So, consider the ideal $J = \bigcap_k ID_kI$, where the intersection runs over all such D_k. Then J has a finite matrix structure $J = \bigcup_{i,j=1}^{q} J_{ij}$, where the components with nonzero multiplication are $J_{11}, \ldots, J_{qq}$ (because $\widehat{I}$ is a Brandt semigroup) and they are 0-cancellative.

Next we deal in a similar way with the remaining matrix components of J. Fix some J_{ij} such that $J_{ij}^2 = \{0\}$ and fix any other J_{pk}. Let $B = \sum(a - b)K_0[S]$ where summation runs over all $a, b \in J_{ij}$ such that $ta = tb \neq 0$ for some p and some $t \in J_{pk}$. Then, since $J_{pk}J_{ij} \neq 0$, we must have $k = i$. Moreover, $\{0\} \neq J_{kp}J_{pk} \subseteq J_{kk}$. Therefore, for every $a - b$ as above there exists $c \in J_{kk}$ such that $ca = cb \neq 0$. Clearly, B is a finitely generated right ideal of $K_0[S]$ because the latter is right Noetherian. Using the fact that every two left ideals of $T_{kk} = J_{kk} \setminus \{0\}$ intersect nontrivially, as in the proof of the fact that ρ' is a congruence on T, one shows that $M = \{x \in J_{kk} \mid xB = \{0\}\}$ is nonzero and also that B is a left $K_0[J_{kk}]$-module. Therefore M is an ideal of J_{kk}. As before, using Lemma 4.1.9 we can find an element $0 \neq w \in M$ such that $wI_{kk}^2 = I_{kk}wI_{kk}$. Hence $F_k = wI_{kk}^2 \subseteq J_{kk}$ is an ideal of I_{kk}. Then the matrix pattern on J implies that $JF_kJ \cap J_{kj} \subseteq J_{kk}F_kJ_{kj} \subseteq F_kJ_{kj} = wI_{kk}^2J_{kj} \subseteq wJ_{kj}$. So, if $a - b \in B$ for some $a, b \in JF_kJ \cap J_{kj}$, then $a = wx, b = wy$ for some $x, y \in J_{kj}$. Clearly $x - y \in B$ and hence $a = b$ because $w \in M$. Hence, again replacing J by a smaller ideal of S we may assume that $ta = tb \neq 0$ implies $a = b$, for $a, b \in J_{kj}$ and $t \in J_{pk}$. A symmetric argument allows us to assume that $at = bt \neq 0$ implies $a = b$ for $a, b \in J_{jk}$ and $t \in J_{kp}$. Moreover, this can be done for all pairs of indices k, j and all pairs p, k.

It is now clear that the ideal $J \subseteq I$ of S obtained in this way satisfies the hypotheses of Theorem 2.1.5. Therefore J has a completely 0-simple semigroup $\widehat{J}$ of (two-sided) quotients. Since every $J_{kk} \setminus \{0\}, k = 1, \ldots, q$, satisfies the left and right Ore conditions, it is clear that $J_{kk} \setminus \{0\}$ is contained in a maximal subgroup of $\widehat{J}$. Moreover, $J_{kk}J_{mm} = 0$ for $m \neq k$ implies that $J_{kk} \setminus \{0\}, k = 1, \ldots, q$, are contained in different $\mathcal{H}$-classes of $\widehat{J}$. It follows that $\widehat{J}$ is a Brandt semigroup. In particular, J is a uniform subsemigroup of $\widehat{J}$, which contradicts our supposition on the nonexistence of such ideals in S. This completes the proof of the theorem. □

Notice that every monoid S satisfying an identity of the form $xwy = yvx$, where w, v are some words in x, y, fulfills the hypothesis on the 2-generated subsemigroups of S used in Theorem 5.2.3. For example, Malcev nilpotent semigroups are defined by an identity that implies an identity of this type, see Section 3.5 and [123]. The proof of Theorem 5.2.3 also shows that for Malcev nilpotent semigroups we do not have to assume that $\mathrm{GK}(K[S]) < \infty$ (notice however that Theorem 5.4.3 will show that the latter is a consequence of the remaining assumptions). So, if $K[S]$ is right and left Noetherian and S is Malcev nilpotent, then we get an ideal chain as in Theorem 5.2.3.

5.3 Monomial semigroups and sufficient conditions

Our next aim is to study certain Noetherian algebras determined by semigroups related to polycyclic-by-finite groups G, but more general than those described in Section 4.4. Namely, we consider the class of Noetherian algebras $K[S]$ for monomial monoids, that is submonoids S of $\mathcal{M}_n(G), n \geq 1$. On one hand, the first main result of this section extends Theorem 4.1.6. On the other hand, it will be a key ingredient of the proof of Theorem 5.3.7, which provides certain sufficient conditions for a semigroup algebra to be Noetherian, and which is also a partial converse to Theorem 5.2.3.

We start with some auxiliary results on monomial semigroups and their algebras. The first can be viewed as an analogue of Lemma 2.3.6 and the standard proof will be omitted.

Lemma 5.3.1. *Let G be a group. Then $\mathcal{M}_n(G)$ is an inverse semigroup with the only ideals*

$$\{0\} = \mathcal{M}_0 \subset \mathcal{M}_1 \subset \cdots \subset \mathcal{M}_n = \mathcal{M}_n(G),$$

where $\mathcal{M}_j = \{s \mid s \text{ has at most } j \text{ nonzero entries }\}$. Moreover, for $j > 0$, $\mathcal{M}_j/\mathcal{M}_{j-1} \cong \mathcal{M}(G_j, \binom{n}{j}, \binom{n}{j}; I)$, where G_j is an extension of G^j by the symmetric group Sym_j and I is the identity matrix. A power of every element of $\mathcal{M}_n(G)$ is diagonal, so that all idempotents are diagonal.

The elements of $\mathcal{M}_j \setminus \mathcal{M}_{j-1}$ are referred to as matrices of rank j. Because of the above lemma, as is the case of linear semigroups, Theorem 2.4.3, we get that in every subsemigroup $S \subseteq \mathcal{M}_n(G)$ one can define uniform components.

This is because every subsemigroup T of a completely 0-simple semigroup Z, in particular $T = (S \cap \mathcal{M}_j)/(S \cap \mathcal{M}_{j-1}) \subseteq Z = \mathcal{M}_j/\mathcal{M}_{j-1}$, is a 0-disjoint union of its subsemigroups $\bigcup_\alpha U_\alpha \cup N$ such that N is the largest nil ideal of T and each U_α is na uniform subsemigroup of a completely 0-simple subsemigroup Z_α of Z and Z_α, Z_β intersect different $\mathcal{R}$-classes and different $\mathcal{L}$-classes of Z if $\alpha \neq \beta$. All such U_α, coming from all possible ranks of matrices in S, are called the uniform components of S. In particular, since $\mathcal{M}_n(G)$ has finitely many maximal subgroups, it follows also that S has an ideal chain $I_1 \subset \cdots \subset I_t = S$ such that all factors are either uniform subsemigroups of Brandt semigroups or nil. Moreover, the nil factors must be nilpotent.

The following lemma allows an inductive approach to subsemigroups of $\mathcal{M}_n(G)$, just as exterior power does in the case of linear semigroups over a field, see Lemma 2.6.6.

Lemma 5.3.2. *Assume that* $1 \leq j \leq n$. *Then there exists a homomorphism* $\varphi : \mathcal{M}_n(G) \longrightarrow \mathcal{M}_{\binom{n}{j}}(G_j)$ *such that* $\varphi(\mathcal{M}_{j-1}) = \{0\}$ *and* φ *is one-to-one on* $\mathcal{M}_n \backslash \mathcal{M}_{j-1}$. *So,* $\mathcal{M}_n/\mathcal{M}_{j-1}$ *is isomorphic with a subsemigroup of* $\mathcal{M}_{\binom{n}{j}}(G)$.

Proof. Since all $\mathcal{M}_j/\mathcal{M}_{j-1}$ are Brandt semigroups with finitely many idempotents, from Lemma 2.5.7 it follows that

$$\begin{aligned} K_0[\mathcal{M}_n(G)] &\cong K_0[\mathcal{M}_n/\mathcal{M}_{n-1}] \oplus \cdots \oplus K_0[\mathcal{M}_2/\mathcal{M}_1] \oplus K_0[\mathcal{M}_1] \\ &\cong \bigoplus_{j=1}^{n} \mathrm{M}_{\binom{n}{j}}(K[G_j]) \end{aligned}$$

The corresponding natural homomorphism

$$\sigma : K_0[\mathcal{M}_n(G)] \longrightarrow \mathcal{M}_{\binom{n}{j}}(K[G_j])$$

can be described as follows. First factor out $K_0[\mathcal{M}_{j-1}]$. Next notice that any nonzero $\mathcal{L}$-class L of $\mathcal{M}_j/\mathcal{M}_{j-1}$ spans a left ideal of $K_0[\mathcal{M}_j/\mathcal{M}_{j-1}] \cong \mathrm{M}_{\binom{n}{j}}(K[G_j])$ that is a free right $K[G_j]$-module of rank $\binom{n}{j}$. For a $K[G_j]$-basis E contained in L, define $\sigma(s)$ as the matrix of left multiplication by $s \in \mathcal{M}_n(G) \setminus \mathcal{M}_{j-1}$ on L. Extend this linearly to $K_0[\mathcal{M}_n(G)]$. So this is defined as in the proof of Proposition 2.5.6.

We claim that every $\sigma(s)$ is a monomial matrix over G_j. From Proposition 2.5.6 we know that every column of $\sigma(s)$ has at most one nonzero entry. Notice that $L = \{x_1, \ldots, x_{\binom{n}{j}}\}$ with all x_i in different $\mathcal{R}$-classes of $\mathcal{M}_j$. If, for some i, k, the elements sx_i, sx_k are in $\mathcal{M}_j \setminus \mathcal{M}_{j-1}$ and they are in the same

$\mathcal{H}$-class of $\mathcal{M}_j/\mathcal{M}_{j-1}$, we get $(sx_i)y \neq 0 \neq (sx_k)y$ for some $y \in \mathcal{M}_j/\mathcal{M}_{j-1}$. This implies $i = k$ because $\mathcal{M}_j/\mathcal{M}_{j-1}$ is a Brandt semigroup. So, indeed $\sigma(s)$ is a monomial matrix. Hence, $\sigma(S) \subseteq \mathcal{M}_{\binom{n}{j}}(G_j)$.

Finally, suppose $\sigma(s) = \sigma(t)$ for some $s, t \notin \mathcal{M}_{j-1}$. Then, for every x_i, either $sx_i = tx_i \notin \mathcal{M}_{j-1}$ or $sx_i, tx_i \in \mathcal{M}_{j-1}$. Since L generates $\mathcal{M}_j \setminus \mathcal{M}_{j-1}$ as a right ideal, the same holds for every sx, tx with $x \in \mathcal{M}_j \setminus \mathcal{M}_{j-1}$. It follows that every j nonzero columns of s are equal to the corresponding j columns of t. This implies $s = t$, as desired. □

Before proving Theorem 5.3.5 we describe in further detail monomial monoids satisfying the ascending chain condition. For this, we introduce more notation. Let G be a group and $S \subseteq \mathcal{M}_n(G)$ a monomial monoid. Let $e_1, \ldots, e_n$ denote the diagonal idempotents of rank one in $\mathcal{M}_n(G)$. Put $D(S) = \{s \in S \mid e_i s = s e_i, \text{ for } i = 1, \ldots, n\}$. So $D(S)$ is the diagonal part of S. Let $\pi(S) = \bigcup_H e_H D(S)$, where the union runs over all maximal subgroups H of $\mathcal{M}_n(G)$ and $e_H = e_H^2 \in H$. So each e_H is a diagonal idempotent and, if $j = \operatorname{rank}(e_H) - 1 \geq 0$, then $e_H D(S) \cap H$, or $(e_H D(S) \cap H) \cup \{0\}$, is an ideal of the homomorphic image of $D(S)$ obtained by factoring out $\mathcal{M}_j$. Moreover, since e_H is central in $D(S)$, we have natural homomorphisms $D(S) \longrightarrow D(S)/(D(S) \cap \mathcal{M}_j) \longrightarrow (e_H D(S) \cap H) \cup \{0\}$. In particular, $e_H D(S) \cap H$ is a monoid. Clearly $D(S) \subseteq \pi(S)$ and $\pi(S) = \bigcup_H e_H D(S) \cap H$ is also a monoid. If $T, U \subseteq S$ are subsemigroups such that $UT \subseteq U$ then we say that U is a right T-module.

Lemma 5.3.3. *Let G be a polycyclic-by-finite group and S a submonoid of $\mathcal{M}_n(G)$. Then the following properties hold.*

1. *$D(S)$ satisfies the ascending chain condition on right ideals if and only if $\pi(S)$ satisfies the ascending chain condition on right ideals.*

2. *$D(S)$ satisfies the ascending chain condition on right ideals if and only if $K[D(S)]$ is right Noetherian.*

3. *$D(S)$ satisfies the ascending chain condition on right ideals if and only if $D(S)$ satisfies the ascending chain condition on left ideals.*

Proof. We may assume that $0 \in S$. Every $e_H D(S)$ is a homomorphic image of $D(S)$. Therefore, if $D(S)$ satisfies the ascending chain condition on right ideals, then $\pi(S) = \bigcup_H e_H D(S)$ also has this property. Now, assume that $\pi(S)$ satisfies the ascending chain condition on right ideals. Clearly, every $(D(S) \cap H) \cup \{0\}$ (viewed as an ideal of a homomorphic image of $D(S)$) is a right $\pi(S)$-module and has the same lattice of right $D(S)$-submodules as the lattice of right $\pi(S)$-submodules. Since $(D(S) \cap H) \cup \{0\}$ is an ideal of $(\pi(S) \cap H) \cup \{0\}$, it follows that $(D(S) \cap H) \cup \{0\}$ is satisfies the ascending chain condition on right $D(S)$-submodules. As $D(S) = \bigcup_H D(S) \cap H$, the first part of the lemma follows easily.

Suppose $\pi(S)$ satisfies the ascending chain condition on right ideals. Since $\pi(S)$ has an ideal chain with factors isomorphic to $D(S)_H = (e_H D(S) \cap H) \cup \{0\}$ and since all $D(S)_H$ are monoids, by Lemma 2.5.7 we get

$$K_0[\pi(S)] \cong \bigoplus_H K_0[D(S)_H].$$

Since G is polycyclic-by-finite, every H is polycyclic-by-finite by Lemma 5.3.1. As $D(S)_H \subseteq H^0$ and H is polycyclic-by-finite, from Theorem 3.4.1 it follows that each $K_0[D(S)_H]$ is right Noetherian. This proves that $K[\pi(S)]$ is right Noetherian. As in the proof of (1) this implies that $K[D(S)]$ also is right Noetherian. Statement (2) now follows

The above proof yields also that $\pi(S)$ satisfies the ascending chain condition on right ideals if and only if each (nonempty) submonoid $e_H D(S) \cap H$ of H satisfies this ascending chain condition. It was shown in Theorem 4.4.7 that the ascending chain condition on right ideals is equivalent with the ascending chain condition on left ideals for such monoids. Hence (3) follows. □

In Theorem 4.4.7 we have given a structural characterization of submonoids of a polycyclic-by-finite group G that satisfy the ascending chain condition for right ideals. So the structure of $\pi(S)$ in the lemma above is well understood.

Lemma 5.3.4. *Let G be a group and S a submonoid of $\mathcal{M}_n(G)$. If S satisfies the ascending chain condition on right ideals, then the following properties hold.*

1. *S is a finitely generated right $D(S)$-module.*

2. *$D(S)$ satisfies the ascending chain condition on right ideals.*

Proof. Suppose that

$$a_1 D(S) \subset a_1 D(S) \cup a_2 D(S) \subset \cdots$$

for some $a_i \in S$. Because S satisfies the ascending chain condition on right ideals, as in the proof of Lemma 4.1.4 we see that there exists a subsequence a_{j_n} such that

$$a_{j_1} S \supseteq a_{j_2} S \supseteq a_{j_3} S \supseteq \cdots .$$

Replacing a_i by a_{j_i} we may assume that $a_1S \supseteq a_2S \supseteq \cdots$. So $a_{i+1} = a_is_i$ for some $s_i \in S$ and

$$a_{j+1} = a_it, t = s_i \cdots s_j, \text{ for } i < j.$$

Consider the natural homomorphism $\phi : \mathcal{M}_n(G) \longrightarrow \mathcal{M}_n(\{1\})$. The inverse image, under ϕ, of an idempotent is the diagonal part $D(F)$ of a maximal subgroup F of $\mathcal{M}_n(G)$. Hence from Theorem 2.4.8, it follows that there exist $i < j$ such that $t = s_i \cdots s_j \in D(F)$ for some maximal subgroup F of $\mathcal{M}_n(G)$. Then $t \in D(S)$ and this implies that $a_{j+1} = a_it \in a_iD(S)$, a contradiction. This proves assertion (1). The same argument, applied to a sequence $a_i \in D(S), i = 1, 2, \ldots$, also shows that every right ideal of $D(S)$ must be finitely generated. Hence (2) also follows. □

We are now ready to prove the first main result of this section.

Theorem 5.3.5. *Let G be a polycyclic-by-finite group and let K be a field. The following conditions are equivalent for a submonoid S of $\mathcal{M}_n(G)$.*

1. *S satisfies the ascending chain condition on right ideals.*
2. *$D(S)$ satisfies the ascending chain condition on right ideals and S is a finitely generated right $D(S)$-module.*
3. *$K[S]$ is right Noetherian.*

Proof. From Lemma 5.3.4 it follows that (1) implies (2). Lemma 5.3.3 proves that (3) is a consequence of (2). That (3) implies (1) is clear. □

Notice that we do not have the left-right symmetry as in Theorem 4.4.7. For example, if G is any infinite group and $S \subseteq \mathcal{M}_2(G)$ consists of all matrices of the form $\begin{pmatrix} 1 & 0 \\ 0 & g \end{pmatrix}$ and $\begin{pmatrix} 0 & g \\ 0 & 0 \end{pmatrix}$, then S satisfies the ascending chain condition on right ideals but not on left ideals.

Lemma 5.3.6. *Assume that S is a monoid with zero satisfying the ascending chain condition on right ideals. Let N be a nilpotent ideal of S. If $K_0[S/N]$ is right Noetherian, then $K_0[S]$ is right Noetherian.*

Proof. We know that $N^r = \{0\}$ for some $r \geq 1$. We prove the assertion by induction on r. The case $r = 1$ is obvious. Assume now that $r > 1$. Then, by the induction hypothesis, $K_0[S/N^{r-1}] \cong K[S]/K[N^{r-1}]$ is right Noetherian. But, as S satisfies the ascending chain condition on right ideals, the ideal $K[N^{r-1}]$ of $K[S]$ is finitely generated as a right $K[S]$-module.

Hence $K_0[N^{r-1}]$ is finitely generated as a right $K_0[S/N]$-module. Since the latter algebra, by assumption, is right Noetherian, we obtain that $K_0[N^{r-1}]$ is a right Noetherian $K[S]$-module. Consequently $K_0[S]$ is right Noetherian. □

We are now ready to show that Theorem 5.2.3 admits a surprisingly general partial converse. The proof will exploit monomial semigroups considered above.

Theorem 5.3.7. *Assume that S is a finitely generated monoid with an ideal chain $S_1 \subseteq S_2 \subseteq \cdots \subseteq S_n = S$ such that S_1 and every factor S_j/S_{j-1} is either nilpotent or a uniform subsemigroup of a Brandt semigroup. Let K be a field. If S satisfies the ascending chain condition on right ideals and* $\mathrm{GK}(K[S])$ *is finite, then $K[S]$ is right Noetherian.*

Proof. We may assume S has a zero element θ, adjoining it to S, if necessary. Write $S_0 = \{\theta\}$. By induction on j we will show that $K[S/S_{n-j}]$ is right Noetherian. With $j = n$ this will yield the assertion. The claim is clear for $j = n$.

Suppose that we have already shown that $K_0[S/S_i]$ is right Noetherian, for some $i > 0$. Define $T = S/S_{i-1}$. First consider the case where S_i/S_{i-1} is nilpotent. Then the inductive assertion follows from Lemma 5.3.6. Next, assume that $J = S_i/S_{i-1}$ is a uniform subsemigroup of a Brandt semigroup. So we know that $J \subseteq \widehat{J} \cong \mathcal{M}(G, X, X; I)$ for a group G, a set X and the identity $X \times X$ matrix I. Since S satisfies the ascending chain condition on right ideals, X is finite, say $|X| = m$. Moreover every maximal subgroup of $\widehat{J}$ is generated by its intersection with J. Then the disjoint union $\widehat{T} = (T \setminus J) \cup \widehat{J}$ has a natural semigroup structure, extending that of T, see Lemma 2.5.1. Since G is isomorphic to a group generated by a subsemigroup of S and $\mathrm{GK}(K[S]) < \infty$, we also have $\mathrm{GK}(K[G]) < \infty$ by Theorem 3.5.6. Moreover, by Proposition 2.5.6, G is finitely generated because S is finitely generated. Then Theorem 3.5.6 implies that G is nilpotent-by-finite. Now $K_0[J] \subseteq K_0[\widehat{J}]$ and, by Lemma 2.5.7, the latter is naturally isomorphic to $\mathrm{M}_m(K[G])$. Moreover, the map $\phi : K_0[T] \longrightarrow K_0[\widehat{J}] \cong \mathrm{M}_m(K[G])$ given by $x \mapsto xE$, where E is the identity of $K_0[\widehat{J}]$, agrees with the action of $K_0[T]$ by left multiplication on $K_0[\widehat{J}]$. If $x \in T$ then $a(xE), (xE)a \in J$ for every $a \in J$. It follows easily that the matrix xE has at most one nonzero entry in each row and in each column. Therefore $\phi(T)$ is a monomial monoid over G. From Theorem 5.3.5 it follows that the algebra $K_0[\phi(T)]$ is right Noetherian. Since $\phi(J) \cong J$ is a finitely generated right ideal of $\phi(T)$, it follows that $K_0[J]$ is a Noetherian right $K_0[\phi(T)]$-module, hence also a Noetherian right $K_0[T]$-module. But $K_0[T/J]$ is right Noetherian by the induction hypothesis, so that $K_0[T]$ also is right Noetherian. This completes the proof of the theorem. □

The same proof would also work without the hypothesis $\mathrm{GK}(K[S]) < \infty$, if we knew that all involved groups G are polycyclic-by-finite, see Theorem 4.4.7. While there are no known Noetherian group algebras of other types, there is no natural general condition that would allow us to soften the requirement $\mathrm{GK}(K[S]) < \infty$. If the ideal chain of Theorem 5.3.7 has no nilpotent factors, one can prove more.

Proposition 5.3.8. *Let S be a semigroup with an ideal chain $\{0\} = S_0 \subseteq S_1 \subseteq S_2 \subseteq S_n = S$ such that all factors S_i/S_{i-1} are uniform subsemigroups of Brandt semigroups $\mathcal{M}(G_i, k_i, k_i; I)$ over groups G_i, with $1 \leq i \leq n$. Then S is a subsemigroup of the monomial monoid $\mathcal{M}_m(\prod_{i=1}^n G_i)$, where $m = k_1 + \cdots + k_n$. So, S is a monomial semigroup. Moreover the algebra $K_0[S]$ embeds into the product $\bigoplus_{i=1}^n M_{k_i}(K[G_i])$.*

If, furthermore, S is a finitely generated monoid and all groups G_i are polycyclic-by-finite, then $K[S]$ is right Noetherian if and only if S satisfies the ascending chain condition on right ideals.

Proof. We prove the first part of the result by induction on the length n of the chain. If $n = 1$, then the assertion is clear. Thus, assume that $n > 1$. Let $J = S_1$, $G = G_1$ and $k = k_1$. Then J is an ideal of S that is a uniform subsemigroup of a Brandt semigroup $\widehat{J} = \mathcal{M}(G, k, k; I)$. As in the proof of Theorem 5.3.7, we get that $K_0[\widehat{J}] \cong M_k(K[G])$ and this algebra has an identity, say e. Moreover, $S' = (S \setminus J) \cup \widehat{J}$ has a semigroup structure that extends the operation on S. So $K_0[\widehat{J}]$ is an ideal of $K_0[S]$ and we must have $K_0[S'] \cong M_k(K[G]) \oplus K_0[S/J]$. Also, eS is a subsemigroup of the monomial semigroup $\mathcal{M}_k(G)$. Therefore, the induction hypothesis applied to the semigroup S/J yields that S embeds into

$$\begin{aligned} eS \times S/J \;\subseteq\; & \mathcal{M}_{k_1}(G_1) \times S/J \subseteq \cdots \subseteq \\ & \subseteq \mathcal{M}_{k_1}(G_1) \times \cdots \times \mathcal{M}_{k_n}(G_n) \subseteq \mathcal{M}_m(G_1 \times \cdots \times G_n) \end{aligned}$$

with $m = k_1 + \cdots + k_n$ and

$$K_0[S] \subseteq K_0[eS] \oplus K_0[S/J] \subseteq M_{k_1}(K[G_1]) \oplus M_{k_2+\cdots+k_n}(K[\prod_{i=2}^n G_i]).$$

The first part of the result thus follows. The second part is a consequence of the comment after Theorem 5.3.7. □

There are many examples of algebras satisfying conditions of Theorem 5.3.7. One of the interesting classes consists of certain Rees factors of cancellative semigroups, see Theorem 4.5.2. Ideal chains used in Theorem 5.3.7 will also play an important role in Chapter 9.

Corollary 5.3.9. *Assume that S is a Malcev nilpotent monoid. Then $K[S]$ is right and left Noetherian if and only if S is finitely generated, satisfies the ascending chain condition on one-sided ideals and it has an ideal chain $S_1 \subseteq S_2 \subseteq \cdots \subseteq S_n = S$ such that S_1 and every factor S_j/S_{j-1} is either nilpotent or a uniform subsemigroup of a Brandt semigroup.*

Proof. Assume that S is finitely generated, satisfies the ascending chain condition on right ideals and has an ideal chain of the described type. Then, in view of Theorem 3.5.9, the proof of Theorem 5.3.7 remains valid, so it follows that $K[S]$ is right Noetherian. Similarly, $K[S]$ is left Noetherian.

The necessity of the stated conditions follows from Theorem 5.1.5 and from Theorem 5.2.3 and the comment following it. □

It is clear that, if I is a finite ideal of S and $K_0[S/I]$ is right Noetherian, then $K[S]$ also is right Noetherian. So, the conditions in Theorem 5.3.7 are far from necessary. On the other hand, let $S = \{1\} \cup I$, where $I = \mathcal{M}(G, X, Y; P)$ (without zero) with $X = \{1\}, Y = \{1, 2\}$, the 2×1 matrix P with entries $p_{11} = p_{21} = 1$ and G standing for the infinite cyclic group. Then $K[S]$ is not right Noetherian since the left annihilator of $K[I]$ is infinite dimensional. Namely, $((g, 1, 1) - (g, 1, 2))I = 0$ for every $g \in G$. One can represent S as a submonoid of $M_2(\mathbb{Q})$ with I consisting of all matrices of the forms

$$\begin{pmatrix} 2^n & 0 \\ 0 & 0 \end{pmatrix} \quad \text{and} \quad \begin{pmatrix} 2^n & 2^n \\ 0 & 0 \end{pmatrix},$$

where n runs over the set of integers. In particular, this shows that Theorem 6 in [142] is not correct. Moreover, this suggests that the ascending chain condition on right congruences, rather than the ascending chain condition on right ideals, might be of importance for a possible extension of Theorem 5.3.7. Clearly, if $K[S]$ is right Noetherian, then S satisfies the ascending chain condition on right congruences. This finiteness condition was studied in [64, 65, 98].

5.4 Gelfand-Kirillov dimension

In this section we study the Gelfand-Kirillov dimension of Noetherian semigroup algebras $K[S]$. The main idea is to show that the cancellative subsemigroups T of S carry the information on the finiteness of $\mathrm{GK}(K[S])$ and knowing the dimensions $\mathrm{GK}(K[T])$ is sometimes sufficient to determine the dimension of $K[S]$.

We start with a partial converse of Lemma 5.2.2.

Lemma 5.4.1. *Assume that T is a cancellative subsemigroup of a semigroup S. Let $\overline{T}$ denote the image of T under the natural homomorphism $\phi : K[S] \longrightarrow K[S]/\mathcal{B}(K[S])$.*

1. *If $\mathrm{char}(K) = 0$ then ϕ determines an isomorphism of $K[T]$ and $K[\overline{T}]$.*
2. *If $\mathrm{char}(K) = p > 0$ and $K[S]/\mathcal{B}(K[S])$ is a right Goldie ring then there exists $t \in \overline{T}$ such that $t\overline{T}t$ is a cancellative semigroup.*

Proof. If $\mathrm{char}(K) = 0$ then Theorem 3.2.8 implies that $K[T]$ is a semiprime ring. Since

$$\ker(\phi) \cap K[T] \subseteq \mathcal{B}(K[S]) \cap K[T] \subseteq \mathcal{B}(K[T]),$$

(1) follows.

Assume that $\mathrm{char}(K) = p > 0$. By Theorem 3.1.5, the image $\overline{S}$ of S in the right Goldie ring $K[S]/\mathcal{B}(K[S])$ embeds into $\mathrm{M}_n(D)$, for a division algebra D and $n \geq 1$. Let $t \in \overline{T}$ be an element of the minimal nonzero rank, as a matrix in $\mathrm{M}_n(D)$, and let $s \in S$ be an inverse image of t in T. Then $t \neq 0$ because otherwise $s \in \mathcal{B}(K[S]) \cap S$, a contradiction, as the latter set is empty. It follows also that t is not nilpotent. If H denotes the $\mathcal{H}$-class of t in $\mathrm{M}_n(D)$ then $t\overline{T}t$ is a subsemigroup of $\mathrm{M}_n(D)$ contained in H. Since $H^2 \neq 0$, we get that H is a group. Then $t\overline{T}t \subseteq H$ is cancellative, which proves the assertion. □

Proposition 5.4.2. *Assume that the semigroup algebra $K[S]$ is right Noetherian and S does not have free nonabelian subsemigroups. Let $\overline{S}$ denote the image of S in $K[S]/\mathcal{B}(K[S])$.*

1. *Every cancellative subsemigroup T of S has a finitely generated group of right quotients.*
2. *We have $\sup\{\mathrm{GK}(K[T]) \mid T$ is a cancellative subsemigroup of $S\} = \sup\{\mathrm{GK}(K[T]) \mid T$ is a cancellative subsemigroup of $\overline{S}\}$. Moreover, this is an integer or ∞.*
3. *$\mathrm{GK}(K[T]) \leq \mathrm{GK}(K[S]/\mathcal{B}(K[S]))$ for every cancellative subsemigroup T of S.*

4. *The set of isomorphism classes of groups of quotients of the maximal cancellative subsemigroups of $\overline{S}$ is finite.*

Proof. By Lemma 2.1.3, any cancellative subsemigroup $T \subseteq S$ has a group of quotients G. Since S satisfies the ascending chain condition on right congruences, G satisfies the ascending chain conditions on subgroups by Proposition 2.4.11. So G must be finitely generated.

Let T' be a cancellative subsemigroup of $\overline{S}$ and let T be the inverse image of T' in S. By Lemma 5.2.2 there exists $y \in T$ such that yTy is cancellative. If y' is the image of y in T' then the group of quotients of T', existing by Lemma 2.1.3, can be described as $(T')(T')^{-1} = (y'Ty)('y'Ty')^{-1}$. From Theorem 3.5.6 it then follows that $\mathrm{GK}(K[T']) = \mathrm{GK}(K[y'T'y']) \leq \mathrm{GK}(K[yTy])$ and the former is an integer or ∞.

On the other hand, let T be a cancellative subsemigroup of S and let T' be the image of T in $\overline{S}$. By Lemma 5.4.1, $tT't$ is cancellative for some $t \in T'$. As above, $\mathrm{GK}(K[T]) = \mathrm{GK}(K[yTy])$, where $y \in T$ is an inverse image of t. To establish the equality in (2) it is then sufficient to show that $\mathrm{GK}(K[U]) \leq \mathrm{GK}(K[U'])$ for every finitely generated subsemigroup U of yTy and its image U' in $\overline{S}$. We may assume that $\mathrm{GK}(K[U']) < \infty$. Since the kernel of the natural map $K[U] \longrightarrow K[U']$ is contained in the nilpotent ideal $\mathcal{B}(K[S])$, by Theorem 3.5.3 we know that $\mathrm{GK}(K[U]) < \infty$. So Theorem 3.5.6 implies that the finitely generated group $G = UU^{-1}$ is nilpotent-by-finite and $\mathrm{GK}(K[G]) = \mathrm{GK}(K[U])$. By Proposition 3.3.3 G has a torsion free normal subgroup N of finite index. Then $N = \mathrm{gr}(N \cap U)$ by Lemma 4.1.2 and Theorem 3.2.8 implies that $K[N \cap U]$ is prime. Hence $K[N \cap U] \cap \mathcal{B}(K[S]) = 0$ and we must have $N \cap U \cong (N \cap U)' \subseteq U'$. It follows that $\mathrm{GK}(K[U]) = \mathrm{GK}(K[N]) = \mathrm{GK}(K[N \cap U]) = \mathrm{GK}(K[(N \cap U)']) \leq \mathrm{GK}(K[U'])$. Therefore $\mathrm{GK}(K[U]) = \mathrm{GK}(K[U'])$ and assertion (2) follows.

The preceding paragraph shows that $K[N \cap U]$ embeds into the algebra $K[S]/\mathcal{B}(K[S])$, whence $\mathrm{GK}(K[U]) = \mathrm{GK}(K[N \cap U]) \leq \mathrm{GK}(K[S]/\mathcal{B}(K[S]))$. Thus assertion (3) follows.

We view $\overline{S}$ as a subsemigroup of a simple Artinian algebra $\mathrm{M}_n(D)$. Then $\overline{S}$ intersects finitely many $\mathcal{R}$-classes of $\mathrm{M}_n(D)$ by Proposition 5.1.1. Therefore, from Theorem 2.4.3 it follows that $\overline{S}$ has only finitely many uniform components. Let T be a maximal cancellative subsemigroup of $\overline{S}$. The set I consisting of matrices of the least rank in T is an ideal of T. Since I is left and right Ore, it is contained in an $\mathcal{H}$-class of the monoid $\mathrm{M}_n(D)$. Then I is contained in a uniform component of S. Moreover, $II^{-1} \cong TT^{-1}$. Since the groups of quotients of any two maximal cancellative subsemigroups of a uniform component of $\overline{S}$ are isomorphic by Proposition 2.2.1, assertion (4) follows. □

The first way of approaching the dimension of Noetherian algebras $K[S]$ is motivated by Theorem 5.2.3.

Theorem 5.4.3. *Let K be a field and let S be a semigroup with a finite chain of ideals $S_n \subseteq S_{n-1} \subseteq \cdots \subseteq S_0 = S$ such that S_n and every factor S_i/S_{i+1} is either nilpotent or a uniform subsemigroup of a completely 0-simple semigroup having finitely many $\mathcal{R}$-classes. Then the following conditions are equivalent:*

1. $\mathrm{GK}(K[S]) < \infty$,
2. $\mathrm{GK}(K[T]) < \infty$ *for every cancellative subsemigroup T of S.*

Moreover, in this case, $\mathrm{GK}(K[S]) < 2^p kq$ where q denotes the supremum of the GK-dimensions of the semigroup algebras of cancellative subsemigroups of S, k is the product of the nilpotency indices of the nilpotent factors $S_n, S_i/S_{i+1}$ and p is the number of uniform factors.

Proof. It is clear that (2) is a consequence of (1). Assume that (2) holds. We show that $\mathrm{GK}(K[S]) < \infty$, proceeding by induction on n. We may assume that S has a zero element and that $\mathrm{GK}(K[T]) < \infty$ for every maximal cancellative subsemigroup T of S. If $n = 1$ then S is nilpotent or uniform. In the former case $\mathrm{GK}(K[S]) = 0$, so consider the latter case. Let us fix a maximal cancellative $T \subseteq S$. Then T has a group of quotients G by Lemma 2.1.3 and S embeds into a completely 0-simple semigroup M whose maximal subgroups are isomorphic to G (notice that T is contained in a maximal subgroup of M because it satisfies the right and left Ore conditions). Then the group ring $K[G]$ can be viewed as a corner in the algebra $K_0[M]$ in the sense that $K[G]K_0[M]K[G] \subseteq K[G]$. Moreover $M = MGM$. From Theorem 3.5.3 it then follows that $\mathrm{GK}(K_0[M]) = \mathrm{GK}(K_0[M]K[G]^2K_0[M]) = \mathrm{GK}(K[G])$. Since we also have $\mathrm{GK}(K[T]) = \mathrm{GK}(K[G])$ by Theorem 3.5.6, this implies that $\mathrm{GK}(K[S]) = \mathrm{GK}(K_0[S]) = \mathrm{GK}(K[T])$ and the assertion follows.

Assume that $n > 1$. By the inductive hypothesis $\mathrm{GK}(K[S/S_n]) < \infty$. If S_n is nilpotent, then S has a zero element and $K_0[S/S_n] \cong K_0[S]/K_0[S_n]$ implies that $\mathrm{GK}(K[S]) = \mathrm{GK}(K_0[S]) \leq r\,\mathrm{GK}(K_0[S/S_n]) < \infty$, where r is the nilpotency index of S_n, see Theorem 3.5.3. So assume that S_n is uniform. Then S_n has a completely 0-simple closure $J = \mathcal{M}(G, I, M; P)$ over a group G that is isomorphic to the group of quotients of every maximal nonzero cancellative subsemigroup of S_n. Theorem 2.1.5 implies that J is a semigroup of quotients of S_n. By Lemma 2.5.1 S embeds into the semigroup $S' = S \cup J$, the disjoint union. Moreover, $|I| < \infty$ by the hypothesis. Let $\phi : K[S'] \longrightarrow M_{|I|}(K[G])$ be the homomorphism defined in Proposition 2.5.6. We know that $L = \ker(\phi) \cap K_0[J]$ is the left annihilator of $K_0[J]$ in $K_0[J]$. Then we may define a homomorphism $\eta : K_0[S']/L \longrightarrow K_0[S'/J] \oplus M_{|I|}(K[G])$, by the rule $\eta(a + L) = \pi(a) + \phi(a)$ for $a \in S'$, where $\pi : K_0[S] \longrightarrow K_0[S'/J]$ is the natural map. Clearly, η is an embedding. Since $L^2 = \{0\}$, we get

$$\begin{aligned}\mathrm{GK}(K_0[S]) &\leq \mathrm{GK}(K_0[S']) \leq 2\,\mathrm{GK}(K_0[S']/L)\\ &= 2\max\{\mathrm{GK}(K[G]), \mathrm{GK}(K_0[S'/J])\}.\end{aligned}$$

Now $S'/J = S/S_n$ and $\mathrm{GK}(K[S/S_n]) < \infty$ by the induction hypothesis. Moreover, a maximal cancellative subsemigroup U of S_n satisfies $UU^{-1} \cong G$, so that $\mathrm{GK}(K[G]) = \mathrm{GK}(K[U])$ by Theorem 3.5.6. Therefore assertion (1) follows.

The stated bound on $\mathrm{GK}(K[S])$ is an easy consequence of the proof. □

In many cases, a better bound for $\mathrm{GK}(K[S])$ can be obtained from the following theorem. Recall that it is still an open question whether $\mathrm{GK}(R) = \mathrm{GK}(R/\mathcal{B}(R))$ for a right Noetherian algebra R.

Theorem 5.4.4. *Assume that K is a field and S is a semigroup such that $K[S]$ is right Noetherian. Then the following conditions are equivalent:*

1. $\mathrm{GK}(K[S]) < \infty$,
2. $\mathrm{GK}(K[T]) < \infty$ *for every cancellative subsemigroup T of S,*
3. *every cancellative subsemigroup of S has a nilpotent-by-finite group of quotients.*

Moreover, $\mathrm{GK}(K[S]/\mathcal{B}(K[S])) = \mathrm{GK}(K[T])$ *for a cancellative subsemigroup T of S and* $\mathrm{GK}(K[S]) \leq r\,\mathrm{GK}(K[T])$, *where r is the nilpotency index of* $\mathcal{B}(K[S])$.

Proof. Since each of the conditions implies that S has no free nonabelian subsemigroups, Proposition 5.4.2 allows us to assume that every cancellative subsemigroup of S has a group of quotients that is finitely generated. Therefore, (2) and (3) are equivalent by Theorem 3.5.6 and to establish the equivalence of the three conditions, it is enough to show that (1) is a consequence of (2). So assume that (2) holds. To complete the proof it is now sufficient to show that $\mathrm{GK}(K[S]/\mathcal{B}(K[S])) = \mathrm{GK}(K[T])$ for a cancellative subsemigroup T of S.

We may assume S has a zero element. Then $\mathrm{GK}(K[S]/\mathcal{B}(K[S])) = \mathrm{GK}(K_0[S]/\mathcal{B}(K_0[S])) = \mathrm{GK}(K_0[S]/P)$ for a prime ideal P of $K_0[S]$, see Theorem 3.5.3. Let $\mathrm{M}_n(D)$ be the right quotient ring of $K_0[S]/P$. Then, for the image S_P of S in $K_0[S]/P \subseteq \mathrm{M}_n(D)$, we choose an ideal uniform component U, its completely 0-simple closure $J \subseteq M_n(D)$ and from Lemma 2.5.2 we know that the disjoint union $S' = S_P \cup J$ is a subsemigroup of $\mathrm{M}_n(D)$. If L is the left annihilator of $K_0[J]$ in $K_0[S']$ then by the primeness of P we get $L \subseteq P$. However, from Proposition 2.5.6 we get an embedding $\psi : K_0[S]/L \longrightarrow \mathrm{M}_t(K[G])$ for some $t \geq 1$. Hence, if F is a maximal cancellative subsemigroup of U then as in the proof of Theorem 5.4.3 it follows that

$$\mathrm{GK}(K_0[S]/P) \leq \mathrm{GK}(K_0[S]/L) \leq \mathrm{GK}(\mathrm{M}_t(K[G])) = \mathrm{GK}(K[F]).$$

Therefore Proposition 5.4.2 implies that $\mathrm{GK}(K_0[S]/P) \leq \mathrm{GK}(K[T])$ for a cancellative subsemigroup T of S. So we get $\mathrm{GK}(K[S]/\mathcal{B}(K[S])) \leq \mathrm{GK}(K[T])$. Then, again Proposition 5.4.2 implies that $\mathrm{GK}(K[S]/\mathcal{B}(K[S])) = \mathrm{GK}(K[T])$. □

The computation of the dimension of $K[S]$ simplifies significantly in certain important special cases. For example, if $K[S]$ is semiprime and right Goldie then $\mathrm{GK}(K[S]) = \mathrm{GK}(K[T])$ for some T and this is an integer or ∞. This is shown as in the proof of Theorem 5.4.4. Our final result will be applicable in the case of algebras studied in Chapter 9.

Corollary 5.4.5. *Assume that $K[S]$ is right Noetherian and satisfies a polynomial identity. Then $\mathrm{GK}(K[S]) = \mathrm{GK}(K[F])$ for a free abelian subsemigroup F of S, and it is an integer.*

Proof. By Theorem 5.1.5 we know that S is finitely generated. Hence, Theorem 3.5.2 implies that $\mathrm{GK}(K[S]) = \mathrm{GK}(K[S]/\mathcal{B}(K[S]))$ and it is an integer. Therefore, from Theorem 5.4.4 it follows that $\mathrm{GK}(K[S]) = \mathrm{GK}(K[T])$ for a cancellative subsemigroup T of S. Finally, Theorem 3.5.6 implies that $\mathrm{GK}(K[U]) = \mathrm{rk}(U)$ for every finitely generated subsemigroup U of T. Therefore $\mathrm{GK}(K[T]) = \mathrm{rk}(T)$ and the remaining assertion follows. □

5.5 Comments and problems

The results on finite generation of $K[S]$, proved in Section 5.1, come from [119] and [79]. The necessary and sufficient conditions for $K[S]$ to be

Noetherian were obtained in [124]. The case of monomial semigroups, Theorem 5.3, and some of its applications were considered in [80]. We followed [121] for the results on the Gelfand-Kirillov dimension of $K[S]$. Theorem 5.1.6 comes from [78]. Some of the results of this chapter can also be proved under the formally weaker assumption that S satisfies the ascending chain condition on right ideals. In particular, it has been shown by Hotzel [64] that, for some special classes of semigroups S with the latter property, S necessarily is finitely generated. However, it is unknown whether this holds in general. Some other consequences of this finiteness condition, related to our results, can be found in [64, 65, 98].

Problems

1. Is S finitely generated whenever $K[S]$ is right Noetherian? This is not known in general, even if S is a submonoid of a group. See Theorem 5.1.5.

2. If $K[S]$ is right Noetherian and $\mathrm{GK}(K[S]) < \infty$, does S have a finite ideal chain with uniform or nilpotent factors? See Theorem 5.2.3.

3. Do we have $\mathrm{GK}(K[S]) = \mathrm{GK}(K[S]/\mathcal{B}(K[S])$ for every right Noetherian semigroup algebra $K[S]$? In this case, it follows that $\mathrm{GK}(K[S]) = \mathrm{GK}(K[T])$ for a cancellative subsemigroup T of S. See Theorem 5.4.4.

4. Assume that S is a Malcev nilpotent monoid. If S is finitely generated and satisfies the ascending chain condition on right ideals, is $K[S]$ right Noetherian?

Chapter 6

Principal ideal rings

In this chapter we study semigroup algebras $K[S]$ that are principal right ideal rings. First, we show that these are finitely generated PI algebras of Gelfand-Kirillov dimension at most 1. Then we completely characterize finite semigroups S for which $K[S]$ is a principal right ideal ring. It turns out that the algebra $K[S]$ is a finite direct product of matrix algebras over principal right ideal semigroup algebras of a very special type. Moreover, this decomposition comes from an ideal chain of S. In the last section, this result is extended to the class of arbitrary principal right and left ideal rings. So, a complete description of principal (right and left) ideal semigroup algebras $K[S]$, and of the underlying semigroups S, is obtained. It follows that $K[S]$ is a semiprime principal right ideal ring if and only if it is a principal left ideal ring; also a description of this class of rings is obtained. It is also shown that every prime contracted semigroup algebra $K_0[S]$ which is a principal ideal ring must be of the form $M_n(K)$, $M_n(K[X])$ or $M_n(K[X,X^{-1}])$ and the structure of S is completely determined.

The results show in particular that in many cases principal ideal semigroup algebras $K[S]$ are built of blocks that are matrix rings over algebras of the type $K[T']$, where T' is a homomorphic image of a cancellative monoid T so that $K[T]$ is a principal ideal ring.

Recall that a ring R (with identity) is said to be a principal right ideal ring if every right ideal $I$of R is generated by one element, that is, $I = rR$ for some $r \in R$. Clearly such rings are right Noetherian and this property is inherited by epimorphic images. In contrast with other chapters we work with semigroups that are not necessarily monoids, the reasons being that we can resolve the problem for arbitrary semigroups but also that this more general case shows up naturally in the reduction steps used in the proofs. Of course we assume that semigroup algebras have an identity.

6.1 Group algebras

In this section we give a characterization of group algebras that are principal ideal rings. This will be a first step to deal with the problem for arbitrary semigroup algebras. However we therefore are forced to deal with group rings $R[G]$ over $n \times n$ matrices $R = \mathrm{M}_n(K)$ over a field K. Note that the transpose map on R and the classical involution on G, defined by mapping $g \in G$ onto g^{-1}, extend naturally to an involution on $R[G]$. It follows that $R[G]$ is right Noetherian if and only if it is left Noetherian. Also $R[G]$ is a principal right ideal ring if and only if it is a principal left ideal ring.

We begin with several technical lemmas.

Lemma 6.1.1. *Let R be a prime Noetherian ring. Assume I is a proper ideal of R with $I = rR$ for some $r \in R$ and J is another ideal of R. The following properties hold.*

1. *If $IJ = J$ then $J = \{0\}$.*
2. *If J is a prime ideal and $J \subset I$ then $J = \{0\}$.*

Proof. To prove the first part, suppose $IJ = J$. Suppose $J \neq \{0\}$. Because R is prime and right Noetherian we know by Proposition 3.1.4 that J contains a regular element, say x. Hence, because R also is left Noetherian, there exists a principal left ideal Ry that is maximal amongst the left ideals Rz generated by a regular element z that belongs to J. So, $y \in J = IJ = rJ$ and thus $y = ry_1$ for some $y_1 \in J$. Since y is regular it is clear that $\mathrm{ra}_R(y_1) = \{0\}$. So, by Proposition 3.1.4, also y_1 is regular. Furthermore, $Ry \subseteq Ry_1$. The maximality condition yields that $Ry = Ry_1$ and thus $y_1 = r_1 y$ for some $r_1 \in R$. It follows that $y = ry_1 = rr_1 y$ and hence $(1 - rr_1)y = 0$. Since y is regular we get that $1 = rr_1$ and hence, again by Proposition 3.1.4, r is invertible in R. So $I = rR = R$, a contradiction.

To prove the second part, suppose J is a prime ideal and J is properly contained in I. Obviously, $\overline{R} = R/J$ is a Noetherian prime ring. Let $\overline{r}$ and $\overline{I}$ denote the natural images in $\overline{R}$ of r and I respectively. Clearly $\overline{I} = \overline{r}\overline{R}$ is a nonzero ideal of $\overline{R}$. Thus the primeness of $\overline{R}$ yields that $\mathrm{la}_{\overline{R}}(\overline{r}) = \{0\}$ and hence $\overline{r}$ is a regular element of $\overline{R}$. Because $J \subset I = rR$, we have that for any $x \in J$ there exists $y \in R$ so that $x = ry$. Hence $0 = \overline{x} = \overline{ry}$. Thus $\overline{y} = 0$, that is, $y \in J$. We therefore have shown that $J = rJ = IJ$. The first part hence yields that $J = \{0\}$, as desired. $\square$

Let K be a field and G a group. Recall that $\omega(K[G])$ denotes the augmentation ideal of $K[G]$. For any positive integer n the dimension subgroup $D_n(K[G])$ is defined by

$$D_n(K[G]) = \{x \in G \mid x - 1 \in \omega(K[G])^n\}.$$

It is well known ([129, Section 1.3.3]) that the dimension subgroups form a decreasing sequence of normal subgroups of G starting with $D_1(K[G]]) = G$ and $[D_n(K[G]), D_m(K[G])] \subseteq D_{n+m}(K[G])$. Furthermore, if $\mathrm{char}(K) = 0$ then $G/D_n(K[G])$ is torsion free for all n. If $\mathrm{char}(K) = p > 0$ then $g^p \in D_{np}(K[G])$ for any $g \in D_n(K[G])$; in particular, each $G/D_n(K[G])$ is a p-group.

Recall that a group G is said to be a p'-group if G does not contain elements of order p.

Lemma 6.1.2. *Let G be a finite group, K a field with $\mathrm{char}(K) = p > 0$ and $R = \mathrm{M}_n(K)$. If $\omega(R[G])$ is principal as a right ideal then G contains a normal p'-subgroup N so that G/N is a cyclic p-group.*

Proof. Let $I = \omega(R[G])$ and $T = R[G]$. By assumption $I = \alpha T$ for some $\alpha \in T$. Since I is an ideal we have $T\alpha \subseteq \alpha T$.

We first show that $T\alpha = \alpha T$. Let $\widehat{G} = \sum_{g\in G} g \in T$. It is readily verified that $\mathrm{la}_T(\alpha) = \mathrm{la}_T(I) = R\widehat{G}$ and thus $\dim_K(\mathrm{la}_T(\alpha)) = n^2$. Since the K-linear epimorphism $f : T \longrightarrow T\alpha$ defined by $f(t) = t\alpha$ has kernel $\mathrm{la}_T(\alpha)$ we get that $n^2|G| = \dim_K(T) = \dim_K(\ker(f)) + \dim_K(T\alpha)$. Thus $\dim_K(T\alpha) = n^2(|G| - 1)$. On the other hand $\dim_K(I) = \dim_K(\mathrm{M}_n(\omega(K[G])) = n^2(|G| - 1)$. So $\dim_K(T\alpha) = \dim_K(\alpha T)$. Since also $T\alpha \subseteq \alpha T$ we indeed obtain that $T\alpha = \alpha T$. It follows that $T\alpha^m = \alpha^m T$ for all positive integers m.

Let N be the last term in the descending chain of the dimension subgroups. So N is a normal subgroup of G and G/N is a p-group, because $\mathrm{char}(K) = p > 0$. We next show that N is a p'-subgroup of G. Because $K[G]$ is finite dimensional there exists a positive integer k so that $\omega(K[G])^k = \omega(K[G])^{k+1}$. Since T can be naturally identified with $\mathrm{M}_n(K[G])$ and thus $I = \mathrm{M}_n(\omega(K[G]))$ it follows that $T\alpha^k = I^k = I^{k+1} = \alpha^{k+1}T$. Thus $\alpha^k = \alpha^{k+1}\beta$ for some $\beta \in T$. So $\alpha^k(1-\alpha\beta) = 0$ and hence $I^k(1-\alpha\beta) = \{0\}$. Now by definition of N we have that $\omega(K[N]) \subseteq \omega(K[G])^k$. The latter is, via the diagonal embedding, contained in $\omega(R[G])^k = I^k$. So $\omega(K[N])(1-\alpha\beta) = \{0\}$. Write $1 - \alpha\beta$ as a matrix (δ_{ij}) with each $\delta_{ij} \in K[G]$. Then for any $\gamma \in \omega(K[N])$ we have that

$$\begin{pmatrix} \gamma & 0 & \cdots & 0 \\ 0 & \gamma & \cdots & 0 \\ \vdots & & & \vdots \\ 0 & 0 & \cdots & \gamma \end{pmatrix} (\delta_{ij}) = 0$$

and so $\gamma\delta_{ij} = 0$. Thus each $\delta_{ij} \in \mathrm{ra}_{K[G]}(\omega(K[N])) = \widehat{N}K[G[$. On the other hand, since $\alpha \in I$ we therefore have that $1 - \alpha\beta \notin I$ and thus it follows that $\widehat{N} \notin \omega(K[G])$. This shows that $|N|$ is nonzero in K and hence N is a p'-group.

It remains to show that G/N is cyclic. Suppose the contrary. Since G/N is a finite p-group it then follows that G/N and thus G has a homomorphic image W that is isomorphic to $C_p \times C_p = \mathrm{gr}(w_1) \times \mathrm{gr}(w_2)$, a direct product of two cyclic groups of order p. Note that $\omega(K[W]) = K[W](1 - w_1) + K[W](1 - w_2)$ and $\gamma^p = 0$ for any $\gamma \in \omega(K[W])$. It is readily verified that $(1 - w_1)^{p-1}(1 - w_2)^{p-1} \neq 0$. Hence it follows that $\omega(K[W])$ is a nilpotent ideal of degree $2p - 1$. Because $2p - 1 > p$ we obtain that $\omega(K[W])$ cannot be principal. Now $\omega(R[G]) = \mathrm{M}_n(\omega(K[G]))$ is a principal right ideal. Since $\mathrm{M}_n(\omega(K[W])$ is a homomorphic image of $\mathrm{M}_n(\omega(K[G]))$ it follows that also $\mathrm{M}_n(\omega(K[W]))$ is a principal right ideal of $\mathrm{M}_n(K[W])$. Using the determinant map we thus get that $\omega(K[W])$ is a principal ideal of $K[W]$, a contradiction. Hence we conclude that G/N indeed is cyclic. □

Lemma 6.1.3. *Let G be a nontrivial group, K a field and $R = \mathrm{M}_n(K)$. Assume $R[G]$ is a prime ring. If $R[G]$ is right Noetherian and $\omega(K[G])$ is principal as a right ideal then G is infinite cyclic.*

Proof. Let $I = \omega(R[G])$. By assumption $I = \alpha R[G]$ for some $\alpha \in R[G]$ and $R[G]$ is right and thus also left Noetherian. So G is finitely generated. We first show that G/G' is infinite cyclic, where $G' = [G, G]$ is the commutator subgroup of G. Note that $G' \subseteq D_2(K[G])$. Since $R[G]$ is prime, Lemma 6.1.1 yields that $I^{m+1} \subset I^m$ for any positive integer m. Hence also $\omega(K[G])^{m+1} \subset \omega(K[G])$ and, in particular, $\omega(K[G])^2 \subset \omega(K[G])$. So $D_2(K[G]) \subset G$. If $\mathrm{char}(K) = 0$ we obtain then that $G/D_2(K[G])$ is torsion free and abelian. Hence $G/D_2(K[G])$ and thus also G/G' is infinite.

If $\mathrm{char}(K) = p > 0$ then we put $N = \bigcap_{i=1}^{\infty} D_i(K[G])$. Clearly $I = \omega(K[N])K[G] \subseteq \omega(K[G])^m$ for any positive integer m. Hence, it follows that $I \subseteq \omega(K[G])^{m+1} \subset \omega(K[G])^m$ for any $m > 0$. It follows that $|G/N| = \dim_K(K[G]/\omega(K[N])K[G]) = \infty$. Because G is finitely generated and $\mathrm{char}(K) = p > 0$ the group $G/D_m(K[G])$ is a finite p-group. Lemma 6.1.2

yields that this group is also cyclic. Thus $G' \subseteq D_m(K[G])$ for all m. So $G' \subseteq N$ and again G/G' is infinite.

Therefore, we have shown that always G/G' is an infinite finitely generated group. Hence there exists a normal subgroup W of G so that G/W is infinite cyclic. Put $J = \mathrm{M}_n(\omega(K[W]))$, a prime ideal of the ring $\mathrm{M}_n(K[G]) = R[G]$. Clearly $J \subset I$ and therefore Lemma 6.1.1 gives that $J = \{0\}$. So $W = \{1\}$ and thus G is infinite cyclic. □

Theorem 6.1.4. *Let G be a group and $R = \mathrm{M}_n(K)$ for some n and some field K. The following conditions are equivalent.*

1. *$R[G] = \mathrm{M}_n(K[G])$ is a principal right ideal ring.*
2. *$R[G]$ is right Noetherian and the augmentation ideal $\omega(R[G])$ is a principal right ideal.*
3. *If $\mathrm{char}(K) = 0$ then G is finite or finite-by-(infinite cyclic).*
   If $\mathrm{char}(K) = p > 0$ then G is (finite p')-by-(cyclic p-group) or G is (finite p')-by-(infinite cyclic).

Proof. Clearly (1) implies (2). To prove that (2) implies (3), suppose that $R[G]$ is right Noetherian and $\omega(R[G])$ is a principal right ideal. Then all subgroups of G are finitely generated and, in particular, the torsion subgroup $N = \Delta^+(G)$ of the finite conjugacy center $\Delta(G)$ is a finite normal subgroup. Since G/N has no nontrivial finite subgroups, Theorem 3.2.8 implies that $K[G/N]$, and thus also $R[G/N]$, is a prime ring. Lemma 6.1.3 then implies that G/N is either trivial or infinite cyclic. Hence G is either finite or finite-by-(infinite cyclic). Thus the characteristic zero case is shown.

Assume now that $\mathrm{char}(K) = p > 0$. If G is finite then the result follows from Lemma 6.1.2. If G is infinite then by the above $G = \langle N, x\rangle$ for some $x \in G$ of infinite order modulo N. Clearly x^m centralizes N for some positive integer m. Thus $x^m \in \mathrm{Z}(G)$. Consider the finite epimorphic image $\overline{G} = G/\langle x^{mp}\rangle$. Because of Lemma 6.1.2 the Sylow p-subgroup of $\overline{G}$ is cyclic. Furthermore, since the image $\overline{x}^m \in \overline{G}$ is an element of order p not contained in the normal subgroup $\overline{N}$ it follows that $\overline{N}$ is a p'-group. As $N \cong \overline{N}$ the group N also is a p'-group. So G satisfies condition (3).

We now prove that (3) implies (1). So assume $G = \langle N, x\rangle$ with N a finite normal subgroup such that $|N|$ is nonzero in K and $x \in G$ has infinite order modulo N. It is easily seen that $R[G] = (R[N])[x, x^{-1}, \sigma]$, a skew Laurent polynomial ring over the group ring $R[N]$. Because of Theorem 3.4.5 the ring $R[N]$ is semisimple Artinian. Theorem 3.7.1 therefore yields that $R[G]$ is a principal ideal ring. It is clear that every group G_1 satisfying the properties listed in (3) is an image of such a group G. Indeed, if H is a normal and finite p'-subgroup of G_1 so that G_1/H is cyclic of order p^n then any p-Sylow subgroup F of G_1 has order p^n. So G_1 is a semidirect product of H and F. Therefore the action of F on H yields an action of the infinite cyclic group C on H, so indeed G_1 is a homomorphic image of a semidirect product of H and C. It follows that the group rings of all groups listed in (3) are principal ideal rings. Hence (1) has been proven. □

6.2 Matrix embedding

Let S be a semigroup and let K be a field. In this section we prove that if $K[S]$ is a principal right ideal ring then $K[S]$ satisfies a polynomial identity. The proof relies, among other things, on a reduction similar to that used in the proof of Theorem 5.1.6.

In the proof we will make use of the following notation. Let Q be a prime ideal of $K[S]$ and T a subset of $K[S]/Q$. Recall that the subalgebra of $K[S]/Q$ generated by T is denoted $K\{T\}$. By $\sim_Q$ we denote the congruence relation on S defined by the ideal Q, that is $s \sim_Q t$ if $s - t \in Q$, for $s, t \in S$. Then $S/\sim_Q$ can be identified with the image of S in $K[S]/Q$ and we have a natural morphism $K[S/\sim_Q] \longrightarrow K[S]/Q$. Moreover, if $K[S]/Q$ is a right Goldie ring, then $S/\sim_Q$ embeds into a matrix ring $\mathrm{M}_n(D)$ over a division ring D.

Theorem 6.2.1. *Let K be a field and S a semigroup. If $K[S]$ is a principal right ideal ring then $K[S]$ satisfies a polynomial identity.*

Proof. By Noetherian induction, we may assume that $K[S/\sim]$ satisfies a polynomial identity for every proper congruence $\sim$ on S. Because of Theorem 3.1.1 and Theorem 3.1.2, $K[S]$ has nilpotent prime radical and has only finitely many minimal prime ideals. Hence it is enough to deal with the case that $S = S/\sim_Q$ for a prime ideal Q of $K[S]$. Considering S as a subsemigroup of the classical ring of right quotients $\mathrm{Q}_{cl}(K[S]/Q)$, from Lemma 5.1.2 we get that S has an ideal uniform component I with a completely 0-simple closure $\widehat{I} = \mathcal{M}(G, X, Y; P)$ such that $\widehat{S} = S \cup \widehat{I}$ is a subsemigroup of $\mathrm{Q}_{cl}(K[S]/Q)$. Furthermore, if we identify G with a fixed maximal subgroup of $\widehat{I}$, then G is the group of right quotients of $G \cap I$ and $\widehat{I}$ is an ideal of $\widehat{S}$, Proposition 5.1.3. Note that for any 'row' C of $\widehat{I}$, the set $K_0[C \cap I]$ is a right ideal of the right Noetherian ring $K_0[S]$. Hence X is a finite set, say $r = |X| < \infty$.

In view of Proposition 5.1.3, in order to prove the theorem it is now sufficient to show that $K[G]$ is a PI algebra. First note that, because of Proposition 5.1.3, $K[G]$ is (right) Noetherian and G is finitely generated. Next we show that we may assume that $K[G]$ is prime. Suppose the contrary, then by Theorem 3.2.8 the group G has a nontrivial finite normal subgroup. Since $K[G]$ is Noetherian, there exists a maximal finite normal subgroup N of G. Then the natural group epimorphism $G \longrightarrow G/N$ determines a semigroup homomorphism from $\widehat{I} = \mathcal{M}(G, X, Y; P)$ to the corresponding completely 0-simple semigroup over G/N. This then determines a congruence relation on $\widehat{I}$. Extending it via the identity relation on $\widehat{S} \setminus \widehat{I}$ we obtain a relation $\sim$ on $\widehat{S}$ and thus on S. It is readily verified that $\sim$ is a congruence relation. Hence we have a semigroup homomorphism $S \longrightarrow S_N = S/\sim$. Note that $\sim$ is not trivial. Indeed let $1 \neq n \in N$. Since G is the group of right quotients of $S \cap G$, there exist $s, t \in S \cap G$ such that $n = ts^{-1}$, and thus $ns = t$. Therefore $s \sim t$, in particular $\sim$ is nontrivial. Hence our inductive hypothesis implies that $K[S_N]$ is a PI algebra. It contains $K[T]$ for a subsemigroup T of S_N that generates G/N. Therefore Theorem 3.1.9 implies that $K[G/N]$ is PI and G/N is finite-by-abelian-by-finite. Since G is finitely generated, this implies that G is abelian-by-finite. Hence, again by Theorem 3.1.9, $K[G]$ is a PI algebra; proving the claim. So, we indeed may assume that $K[G]$ is prime. Therefore $K[S \cap G]$ is prime by Theorem 3.2.8.

Let Q' be an ideal of $K_0[S]$ which is maximal with respect to the property $Q' \cap K[G] = \{0\}$. Then Q' is maximal with respect to $Q' \cap K[S \cap G] = \{0\}$ because G is the group of right quotients of $S \cap G$. Because $K[G]$ is prime it follows that Q' is a prime ideal of $K_0[S]$. Let $\varphi : K_0[S] \longrightarrow K_0[S]/Q'$ be the natural epimorphism. Note that $K_0[S]/Q'$ is a right order in a simple Artinian ring, say $\mathrm{M}_n(D)$ for a division algebra D. Suppose now first that $\sim_{Q'}$ is not trivial. Then, by the induction hypothesis, $K_0[S/\sim_{Q'}]$ is a PI algebra. Hence $K_0[S]/Q'$ is PI as well. Since $Q' \cap K[G] = \{0\}$, the subring $K[S \cap G]$ is also a PI algebra. Hence, because G is the group of right quotients of $S \cap G$, it follows from Theorem 3.1.9 that $K[G]$ is PI; proving the claim. So we may assume that $\sim_{Q'}$ is trivial, in particular $S \cong S' = \varphi(S) \subseteq \mathrm{M}_n(D)$. Let J be the set of matrices of minimal rank in the cancellative semigroup $\varphi(S \cap G) \cong S \cap G$. Since we know that G is the group of right quotients of $S \cap G$, the semigroup $\varphi(S \cap G)$ has a group of right quotients isomorphic to G. Because J is an ideal of $\varphi(S \cap G)$, the group G is isomorphic to the group of right quotients of J. From Lemma 2.1.8 it follows that J is contained in a maximal subgroup of the corresponding principal factor of $\mathrm{M}_n(D)$. Clearly, the group $G' \subset \mathrm{M}_n(D)$ generated by J is its group of right quotients; in particular $G' \cong G$. Let I' be the uniform component of

$S' \subseteq M_n(D)$ containing J. Suppose that there exists a nonzero ideal L in S' such that $L \cap I' = \emptyset$ or $L \cap I' = \{0\}$ (the latter is possible if I' is an ideal uniform component of S' and $0 \in S'$). Then, by the induction hypothesis, $K_0[S'/L]$ is a PI algebra. Since $K_0[J] \subseteq K_0[S'/L]$, we therefore obtain that $K[J]$ is PI. Hence, by Theorem 3.1.9 $K[G'] \cong K[G]$ is a PI algebra as well; again proving the claim. So we may assume that every nonzero ideal L of S' intersects I' nontrivially (that is $L \cap I' \neq \emptyset$ and $L \cap I' \neq \{0\}$). This implies that I' is the unique ideal uniform component of S'. Lemma 2.5.2 allows us to construct the semigroup $\widehat{S'} = S' \cup \widehat{I'} \subseteq M_n(D)$, in which $\widehat{I'}$ is an ideal. So, replacing Q by Q' and I, G by I', G', we may assume that $Q \cap K[G] = \{0\}$. Moreover we may identify S and S'. We then have a natural homomorphism $\varphi : K_0[\widehat{S}] \longrightarrow M_n(D)$.

Put $A = K_0[\widehat{I}]$. Then $A = \mathcal{M}(K[G], X, Y; P)$ is a Munn algebra over $K[G]$, see Section 2.1. So, if necessary, we consider the elements of A as $X \times Y$ matrices with entries in $K[G]$. Since $K[S]$ is a principal right ideal ring, $K_0[I] = aK_0[S]$ for some $a \in K_0[I]$. Because $\widehat{I} = I\widehat{I}$, we obtain that $A = K_0[I]A = aA$. This means that $a \circ P \circ A = A$, where $\circ$ denotes the classical matrix multiplication. Hence $a' \circ P' \circ B = B$, where $B = \mathcal{M}(K, X, Y; P')$ is the image of A coming from the natural epimorphism $K[G] \longrightarrow K$ and a' is the image of a. If $|X| > |Y|$, then by Lemma 2.5.7, $c \circ a' = 0$ for a nonzero $Y \times X$ matrix c over K. Hence $c \circ B = c \circ a' \circ P' \circ B = \{0\}$, which is of course impossible because B is the set of all $X \times Y$ matrices over K. Therefore $r = |X| \leq |Y|$. So let Y' be a subset of Y with $|Y'| = r$. Since $aA = A$, there exists $x \in A$ such that $a \circ P \circ x$ is the identity $X \times Y'$ matrix (we impose an order on the set Y and also on X). Then x can be chosen so that $x \in \tilde{A}$, where the latter is defined as the algebra of matrix type consisting of columns of A indexed by Y'. The elements of $\tilde{A}$ can be identified with those of $M_r(K[G])$. Viewing x in this way we therefore have $(a \circ P) \circ x = 1$ in $M_r(K[G])$. Since $K[G]$ and thus $M_r(K[G])$ is Noetherian, both x and $a \circ P$ are invertible in $M_r(K[G])$, so that $x \circ a \circ P = 1$ in $M_r(K[G])$. Moreover, for any $y \in A$, we obtain $x \circ a \circ P \circ y = y$, and thus $x \circ a$ is a left identity of A. In particular,

$$A = K_0[\widehat{I}] \longrightarrow M_r(K[G]) : y \mapsto y \circ P$$

is a ring epimorphism. Since A has a left identity, it is also clear that the left annihilator $\mathrm{la}_A(A)$ of A in A satisfies $\mathrm{la}_A(A) = \{y \in A \mid y \circ P = 0\} = \{y \in A \mid P \circ y \circ P = 0\}$, and $\mathrm{la}_A(A)$ is the kernel of the above map. In particular, $\mathrm{la}_A(A)$ is a prime ideal of A because $M_r(K[G])$ is a prime algebra.

We come back to the natural homomorphism $\varphi : K_0[\widehat{S}] \longrightarrow M_n(D)$. For any subset V of $\widehat{S}$ we denote by $K\{V\}$ the subalgebra of $M_n(D)$ generated by $\varphi(V)$. The restriction of the map φ to $A = K_0[\widehat{I}]$ is denoted by ϕ. Since $K\{\widehat{I}\}$ is prime (because $K\{I\} \subseteq K\{\widehat{I}\}$ is a right order in $\mathrm{M}_n(D)$), we get that $\ker(\phi)$ is a prime ideal of A. Since $\mathrm{la}_A(A)$ is a nilpotent ideal of A, we get $\mathrm{la}_A(A) \subseteq \ker(\phi)$. We know that $\ker(\varphi) \cap K[G] = Q \cap K[G] = \{0\}$. Therefore, from Proposition 2.1.2 it follows that $\ker(\phi)$ is nilpotent and consequently $\mathrm{la}_A(A) = \ker(\phi)$, because $\mathrm{la}_A(A)$ is a prime ideal of A. Hence, $\phi(A) = K\{\widehat{I}\} \cong M_r(K[G])$ and we obtain the following natural embeddings:

$$K\{I\} \subseteq K\{\widehat{I}\} \cong \mathrm{M}_r(K[G]) \subseteq \mathrm{M}_r(D'),$$

where D' is the simple Artinian ring of right quotients of $K[G]$. Note that $\mathrm{M}_r(D')$ is the classical ring of right quotients of $K\{\widehat{I}\}$. But $\mathrm{M}_n(D)$ is the ring of right quotients of $K\{S\}$, whence of its ideal $K\{I\}$. Since $K\{\widehat{I}\}$ is an ideal of the prime ring $K\{\widehat{S}\}$ and since it has an identity, it follows that $K\{S\} \subseteq K\{\widehat{S}\} = K\{\widehat{I}\}$. Therefore $\mathrm{M}_n(D)$ is also the ring of right quotients of the intermediate ring $K\{\widehat{I}\}$. Hence $n = rk$ and $D' \cong \mathrm{M}_k(D)$ for some $k \geq 1$. Under the natural identification, we get

$$K\{I\} \subseteq K\{S\} \subseteq K\{\widehat{I}\} = \mathrm{M}_r(K[G]) \subseteq \mathrm{M}_r(D').$$

From Theorem 3.7.3 we know also that $K\{S\} \cong \mathrm{M}_q(C)$ for a right Noetherian domain C and some $q \geq 1$. The uniqueness of the classical ring of quotients implies that $q = rk$ and D is the quotient ring of C. Our aim is to show that $\mathrm{M}_r(K[G])$ is a right localization of $K\{S\}$.

Consider the homomorphism $K_0[\widehat{I}] \xrightarrow{\eta} \mathrm{M}_r(K[G]) \xrightarrow{\omega} \mathrm{M}_r(K)$, where the first map is given by $y \mapsto y \circ P$ and the second is determined by the augmentation map $K[G] \longrightarrow K$, that is $g \mapsto 1$ for $g \in G$. We know that each of these homomorphisms is onto. Also, consider $K_0[\widehat{I}] \xrightarrow{\omega'} K_0[\mathcal{M}(\{1\}, X, Y, P')] = \mathcal{M}(K, X, Y, P') \xrightarrow{\eta'} \mathrm{M}_r(K)$, where the first map is determined by $(g, x, y) \mapsto (1, x, y)$ for $g \in G$ and the second is defined by $y' \mapsto y' \circ P'$, with P' obtained from P by applying the augmentation map to the entries of P. Clearly, $\omega\eta = \eta'\omega'$. So $\eta'\omega'(K_0[\widehat{I}]) = \mathrm{M}_r(K)$. Since $\omega'(\widehat{I}) = \omega'(I)$, we get $\omega(K\{I\}) = \mathrm{M}_r(K)$. But $\ker(\omega) \cap K\{I\}$ is an ideal of $K\{S\}$. As $K\{S\} \cong \mathrm{M}_{rk}(C)$, it follows that $\mathrm{M}_r(K) \cong K\{I\}/(\ker(\omega) \cap K\{I\}) \cong \mathrm{M}_{rk}(C/T)$ for an ideal T of C. Therefore $k = 1$ and hence $D = D'$ is a division ring of quotients of $K[G]$.

Now, since P is left invertible, the rank of P (as a matrix over D) is equal to r. Therefore P contains an $r \times r$ submatrix that is invertible. Let P_J be such a submatrix and let J be the corresponding subsemigroup of I. So $J = I \cap \widehat{J}$, where $\widehat{J}$ is the union of $\mathcal{L}$-classes of $\widehat{I}$ intersecting J. Then

$\widehat{J}$ is the completely 0-simple closure of J in $\widehat{I}$ and $\mathcal{M}(G,r,r,P_J) \cong \widehat{J} \subseteq \widehat{I}$. The identity of $K_0[\widehat{J}] \cong \mathrm{M}_r(K[G])$, see Lemma 2.5.7, is a left identity of $K_0[\widehat{I}]$ because $\widehat{J}\widehat{I} = \widehat{I}$. So η maps the identity of $K_0[\widehat{J}]$ onto the identity of the prime algebra $\mathrm{M}_r(K[G])$. It follows that $K\{\widehat{J}\} = K\{\widehat{I}\}$. Since $\ker(\eta)$ is nilpotent, $\eta_{|K_0[\widehat{J}]} : K_0[\widehat{J}] \longrightarrow K\{\widehat{J}\}$ is an isomorphism. Therefore we get

$$K\{J\} \subseteq K\{I\} \subseteq K\{S\} \subseteq K\{\widehat{I}\} = K\{\widehat{J}\} = \mathrm{M}_r(K[G])$$

and $\mathrm{M}_r(K[G])$ is a right localization of $K\{J\}$ because $K_0[\widehat{J}]$ is a right localization of $K_0[J]$ by Lemma 4.5.6. Then $\mathrm{M}_r(K[G])$ is also a right localization of the principal right ideal ring $K\{S\}$. Hence $\mathrm{M}_r(K[G])$ is a prime principal right ideal ring. Theorem 6.1.4 therefore yields in view of Theorem 3.2.8 that G is cyclic. This completes the proof of the claim that $K[G]$ is a PI algebra; and hence finishes the proof of the theorem. □

We derive some consequences of the theorem and of its proof.

Corollary 6.2.2. *Let K be a field and S a semigroup. If $K[S]$ is a principal right ideal ring, then S is finitely generated and $K[S]$ embeds into a matrix ring $\mathrm{M}_n(F)$ over a field extension F of K. In particular, S is a linear semigroup. The groups associated to S are either finite or finite-by-(infinite cyclic).*

Proof. By Theorem 5.1.5 $K[S]$ is finitely generated because it is a right Noetherian PI algebra. Theorem 3.5.2 implies that a finitely generated right Noetherian PI algebra embeds into a matrix ring over a field. This proves the first part of the result.

To prove the second part, let H be a maximal subgroup of the multiplicative monoid $\mathrm{M}_n(F)$ that satisfies $S \cap H \neq \emptyset$. We have to show that $S \cap H$ has a group of fractions that is either finite or finite-by-(infinite cyclic). So, it is enough to prove this for every cancellative subsemigroup T of S. By Noetherian induction we may assume that cancellative subsemigroups of every proper homomorphic image of S have this property. Fix some T. Then we may assume that every nonzero ideal of S intersects T. Hence, the structure theorem for S (viewed as a linear semigroup, see Theorem 2.4.3) implies that S has a unique ideal uniform component I. It is easy to see that $T \cap I$ is contained in a maximal subgroup of $\widehat{I}$ because by Theorem 6.2.1 and Lemma 2.1.3 it satisfies the right and left Ore conditions. Since the group of quotients of $T \cap I$ is isomorphic to the group of quotients of T, it is enough to show that the maximal subgroup G of $\widehat{I}$ is of the desired type. We continue as in the proof of Theorem 6.2.1, showing that $K[G]$ is

Noetherian. Next, factoring by the congruence determined by a maximal finite normal subgroup N of G allows us to assume that $K[G]$ is prime. The proof of Theorem 6.2.1 shows then that $K[G]$ is a prime principal right ideal ring and G is cyclic. The result follows. $\square$

Remark 6.2.3. Assume that $K[S]$ is a principal right ideal ring. Let I be an ideal uniform component of S (treated as a linear semigroup). If $\widehat{I}$ is a 'square' completely 0-simple semigroup, then, since one-sided invertible elements of the Noetherian ring $\mathrm{M}_r(K[G])$ are invertible, it follows as in the proof of Theorem 6.2.1 that the sandwich matrix P is invertible in $\mathrm{M}_r(K[G])$. (First, one has to use Lemma 2.5.1 in order to construct the semigroup $\widehat{S}$, in which $\widehat{I}$ is an ideal, but a reduction to the prime case is not needed). Consequently, we obtain that $K_0[\widehat{I}] \cong \mathrm{M}_r(K[G])$.

Corollary 6.2.4. *If $K[S]$ is a principal right ideal ring, then the Gelfand-Kirillov dimension of $K[S]$ is equal to its classical Krull dimension and it is 0 or 1 (this also holds for any epimorphic image of $K[S]$). In the former case S is finite. Moreover, every right Artinian homomorphic image of $K[S]$ is finite dimensional over K.*

Proof. Because of Corollary 6.2.2, S is finitely generated and so $\mathrm{GK}(K[S]) = 0$ if and only if S is finite. From Theorem 6.2.1 we know that $K[S]$ is a PI algebra. Theorem 3.2.1 therefore implies that $K[S]/\mathcal{B}(K[S])$ is also left Noetherian. Then from Lemma 6.1.1 it follows that the classical Krull dimension of $K[S]$ is at most 1. Theorem 3.5.2 thus yields that $\mathrm{GK}(K[S]) = \mathrm{clKdim}(K[S]) \leq 1$ and the same holds for all homomorphic images of $K[S]$.

Finally, suppose that Q is an ideal of $K[S]$ such that $K[S]/Q$ is right Artinian. Then $\mathrm{GK}(K[S]/Q) = \mathrm{clKdim}(K[S]/Q) = 0$ implies that $K[S]/Q$ is finite dimensional over K because it is finitely generated. $\square$

6.3 Finite dimensional case

Recall from Theorem 3.7.3 that a finite dimensional algebra is a principal right ideal ring if and only if it is a principal left ideal ring. In this section we characterize semigroup algebras of finite semigroups that are principal ideal rings. Of course the finite groups listed in Theorem 6.1.4 provide a class of examples. We now first enlarge this class and later we show that these are the building blocks in the characterization. Because of their use in the solution of the problem for general semigroups we also include some infinite examples.

Proposition 6.3.1. *Let K be a field and let G be a finite group so that $|G|$ is nonzero in K. Let T be a monoid with group of units G such that $T = \bigcup_{i \geq 0} Gx^i$ for some $x \in T$ and $Gx = xG$. If the central idempotents of $K[G]$ commute with x then $K[T]$ is a principal ideal ring.*

Proof. Because of Theorem 3.2.8 and the assumptions on G we know that $K[G]$ is semisimple Artinian. So $K[G] = A_1 \oplus \cdots \oplus A_m$, where each A_i is a simple Artinian ring. Since, by assumption, the central idempotents of $K[G]$ commute with x and because $K[G]x = xK[G]$ we obtain that $Ix = xI$ for any ideal I of $K[G]$. Now for any $\alpha \in K[G]$ it is clear that $\alpha x = 0$ if and only if $K[G]\alpha K[G]x = 0$. Hence we get that

$$A = \{\alpha \in K[G] \mid \alpha x = 0\} = \{\alpha \in K[G] \mid x\alpha = 0\}$$

is an ideal of $K[G]$. Re-numbering the simple components, if needed, we thus may write $A = A_1 \oplus \cdots \oplus A_l$, with $l \leq m$. As $1 \in K[G]$ and $1 \notin A$ we also have that $l < m$. Let $B = A_{l+1} \oplus \cdots \oplus A_m$. It follows that $xb \neq 0$ and $bx \neq 0$ for every nonzero $b \in B$. Again since $K[G]x = xK[G]$, we thus get that for every $0 \neq b \in B$ there exists a unique $b' \in B$ such that $bx = xb' \neq 0$. This yields an algebra homomorphism $\sigma : B \longrightarrow B$ defined by mapping b onto b'. This map is an automorphism as also for every $0 \neq b \in B$ there exists a unique $b'' \in B$ so that $xb = b''x \neq 0$.

Clearly

$$K[T] = A \oplus \left(\sum_{i \geq 0} Bx^i \right),$$

a direct product. Again since central idempotents of $K[G]$ commute with x, the automorphism σ fixes all central idempotents. Hence

$$K[T] = A \oplus (\bigoplus_e B_e),$$

where $B_e = \sum_{i \geq 0} Bex^i$ is a subring of $K[T]$, and where the sum runs over the set of all primitive central idempotents of B. Obviously B_e is an epimorphic image of the skew polynomial algebra $Be[X, \sigma]$. Because Be is a simple Artinian ring, Theorem 3.7.1 yields that $Be[X, \sigma]$ is a principal ideal ring. It follows that also $K[T]$ is a principal ideal ring because it is an epimorphic image of $A \oplus \bigoplus_e (Be[X, \sigma])$. □

A few remarks on the semigroups considered in Proposition 6.3.1. If T is infinite then it follows that $Gx^i \cap Gx^j = \emptyset$ if $i \neq j$. With notations as in the proof of Proposition 6.3.1, it follows in this case that $B_e = \bigoplus_{i \geq 0} Bex^i$. Of course $K[T]$ is semiprime if and only if each B_e is semiprime. Since B_e is a $\mathbb{Z}$-graded ring, Theorem 3.4.2 yields that the prime radical of B_e is $\mathbb{Z}$-homogeneous. Since Be is simple it then easily is seen that B_e is semiprime if and only if either $ex = 0$ or $ex^i \neq 0$ for all $i \geq 1$ (or equivalently, either $B_e = Be$ or $B_e = (Be)[x, \sigma]$, a skew polynomial ring). Hence we have shown that $K[T]$ is semiprime if and only if for every central idempotent $e \in K[G]$ either $ex = 0$ or $ex^i \neq 0$ for all $i \geq 1$. It also readily follows that $K_0[T]$ is prime if and only if $G = \{1\}$, and thus $T = \langle x \rangle$, an infinite cyclic monoid.

It also is easily verified that if T is infinite then T is cancellative if and only if $|Gx^i| = |G|$ for every $i \geq 0$. In this case $K[T] = (K[G])[x, \sigma]$, a skew polynomial ring over the group algebra $K[G]$.

If T is finite then the cyclic semigroup $\prec x \succ$ is finite and thus contains a unique idempotent, say $e = x^n$. Hence eT is an ideal of T. Furthermore, eT is a finite group with identity e and if $eT \neq T$ (that is, if $x \notin G$) then the Rees factor T/eT can be written as a union $\bigcup_{0 \leq i \leq n} Gx^i$ but with $x^n = \theta$, where θ is the zero of T/eT. It will be shown that, apart from the groups listed in Theorem 6.1.4, the monoids of the first and latter type are the building blocks of semigroup algebras that are principal ideal rings.

Let K be a field and let S be a finite semigroup. If the finite dimensional algebra $K[S]$ is a principal ideal ring then it follows from Theorem 3.7.3 $K_0[S] = R_1 \oplus \cdots \oplus R_m$, a direct product of matrix rings R_i over a chain ring (of course the rings R_i also are principal ideal rings). So, the rings R_i have a unique maximal ideal, say M_i, and every other ideal is of the type M_i^k, for some k. Furthermore, R_i/M_i is a matrix ring over a division algebra and $M_i = R_i m_i = m_i R_i$ for some $m_i \in M_i$. The identity of each R_i is denoted by e_i. This notation will be used throughout without any further reference.

It will be shown that such an S has an ideal chain whose factors are of a very special type and they lead to a direct product decomposition of the algebra $K[S]$. Therefore, we first deal with finite semigroups that have an ideal chain $N \subseteq S$ so that N is a nilpotent ideal (as before, the case $N = \emptyset$ is allowed by putting $S/N = S$; of course it is implicitly assumed that S has a zero element in case $N \neq \emptyset$) and S/N is completely 0-simple. Since $K[S/N]$ has an identity element, Lemma 2.5.7 yields that $S/N \cong \mathcal{M}(G, n, n; P)$, where G is a maximal subgroup of S/N and the sandwich matrix P is invertible over $K[G]$. In the next three lemmas we investigate the particular case that $n = 1$. Note that then $S = G \cup N$ (a disjoint union) is a monoid with identity the idempotent of G. Indeed, since $K[G] = K[S]/K[N] = K_0[S/N]$, it is clear that $1 = e + \alpha$ for some $\alpha \in K[N]$ and e the identity of G. So, for any $x \in N \setminus N^2$, we get that $x = 1x = x1 = xe + x\alpha = ex + \alpha x$.

As $x\alpha, \alpha x \in K[N^2]$ and $xe, ex \in N$ we obtain that $x = xe = ex$. It follows that $1 = e$, as desired.

Lemma 6.3.2. *Assume S is a finite monoid so that $S = G \cup N$, with G a group and N a nilpotent ideal. If $K_0[S]$ is a principal ideal ring then $e \in K_0[S]$ is a central idempotent in $K_0[S]$ if and only if $e \in K[G]$ is a central idempotent in $K[G]$.*

Proof. Let $e \in K_0[S]$ be a central idempotent. Let $\alpha, \beta \in K_0[S]$ be so that $e = \alpha + \beta$, $\text{supp}(\alpha) \subseteq G = S \setminus N$ and $\text{supp}(\beta) \subseteq N$. Since N is an ideal and because both $\text{supp}(\alpha)$ and $\text{supp}(\alpha^2)$ are contained in the group G, the equality $\alpha + \beta = e = e^2 = \alpha^2 + \beta\alpha + \alpha\beta + \beta^2$ implies that $\alpha^2 = \alpha$. Since central idempotents of each R_i are trivial, $\alpha e_i + \beta e_i = e_i$ or $\alpha e_i + \beta e_i = 0$, where e_i denotes the identity of R_i. Hence, if $\alpha e_i + \beta e_i = 0$, then $\alpha = \alpha^2$ implies that

$$\alpha e_i = (\alpha e_i)^l = (-\beta e_i)^l = (-\beta)^l e_i,$$

for any $l > 0$. Because $K_0[N]$ is a nilpotent ideal of $K_0[S]$ it follows in this case that $\alpha e_i = 0$, and thus $\beta e_i = 0$. On the other hand $e_i = \alpha e_i + \beta e_i$ implies that $\alpha e_i = (1 - \beta) e_i$ is an invertible element in R_i. Recall from above that α is an idempotent and thus αe_i is an idempotent as well. Hence it follows that $\alpha e_i = e_i$ and thus again $\beta e_i = 0$. So $\beta e_i = 0$ for every e_i and hence $\beta = 0$. Thus $e = \alpha \in K[G]$. So we have shown that indeed all central idempotents of $K_0[S]$ belong to $K[G]$.

In particular, all central idempotents e_i belong to $K[G]$. As $K[G] \cong K_0[S]/K_0[N] = R_1/K_0[N]e_1 \oplus \cdots \oplus R_m/K_0[N]e_m$, the group algebra $K[G]$ has precisely as many primitive central idempotents as $K_0[S]$. So these must be $e_1, \ldots, e_m$. It follows that if $e \in K[G]$ is a central idempotent in $K[G]$ then e is a sum of some of the idempotents $e_1, \ldots, e_m$. Thus e is central in $K_0[S]$. □

Lemma 6.3.3. *Assume S is a finite monoid so that $S = G \cup N$, with G a group and N a nilpotent ideal. Let K be a field of characteristic p and assume $K[S]$ is a principal ideal ring. If $N \neq \emptyset$ and $N \neq \{\theta\}$ then the following properties hold.*

1. *$K[G]$ is semisimple, that is either $p = 0$ or $p \not| \, |G|$.*

2. *There exists $t \in S$ such that $t^k = \theta$, $S = G \cup Gt \cup \cdots \cup Gt^{k-1} \cup \{\theta\}$ and $Gt = tG$.*

Proof. Let H be a normal subgroup of G so that $|H| \neq 0$ in K. It is well known, and easily verified, that

$$\widehat{H} = \frac{1}{|H|} \sum_{h \in H} h$$

is a central idempotent of $K[G]$. Hence, by Lemma 6.3.2, $\widehat{H}$ is a central idempotent of $K_0[S]$. In particular, $\widehat{H}s = s\widehat{H}$ and thus $\mathrm{supp}(\widehat{H}s) = \mathrm{supp}(s\widehat{H})$ for any $s \in S$. Let $L_s = \{h \in H \mid hs = s\}$. Then L_s is a subgroup of H (recall the identity of G is the identity of S). Also note that $0 \neq |L_s|$. Write $H = \bigcup_{m \in M} mL_s$, a disjoint union, with $M \subseteq H$. Then $\hat{H}s = \sum_{m \in M} \frac{|L_s|}{|H|} ms$ and $ms \neq m's$ for $m \neq m'$. Hence $\mathrm{supp}(\widehat{H}s) = Ms = Hs$. Similarly $\mathrm{supp}(s\widehat{H}) = sH$. Consequently, $Hs = \{hs \mid h \in H\} = \{sh \mid h \in H\} = sH$. Note that if $s_1, s_2 \in S$ and $s_1H \cap s_2H \neq \emptyset$ then $s_1H = s_2H$. The classical arguments in group theory yield that $S/H = \{sH \mid s \in S\}$ is a monoid for the multiplication $(sH)(tH) = stH$, where $s, t \in S$. Clearly $K_0[S/H]$ is a natural epimorphic image of $K_0[S]$. Therefore $K_0[S/H]$ is a principal ideal ring as well.

We now prove the first part of the statement. For this it clearly is sufficient to deal with the case that $p > 0$. Clearly $K[G]$ is an epimorphic image of $K_0[S]$. Hence it also is a principal ideal ring and thus, by Theorem 6.1.4, we know that G has a normal subgroup H_1 such that G/H_1 is a cyclic p-group and H_1 is a p'-group. We have to prove $G = H_1$. Taking $H = H_1$ in the first part of the proof, it is clear that we may replace S by S/H and thus we may assume that G is a cyclic p-group, say with generator g. We then have to prove that G is trivial. Suppose the contrary, that is assume $g \neq 1$. Since $G = \mathrm{gr}(g)$ is a cyclic p-group and because $\mathrm{char}(K) = p$, the ideal $(1-g)K[G]$ is nilpotent and it is the unique maximal ideal of $K[G]$. Hence central idempotents of $K[G]$ are trivial. Thus, by Lemma 6.3.2, $K_0[S]$ has only trivial central idempotents. Consequently $K_0[S]$ is a matrix ring over a chain ring which has finite dimension over K. Let W be the ideal of $K_0[S]$ generated by $1 - g \in K[G]$ and let $V = K_0[N]$. As $W \not\subseteq V$ we have $V \subseteq W$. Hence $K_0[S]/W \cong K[G]/K[G](1-g) \cong K$ and thus W is the unique maximal ideal of $K_0[S]$. Therefore $V = W^l$ for some $l > 0$. By assumption $V \neq \{0\}$. Hence there exists $t \geq l$ such that $W^t \neq \{0\}$ and $W^{t+1} = \{0\}$. Choose $0 \neq \alpha \in W^t$ and $\alpha_1, \ldots, \alpha_{t+1} \in K_0[S]$ so that

$$0 \neq \alpha = \alpha_1(1-g)\alpha_2(1-g)\cdots\alpha_t(1-g)\alpha_{t+1} \in K_0[N].$$

Since $1 - g \in W$ and $W^t \subseteq K_0[N]$, we may assume that $\mathrm{supp}(\alpha_i) \subseteq G$ for all i. But then $\mathrm{supp}(\alpha) \subseteq G \cap N = \emptyset$, a contradiction. So we have shown that indeed G is trivial. This finishes the proof of the first part.

Next we prove the second part of the statement. Note that by assumption $\{\theta\} \subset N$. Taking $H = G$ in the first part of the proof it is then clear that we also may replace S by S/G. So we may assume that G is trivial. Hence Lemma 6.3.2 yields that then $K_0[S]$ has only one nonzero central idempotent and thus $K_0[S]$ is a matrix ring over a chain ring. Because $|N| > 1$ there exists $k > 1$ so that $N \supset N^2 \supset \cdots \supset N^k = \{\theta\}$ and $N^{k-1} \neq \{\theta\}$. Hence $StS \setminus N^2 = \{t\}$ for any $t \in N \setminus N^2$. Since ideals are ordered, it follows that $N \setminus N^2 = \{t\}$ for some $t \in S$. Clearly this implies that $N^i \setminus N^{i+1} = \{t^i\}$, for any $0 < i < k$. So the result follows. □

We are now in a position to describe principal ideal rings of finite semigroups of the type $S = G \cup N$ with $N \neq \{\theta\}$ and $N \neq \emptyset$.

Proposition 6.3.4. *Let K be a field of characteristic p, and let S be a finite monoid such that $S = G \cup N$, for some subgroup G of S and some nonempty and nonzero nilpotent ideal N of S. Then $K[S]$ is a principal ideal ring if and only if the following conditions are satisfied.*

1. *Either $p = 0$ or $p \not| \, |G|$.*
2. *There exist $t \in S$ and $k > 0$ such that $t^k = \theta$, $Gt = tG$ and $S = \bigcup_{0 \leq i \leq k} Gt^i$.*
3. *Central idempotents of $K[G]$ commute with t (that is, they are central in $K[S]$).*

Proof. That the listed conditions are necessary follows from Lemma 6.3.2 and Lemma 6.3.3. That they are also sufficient follows from Proposition 6.3.1. □

Next we deal with a finite semigroup S that contains an ideal N so that S/N is completely 0-simple and N is nilpotent or $N = \emptyset$.

Proposition 6.3.5. *Let K be a field of characteristic p and let S be a finite semigroup. Suppose N is an ideal of S so that S/N is completely 0-simple and N is nilpotent or $N = \emptyset$. Then $K[S]$ is a principal ideal ring if and only if one of the following conditions is satisfied.*

1. *$S \cong \mathcal{M}(G, n, n; Q)$ for a finite group G satisfying the conditions of Theorem 6.1.4 and for a sandwich matrix Q that is invertible over $K[G]$.*

2. *$S \cong \mathcal{M}(T,n,n;Q)$ for a monoid T satisfying the conditions of Proposition 6.3.4 and for a sandwich matrix Q that is invertible over $K_0[T]$.*

Proof. Note that if S has a zero element then $K[S] \cong K_0[S] \oplus K$ and thus $K[S]$ is a principal ideal ring if and only if $K_0[S]$ is a principal ideal ring.

To prove that the listed conditions are sufficient, one first notes that they imply that $K_0[S] \cong \mathrm{M}_n(K[G])$ or $\mathrm{M}_n(K_0[T])$. Because of Theorem 6.1.4 and Proposition 6.3.4 both $K[G]$ and $K_0[T]$ are principal ideal rings. Theorem 3.7.1 therefore yields $K_0[S]$ is a principal ideal ring.

We now prove that the conditions are necessary. So assume $K_0[S]$ is a principal ideal ring and S/N is completely 0-simple. So $K_0[S] = R_1 \oplus \cdots \oplus R_m$, a direct product of matrix rings R_i over a chain ring. Furthermore, as explained earlier in this section, $S/N \cong \mathcal{M}(G,n,n;P)$ with P invertible over $K[G]$ and thus $K_0[S/N] \cong \mathrm{M}_n(K[G])$. As we have done often, throughout the proof we will several times identify the elements of $V = S \setminus N$ with the nonzero elements of S/N.

Let $e \in V$ be an idempotent, that is, e belongs to a maximal subgroup G of S that is contained in V. Since $K_0[S]$ is finite dimensional, Corollary 3.7.4 yields that $eK_0[S]e = K_0[eSe]$ is a principal ideal ring. Under the isomorphism $K_0[S/N] = \mathcal{M}(K[G],n,n;P) \longrightarrow \mathrm{M}_n(K[G])$, defined by $x \mapsto x \circ P$, the element e maps to the idempotent matrix $e \circ P$ over $G \cup \{0\}$ with only one nonzero row. So, conjugating in $\mathrm{M}_n(K[G])$, we may assume that $e \circ P$ is a matrix unit in $\mathrm{M}_n(K[G])$. In particular, we may assume that $e = (1,1,1) \in S/N = \mathcal{M}(G,n,n;P)$ and identify the natural image of e in $\mathrm{M}_n(K[G])$ with the matrix unit $e_{11} \in \mathrm{M}_n(K[G])$. Clearly $eSe = G \cup M$ with either $M = \emptyset$ or M a nilpotent ideal of eSe. In the former case $eK_0[S]e = K[G]$ and the group G satisfies the conditions of Theorem 6.1.4. In the latter case the monoid eSe satisfies the conditions of Proposition 6.3.4. The result thus follows if $N = \emptyset$.

In the remainder of the proof we thus may assume that N is a nilpotent ideal. Let $s \in S$. Clearly $s = 1s$ in $K_0[S]$ implies that $s \in Ss$. Hence $s = s_1 s$ for some $s_1 \in S$. So $s = s_1^r s$ for every $r \geq 1$. If $s \neq \theta$ then, since N is nilpotent, it follows that $s \in VS$ and consequently $S = VS$. A symmetric argument shows then that $S = VSV$. Choose elements $a_{i1} = (g_i, i, 1) \in V$ and $a_{1j} = (h_j, 1, j) \in V$ for $i,j = 1,\ldots,n$. Because S/N is completely 0-simple it follows that

$$S = VSV = \bigcup_{i,j=1,\ldots,n} a_{i1} S a_{1j} = \bigcup_{i,j=1,\ldots,n} a_{i1} eSe a_{1j}.$$

Define the $n \times n$ matrix $Q = (q_{ji})$ over eSe by $q_{ji} = a_{1j}a_{i1}$. Let $\mathcal{M} = \mathcal{M}(eSe, n, n; Q)$ be the corresponding semigroup of matrix type over the monoid eSe.

Let $\psi : \mathcal{M} \longrightarrow S$ be the map given by $\psi((x, i, j)) = a_{i1}xa_{1j}$ for $x \in eSe$. It is easy to see that ψ is an epimorphism. We claim that it also is injective and thus an isomorphism. In order to prove this, recall that $K_0[S]/\mathcal{B}(K_0[S]) \cong \mathrm{M}_n(K[G]/\mathcal{B}(K[G]))$ and that the image of the idempotent e in $\mathrm{M}_n(K[G]/\mathcal{B}(K[G]))$ is a matrix unit. Hence Proposition 3.2.2 (and the subsequent comments) implies that $K_0[S] \cong \mathrm{M}_n(A)$, where $A \cong eK_0[S]e$. Therefore

$$|S \setminus \{\theta\}| = \dim_K K_0[S] = n^2 \dim_K K_0[eSe] = n^2\, |eSe \setminus \{\theta\}| = |\mathcal{M} \setminus \{\theta\}|.$$

So $|\mathcal{M}| = |S|$ and thus φ is indeed injective. Hence $\mathcal{M} \cong S$. Since $K_0[S]$ has an identity it follows from Lemma 2.5.7 that Q is invertible over $K_0[eSe]$. So the result follows. □

We now prove the main result of this section. Together with Proposition 6.3.5 it describes all finite dimensional principal ideal semigroup algebras.

Theorem 6.3.6. *Let K be a field and let S be a finite semigroup. Then $K[S]$ is a principal ideal ring if and only if S has an ideal chain $I_1 \subset I_2 \subset \cdots \subset I_r = S$ such that I_1 and every factor I_j/I_{j-1} is of the type described in Proposition 6.3.5.*

Proof. To prove the sufficiency of the conditions note that if the ideal $K_0[I_1]$ is a principal ideal ring then its identity is a central idempotent of $K_0[S]$ and thus $K_0[I_1]$ is a ring direct summand. Hence $K_0[S] \cong K_0[I_1] \oplus K_0[S/I_1]$. Since S/I_1 has a shorter ideal chain of the described type, an induction argument shows that $K_0[S]$ is the direct product of the rings $K_0[I_1]$ and $K_0[I_j/I_{j-1}]$, $j = 2, \ldots, r$. Hence $K_0[S]$, and thus also $K[S]$ is a principal ideal ring.

To prove the converse, assume that $K[S]$ is a principal ideal ring. We may assume that S has a nonzero idempotent. Indeed, for otherwise, since S is finite, it follows that S is nil and thus nilpotent. However, as $K[S]$ has an identity element, this yields that $S = \{\theta\}$ and thus $K[S] \cong K$ is a principal ideal ring. So, choose a nonzero idempotent $e \in S$ that is minimal with respect to the $\mathcal{J}$-order in S. Let $I = SeS$. Then either S has a nilpotent ideal $N \subseteq I$ such that $U = I/N$ is completely 0-simple or I itself is completely 0-simple. In the latter case we can assume that $N = \{\theta\}$, adjoining a zero element to S if necessary. We claim that $K_0[I]$ has an identity element. Hence $K_0[S] \cong K_0[I] \oplus K_0[S/I]$. Clearly $K_0[S/I]$ also is a principal ideal ring. Because of Proposition 6.3.5 the result follows then by an induction argument on the cardinality of S.

To prove the claim, let $e_1, \dots, e_k$ be the primitive central idempotents of $K_0[S]$ so that $K_0[I]e_i \neq \{0\}$. Recall that each $K_0[S]e_i$ is a matrix ring over a local ring that is a principal ideal ring. So, in particular, the only idempotent ideals of each $K_0[S]e_i$ are the trivial ideals. Since I is idempotent, so are the ideals $K_0[I]$ and $K_0[I]e_i$. Hence $K_0[I]e_i = K_0[S]e_i$ and it follows that $K_0[I] = \sum_i K_0[I]e_i = \sum_i K_0[S]e_i = K_0[S]\sum_i e_i$, where the sum runs over all i with $K_0[I]e_i \neq 0$. So $K_0[I]$ indeed has an identity. □

6.4 The general case

In this section we deal with arbitrary semigroup algebras $K[S]$ that are principal (left and right) ideal rings, or more general, principal right ideal rings that satisfy the ascending chain condition on nilpotent left ideals. As in the finite dimensional case we show that such algebras are built of blocks that are semigroup algebras of a very special type. Moreover, the blocks again come from an ideal chain in S. As an application we obtain that a semiprime semigroup algebra $K[S]$ is a principal right ideal ring if and only if it is a principal left ideal ring. Furthermore, a full description of such algebras and the defining semigroups S is obtained. It turns out that semigroup algebras are prime principal ideal rings if and only if they are isomorphic to a matrix algebra over either a field K, a polynomial algebra $K[X]$ or a Laurent polynomial algebra $K[X, X^{-1}]$.

Theorem 6.4.1. *Let K be a field and let S be a semigroup. The following conditions are equivalent.*

1. *$K[S]$ is a principal right ideal ring which satisfies the ascending chain condition on nilpotent left ideals.*

2. *The semigroup S has an ideal chain*

$$I_1 \subset \cdots \subset I_t = S$$

so that I_1 and every factor I_j/I_{j-1} is a matrix semigroup of the type $\mathcal{M}(T, n, n; P)$, with sandwich matrix P invertible over $K_0[T]$, and where T is a group of the type described in Theorem 6.1.4 or T is a monoid of the type described in Proposition 6.3.1.

In case the equivalent conditions are satisfied we get

$$K_0[S] \cong K_0[I_1] \oplus K_0[I_2/I_1] \oplus \cdots \oplus K_0[I_t/I_{t-1}]$$

and $K[S]$ *a principal ideal ring which is a finite module over its center, which is finitely generated.*

Proof. Adjoining a zero element to S, if necessary, we may assume that S has a zero element θ and $S \neq \{\theta\}$.

We first make some remarks in case $K_0[S]$ is a principal right ideal ring. Since $K_0[S]$ has an identity element it follows that $S^2 = S$. Because of Corollary 6.2.2, the semigroup S is finitely generated and linear. Hence, as $K_0[S]$ is right Noetherian, and thus S satisfies the ascending chain condition on right ideals, Proposition 5.1.1 also yields that S has only finitely many $\mathcal{J}$-classes with idempotents and every 0-simple ideal of S is completely 0-simple. In particular, if S' is a 0-simple image of S then S' is completely 0-simple. Since $K_0[S]$ has an identity, Lemma 2.5.7 implies that $S' \cong \mathcal{M}(G, n, n; Q)$ for a group G and a matrix Q that is invertible in $\mathrm{M}_n(K[G])$. Furthermore, $K_0[S'] \cong \mathrm{M}_n(K[G])$. Theorem 6.1.4 then implies that S' is a matrix semigroup of the type stated in the theorem. Consequently, S' has a nonzero idempotent. This applies in particular to $S' = S/J$ for a maximal ideal J of S, whence S has a nonzero idempotent.

To prove that the conditions are sufficient, assume that $K_0[S]$ is a principal right ideal ring with the ascending chain condition on nilpotent left ideals. By the above we may assume that S is not (completely) 0-simple and that S has only finitely many $\mathcal{J}$-classes containing a nonzero idempotent and at least on such class exists.

Because of Corollary 6.2.4, right Artinian homomorphic images of $K_0[S]$ are finite dimensional over K. Since, by assumption, $K_0[S]$ also satisfies the ascending chain condition on nilpotent left ideals, Theorem 3.7.2 and Theorem 3.7.3 therefore yield that $K_0[S] \cong R_1 \oplus \cdots \oplus R_m$, where each R_i is either a prime principal right ideal ring or a matrix ring over a local chain ring which has finite dimension over K. Furthermore, by Theorem 3.7.3 the only nonzero idempotent ideals of $K_0[S]$ are direct products of some factors R_i. In particular, idempotent ideals are principal right ideal rings.

Let e be a nonzero idempotent of S. Then SeS is an idempotent ideal of S and thus $K_0[SeS]$ is an idempotent ideal of $K_0[S]$. Hence, by the previous comment, $K_0[SeS]$ has an identity. Thus $K_0[S] \cong K_0[SeS] \oplus K_0[S/SeS]$. So by induction on the number of $\mathcal{J}$-classes containing a nonzero idempotent we may assume that $S = SeS$ and that S has only one $\mathcal{J}$-class containing a nonzero idempotent. Let M be the ideal of non-generators of S. Since

S is not 0-simple, $M \neq \{\theta\}$. Clearly M has no nonzero idempotents, M is a proper ideal of S and $U = S/M$ is completely 0-simple. Moreover, $K_0[U] \cong K_0[S]/K_0[M]$ is a principal right ideal ring. By the first part of the proof we know that the result holds for U, in particular, $U = \mathcal{M}(G,n,n;P)$ with G a group as in Theorem 6.1.4 and the sandwich matrix is invertible over $K[G]$.

Let r be a positive integer. We claim that if M is not nilpotent then M^{r+1} is properly contained in M^r. Suppose the contrary, that is, $M^r = M^{r+1} \neq \{\theta\}$. Then $M^r = (M^r)^2$ and thus $K_0[M]$ is an idempotent ideal of $K_0[S]$. So again $K_0[M^r]$ has an identity and it is a principal right ideal ring. By the first part of the proof one concludes that M^r has a nonzero idempotent. This contradiction proves the claim.

Let $\overline{M} = \bigcap_{i\geq 1} M^i$. Because of the previous claim, either $M^r = \{\theta\}$ for some $r > 1$ or $M/\overline{M}$ is infinite. Assume the former, that is, suppose $M^r = \{\theta\}$ for some positive integer r. Then $K_0[M]$ is a nilpotent ideal in $K_0[S]$ and thus it is contained in a finite direct sum of R_i's that are finite dimensional K-algebras. Hence $K_0[M]$ is finite dimensional and thus M is a finite set. We claim that then also $U = S/M = \mathcal{M}(G,n,n;Q)$ is finite. Suppose the contrary, that is, assume the group G is infinite and thus G has a finite normal subgroup H so that G/H is infinite cyclic and $K[H]$ is semisimple. Hence $K_0[U] \cong \mathrm{M}_n(K[G]) \cong \mathrm{M}_n(K[H])[X, X^{-1}, \sigma]$. Let Y be an inverse image of X in $K_0[S]$. The natural image of a nonzero K-linear combination $g(Y)$ of powers of Y is regular in $K_0[U]$. Because $K_0[S]$ is a direct sum of prime and finite dimensional algebras and the kernel of the map $K_0[S] \longrightarrow K_0[U]$ is nilpotent it follows that $g(Y)$ is regular in $K_0[S]$. Let m be a nonzero element of M. Because the ideal $K_0[M]$ is finite dimensional, the set $\{Y^i m \in K_0[M] \mid i \geq 1\}$ is K-linearly dependent. Hence there exists a nonzero polynomial $f(Y) \in K[Y]$ so that $f(Y)m = 0$. As $f(Y)$ is regular this yields in $K_0[M]$ that $m = 0$. Hence $m = \theta$, a contradiction. This proves that indeed U is finite. Therefore S is finite and the desired form of S follows from Proposition 6.3.5.

Thus, assume now that $M/\overline{M}$ is infinite. Let $K_0[M] = J_1 \oplus \cdots \oplus J_m$ and $K_0[\overline{M}] = \overline{J}_1 \oplus \cdots \oplus \overline{J}_m$ for some ideals $J_j, \overline{J}_j$ of R_j, $j = 1,\ldots,m$. Choose j such that $J_j/\overline{J}_j$ (and thus also $R_j/\overline{J}_j$) is of infinite dimension over K. Because of Corollary 6.2.4 every proper epimorphic image of R_j is finite dimensional over K. Hence $\overline{J}_j = \{0\}$. On the other hand, if $J_j/\overline{J}_j$ is finite dimensional, then a power of J_j is idempotent, and thus it follows that $J_j = \overline{J}_j = R_j$ or $\overline{J}_j = \{0\}$. Consequently, $K_0[\overline{M}] = R_{i_1} \oplus \ldots \oplus R_{i_q}$ for some $i_1, \ldots, i_q$ and thus $K_0[\overline{M}]$ is a principal right ideal ring. Again by the first part of the proof we obtain that $\overline{M}$ has nonzero idempotents if $\overline{M} \neq \{\theta\}$. Therefore we must have $\overline{M} = \{\theta\}$.

We know that $S = SeS$ and $S/M = \mathcal{M}(G, n, n; P)$ for a group G and an invertible matrix P. We may assume that e corresponds to the element $(1, 1, 1) \in S/M$. For any $i > 1$, the semigroup S/M^i has a nonzero nilpotent ideal M/M^i and only one $\mathcal{J}$-class of S containing nonzero idempotents. Hence, the previous case can be applied. So all semigroups S/M^i are finite. In particular, G is a finite group. Moreover, for every $i \geq 1$, $S/M^i \cong \mathcal{M}(T_i, n_i, n_i; P_i)$ for a monoid $T_i = G_i \cup N_i$, where G_i is the group of units of T_i, N_i is either empty or a nilpotent ideal of T_i, and the matrix P_i is invertible over $K_0[T_i]$.

For every $i \geq 1$ consider the natural homomorphisms $\psi_{i+1} : S/M^{i+1} \longrightarrow S/M^i$ and $\alpha_i : S \longrightarrow S/M^i$. We shall identify the elements e and $\alpha_i(e)$. Since M^i/M^{i+1} is a nilpotent ideal of S/M^{i+1} it is straightforward to verify that the unit group G_{i+1} of $e(S/M^{i+1})e$ maps isomorphic onto the unit group G_i of $e(S/M^i)e$ under ψ_{i+1}. Therefore we may identify G and all G_i, $i \geq 1$. Let $S_i = \mathcal{M}(G_i, n_i, n_i, \overline{P}_i)$ be the natural image of S/M^i determined by the natural epimorphism $T_i \longrightarrow G_i^0$. Since P_i is invertible, $\overline{P}_i$ is invertible over $K[G_i]$. Therefore, S_i is completely 0-simple and clearly this is the largest completely 0-simple homomorphic image of S/M^i (because $\mathcal{M}(N_i, n_i, n_i, \overline{P}_i)$ is a nilpotent ideal of S/M^i). Hence we must have $S/M \cong (S/M^i)/(M/M^i) \cong S_i$ and consequently $n = n_i$ for $i \geq 1$.

Again we identify the elements of $V = S \setminus M$ with the nonzero elements of S/M. Let $s \in S$ and assume $\theta \neq s$. Since $\overline{M} = \{\theta\}$ we have that either $s \in V$ or $s \in M^i \setminus M^{i+1}$ for some i. Because $S = S^2$ one obtains that $s = gsf$ for some $g, f \in V = S \setminus M$. It follows that $S = VSV$. Thus, since $S/M = \mathcal{M}(G, n, n; P)$ is completely 0-simple and $e = (1, 1, 1) \in V$,

$$S = \bigcup_{1 \leq k, j \leq n} q_{k1} e S e q_{1j}$$

for some elements $q_{k1} = (h_k, k, 1)$ and $q_{1j} = (g_j, 1, j)$ in V. Clearly $q_{11}e = eq_{11}$ and thus one may take $q_{11} = e$. Let $Q = (q_{jk})$ be the $n \times n$ matrix with $q_{jk} = q_{1j}q_{k1} \in eSe$ and let $\mathcal{M} = \mathcal{M}(eSe, n, n; Q)$ be the corresponding semigroup of matrix type over eSe.

Then the mapping $\delta : \mathcal{M} \longrightarrow S$ defined by $\delta((z, k, j)) = q_{k1} z q_{1j}$ is a semigroup epimorphism. Clearly we have the natural epimorphisms $\delta_i : \mathcal{M} \longrightarrow \mathcal{M}_i = \mathcal{M}(\alpha_i(eSe), n, n; \alpha_i(Q))$ with $\alpha_i(Q) = (\alpha_i(q_{jk}))$. Because $n_i = n$ and $e = \alpha_i(e)$ is a nonzero idempotent in S/M^i, it follows from Proposition 6.3.5 that S/M^i is a semigroup of matrix type over $\alpha_i(eSe)$. It then easily is verified that $\mathcal{M}_i \cong \alpha_i(S)$ and that the diagram

$$\begin{array}{ccc} \mathcal{M} & \xrightarrow{\delta} & S \\ \downarrow \delta_i & & \downarrow \alpha_i \\ \mathcal{M}_i & \xrightarrow{\cong} & \alpha_i(S) \end{array}$$

commutes for every i. It then also easily follows that δ is injective. Hence $\mathcal{M} \cong S$.

Consequently, $K_0[S] = K_0[\mathcal{M}(eSe, n, n; Q)] \cong \mathcal{M}(K_0[eSe], n, n; Q)$, a Munn algebra. Lemma 2.5.7 implies that Q is invertible over $K_0[eSe]$.

It remains to show that $T = eSe \setminus \{\theta\}$ is a semigroup of the desired type. Choose $x = (z, 1, 1) \in eSe \cap (M \setminus M^2)$. Because $\overline{M} = \bigcap_{i \geq 1} M^i = \{\theta\}$ it is clear that $T = (eSe \setminus M) \cup \bigcup_i (eM^ie \setminus M^{i+1})$. Proposition 6.3.5 applied to all S/M^i, $i \geq 1$, yields that $T = \bigcup_{i \geq 0} Gx^i$, central idempotents of $K[G]$ are central in $K[T]$, the order of G is not divisible by the characteristic of K and $Gx = xG$. This completes the proof of the necessity of the conditions.

To prove the conditions are sufficient assume that condition (2) holds. Then each $K_0[I_j/I_{j-1}]$ is an ideal of $K_0[S/I_{j-1}]$ and it has an identity element. Hence, $K_0[S] \cong K_0[I_1] \oplus K_0[I_2/I_1] \oplus \cdots \oplus K_0[I_t/I_{t-1}]$. Since each $K_0[I_j/I_{j-1}]$ is a principal ideal ring, it follows that $K[S]$ is a principal ideal ring as well.

Finally we prove the last assertion of the theorem. So assume $K[S]$ is a principal right ideal ring with the ascending chain condition on nilpotent left ideals. Because of Corollary 6.2.2 and Corollary 6.2.4, $K[S]$ is a finitely generated K-algebra with Gelfand-Kirillov dimension 0 or 1. Since $K[S]$ is a direct product of a semiprime ring and a finite dimensional algebra, Theorem 3.5.4 then yields that $K[S]$ is a finite module over its center. Because of Theorem 3.1.8 the center is a finitely generated K-algebra.

We note that the last assertion also easily can be derived directly from the structure of the semigroup rings $K_0[I_j/I_{j-1}]$. □

We finish this chapter with some consequences. The first one shows that a semiprime semigroup algebra is a principal right ideal ring if and only if it is a principal left ideal ring.

Corollary 6.4.2. *Let S be a semigroup and K a field of characteristic p. The following conditions are equivalent.*

1. *$K[S]$ is a semiprime principal right ideal ring.*
2. *$K[S]$ is a semiprime principal left ideal ring.*

3. *S has an ideal chain*

$$I_1 \subset \cdots \subset I_t = S$$

such that I_1 *and every factor* I_j/I_{j-1} *is of the form* $\mathcal{M}(T, n, n; P)$ *for an invertible over* $K_0[T]$ *sandwich matrix* P *and a monoid* T *that satisfies one of the following conditions.*

(a) T *is a group with a finite normal subgroup* N *so that* $T = N$ *or* T/N *is an infinite cyclic group, and if* $\text{char}(K) = p > 0$ *then* N *is a* p'*-group.*

(b) T *is a monoid with a finite group of units* H *that is a* p'*-group in case* $\text{char}(K) = p > 0$, $T = \bigcup_{i \geq 0} Hx^i$, *a disjoint union,* $Hx = xH$, *and for every primitive central idempotent* e *of* $K[H]$ *we have that* e *is central in* $K[T]$ *and either* $ex = 0$ *or* $ex^i = 0$ *for all* $i \geq 1$.

Proof. Suppose $K[S]$ is a semiprime principal right ideal ring. Because of Theorem 6.2.1 the algebra $K[S]$ satisfies a polynomial identity. Since it also is right Noetherian, Theorem 3.2.1 yields that $K[S]$ is left Noetherian. Hence Theorem 6.4.1 applies and thus $K[S]$ is a principal left ideal ring. Hence (1) implies (2), and by symmetry the two conditions are equivalent. Because of Theorem 6.4.1 (we use the notation of the statement), in order to prove the equivalence of (1) and (3) it is sufficient to show that $K_0[\mathcal{M}(T, n, n; P)]$ is semiprime, or equivalently $K_0[T]$ is semiprime, if and only if T is described as in the statement of (3). In case T is a group and thus of the type described in Theorem 6.1.4 then, also using the description of semiprime group algebras (see Theorem 3.2.8), it follows that indeed T has a finite normal subgroup N so that $T = N$ or T/N is an infinite cyclic group. Furthermore, if $\text{char}(K) = p > 0$ then N is a p'-group. If T is not a group then T is a monoid of the type described in Proposition 6.3.1. So if H is the group of units of T then $|H|$ is nonzero in K, $T = \bigcup_{i \geq 0} Hx^i$, $Hx = xH$ and every primitive central idempotent of $K[H]$ is central in $K[T]$. We claim that T is infinite if $K_0[T]$ is semiprime. Suppose the contrary. Then by the comments after Proposition 6.3.1, x^n is an idempotent for some $n \geq 1$. Hence $K_0[T]x^n = x^n K_0[T]$ is an ideal of $K_0[T]$. Clearly x^n is the identity of this ideal and hence x^n is a central idempotent. Then $K_0[T] = K_0[T/x^nT] \oplus K_0[x^nT]$, a direct product. But the former has a nilpotent ideal except when $T = x^nT$. In this case one gets that $x^nT = T$, and thus some power of x is in the group H, and so $T = H$; contradicting that T is not a group. Hence if $K_0[T]$ is semiprime then we get that T is infinite. So, by the remarks following Proposition 6.3.1, the cosets Tx^i

are disjoint and for every central idempotent e of $K[H]$ we have that either $ex = 0$ or $ex^i \neq 0$ for every $i \geq 1$. Conversely, if T is such a monoid then, again by the remarks given after Proposition 6.3.1, $K_0[T]$ is semiprime. The result thus follows. □

Corollary 6.4.3. *Let S be a semigroup with zero element and K a field. The following conditions are equivalent for the contracted semigroup algebra $K_0[S]$.*

1. *$K_0[S]$ is a prime principal right ideal ring.*
2. *$K_0[S]$ is a prime principal left ideal ring.*
3. *S is a semigroup of one of the following types.*

   (a) $\mathcal{M}(\{1\}, n, n; Q)$,
   (b) $\mathcal{M}(\langle X\rangle, n, n; Q)$,
   (c) $\mathcal{M}(\langle X, X^{-1}\rangle, n, n; Q)$,

   where Q is invertible in $\mathrm{M}_n(K)$, $\mathrm{M}_n(K[X])$ or $\mathrm{M}_n(K[X, X^{-1}])$, respectively.

Moreover, if $K_0[S]$ satisfies one of the equivalent conditions then $K_0[S] \cong \mathrm{M}_n(K)$, $\mathrm{M}_n(K[X])$ or $\mathrm{M}_n(K[X, X^{-1}])$ respectively.

Proof. Because $K[S] \cong K_0[S] \oplus K$ it follows from Corollary 6.4.2 that (1) and (2) are equivalent. The equivalence with (3) is then obtained at once from Corollary 6.4.2, the decomposition of the semigroup algebra via the ideal chain of S stated in Theorem 6.4.1, the remarks given after Proposition 6.3.1 and the description of prime group algebras given in Theorem 3.2.8. □

It follows easily from the comments given after Proposition 6.3.1 that the monoids T in the statement of Theorem 6.4.1 are homomorphic images of cancellative monoids of the same type. From this point of view the following extension of Theorem 6.1.4 is of special interest.

Corollary 6.4.4. *Let T be a cancellative monoid and K a field. The following conditions are equivalent.*

1. *$K[T]$ is a principal right ideal ring.*
2. *$K[T]$ is a principal left ideal ring.*

3. T is a semigroup satisfying one of the following conditions.

(a) T is a group satisfying the conditions of Theorem 6.1.4.

(b) T contains a finite subgroup H and an element x of infinite order such that $xH = Hx$, $T = \bigcup_{i\in\mathbb{N}} Hx^i$, H is a p'-group if $\mathrm{char}(K) = p > 0$, and the central idempotents of $K[H]$ are central in $K[T]$.

Proof. We first note that if $K[T]$ is a principal right ideal ring then $K[T]$ is right and left Noetherian. Of course it is right Noetherian. Since T is cancellative, it follows from Lemma 4.1.5 that T has a group of right quotients $G = TT^{-1}$. Because right ideals of $K[G]$ are generated by right ideals of $K[T]$, also $K[G]$ is a principal right ideal ring. Hence, by Theorem 6.1.4, the group G is polycyclic-by-finite. Therefore Theorem4.4.7 yields that $K[T]$ indeed is left Noetherian. Similarly, if $K[T]$ is a principal left ideal ring then $K[T]$ is left and right Noetherian.

Since T has only one idempotent, the result then follows at once from Theorem 6.4.1 and the remarks given after Proposition 6.3.1. □

6.5 Comments and problems

Section 6.1 on the description of group algebras $K[G]$ that are principal ideal rings is due to Passman [128]. Earlier Fisher and Sehgal [39] dealt with the case of group algebras of nilpotent groups. The other sections of this chapter are taken from [77]. A characterization of commutative semigroup algebras $K[S]$ that are principal ideal rings was obtained earlier by Decruyenaere, Jespers and Wauters in [30]. The reader is referred to Gilmer's book [54] for more background on this problem, as well as on other related arithmetical properties, in the commutative setting. The description of semigroup algebras of cancellative semigroups S that are principal right ideal rings was already shown by Jespers and Wauters in [94]; in case S also is abelian this was done by Gilmer in [54]. The problem of describing principal ideal rings that are commutative semigroup rings $R[S]$ over arbitrary coefficient rings R also has been resolved in [30]; earlier results on this topic can be found in [54, 55, 62].

Problems

1. When does a semigroup algebra satisfy the ascending chain condition on nilpotent left ideals? See Theorem 6.4.1 and Theorem 3.7.2.

2. Describe when an arbitrary semigroup algebra is a principal right ideal ring. Is the ascending chain condition on nilpotent left ideals needed in Theorem 6.4.1?

3. Is a semigroup algebra a principal right ideal ring if and only if it is a principal left ideal ring?

Chapter 7

Maximal orders and Noetherian semigroup algebras

It remains an unsolved problem to characterize when an arbitrary semigroup algebra $K[S]$ over a field K is a prime Noetherian maximal order. In this chapter we describe when a semigroup algebra $K[S]$ is a Noetherian PI domain which is a maximal order. This result will be applied in the context of concrete classes of algebras considered in Chapter 8. Our work relies on the study of the height one primes of $K[S]$ and of the minimal primes of the monoid S and leads to a characterization purely in terms of S. It turns out that the primes P intersecting S play a crucial role, and therefore we reduce the problem to certain 'local' monoids S_P, that is, monoids with only one minimal prime. However, in Section 10.5 we illustrate by examples that such monoids and their algebras are much more complicated than the discrete valuation rings arising in the commutative case. Our work is based on the results of Brown concerning the group ring case and of Chouinard on semigroup algebras of abelian monoids, stated in Section 3.6.

The description of Noetherian algebras $K[S]$ for a submonoid S of a polycyclic-by-finite group, given in Theorem 4.4.7, indicates that the PI case can be crucial for the general problem. So this result together with Theorem 3.1.9, Theorem 3.3.4 and Lemma 4.1.9 yield the following starting point of the investigations.

Theorem 7.0.1. *For a monoid S and a field K, the semigroup algebra $K[S]$ is a Noetherian PI domain if and only if S satisfies the ascending chain* *condition on right ideals (or equivalently, left ideals) and S embeds into a finitely generated torsion free abelian-by-finite group. Equivalently, S has a torsion free group of quotients G and $S \cap A$ is finitely generated for an abelian normal subgroup A of finite index in G. In this case $G = SS^{-1} = S\,\mathrm{Z}(S)^{-1}$.*

7.1 Maximal orders and monoids

In this section some properties are deduced for cancellative semigroups T that are a maximal order in a finitely generated torsion free abelian-by-finite group G. The focus mainly is on prime ideals P of T and several types of localization of T with respect to P. Note that, because of Lemma 4.1.9, the monoid T has a group of fractions $TT^{-1} = T^{-1}T = T\,\mathrm{Z}(T)^{-1}$.

So, throughout this section we assume G is a finitely generated torsion free group and A is an abelian subgroup A of finite index that is normal in G. Further, we assume T is a submonoid of G and $G = T\,\mathrm{Z}(T)^{-1}$. If $g \in G$ then C_g denotes the conjugacy class of g in G.

For a prime ideal P in T consider the following localizations.

$$\begin{aligned} T_P &= \{g \in G \mid Cg \subseteq T \text{ for some } G-\text{conjugacy class } C \text{ of } G \\ &\qquad \text{contained in } T, \text{ with } C \not\subseteq P\} \\ T_{(P)} &= T_{(P)}(A) \\ &= \{g \in G \mid Cg \subseteq T \text{ for some } G-\text{conjugacy class } C \\ &\qquad \text{of } G \text{ contained in } T \cap A, \text{ with } C \not\subseteq P\} \\ T_{[P]} &= \{g \in G \mid Jg \subseteq T \text{ for some ideal } J \text{ of } T \text{ with } J \not\subseteq P\} \end{aligned}$$

In case T is abelian the above constructions agree with the classical localization of T with respect to the multiplicative set $T \setminus P$.

Because P is a prime ideal it follows at once that $T_{[P]}$ is a submonoid of G containing T. That also T_P and $T_{(P)}$ are submonoids of G follows from the following two facts. First if C is a conjugacy class of G then $Cg = gC$ for all $g \in G$. Second, if $g_1, g_2 \in G$ and C_1, C_2 are conjugacy classes of G then $g_1g_2C_1C_2 = g_1C_1g_2C_2$ and C_1C_2 is a union of conjugacy classes of G.

So we get a chain of submonoids

$$T \subseteq T_{(P)} \subseteq T_P \subseteq T_{[P]} \subseteq G.$$

Further,

$$\mathrm{N}(T) \subseteq \mathrm{N}(T_P).$$

Indeed, let $n \in \mathrm{N}(T)$ and $x \in T_P$. Suppose C is a conjugacy class of G contained in T so that $Cx \subseteq T$ and $C \not\subseteq P$. Since $n \in \mathrm{N}(T)$ and $Cg = gC$ for any $g \in G$, it follows that

$$(nxn^{-1})C = nCxn^{-1} \subseteq nTn^{-1} \subseteq T.$$

Hence $nxn^{-1} \in T_P$ and therefore $nT_P \subseteq T_P n$. Similarly, $T_P n \subseteq nT_P$.

We now give a series of technical lemmas. The first one yields some consequences on $T \cap A$ when T is a maximal order.

Lemma 7.1.1. *The following properties hold.*

1. *If T is a maximal order, then $T \cap A$ is maximal order.*
2. *If $T \cap A$ is a maximal order then $T \cap A$ is G-invariant.*
3. *If $T \cap A$ is G-invariant, then $T_{(P)} \cap A$ is G-invariant, for any prime ideal P in T.*

Proof. To prove the first part, suppose T is a maximal order. Let I be an ideal of $T \cap A$ and let $a \in A = \mathrm{gr}(T \cap A)$ be such that $aI \subseteq I$. Then $aJ \subseteq J$ with $J = \prod_{f \in F} f^{-1} I f$, where F is a transversal for A in G. Note that $J \subseteq A$ because A is normal in G. Since $G = T\,\mathrm{Z}(T)^{-1} = G(\mathrm{Z}(T) \cap A)^{-1}$ and because F is finite, there exists $b \in \mathrm{Z}(T) \cap A$ with $bJ \subseteq T \cap A$. Hence J is a fractional ideal of $T \cap A$. Furthermore $TJ = JT$, because $g^{-1}Jg = J$ for all $g \in G$. Also $bJT \subseteq T$ and thus JT is a fractional ideal of T. As $aJT \subseteq JT$ and T is a maximal order, one obtains that $a \in T \cap A$, as desired.

For the second part, suppose $T \cap A$ is a maximal order. Let $g \in G$. Since $G = T\,\mathrm{Z}(T)^{-1}$ there exists $z \in \mathrm{Z}(T) \cap A$ so that $zg, zg^{-1} \in T$. Consider the monoid $M = (T \cap A)g^{-1}(T \cap A)g$. Clearly $T \cap A \subseteq M \subseteq A$ and $z^2 M \subseteq T \cap A$. Because, by assumption, $A \cap T$ is a maximal order, it follows that $M = T \cap A$. Hence $g^{-1}(T \cap A)g \subseteq T \cap A$ for any $g \in G$. So indeed $T \cap A$ is G-invariant.

To prove the third part, suppose $T \cap A$ is G-invariant. Let $a \in T_{(P)} \cap A$. Then there exists a G-conjugacy class $C \subseteq T \cap A$ so that $C \not\subseteq P$ and $Ca \subseteq T \cap A$. Hence, for any $g \in G$, $Cg^{-1}ag = (g^{-1}Cg)(g^{-1}ag) = g^{-1}(Ca)g \subseteq (T \cap A)^g$. Because $T \cap A$ is G-invariant, this yields that $Cg^{-1}ag \subseteq T \cap A$ and hence $g^{-1}ag \in T_{(P)}$. So, $T_{(P)} \cap A$ is G-invariant. $\square$

For any prime ideal P of T put

$$\begin{aligned} I(P) &= \{g \in T_P \mid gC \subseteq P \text{ for some} \\ &\qquad G-\text{conjugacy class } C \subseteq T \text{ with } C \not\subseteq P\} \end{aligned}$$

The following two lemmas deal with the structure of T_P.

Lemma 7.1.2. *If P is a prime ideal of T then $I(P)$ is a prime ideal of T_P and $I(P) \cap T = P$.*

Proof. Since $Cg = gC$ for any conjugacy class C and $g \in G$, it follows at once that $I(P)$ is a two-sided ideal of T_P and that $I(P) \cap T = P$. To prove it is a prime ideal, let $s, t \in T_P \setminus I(P)$ and suppose $sT_Pt \subseteq I(P)$. Let C_1 and C_2 be G-conjugacy classes contained in T but not contained in P so that $C_1 s \subseteq T$ and $C_2 t \subseteq T$. Then

$$(C_1 s)T(C_2 t) \subseteq I(P) \cap T = P.$$

Because P is a prime ideal of T it follows that $C_1 s \subseteq P$ or $C_2 t \subseteq P$ and thus $s \in I(P)$ or $t \in I(P)$. □

Lemma 7.1.3. *Suppose P be a prime ideal in T.*

1. *If $T \cap A$ is G-invariant, then $T_{(P)} = T_P$.*
2. *If T is a maximal order that satisfies the ascending chain condition on right ideals (or equivalently, on left ideals) then T_P is a maximal order.*

Proof. To prove the first part, assume that $T \cap A$ is G-invariant. Suppose $T_{(P)} \neq T_P$. Then there exists $x \in T_P \setminus T_{(P)}$. Let C be a G-conjugacy class contained in T so that $C \not\subseteq P$ and $Cx \subseteq T$.

Let $n = [G : A]$. We claim that $t^n \in P$ for every $t \in C$. Suppose the contrary. Then let $t \in C$ be so that $t^n \notin P$. The assumption that $T \cap A$ is G-invariant implies that the conjugacy class C_{t^n} of t^n is contained in $T \cap A$. Of course, $C_{t^n} \subseteq C^n$ and therefore $C_{t^n} x \subseteq T$. Hence, as $C_{t^n} \not\subseteq P \cap A$ one obtains that $x \in T_{(P)}$, a contradiction. This proves the claim.

Now, for any conjugacy class $D \subseteq T$, we have $DC^n x \subseteq T$ and $DC^n = \bigcup_{i=1}^{l} C_i$, each C_i a conjugacy class contained in T. Hence, $C_i x \subseteq T$ and thus by the above, C_i is nil of exponent at most n modulo P. So DC^n is nil of exponent at most n modulo P.

As $(tC)^n \subseteq C_{t^n} C^n$ it follows that tC is nil modulo P, for any $t \in T$. Hence the ideal TC is nil modulo P. Let K be a field. Then there exists an ideal Q of $K[T]$ maximal with respect to $Q \cap T = P$. Since P is prime in T it follows that Q is a prime ideal in $K[T]$. Because G is abelian-by-finite, Theorem 3.1.9 then implies that $K[T]/Q$ is a prime PI algebra. So any nil subsemigroup of $K[T]/Q$ is nilpotent by Proposition 2.4.9, Theorem 3.1.7 and Theorem 3.1.3. Therefore TC is nilpotent modulo P. As $C \not\subseteq P$ and because P is prime in T this yields a contradiction. This finishes the proof of the first part.

To prove the second part assume that T is a maximal order that satisfies the ascending chain condition on left and right ideals (see Theorem 4.4.7). Let $g \in G$ and J an ideal of T_P so that $Jg \subseteq J$. Since $T \cap J$ is finitely generated as a left ideal in T, we get that $T \cap J = \bigcup_{i=1}^{m} Ta_i$, for some $a_i \in T \cap J$. Since $a_i g \in T_P$, there exists a G-conjugacy class $C_i \subseteq T$ so that $C_i \not\subseteq P$ and $C_i a_i g \subseteq T$. Hence $(T \cap J)C_1 \cdots C_m g = C_1 \cdots C_m (T \cap J) g \subseteq (T \cap J)$. As, by assumption, T is a maximal order, we therefore obtain that $C_1 \cdots C_m g \subseteq T$. Because $C_1 \cdots C_m$ is a union of conjugacy classes contained in T, but not all contained in P, it follows that $g \in T_P$. Therefore $(J :_r J) = T_P$, as desired. Similarly one proves that $(J :_l J) = T_P$. Hence T_P is a maximal order. □

In the next lemma height one primes of $K[T]$ are investigated.

Lemma 7.1.4. *If $Q \in \operatorname{Spec}^1(K[T])$ and $T \cap A$ is G-invariant, then the following properties hold.*

1. *$Q \cap K[A] = Q_1 \cap \cdots \cap Q_m$, where $Q_1, \ldots, Q_m$ are all the height one prime ideals of $K[T \cap A]$ containing $Q \cap K[A]$. Moreover, all Q_i are G-conjugate.*

2. *If P is an ideal of T so that $K[P] \in \operatorname{Spec}^1(K[T])$ and $Q \cap T = \emptyset$, then $\overline{Q} = \bigcap_{g \in G} g^{-1}(Q \cap K[A])g \not\subseteq K[P \cap A]$.*

Proof. Consider $K[T]$ as an algebra which is naturally graded by the finite group G/A. So the homogeneous component indexed by the identity of G/A is the algebra $K[T \cap A]$. Because of Theorem 3.4.8,

$$Q \cap K[A] = Q_1 \cap Q_2 \cap \cdots \cap Q_m,$$

where $Q_1, \ldots, Q_m$ are all height one primes of $K[T \cap A]$ containing $Q \cap K[A]$. Let $\overline{Q_i} = \bigcap_{f \in F} f^{-1} Q_i f$, where F is a transversal for A in G. Since, by assumption, $T \cap A$ is G-invariant, each $f^{-1} Q_i f$ is a height one prime ideal of $K[T \cap A]$ and $\overline{Q_i}$ is a G-invariant ideal of $K[T \cap A]$. Hence $\overline{Q_i} K[T] = K[T]\overline{Q_i}$ is an ideal of $K[T]$. Clearly, $\prod_{i=1}^{m} (\overline{Q_i} K[T]) \subseteq Q$ and thus, as Q is prime, $\overline{Q_i} K[T] \subseteq Q$ for some i. Hence $\overline{Q_i} \subseteq Q \cap K[A]$, and therefore

$$\overline{Q_i} = \bigcap_{f \in F} f^{-1} Q_i f \subseteq Q_1 \cap \cdots \cap Q_m.$$

So, for every j, $\bigcap_{f \in F} f^{-1} Q_i f \subseteq Q_j$ and hence there exists $f(j) \in F$ such that $f(j)^{-1} Q_i f(j) \subseteq Q_j$. As $Q_j, f(j)^{-1} Q_i f(j) \in \operatorname{Spec}^1(K[T \cap A])$ it follows that $f(i)^{-1} Q_i f(i) = Q_j$. This finishes the proof of the first part of the result.

We now prove the second part. So assume that P is an ideal of T so that $K[P]$ is a height one prime of $K[T]$. Applying the first part to this prime one gets that there exists $P_1 \in \operatorname{Spec}^1(K[T \cap A])$, with $P \cap A \subseteq P_1$, so that $K[P \cap A] = P_1 \cap g_1^{-1} P_1 g_1 \cap \cdots \cap g_r^{-1} P_1 g_r$ for some $g_1, \ldots, g_r \in G$ and $r \geq 1$. Also from the first part one obtains that $\overline{Q} = \bigcap_{g \in G} g^{-1} Q_1 g = \bigcap_{f \in F} f^{-1} Q_1 f$, where Q_1 is a height one prime of $K[T \cap A]$ containing $Q \cap K[A]$. Assume now that $\overline{Q} \subseteq K[P \cap A]$. Then $\overline{Q} \subseteq P_1$. Thus $g^{-1} Q_1 g \subseteq P_1$ for some $g \in G$. Because both primes are of height one in $K[T \cap A]$ we thus have that $g^{-1} Q_1 g = P_1$. Hence $\overline{Q} = \bigcap_{f \in F} f^{-1} P_1 f \supseteq \bigcap_{f \in F} f^{-1} (P \cap A) f \neq \emptyset$. As $\bigcap_{f \in F} f^{-1} (P \cap A) f \subseteq T$ and $\overline{Q} \subseteq Q$ it follows that $Q \cap T \neq \emptyset$. This contradiction finishes the proof of the second part. □

Finally we deduce some properties on T_P in case this monoid has a unique minimal prime ideal.

Lemma 7.1.5. *Suppose T satisfies the ascending chain condition on right ideals. Assume $P \in \operatorname{Spec}^0(T)$ and $\operatorname{Spec}^0(T_P) = \{M\}$. If $T \cap A$ is G-invariant then the following properties hold.*

1. *$T \cap M = P$.*
2. *$M = I(P)$.*
3. *$M \cap A$ is G-invariant.*
4. *$P \cap A$ is G-invariant.*
5. *$K[M]$ is the only height one prime ideal of $K[T_P]$ that intersects T_P nontrivially.*
6. *$(T_P)_M = T_P$.*

Proof. We begin with two remarks. First, since $G = T_P (T_P)^{-1}$ is finitely generated abelian-by-finite, we know from Theorem 3.5.6 that $\operatorname{clKdim} K[T_P] = \operatorname{rk}(T_P) < \infty$. Since G also is torsion free, Proposition 4.5.8 yields that if V is a prime ideal of T_P then $K[V]$ is a prime ideal of $K[T_P]$. It follows that V cannot contain an infinite descending chain of prime ideals of T_P. So, any prime ideal V of T_P contains a minimal prime ideal.

Second, Proposition 4.5.8 also says that if $Q \in \operatorname{Spec}^1(K[T_P])$ and $Q \cap T_P \neq \emptyset$, then $Q = K[Q \cap T_P]$ with $Q \cap T_P \in \operatorname{Spec}^0(T_P)$. Since, by assumption T_P has a unique minimal prime, we thus obtain that $Q \cap T_P = M$.

Consider the natural G/A-gradation on $K[T_P]$, with identity component $K[T_P \cap A]$. We know from Lemma 7.1.2 that $I(P)$ is a prime ideal of T_P with

$I(P) \cap T = P$. Since also by assumption $\mathrm{Spec}^0(T_P) = \{M\}$, the first remark in the proof implies that $M \subseteq I(P)$ and thus also $M \cap T \subseteq P$. Because $G = T\,\mathrm{Z}(T)^{-1}$, there exists $m \in M \cap \mathrm{Z}(T) \subseteq I(P) \cap T = P$. Let $C = \mathrm{Z}(K[T]) \setminus K[P]$. This is a multiplicatively closed set as $K[P]$ is a height one prime ideal by Proposition 4.5.8. Since T satisfies the ascending chain condition on right ideals, we know from Theorem 7.0.1 that $K[T]$ is a Noetherian domain. It follows that also the localization $K[T]\,C^{-1}$ is a Noetherian domain. Note that $T_P \subseteq K[T]C^{-1}$ and thus $K[T]C^{-1} = K[T_P]\,C^{-1}$. As $m \in \mathrm{Z}(T) \subseteq \mathrm{Z}(T_P)$ we therefore obtain from the principal ideal theorem (Theorem 3.2.4) that $K[T_P]C^{-1}m \subseteq Q$ for some $Q \in \mathrm{Spec}^1(K[T_P]C^{-1})$. Since the primes of $K[T_P]C^{-1}$ are in one-to-one correspondence with the primes of $K[T_P]$ not intersecting C, we get that $Q \cap K[T_P]$ is a height one prime of $K[T_P]$ intersecting T_P. Similarly, $Q \cap K[T] \in \mathrm{Spec}^1(K[T])$. Hence, by the second remark in the proof, $Q \cap K[T_P] = K[M] \in \mathrm{Spec}^1(K[T_P])$ and $K[M]$ is the only height one prime of $K[T_P]$ intersecting T_P nontrivially. This proves part (5). Because also $M \cap T \subseteq P$, we get

$$Q \cap K[T] = K[M] \cap K[T] = K[M \cap T] \subseteq K[P].$$

As $K[P]$ is a height one prime ideal of $K[T]$ and $K[M \cap T] = Q \cap K[T]$ is prime in $K[T]$, it follows that $M \cap T = P$. This proves part (1).

To prove part (2), take $g \in I(P)$. Then there exists a G-conjugacy class $C \subseteq T$ with $C \not\subseteq P$ and $Cg \subseteq P \subseteq M$. As $M \cap T = P$ we get $C \not\subseteq M$. Hence, by the primeness of M, $g \in M$. So $I(P) \subseteq M$. Because of Lemma 7.1.2 we know that $I(P)$ is a prime ideal of T_P. Again using the assumption that M is the only minimal prime ideal of T_P, we obtain that $M = I(P)$ and thus (2) has been proven.

Next we prove (3). Since, $T \cap A$ is G-invariant we get from Lemma 7.1.1 and Lemma 7.1.3 that also $T_P \cap A$ is G-invariant. Now, from the above we know that $K[M] \in \mathrm{Spec}^1(K[T_P])$. Therefore Lemma 7.1.4 yields that $K[M] \cap K[A] = M_1 \cap \cdots \cap M_m$, where $M_1, \ldots, M_m$ are all height one primes of $K[T_P \cap A]$ containing $M \cap A$, and all the primes M_i are G-conjugate. To prove that $M \cap A$ is G-invariant, it is now sufficient to show that all G-conjugates of M_1 are in the set $\{M_1, \ldots, M_m\}$. Therefore let M' be a G-conjugate of M_1. Since $T_P \cap A$ is G-invariant one gets that $M' \in \mathrm{Spec}^1(K[T_P \cap A])$ and $M' \cap T_P \cap A \neq \emptyset$. So by Theorem 3.4.8 and Lemma 7.1.4, there exists a height one prime M^e of $K[T_P]$ so that

$$M^e \cap K[A] = M^e \cap K[T_P \cap A] = M' \cap J,$$

where J itself is an intersection of (finitely many) height one primes of $K[T_P \cap A]$, all intersecting $T_P \cap A$. Then $M^e \cap T_P \neq \emptyset$. Hence $M^e =$

$K[M]$, as $K[M]$ is the only height one prime ideal of $K[T_P]$ intersecting T_P nontrivially. It follows that $K[M] \cap K[A] \subseteq M'$ and thus M' is one of the primes $M_1, \ldots, M_n$, as desired.

Because of (1), $P \cap A = (T \cap M) \cap A = (T \cap A) \cap (M \cap A)$. Hence the assumption and (3) yield that also $P \cap A$ is G-invariant. This proves (4).

Finally we prove (6). So let $x \in (T_P)_M$. Then there exists a G-conjugacy class $C \subseteq T_P$ so that $C \not\subseteq M$ and $Cx \subseteq T_P$. As C is a finite set, there exists a G-conjugacy class $C' \subseteq T$ so that $C' \not\subseteq P$ and $C'(Cx) \subseteq T$. As $C'C \subseteq T_P$ is a finite set as well, there also exists a G-conjugacy class $C'' \subseteq T$ with $C'' \not\subseteq P$ and $C''C'C \subseteq T$. Because of (1), each of C'', C' and C is not contained in M and thus $C''C'C \not\subseteq M$. Therefore $C''C'C \not\subseteq P$. As $C''C'Cx \subseteq T$ we obtain that $x \in T_P$, as desired. □

We finish this section with one more remark that is of interest in view of Lemma 7.1.5 and it will be relevant for some results in the next section. If T is a Krull order with only one minimal prime ideal, then $\mathrm{N}(T) = \mathrm{U}(T)\langle n \rangle$ for some $n \in \mathrm{N}(T)$. Indeed, since there only is one minimal prime, the integral divisorial ideals of T form a chain. Because a normal element of T generates an integral divisorial ideal, it follows that there exists $x \in \mathrm{N}(T)$ so that Tx is the only maximal ideal among the ideals generated by a normal element. Hence, if $y \in \mathrm{N}(T)$, then there exists $v_1 \in \mathrm{N}(T)$ so that $y = v_1 x$. If $v_1 \notin \mathrm{U}(T)$ then $v_1 = v_2 x$ for some $v_2 \in \mathrm{N}(T)$. Clearly $v_1 T \subset v_2 T$. After a finite number of steps and because of the ascending chain condition on divisorial ideals, we get that $y \in \mathrm{U}(T)\langle x \rangle$.

7.2 Algebras of submonoids of abelian-by-finite groups

In this section we characterize when a Noetherian semigroup algebra is a maximal order that is a domain and satisfies a polynomial identity. The results obtained rely on the characterization of the group algebra case (Theorem 3.6.4) and the commutative semigroup algebra case (Theorem 3.6.5).

Throughout this section K is a field, S is a submonoid of a finitely generated torsion free abelian-by-finite group. Let G denote its group of quotients and let A be an abelian normal subgroup of finite index in G. So the semigroup algebra $K[S]$ is a PI domain.

Lemma 7.2.1. *If $K[S]$ is a maximal order then S is a maximal order. Furthermore, if $K[S]$ is a Krull order then S is a Krull order.*

Proof. Assume $K[S]$ is a maximal order. Let I be an ideal of S. Clearly $K[I]$ is an ideal of $K[S]$ and $\{g \in G \mid gI \subseteq I\} \subseteq (K[I] :_l K[I])$. Since, by assumption $K[S]$ is maximal order, it follows at once that $\{g \in G \mid gI \subseteq I\} = (I :_l I) = S$. Similarly $(I :_r I) = S$ and thus S is a maximal order.

Assume now that $K[S]$ is a Krull order and let I be a fractional ideal of S. Then $K[I]$ is a fractional ideal of $K[S]$ and $K[(S : I)] = (K[S] : K[I])$. One inclusion is obvious. In order to prove the other inclusion let α be an element in the classical ring of quotients of $K[S]$ and suppose $\alpha \in (K[S] : K[I])$. Then $\alpha K[I] \subseteq K[S]$. Since $K[I]K[G] = K[G]$, it follows that $\alpha K[G] = \alpha K[I]K[G] \subseteq K[S]K[G] = K[G]$. So $\alpha \in K[G]$. Write $\alpha = \sum_{g \in G} k_g g$ with all $k_g \in K$. As $\alpha \in (K[S] : K[I])$ and thus $\alpha i \in K[S]$ for any $i \in I$, we obtain that $gi \in S$, for any $g \in G$ with $k_g \neq 0$. Thus $\alpha \in K[(S : I)]$, as claimed.

So, for an integral divisorial ideal I of S we obtain that $K[I] = K[(S : (S : I))] = (K[S] : K[(S : I)]) = (K[S] : (K[S] : K[I])) = (K[I])^*$. Hence $K[I]$ is an integral divisorial ideal of $K[S]$. Since by assumption $K[S]$ satisfies the ascending chain condition on integral divisorial ideals it follows that also S satisfies the ascending chain condition on integral divisorial ideals. The result thus follows. □

Lemma 7.2.2. *Suppose $K[S]$ is a Krull order and $P \in \operatorname{Spec}^0(S)$. The following properties hold.*

1. *If J is an ideal of S, then $J \not\subseteq P$ if and only if there exists a G-conjugacy class C contained in J so that $C \not\subseteq P$.*
2. *S is a Krull order and $S_{[P]} = S_P = S_{(P)}$.*

Proof. Suppose $K[S]$ is a Krull order. Since $K[S]$ is a PI algebra, Theorem 3.6.1 yields that for any $Q \in \operatorname{Spec}^1(K[S])$ and any ideal I of $K[S]$, $I \not\subseteq Q$ if and only if $I \cap \mathrm{Z}(K[S]) \not\subseteq Q \cap \mathrm{Z}(K[S])$.

Assume $P \in \operatorname{Spec}^0(S)$. Then by Proposition 4.5.8, $K[P] \in \operatorname{Spec}^1(K[S])$. Hence for any ideal J of S, if $J \not\subseteq P$ then $K[J] \not\subseteq K[P]$ and thus $K[J] \cap \mathrm{Z}(K[S]) \not\subseteq K[P] \cap \mathrm{Z}(K[S])$. Since the center of $K[S]$ is spanned by full G-conjugacy class sums contained in $K[S]$, statement (1) follows.

Because $K[S]$ is a Krull order, Lemma 7.2.1 yields that S is a Krull order. Hence, by Lemma 7.1.1, $S \cap A$ is G-invariant and thus by Lemma 7.1.3, $S_{(P)} = S_P$. That $S_{[P]} = S_P$ follows at once from (1). □

Lemma 7.2.3. *If $K[S]$ is a Noetherian maximal order and $P \in \operatorname{Spec}^0(S)$ then $\operatorname{Spec}^0(S_P) = \{I(P)\}$.*

Proof. Assume $K[S]$ is a Noetherian maximal order and $P \in \mathrm{Spec}^0(S)$. Let $Z = \mathrm{Z}(K[S]) \setminus K[P]$. By Proposition 4.5.8, $K[P] \in \mathrm{Spec}^1(K[S])$. Thus, by Theorem 3.6.1, $R = K[S]Z^{-1}$ has a unique maximal ideal $M = K[S]Z^{-1}P$ and any nonzero ideal of R can be written uniquely as M^k with $k \geq 0$. So, for every $n \in \mathrm{N}(S) \cap P$ there exists $k > 0$ so that $Rn = M^k = RP^k$. Hence, for any $p \in P^k$ there exists $\gamma \in Z$ so that $\gamma p \in K[S]n$. Therefore $Cp \subseteq Sn$ for some G-conjugacy class C with $C \not\subseteq P$ and $C \subseteq S$. Thus $pn^{-1} \in S_P$. Since p is arbitrary in P^k, we have shown that

$$P^k \subseteq S_P n \quad \text{for some } k \geq 1. \tag{7.1}$$

Let $Q \in \mathrm{Spec}(S_P)$. Because $Q \cap S$ is an ideal of S and $(Q \cap S)G = G = S\,\mathrm{Z}(S)^{-1}$, there exists $m \in Q \cap \mathrm{Z}(S)$. Note that $m \in P$ for otherwise $m^{-1} \in S_P$ and thus $1 = m^{-1}m \in Q$, a contradiction. Consequently, by(7.1), $P^k \subseteq S_P m$, for some $k > 0$. We now show that this implies

$$I(P)^k \subseteq S_P m. \tag{7.2}$$

In order to prove this, let $a \in I(P)^k$. Write $a = a_1 \cdots a_k$ with $a_i \in I(P)$. Thus for some G-conjugacy classes C_i with $C_i \subseteq S$ and $C_i \not\subseteq P$, $C_i a_i \subseteq I(P) \cap S$. Hence by Lemma 7.1.2, $C_i a_i \subseteq P$. So $Ca \subseteq P^k \subseteq S_P m$, with $C = C_1 \cdots C_k$. Consequently $Cam^{-1} \subseteq S_P$. Therefore, there exists a finite product of G-conjugacy classes $D = D_1 \cdots D_l$ so that each $D_j \subseteq S$ and $D_j \not\subseteq P$, and $DCam^{-1} \subseteq S$. Hence $am^{-1} \in S_P$. This proves (7.2).

Of course (7.2) yields that $I(P)^k \subseteq Q$. Hence, as Q is prime, we obtain that $I(P) \subseteq Q$. So every prime ideal of S_P contains $I(P)$, which is a prime ideal of S_P by Lemma 7.1.2. Hence $\mathrm{Spec}^0(S_P) = \{I(P)\}$. □

Before stating the main result of this section we need one more lemma.

Lemma 7.2.4. *Suppose T is a submonoid of an abelian group. If T is a Krull order with finitely many minimal prime ideals then there exist $t_1, \ldots, t_n \in T$ so that $T = \langle t_1, \ldots, t_n, \mathrm{U}(T) \rangle$.*

Proof. By assumption, $\{P_1, \ldots, P_m\} = \mathrm{Spec}^0(T)$. Since T is a Krull order, we know that the group of divisorial ideals $\mathrm{D}(T)$ is free abelian with basis $\mathrm{Spec}^0(T)$. Recall that $P_1^{n_{1i}} * \cdots * P_m^{n_{mi}} \subseteq P_1^{n_{1j}} * \cdots * P_m^{n_{mj}}$ if and only if $n_{ki} \geq n_{kj}$ for every k with $1 \leq k \leq m$. Hence, for any divisorial ideal I of T it follows that $\bigcap_{n \geq 1} I^n = \emptyset$.

Because of the ascending chain condition on principal ideals, every proper principal ideal is contained in a maximal principal ideal. Suppose that

$Ta_1, Ta_2, \ldots$ are infinitely many such ideals. Obviously Ta_i is a divisorial ideal and thus

$$Ta_i = P_1^{n_{1i}} * \cdots * P_m^{n_{mi}},$$

with each $n_{ji} \geq 0$. It is easily seen that there exists a sequence $k_1 \leq k_2 \leq \ldots$ so that $n_{jk_1} \leq n_{jk_2} \leq \ldots$ for $j = 1, \ldots m$. But then $Ta_{k_2} \subseteq Ta_{k_1}$, a contradiction.

Hence there are only finitely many maximal principal ideals, say, $Tt_1, \ldots,$ Tt_n. Let $t \in T \setminus \mathrm{U}(T)$. Then $t \in Tt_{i_1}$ for some i_1. Now, either $tt_{i_1}^{-1} \in \mathrm{U}(T)$ or $tt_{i_1}^{-1} \in Tt_{i_2}$ for some i_2. In the former case $t \in \langle t_1, \ldots, t_n, \mathrm{U}(T) \rangle$ and in the latter case either $tt_{i_1}^{-1} t_{i_2}^{-1} \in \mathrm{U}(T)$ or $tt_{i_1}^{-1} t_{i_2}^{-1} \in Tt_{i_3}$ for some i_3. Continuing this process one gets that either $t \in \langle t_1, \ldots, t_n, \mathrm{U}(T) \rangle$ or there exists a sequence $i_1, i_2, \ldots$ so that $t \in Tt_{i_1} \cdots t_{i_m}$ for every m. Since the indices i_j belong to a finite set there thus exists k so that $t \in \bigcap_{n \geq 1} (Tt_{i_k})^n = \emptyset$, a contradiction. The result thus follows. $\square$

Theorem 7.2.5. *Let K be a field and S a submonoid of a torsion free finitely generated abelian-by-finite group. The semigroup algebra $K[S]$ is a Noetherian maximal order if and only if the following conditions are satisfied.*

1. *S satisfies the ascending chain condition on left and right ideals.*

2. *For every minimal prime P in S the semigroup S_P is a maximal order with only one minimal prime ideal.*

3. *$\bigcap_{P \in \mathrm{Spec}^0(S)} S_P = S$.*

Proof. Because of Theorem 4.4.7 condition (1) is equivalent with $K[S]$ being Noetherian. To prove the necessity of the conditions assume that $K[S]$ is a Noetherian maximal order. Hence, by Lemma 7.2.1, the semigroup S is a maximal order. So, if $P \in \mathrm{Spec}^0(S)$ then, by Lemma 7.1.3, S_P is a maximal order. Furthermore, because of Lemma 7.2.3, $I(P)$ is the only minimal prime ideal of S_P. Condition (2) thus follows. We now show that condition (3) follows from (1) and the fact that S is a maximal order. Note that these two assumptions imply that the group of divisorial ideals $\mathrm{D}(S)$ is free abelian with basis the set of minimal prime ideals of S. Hence for any ideal J of S that is not contained in any minimal prime ideal we have that $J^* = S$. So, let $g \in \bigcap_{P \in \mathrm{Spec}^0(S)} S_P$. Then, for any $P \in \mathrm{Spec}^0(S)$, there exists an ideal $J(P) \subseteq S$ so that $J(P) \not\subseteq P$ and $J(P)g \subseteq S$ (as $S_P \subseteq S_{[P]}$). In particular, $Jg \subseteq S$, with $J = \bigcup_{P \in \mathrm{Spec}^0(S)} J(P)$. As J is not contained in any minimal prime ideal of S, we know that its divisorial closure J^* equals S. Hence, working in $\mathrm{D}(S)$, we get $(SgS)^* = J^* * (SgS)^* = (JSgS)^* = (JgS)^* \subseteq S^* = S$. Therefore $g \in S$. Hence (3) follows.

In the remainder of the proof we prove the converse. So assume conditions (1), (2) and (3) are satisfied. In particular $K[S]$ and $K[S \cap A]$ are Noetherian algebras.

To prove that $K[S]$ is a maximal order it is sufficient to show that the following properties hold.

(M1) $\bigcap_{P \in \mathrm{Spec}^0(S)} K[S_P] = K[S]$.

(M2) For $P \in \mathrm{Spec}^0(S)$ and $C = \mathrm{Z}(K[S]) \setminus K[P]$ one has that $K[S]C^{-1} \cap K[G] = K[S_P]$.

(M3) If $C = \mathrm{Z}(K[S]) \setminus K[P]$, with $P \in \mathrm{Spec}^0(S)$, then the localization $K[S]C^{-1}$ is a maximal order (note that since $K[S]$ is Noetherian, Proposition 4.5.8, implies that $K[P] \in \mathrm{Spec}(K[S])$). Furthermore, $K[S]C^{-1} = K[S]\,(\mathrm{Z}(K[S]) \cap K[A]) \setminus K[P])^{-1}$.

Indeed, let $\gamma \in \mathrm{Q}_{cl}(K[S])$ and let J be a nonzero ideal of $K[S]$ so that $\gamma J \subseteq J$. Then $\gamma JK[G] \subseteq JK[G]$. Note that $JK[G]$ is two-sided ideal of $K[G]$ as $K[G]$ is a central localization of $K[S]$. Because of Theorem 3.6.4, $K[G]$ is a maximal order. Hence $\gamma \in K[G]$. By (M3), for each $P \in \mathrm{Spec}^0(S)$ the localization $K[S](\mathrm{Z}(K[S]) \setminus K[P])^{-1}$, is a maximal order as well. So (M2) implies that $\gamma \in \bigcap_{P \in \mathrm{Spec}^0(S)} K[S_P]$. From (M1) it then follows that $\gamma \in K[S]$, as required. Similarly, $(J :_r J) = K[S]$, and thus indeed $K[S]$ is a maximal order.

First we show that (M3) holds. Therefore, let $P \in \mathrm{Spec}^0(S)$. Condition (2) says that S_P is a maximal order with only one minimal prime ideal, which we denote by M. Furthermore, Lemma 7.1.5 implies that $M \cap A$ is G-invariant and $K[M]$ is the only height one prime ideal of $K[S_P]$ intersecting S_P nontrivially. Also, because of condition (2) and Lemma 7.1.1, $S_P \cap A$ is G-invariant and it is a maximal order. Note that condition (3) therefore implies that $S \cap A$ is G-invariant as well and it is a maximal order. Hence, because of Lemma 7.1.3, $S_P = S_{(P)}$.

As $S \cap A$ satisfies the ascending chain condition on ideals and since it is a maximal order in a finitely generated torsion free abelian group (because A is finitely generated), we know from Theorem 3.6.5 that $K[S \cap A]$ is a Krull domain. It easily is verified that $K[S_P \cap A] = K[S_{(P)} \cap A]$ is the localization of $K[S \cap A]$ with respect to the multiplicatively closed set of all ideals $K[J]$ of $K[S_P \cap A]$, where J is an ideal of $S_P \cap A$ so that there exists a G-conjugacy class $D \subseteq J$ with $D \not\subseteq P$. Theorem 3.6.1 therefore yields that the localization $K[S_P \cap A]$ also is a Krull domain and thus, by Lemma 7.2, $S_P \cap A$ is a Krull order. So by Proposition 4.5.8, the height one primes of $K[S_P \cap A]$ intersecting $S_P \cap A$ nontrivially are precisely the ideals $K[Q]$ with Q a minimal prime ideal of $S_P \cap A$. Consider $K[S_P]$ as a natural G/A-graded ring. Then its identity component is $K[S_P \cap A]$. Because of Theorem 3.4.8, for every such minimal prime Q there exists a height one prime ideal J of $K[S_P]$ so that $J \cap K[S_P \cap A] = K[Q_1] \cap \cdots \cap K[Q_m]$ where $K[Q_1] = K[Q], \ldots, K[Q_m]$ are all the height one prime ideals of $K[S_P \cap A]$ containing $J \cap K[S_P \cap A]$. Clearly J intersects S_P nontrivially. Hence, by the above, $J = K[M]$. It follows that $Q_1, \ldots, Q_m$ are all the minimal primes of $S_P \cap A$. In particular, $S_P \cap A$ has only finitely many minimal prime ideals and they all contain $M \cap A$. Since $\mathrm{U}(S_P \cap A)$ is a finitely generated group (as a subgroup of A), we thus get from Lemma 7.2.4 that $S_P \cap A$ is a finitely generated semigroup and thus $K[S_P \cap A]$ is Noetherian. Lemma 4.1.3 then implies that also $K[S_P]$ is Noetherian.

Let $Q \in \mathrm{Spec}^1(K[S_P])$ so that $Q \cap S_P = \emptyset$. It follows from Lemma 7.1.4 that the ideal

$$\overline{Q} = \bigcap_{g \in G} g^{-1}(Q \cap K[A])g$$

is not contained in $K[M \cap A]$. As $\overline{Q}$ is G-invariant, $K[S_P]\overline{Q}$ is an ideal of $K[S_P]$. So

$$\left(K[S_P]\overline{Q} + K[M]\right)/K[M]$$

is a nonzero ideal of the Noetherian ring $R = K[S_P]/K[M] = K[S_{(P)}]/K[M]$. Since R_e is commutative Noetherian semiprime ring (in particular a Goldie ring), Proposition 3.4.6 then yields that the ring R has a classical ring of quotients which is obtained by inverting the regular elements of R_e. Hence the ideal $\left(K[S_P]\overline{Q} + K[M]\right)/K[M]$ of $K[S_P]/K[M]$ contains a regular element $\overline{\alpha}$ that is contained in R_e. Since $K[S_P \cap A]$ and $K[M \cap A]$ are G-invariant, conjugation induces an action of G on R_e. Hence the product of the finitely many G-conjugates of $\overline{\alpha}$ also belongs to $R_e \cap \left(K[S_P]\overline{Q} + K[M]\right)/K[M])$. Since this element is central, we thus may assume that $\overline{\alpha}$ also is central in R. As

$$((K[S_P]\overline{Q} + K[M])/K[M]) \cap R_e = \left(K[S_P \cap A]\overline{Q} + K[M \cap A]\right)/K[M \cap A]$$

and $\overline{Q}$ is an ideal of $K[S_P \cap A]$, we get that

$$\overline{\alpha} \in \left(\overline{Q} + K[M \cap A]\right)/K[M \cap A].$$

Consequently, there exists $\gamma \in \overline{Q}$ and $\omega \in K[M \cap A]$ so that $g^{-1}\delta g - \delta \in K[M \cap A]$ for all $g \in G$, where $\delta = \gamma + \omega$ and $\delta + K[M \cap A] = \overline{\alpha}$. Clearly,

$g^{-1}\gamma g - \gamma = (g^{-1}\delta g - \delta) - (g^{-1}\omega g - \omega)$ for all $g \in G$. As $M \cap A$ is G-invariant we get

$$g^{-1}\gamma g - \gamma \in K[M \cap A],$$

for all $g \in G$. Let

$$\beta = \prod_{f \in F} f^{-1}\gamma f,$$

where F is a transversal for A in G. Obviously $\beta \in \mathrm{Z}(K[S_P]) \cap K[A]$ and $\beta \in Q$. Moreover, $\beta \notin K[M \cap A]$. Indeed, for otherwise,

$$K[M \cap A] = \beta + K[M \cap A] = \prod_{f \in F} f^{-1}\gamma f + K[M \cap A] = \gamma^{|F|} + K[M \cap A].$$

Hence, as $K[M \cap A]$ is a semiprime ideal of the commutative ring $K[S_P \cap A]$, $\gamma \in K[M \cap A]$ and thus $\delta \in K[M \cap A]$, a contradiction.

So we have shown that

$$Q \cap C' \neq \emptyset,$$

where $C' = (\mathrm{Z}(K[S_P]) \cap K[A]) \setminus K[M]$, and this for any height one prime ideal Q in $K[S_P]$ with $Q \cap S_P = \emptyset$. Hence the only height one prime ideals of the localization $K[S_P](C')^{-1}$ are generated by height one prime ideals of $K[S_P]$ intersecting S_P nontrivially. As $K[M]$ is the only such prime ideal it has been shown that

$$\mathrm{Spec}^1(K[S_P](C')^{-1}) = \{K[S_P](C')^{-1}M\}.$$

Also $K[S_P](C')^{-1}$ is a natural G/A-graded ring, with identity component $K[S_P \cap A](C')^{-1}$. Hence, by Theorem 3.4.8, $K[S_P \cap A](C')^{-1}$ has only finitely many height one prime ideals. Hence, again by Theorem 3.6.1, the localization $K[S_P \cap A](C')^{-1}$ is a (Noetherian) Krull domain with only finitely many height one prime ideals. In particular it is a principal ideal ring and it has prime dimension one (see Section 3.6). Because of Theorem 3.4.8 we then get that $K[S_P](C')^{-1}$ also has dimension one. Hence every nonzero prime is of height one. Since we already know that $\mathrm{Spec}^1(K[S_P](C')^{-1}) = \{K[S_P](C')^{-1}M\}$ it thus follows that $K[S_P](C')^{-1}M$ is the only nonzero prime ideal of the Noetherian ring $K[S_P](C')^{-1}$. As this algebra also is PI, Theorem 3.1.6 yields that

$$\bigcap_n \left(K[S_P](C')^{-1}M\right)^n = \{0\}. \tag{7.3}$$

Let $N = K[S_P](C')^{-1}M$ and $W = K[S_P](C')^{-1}$. We already know that S_P is a Noetherian maximal order with a unique minimal prime ideal

M. Hence the group of divisorial ideals $\mathrm{D}(S_P)$ is an infinite cyclic group with generator the minimal prime ideal M. Clearly, $(M(S_P : M))^* = M^* * (S_P : M)^* = M * (S_P : M) = S_P$. Therefore the divisorial ideal $(M(S_P : M))^*$ of S_P is not contained in M. Hence, the ideal $M(S_P : M)$ is not contained in M. Because $N \cap S_P = M$ and $N = WM = MW$ is the only maximal ideal of W, we thus also get that $WM(S_P : M)W = W$. So, $N(S_P : M)W = WMW(S_P : M)W = W$. Thus N is an invertible ideal of the ring W.

It follows that every nonzero ideal of W is a power of the unique maximal ideal. Indeed, if V is a nonzero proper ideal of W then by (7.3), $V \subseteq N^n$ and $V \not\subseteq N^{n+1}$, for some $n > 0$. Since N is invertible, we thus get that $V(N')^n \subseteq N^n(N')^n = W$ and $V(N')^n \not\subseteq N$. Since N is the unique maximal ideal of W, we get that $V(N')^n = W$ and thus $V = N^n$, as desired.

So all nonzero ideals of W are invertible and hence it is a maximal order. Thus $K[S_P](C')^{-1}$ is a Noetherian maximal order with a unique height one prime ideal.

Let $C'' = (\mathrm{Z}(K[S]) \cap K[A]) \setminus K[P]$. Obviously, $S_P = S_{(P)} \subseteq K[S](C'')^{-1}$ and, for every element x of C', there exists a product F of G-conjugacy classes in $S \cap A$ such that $F \subseteq (\mathrm{Z}(K[S]) \cap K[A]) \setminus K[P]$ and $Fx \subseteq K[S]$. Hence $K[S_P](C')^{-1} \subseteq K[S](C'')^{-1}$. As the converse inclusion is obvious, it follows that $K[S_P](C')^{-1} = K[S](C'')^{-1}$.

Let $C = \mathrm{Z}(K[S]) \setminus K[P]$. It is then clear that $K[S]C^{-1}$ is a localization of $K[S](C'')^{-1} = K[S_P](C')^{-1}$. Hence by part (3) of Theorem 3.6.1 the ring $K[S]C^{-1}$ is an intersection of the rings that are central localizations of $K[S_P](C')^{-1}$ with respect to height one prime ideals. Since the latter ring is a Noetherian maximal order with a unique height one prime ideal N, it follows that all central elements not contained in N are invertible. Hence $K[S]C^{-1} = K[S_P](C')^{-1} = K[S](C'')^{-1}$. So condition (M3) is proved.

Now we prove condition (M2), that is

$$K[S]C^{-1} \cap K[G] = K[S_P],$$

where $P \in \mathrm{Spec}^0(S)$ and $C = \mathrm{Z}(K[S]) \setminus K[P]$. Clearly, $K[S_P] \subseteq K[S]C^{-1} \cap K[G]$. To prove the converse inclusion, let $\alpha \in K[S]C^{-1} \cap K[G]$. We proceed by induction on the number of elements in $\mathrm{supp}(\alpha)$. If $|\mathrm{supp}(\alpha)| = 1$, then $\alpha = kg$ with $k \in K$ and $g \in G$, and there exists a central element $\gamma \in K[S]$ so that $\gamma \notin K[P]$ and $\gamma g \in K[S]$. Since the center of $K[S]$ is the K-vector space spanned by the G-conjugacy classes, it then easily follows that $g \in S_P$, as desired. Now suppose $|\mathrm{supp}(\alpha)| > 1$. In the proof of (M3) above, it has been shown that $K[S]C^{-1} = K[S](C'')^{-1}$, with $C'' =$

$(\mathrm{Z}(K[S] \cap K[A]) \setminus K[P]$. Hence $K[S]C^{-1} = K[S](C'')^{-1}$ is a G/A-graded ring. It is clear that the G/A-homogeneous components of α also belong to $K[S]C^{-1} \cap K[G]$. Hence, because of the induction hypothesis, we may assume that α is G/A-homogeneous. So write $\alpha = \sum_{i=1}^{n} k_i g_i$, each $g_i \in G$, $k_i \in K$ and all $g_i^{-1} g_j \in A$. Note that this statement holds for any abelian normal subgroup A of G that is of finite index.

Suppose there exists a normal abelian subgroup A' of finite index in G such that $A' \subseteq A$ and $g_i g_j^{-1} \notin A'$ for some $i \neq j$. Then using the G/A'-gradation of $K[S]C^{-1}$ and because of the induction hypothesis we obtain that $\alpha \in K[S]$, as desired. So we may assume that $g_i g_j^{-1} \in A'$ for all i, j and all $A' \subseteq A$, with A' a normal abelian subgroup of finite index in G. As G is residually finite ([140, Theorem 1.1]) it follows that $n = 1$. This proves (M2).

Finally, it is obvious that condition (M1) follows at once from (3). □

We now show that if $K[S]$ is a Noetherian maximal order then so is $K[S_P]$, for each minimal primed ideal P of S.

Proposition 7.2.6. *Let K be a field and S a submonoid of a torsion free finitely generated abelian-by-finite group. If $K[S]$ is a Noetherian maximal order and $P \in \operatorname{Spec}^0(S)$, then $K[S_P]$ is a Noetherian maximal order. In particular, S_P is a maximal order satisfying the ascending chain condition on right and left ideals. Furthermore, for any normal and abelian subgroup A of finite index in SS^{-1} the monoid $S_P \cap A$ is a finitely generated maximal order with finitely many minimal prime ideals.*

Proof. Suppose $K[S]$ is a Noetherian maximal order and let P be a minimal prime ideal of S. It follows from Theorem 7.2.5 and also from its proof that S_P is a maximal order with a unique minimal prime ideal M and $K[S_P]$ is Noetherian. Furthermore, for any normal and abelian subgroup A of finite index in SS^{-1} the monoid $S_P \cap A$ is a finitely generated maximal order with finitely many minimal prime ideals. Because of Lemma 7.1.1 and Lemma 7.1.5, we thus also have that $(S_P)_M = S_P$. So, by Theorem 7.2.5, $K[S_P]$ is a maximal order. □

We finish this section with a reformulation of the main theorem.

Theorem 7.2.7. *Let K be a field and S a submonoid of a torsion free finitely generated abelian-by-finite group. The semigroup algebra $K[S]$ is a Noetherian maximal order if and only if the following conditions are satisfied.*

1. *S satisfies the ascending chain condition on right and left ideals.*

2. *S is a maximal order in its group of quotients.*

3. *For every minimal prime P in S the semigroup S_P has only one minimal prime ideal.*

Proof. The necessity follows from Lemma 7.2.1 and Theorem 7.2.5. So we prove the converse. Because of Theorem 7.2.5 it is sufficient to prove that $\bigcap_{P\in \mathrm{Spec}^0(S)} S_P = S$ and if $P \in \mathrm{Spec}^0(S)$ then S_P is a maximal order. Since, by assumption, S is a maximal order the latter follows from Lemma 7.1.3. The former is proved as in the first part of the proof of Theorem 7.2.5. This finishes the proof. □

In the final lemma we collect some facts on the localizations of algebras and monoids with respect to height one primes.

Lemma 7.2.8. *If $K[S]$ is a Noetherian maximal order and $P \in \mathrm{Spec}^0(S)$, then*

1. $K[S_P] = K[G] \cap K[S]((\mathrm{Z}(K[S]) \cap K[A]) \setminus K[P])^{-1}$,

2. $S_P \cap A = (S\cap A)_{Q_1} \cap \cdots \cap (S\cap A)_{Q_n}$ *and* $\{Q_1, \ldots, Q_n\} = \{g^{-1}Q_1 g \mid g \in SS^{-1}\}$, *where* $Q_1, \ldots, Q_n$ *are all the minimal primes of* $S \cap A$ *containing* $P \cap A$ *(so* $P\cap A = Q_1 \cap \cdots \cap Q_n$*),*

3. $K[S_P \cap A] = K[A] \cap K[S\cap A]((\mathrm{Z}(K[S]) \cap K[A]) \setminus K[P])^{-1} = K[A] \cap \bigcap_{i=1}^{n} K[S\cap A](K[S\cap A] \setminus K[Q_i])^{-1}$.

Proof. The first part follows from the proof of Theorem 7.2.5 (conditions (M2) and (M3)). From (1) we then get

$$\begin{aligned} K[S_P \cap A] &= K[A] \cap K[S]((\mathrm{Z}(K[S]) \cap K[A]) \setminus K[P])^{-1} \\ &= K[A] \cap K[S\cap A]((\mathrm{Z}(K[S]) \cap K[A]) \setminus K[P])^{-1}. \end{aligned}$$

Since $K[S]$ is a Noetherian maximal order we know that S is a maximal order and, by Proposition 4.5.8, $K[P]$ is a height one prime ideal of $K[S]$. Furthermore, the height one primes of $K[S\cap A]$ intersecting $S\cap A$ nontrivially are of the form $K[Q]$ with $Q \in \mathrm{Spec}^0(S\cap A)$. Hence, by Lemma 7.1.1, Lemma 7.1.4 and Lemma 7.1.5, it follows that $P\cap A = Q_1 \cap \cdots \cap Q_n$, where $\{Q_1, \ldots, Q_n\} = \{g^{-1}Q_1 g \mid g \in SS^{-1}\}$ and $Q_1, \ldots, Q_n$ are all the minimal primes of $S \cap A$ containing $P \cap A$. One then gets that

$$K[S_P \cap A]((\mathrm{Z}(K[S]) \cap K[A]) \setminus K[P])^{-1} = \bigcap_{1 \leq i \leq n} K[S \cap A](K[S \cap A] \setminus K[Q_i])^{-1}$$

(as $P \cap A = Q_1 \cap \cdots \cap Q_n$). This shows (3).

For any minimal prime Q of $S \cap A$ containing $P \cap A$ it is readily verified that $A \cap K[S \cap A]((\mathrm{Z}(K[S]) \cap K[A]) \setminus K[Q])^{-1} \subseteq (S \cap A)_Q$. Hence (3) implies that $S_P \cap A \subseteq (S \cap A)_{Q_1} \cap \cdots \cap (S \cap A)_{Q_n}$. To prove the converse inclusion, let $a \in (S \cap A)_{Q_1} \cap \cdots \cap (S \cap A)_{Q_n}$. Then $aJ' \subseteq S \cap A$ for some ideal J' of $S \cap A$ with $J' \not\subseteq Q_i$ for every i with $1 \leq i \leq n$. Let J be the product of all G-conjugates of J'. Recall from Lemma 7.1.1 that $S \cap A$ is G-invariant, and thus that all these conjugates and J are contained in $S \cap A$. So J is an ideal of $S \cap A$ that is G-invariant. Furthermore $J \not\subseteq P \cap A$. Indeed, for otherwise $J \subseteq Q_1$ and thus $g^{-1}(J')g \subseteq Q_1$ for some $g \in G$. Hence $J' \subseteq gQ_1g^{-1}$. Since $gQ_1g^{-1} = Q_j$ for some j, we thus get that $J' \subseteq Q_j$, a contradiction. Hence indeed $J \not\subseteq P \cap A$. Because J is G-invariant, it follows that $SJ = JS$ is an ideal not contained in P. Clearly $aJS \subseteq S$ and thus $a \in S_P \cap A$, as desired. Hence (2) follows. □

7.3 Comments and problems

The results presented in this chapter are taken from [84]. The main theorem deals with submonoids of finitely generated abelian-by-finite groups. Applications will be given in Chapter 8 to semigroup algebras of special classes of semigroups. In [81] a characterization is given of when a contracted semigroup algebra $K_0[S]$ of an arbitrary finitely generated semigroup S is a prime Noetherian maximal order. Attention has also been given to determining when a semigroup algebra $K[S]$ belongs to a special class of maximal orders. For example, in Chapter 6 the class of principal ideal rings has been considered. In [91, 92] it has been investigated when $K[S]$ is a unique factorization ring in case S is a submonoid of a polycyclic-by-finite group or an abelian-by-finite group (not necessarily finitely generated). For commutative semigroup algebras much more results have been obtained; we only list a few relevant references [5, 26, 30, 31, 54]. As mentioned in the introduction of the chapter, Brown in [12, 13, 14] solved the problem of when a group algebra of a polycyclic-by-finite group is a prime maximal order. Earlier results were obtained in [89, 90, 144, 145]. There is an extensive list of publications on related results for group rings and some generalizations, such as skew group rings (see for example [15, 22, 23, 110]). Also in the context of rings graded by abelian groups there is an extensive literature on arithmetical properties (see for example [3, 4, 5, 102, 116]).

Problems

1. Characterize prime Noetherian maximal orders $K[S]$ for a submonoid S of an arbitrary polycyclic-by-finite group G in terms of the properties of S. See Theorem 7.2.5 and Theorem 7.2.7 for the abelian-by-finite case. Because of Brown's result on group algebras (Theorem 3.6.2), a positive answer to the following question would be the best result to hope for. Assume G is dihedral free, $\Delta(G)$ is torsion free, S is a maximal order and $K[S]$ is Noetherian. Is $K[S]$ a maximal order?

2. Suppose $K[S]$ is a prime Noetherian maximal order. Assume that every height one prime ideal P of $K[S]$ is generated by a normal element provided that P intersects S nontrivially. Is then every height one prime ideal of $K[S]$ generated by a normal element? See Proposition 4.5.8 and Theorem 3.6.4.

3. Let S be a submonoid of a torsion free finitely generated abelian-by-finite group. Assume S is a maximal order that satisfies the ascending chain condition on one-sided ideals. If P is a minimal prime ideal of S, does it follow that the minimal prime ideals of S_P are principally generated by a normal element? In Section 10.5 examples are given to illustrate that the local monoid S_P has, in general, a more complicated structure than in the commutative case. However, the answer to the question posed is positive in all examples considered.

Chapter 8

Monoids of I-type

As promised in Chapter 4, we investigate in this chapter an important class of Noetherian semigroup algebras of monoids S that arise in other contexts. The monoids considered are called monoids of (left) I-type and they were introduced by Gateva-Ivanova and Van den Bergh. Their work was inspired by earlier work of Tate and Van den Bergh on Sklyanin algebras. It turns out that these monoids S satisfy the ascending chain condition on left and right ideals and that they have a group of quotients SS^{-1} that is torsion free, finitely generated, solvable and abelian-by-finite; these are called groups of (left) I-type. It follows that the algebras $K[S]$ are Noetherian domains that satisfy a polynomial identity. Furthermore, the monoids S are finitely generated, say by n elements, and they have a presentation defined by $\binom{n}{2}$ homogeneous relations of degree 2. It will be shown in Section 8.1 that these monoids are intimately related with other mathematical notions that are currently of interest, such as set theoretic solutions of the quantum Yang-Baxter equation and Bieberbach groups. In Section 8.3 it is shown that a class of monoids, introduced by Gateva-Ivanova and defined by quadratic homogeneous and square free relations, consists of monoids of I-type. These monoids are called binomial monoids. In Section 8.5 it is shown that the algebras $K[S]$ of monoids of I-type have a rich algebraic structure that resembles that of a polynomial algebra in finitely many commuting generators. In particular $\mathrm{GK}(K[S]) = n$ and $K[S]$ is a maximal order with height one primes that are either S-homogeneous and principal, or that are determined by the height one primes (which are also principal) of the group algebra $K[SS^{-1}]$. In order to prove these results, we describe in Section 8.2, monoids of left I-type as submonoids of a semidirect product $\mathrm{FaM}_n \rtimes \mathrm{Sym}_n$ of a free abelian monoid FaM_n with the symmetric group

Sym_n of degree n. It follows that monoids (groups) of left I-type are also of right I-type. Furthermore, a large class of groups of I-type consists of poly-(infinite cyclic) groups. Examples are given to show that they are not all of this type. It also is shown that such monoids (respectively groups) often can be decomposed as products of monoids (respectively groups) of I-type on less generators. In Section 8.4, it is proved that all non-cyclic monoids of I-type that have a presentation defined via square free relations can be decomposed and are binomial.

8.1 A characterization

Let FaM_n (respectively Fa_n) denote the free abelian monoid (respectively free abelian group) of rank n generated by elements $u_1, \ldots, u_n$. The group Fa_n has a natural degree function deg with respect to the given basis; so $\deg(u_i) = 1$. Throughout we will simply use the terminology "natural degree function" without specifically referring to the chosen basis.

Definition 8.1.1. A monoid S generated by a set $X = \{x_1, \ldots, x_n\}$ is said to be of left I-type if there exists a bijection (called a left I-structure)

$$v : \mathrm{FaM}_n \longrightarrow S$$

such that, for all $a \in \mathrm{FaM}_n$,

$$v(1) = 1 \quad \text{and} \quad \{v(u_1 a), \ldots, v(u_n a)\} = \{x_1 v(a), \ldots, x_n v(a)\}. \tag{8.1}$$

Similarly, one defines the notion of a monoid of right I-type. It is clear that a monoid S is of left I-type if and only if the opposite monoid S^{opp} is of right I-type. Recall that as a set $S^{opp} = S$ and that the product $s_1 \cdot s_2$ of $s_1, s_2 \in S^{opp}$ is defined as the product $s_2 s_1$ in S.

The following lemma states some elementary properties.

Lemma 8.1.2. *The following properties hold for a monoid S of left I-type with generating set $X = \{x_1, \ldots, x_n\}$ and with left I-structure $v : \mathrm{FaM}_n \longrightarrow S$.*

1. *The natural grading by degree on FaM_n induces via v a grading on S such that $\deg(x_i) = 1$.*

2. *For every $a \in \mathrm{FaM}_n$ there exists a permutation $f_a \in \mathrm{Sym}_n$ such that $x_{f_a(i)} v(a) = v(a u_i)$, for all i with $1 \leq i \leq n$. Furthermore, for any $b \in \mathrm{FaM}_n$ and $1 \leq i, j \leq n$,*

$$x_{f_{bu_j}(i)} x_{f_b(j)} = x_{f_{bu_i}(j)} x_{f_b(i)}.$$

3. *Every element of S can be written uniquely in the form*

$$x_{f_{u_1\cdots u_{i_{m-1}}}(i_m)}\cdots x_{f_{u_{i_1}}(i_2)}x_{f_1(i_1)},$$

with $1 \leq i_1 \leq i_2 \leq \cdots \leq i_m \leq n$.

4. *If $s \in S$ has degree two then $s = x_{f_{u_i}(j)}x_{f_1(i)} = x_{f_{u_j}(i)}x_{f_1(j)} = v(u_iu_j) = v(u_ju_i)$ for some $1 \leq i,j \leq n$; and these are all possible expressions for s.*

5. *Right multiplication by $s = v(a) \in S$ induces a bijection between S and $\{v(ab) \mid b \in \mathrm{FaM}_n\}$.*

6. *S is right cancellative.*

7. *The monoid S has a presentation of the form*

$$S = \langle x_1, \ldots, x_n \mid x_{f_{u_i}(j)}x_{f_1(i)} = x_{f_{u_j}(i)}x_{f_1(j)},\ n \geq j > i \geq 1\rangle.$$

In particular, S has a presentation with $\binom{n}{2}$ defining homogeneous relations of degree two.

Proof. (1) and the first part of (2) are obvious. In particular, we get that $v(u_{i_1}) = v(1u_{i_1}) = x_{f_1(i_1)}v(1) = x_{f_1(i_1)}$. Note that every element of S can be written uniquely as $v(u_{i_1}u_{i_2}\cdots u_{i_m})$ with $1 \leq i_1 \leq i_2 \leq \cdots \leq i_m \leq n$. From the first part of (2) we thus also obtain that

$$v(u_{i_1}u_{i_2}\cdots u_{i_m}) = x_{f_{u_1\cdots u_{m-1}}(i_m)}v(u_{i_1}\cdots u_{i_{m-1}})$$

and thus part (3) follows by induction.

Because of (1), it is clear that the elements of S of degree two are of the form $v(u_iu_j) = v(u_ju_i)$ with $1 \leq i,j \leq n$. Hence (4) follows from the first part of (2).

Let $s = v(a) \in S$ with $a \in \mathrm{FaM}_n$. It follows from (8.1) that for any i, with $1 \leq i \leq n$, there exists j with $1 \leq j \leq n$, so that $v(u_ia) = x_jv(a)$. An induction argument then easily yields that, for any $b \in \mathrm{FaM}_n$, there exists $c \in \mathrm{FaM}_n$ so that $v(ba) = v(c)v(a)$. Similarly, for any $d \in \mathrm{FaM}_n$ there exists $e \in \mathrm{FaM}_n$ so that $v(d)v(a) = v(ea)$. Hence, if $\theta : S \longrightarrow S$ denotes the right multiplication by s, then $\theta(S) = \{v(ba) \mid b \in \mathrm{FaM}_n\} = \{v(b) \mid b \in \mathrm{FaM}_n\}\,v(a)$. From (1) it follows that $\deg(\theta(v(b))) = \deg(v(b)) + \deg(v(a))$. So, θ induces a surjective map between the elements of S of a fixed degree k and the elements in $\{v(b) \mid b \in \mathrm{FaM}_n\}\,v(a)$ of degree $k + \deg(v(a))$. Since both sets have the same number of elements, we obtain that both sets are in a bijective correspondence. Hence θ is bijective. Therefore (5) follows and (6) now also is obvious.

We now prove the second part of (2). Let $b \in \mathrm{FaM}_n$ and $1 \leq i, j \leq n$. We already know that

$$\begin{aligned} v(u_i u_j b) &= x_{f_{u_i b}(j)} v(u_i b) \\ &= x_{f_{u_i b}(j)} x_{f_b(i)} v(b). \end{aligned}$$

Interchanging i and j we also get

$$v(u_i u_j b) = x_{f_{u_j b}(i)} x_{f_b(j)} v(b).$$

Because S is right cancellative, it follows that $x_{f_{u_i b}(j)} x_{f_b(i)} = x_{f_{u_j b}(i)} x_{f_b(j)}$, as desired.

In order to obtain part (7), it is sufficient to note that the mentioned relations in (7) hold, and that not more defining relations are needed, as the number of words of a particular degree in S and in FaM_n are the same and possible relations must have the same degree. □

Note that, for any field K, it follows at once from the first statement in Lemma 8.1.2 that $\mathrm{GK}(K[S]) = n$. Furthermore, the semigroup algebra $K[S]$ is of left I-type in the sense of [150].

Clearly, finitely generated free abelian monoids are examples of monoids of left and right I-type. We shall later prove that monoids of left I-type are also of right I-type, and hence they are left and right cancellative. It follows that the only possible nonabelian monoid of left I-type is $\langle x, y \mid x^2 = y^2 \rangle$. We now show that this indeed is a monoid of left and right I-type.

Example 8.1.3. The monoid $S = \langle x, y \mid x^2 = y^2 \rangle$ is of left and right I-type.

Proof. Let $\mathrm{Fa}_2 = \mathrm{gr}(a, b \mid ab = ba)$ denote the free abelian monoid of rank two with generators a and b. Let $\mathrm{Sym}_2 = \{1, \sigma\}$ be the symmetric group on $\{a, b\}$. So $\sigma(a) = b$ and $\sigma(b) = a$. Let $G = \mathrm{Fa}_2 \rtimes_\sigma \mathrm{Sym}_2$ denote the semidirect product, where the action of Sym_2 on Fa_2 is defined naturally via the permutation σ. Thus

$$(c, \alpha)\,(d, \beta) = (c\,\alpha(d),\, \alpha\beta)$$

for $c, d \in \mathrm{Fa}_2$ and $\alpha, \beta \in \mathrm{Sym}_2$.

For $c \in \mathrm{Fa}_2$ we put $\sigma_c = \sigma$, if c has odd length and $\sigma_c = 1$, if c has even length. Let $\mathrm{FaM}_2 = \langle a, b \rangle$ be the free abelian submonoid of Fa_2 generated by a and b. It easily is verified that

$$T = \{(c, \sigma_c) \mid c \in \mathrm{FaM}_2\}$$

is a submonoid of G. We claim that T is a monoid of left and right I-type that is isomorphic with S. Indeed, consider the bijective mapping

$$v : \mathrm{FaM}_2 \longrightarrow T : c \mapsto (c, \sigma_c).$$

Then, for any $c \in \mathrm{FaM}_2$,

$$\begin{aligned} \{v(c)v(a),\ v(c)v(b)\} &= \{(c, \sigma_c)(a, \sigma_a),\ (c, \sigma_c)(b, \sigma_b)\} \\ &= \{(c\,\sigma_c(a), \sigma_c\sigma),\ (c\,\sigma_c(b), \sigma_c\sigma)\} \\ &= \{v(c\sigma_c(a)), v(c\sigma_c(b))\} \\ &= \{v(ca), v(cb)\}. \end{aligned}$$

Since $v(1) = (1, 1)$, the map v defines a right I-structure on T, and thus T is a monoid of right I-type.

Let $x_1 = (a, \sigma)$ and $x_2 = (b, \sigma)$. A direct verification shows that $T = \langle x_1, x_2 \rangle$ and $x_1^2 = (ab, 1) = v(ab) = v(ba) = x_2^2$. Because of the left-right dual of Lemma 8.1.2, it follows that $x_1^2 = x_2^2$ is the only defining relation for T. Hence, T is isomorphic with S. As S is isomorphic with S^{opp}, it follows that S is a monoid of left and right I-type. □

In Lemma 8.1.2, it is shown that a monoid S of left I-type has a presentation

$$S = \langle x_1, \ldots, x_n \mid x_i x_j = x_k x_l \rangle$$

with $\binom{n}{2}$ defining relations, so that every word $x_i x_j$ with $1 \leq i, j \leq n$ appears at most once in one of the relations. So we obtain an associated bijective map

$$r : X \times X \quad \longrightarrow \quad X \times X \tag{8.2}$$

defined by

$$r(x_i, x_j) = (x_k, x_l)$$

if $x_i x_j = x_k x_l$ is a defining relation for S, otherwise we put $r(x_i, x_j) = (x_i, x_j)$. For every $x \in X$, we denote by

$$f_x \ :\ X \longrightarrow X \tag{8.3}$$

and

$$g_x \ : \ X \longrightarrow X \tag{8.4}$$

the mappings defined by $f_x(x_i) = p_1(r(x, x_i))$ and $g_x(x_i) = p_2(r(x_i, x))$, where p_1 and p_2 denote the natural projections onto the first and the second component respectively. So

$$r(x_i, x_j) = (f_{x_i}(x_j), g_{x_j}(x_i)).$$

One says that r (or simply S) is left non-degenerate if g_x is bijective for each $x \in X$. In case f_x is bijective for each $x \in X$, then r (or S) is said to be right non-degenerate.

We now will prove a characterization of monoids of left I-type that relates these monoids to set theoretic solutions of the Yang-Baxter equation. We recall the definition.

Let X be a nonempty set and $r : X^2 \longrightarrow X^2$ a bijective mapping. Let $m \geq 2$ be an integer. If $m = 2$ then we put $r_1 = r$ and if $m > 2$ then for any integer i, with $2 \leq i < m - 2$, put

$$r_i \ = \ \mathrm{id}_{X^{i-1}} \times r \times \mathrm{id}_{X^{m-i-1}} \text{ and } r_{m-1} = \mathrm{id}_{X^{m-2}} \times r, \tag{8.5}$$

a mapping $X^m \longrightarrow X^m$. One says that r is a set theoretic solution of the Yang-Baxter equation if it satisfies on X^3 the braid relation

$$r_1 r_2 r_1 = r_2 r_1 r_2.$$

If V is a vector space over a field K with basis X, then it follows that r induces a linear operator R on $V \otimes_K V$, so that in the automorphism group of $V \otimes_K V \otimes_K V$

$$(R \otimes \mathrm{id}_V)\,(\mathrm{id}_V \otimes R)\,(R \otimes \mathrm{id}_V) = (\mathrm{id}_V \otimes R)\,(R \otimes \mathrm{id}_V)\,(\mathrm{id}_V \otimes V).$$

One says that R is a solution of the Yang-Baxter equation. If we denote by $\tau : X^2 \longrightarrow X^2$ the map defined by $\tau(x, y) = (y, x)$, then it is shown in [37] that r is a set theoretic solution of the Yang-Baxter equation if and only if $\mathcal{R} = \tau \circ r$ is a solution of the quantum Yang-Baxter equation, that is,

$$\mathcal{R}^{12}\mathcal{R}^{13}\mathcal{R}^{23} = \mathcal{R}^{23}\mathcal{R}^{13}\mathcal{R}^{12},$$

where $\mathcal{R}^{ij}$ denotes $\mathcal{R}$ acting on the i-th and j-th component.

The quantum Yang-Baxter equation appeared in a paper by Yang [155] in statistical mechanics and it turned out to be one of the basic equations in mathematical physics. It lies at the foundation of the theory of quantum groups. One of the important problems is to discover all solutions of the quantum Yang-Baxter equation. In general, this still is unsolved. Nevertheless, in recent years, many solutions have been found, and the related algebraic structures, such as Hopf algebras, have been intensively studied (see for example [97]). In [33], Drinfeld posed the question of finding all set theoretic solutions.

The following characterization of monoids of left I-type shows that these monoids yield set theoretic solutions of the Yang-Baxter equation.

Theorem 8.1.4. *Let $X = \{x_1, \dots, x_n\}$. If $S = \langle x_1, \dots, x_n \rangle$ is a monoid of left I-type, then the mapping $r : X^2 \longrightarrow X^2$ defined by (with notations as in Lemma 8.1.2)*

$$r(x_{f_{u_i}(j)}, x_{f_1(i)}) = (x_{f_{u_j}(i)}, x_{f_1(j)})$$

satisfies the following three properties.

(IT.1) $r^2 = \mathrm{id}_{X^2}$.

(IT.2) r *is a set theoretic solution of the Yang-Baxter equation.*

(IT.3) For $i, j \in \{1, \dots, n\}$ there exist unique $k, l \in \{1, \dots, n\}$ such that

$$r(x_k, x_i) = (x_l, x_j),$$

that is, S is left non-degenerate.

Conversely, if a mapping $r : X^2 \longrightarrow X^2$ satisfies (IT.1), (IT.2) and (IT.3), then, for any $f \in \mathrm{Sym}_n$, there exists a unique left I-structure v on the monoid

$$S = \langle x_1, \dots x_n \mid x_i x_j = x_k x_l \text{ if } 1 \leq i, j, \leq n \text{ and } r(x_i, x_j) = (x_k, x_l) \rangle$$

so that, for all i,

$$v(u_i) = x_{f(i)}.$$

Proof. Assume $S = \langle x_1, \dots, x_n \rangle$ is a monoid of left I-type. For each $a \in \mathrm{FaM}_n = \langle u_1, \dots, u_n \rangle$ let f_a denote the permutation on $\{1, \dots, n\}$ introduced in Lemma 8.1.2. Consider the mapping $r : X^2 \longrightarrow X^2$ defined by

$$r(x_{f_{u_i}(j)}, x_{f_1(i)}) = (x_{f_{u_j}(i)}, x_{f_1(j)}).$$

Clearly $r^2 = \mathrm{id}_{X^2}$. Let $U = \{u_1, \dots, u_n\}$ and consider the bijective mapping $w : U^3 \longrightarrow X^3$ defined by

$$w(u_{i_1}, u_{i_2}, u_{i_3}) = (x_{f_{u_{i_2} u_{i_3}}(i_1)}, x_{f_{u_{i_3}}(i_2)}, x_{f_1(i_3)}).$$

Further denote by $\tau_1 : U^3 \longrightarrow U^3$ the mapping that interchanges the first two components, and by $\tau_2 : U^3 \longrightarrow U^3$ the mapping that interchanges the last two components. For any $b \in \mathrm{FaM}_n$ we have shown in Lemma 8.1.2 that $x_{f_{bu_j}(i)} x_{f_b(j)} = x_{f_{bu_i}(j)} x_{f_b(i)}$. It then follows that

$$\begin{aligned} w\tau_1(u_{i_1}, u_{i_2}, u_{i_3}) &= (x_{f_{u_{i_1}u_{i_3}}(i_2)}, x_{f_{u_{i_3}}(i_1)}, x_{f_1(i_3)}) \\ &= r_1(x_{f_{u_{i_2}u_{i_3}}(i_1)}, x_{f_{u_{i_3}}(i_2)}, x_{f_1(i_3)}) \\ &= r_1 w(u_{i_1}, u_{i_2}, u_{i_3}). \end{aligned}$$

Hence $w\tau_1 w^{-1} = r_1$. Similarly one also obtains that $w\tau_2 w^{-1} = r_2$. As $\tau_1\tau_2\tau_1 = \tau_2\tau_1\tau_2$, it follows that $r_1 r_2 r_1 = r_2 r_1 r_2$ and thus r is a set theoretic solution of the Yang-Baxter equation. We have shown that conditions (IT.1) and (IT.2) are necessary. That (IT.3) also holds follows from parts (4) and (7) of Lemma 8.1.2.

Conversely, assume $f \in \mathrm{Sym}_n$ and $r : X^2 \longrightarrow X^2$ is a mapping that satisfies (IT.1), (IT.2) and (IT.3). Let

$$S = \langle x_1, \dots x_n \mid x_i x_j = x_k x_l \text{ if } 1 \leq i, j, \leq n \text{ and } r(x_i, x_j) = (x_k, x_l) \rangle.$$

Put $f_1 = f$. We first show that, for every $a \in \mathrm{FaM}_n = \langle u_1, \dots, u_n \rangle$, with $\deg(a) \geq 1$, there exists a permutation $f_a \in \mathrm{Sym}_n$ so that, if $a = bu_j$ for some j and $b \in \mathrm{FaM}_n$ then

$$r(x_{f_{bu_j}(i)}, x_{f_b(j)}) = (x_{f_{bu_i}(j)}, x_{f_b(i)}), \tag{8.6}$$

for all i with $1 \leq i \leq n$. Note that if such maps f_a exist, then the equalities (8.6), condition (IT.3) and an induction argument yield that these mappings are uniquely determined. So let $a \in \mathrm{FaM}_n$ with $\deg(a) = m \geq 1$. We need to construct f_a. By induction, we may assume that, for all $c \in \mathrm{FaM}_n$ with $\deg(c) \leq m - 1$, we have constructed f_c in such a way that, if $c = bu_j$ for some j and some $b \in \mathrm{FaM}_n$ then the equalities (8.6) are satisfied for all i.

Let $1 \leq i \leq n$. In order to define $f_a(i)$ we consider two cases: either $a \neq u_i^m$ or $a = u_i^m$.

First we deal with $a \neq u_i^m$ (note that in particular $m \geq 2$). So $a = bu_j$ with $i \neq j$ and $f_b \in \mathrm{Sym}_n$ is already defined. Because of (IT.3), there exist uniquely defined elements $p = p(j, i), q \in \{1, \dots, n\}$ so that

$$r(x_p, x_{f_b(j)}) = (x_q, x_{f_b(i)}). \tag{8.7}$$

We now show that $p(j, i)$ is independent of j, and thus we may define $f_a(i) = p(j, i)$. In order to prove this, assume $a = du_j u_k$ with $k \neq i$. So, $b = du_k$.

Put $c = du_j$ and $e = du_i$. As $\deg(c) = m-1$ and $\deg(e) = m-1$, the functions f_c and f_e are already defined. Condition (IT.3) then yields elements $x_{p'}, x_{q'} \in X$ so that

$$r(x_{p'}, x_{f_c(k)}) = (x_{q'}, x_{f_c(i)}). \tag{8.8}$$

We need to show that $x_p = x_{p'}$. By induction, we also know that the following equalities hold:

$$\begin{aligned} r(x_{f_b(j)}, x_{f_d(k)}) &= (x_{f_c(k)}, x_{f_d(j)}) && (8.9)\\ r(x_{f_b(i)}, x_{f_d(k)}) &= (x_{f_e(k)}, x_{f_d(i)}) && (8.10)\\ r(x_{f_c(i)}, x_{f_d(j)}) &= (x_{f_e(j)}, x_{f_d(i)}). && (8.11)\end{aligned}$$

Because r is a set theoretic solution of the Yang-Baxter equation, there exist unique $s, z, t, y \in X$ so that

$$r(x_q, x_{f_c(k)}) = (s, z) \quad \text{and} \quad r(x_p, x_{f_c(k)}) = (t, y). \tag{8.12}$$

We thus get that

$$\begin{aligned} r_1 r_2 r_1 (x_p, x_{f_b(j)}, x_{f_d(k)}) &= r_1 r_2 (x_q, x_{f_b(i)}, x_{f_d(k)})\\ &= r_1 (x_q, x_{f_e(k)}, x_{f_d(i)})\\ &= (s, z, x_{f_d(i)}) \end{aligned}$$

and

$$\begin{aligned} (s, z, x_{f_d(i)}) &= r_2 r_1 r_2 (x_p, x_{f_b(j)}, x_{f_d(k)})\\ &= r_2 r_1 (x_p, x_{f_c(k)}, x_{f_d(j)})\\ &= r_2 (t, y, x_{f_d(j)}). \end{aligned}$$

Hence

$$s = t \text{ and } r(y, x_{f_d(j)}) = (z, x_{f_d(i)}).$$

Because of (8.11), (IT.3) and (8.6) (note that, $c = du_j$ and $\deg(c) = m-1$), we then obtain that $y = x_{f_c(i)}$ and $z = x_{f_e(j)}$. Hence, by (8.12),

$$r(x_p, x_{f_c(k)}) = (t, x_{f_c(i)}).$$

Because of (8.8) and (IT.3), we thus obtain that

$$x_{p'} = x_p.$$

So we have shown that $f_a(i)$ indeed is well defined and $f_a(i) = p$ provided that $\deg(a) = m$ and $a \neq u_i^m$.

Because $i \neq j$ and thus $bu_i \neq u_j^m$, we thus also get that $f_{bu_i}(j)$ is defined. Furthermore, (8.7) and (IT.1) yield that

$$r(x_q, x_{f_b(i)}) = (x_p, x_{f_b(j)}).$$

We thus obtain that $q = f_{bu_i}(j)$. Since, $p = f_a(i)$, equation (8.7) becomes

$$r(x_{f_{bu_j(i)}}, x_{f_b(j)}) = (x_{f_{bu_i}(j)}, x_{f_b(i)}). \tag{8.13}$$

Thus, so far, we have shown that this equality holds for all $b \in \mathrm{FaM}$ provided that either $\deg(b) \leq m-2$, or $\deg(b) = m-1$ and $i \neq j$. We now prove that this equality holds for all $b \in \mathrm{FaM}_n$ with $\deg(b) \leq m-1$ and $bu_j \neq u_i^m$. The only remaining case to be checked is when $j = i$, $\deg(b) = m-1$ and $b \neq u_i^{m-1}$. In this case, we have $b = cu_k$ with $k \neq i$. Because of (IT.2) and (8.13), we get

$$\begin{aligned} r_1(x_{f_{cu_iu_k}(i)}, x_{f_{cu_k}(i)}, x_{f_c(k)}) &= r_2r_1r_2r_1r_2(x_{f_{cu_iu_k}(i)}, x_{f_{cu_k}(i)}, x_{f_c(k)}) \\ &= r_2r_1r_2r_1(x_{f_{cu_iu_k}(i)}, x_{f_{cu_i}(k)}, x_{f_c(i)}) \\ &= r_2r_1r_2(x_{f_{cu_i^2}(k)}, x_{f_{cu_i}(i)}, x_{f_c(i)}) \\ &= r_2r_1(x_{f_{cu_i^2}(k)}, x_{f_{cu_i}(i)}, x_{f_c(i)}) \\ &= r_2(x_{f_{cu_iu_k}(i)}, x_{f_{cu_i}(k)}, x_{f_c(i)}) \\ &= (x_{f_{cu_iu_k}(i)}, x_{f_{cu_k}(i)}, x_{f_c(k)}). \end{aligned}$$

So

$$r(x_{f_{cu_iu_k}(i)}, x_{f_{cu_k}(i)}) = (x_{f_{cu_iu_k}(i)}, x_{f_{cu_k}(i)}).$$

Hence (8.13) holds in all desired cases.

Because of the first case, we know that for any $c \in \mathrm{FaM}_n$ with $\deg(c) = m$ and for any $1 \leq i \leq n$, the element $f_c(i)$ has been determined provided that $c \neq u_i^m$. For such an element c, we claim that $f_c(i) \neq f_c(j)$ if $i \neq j$, $c \neq u_i^m$ and $c \neq u_j^m$. Suppose the contrary. Write $c = bu_l$, for some $b \in \mathrm{FaM}_n$ and some l with $1 \leq l \leq n$. Then, by (8.13), we get

$$\begin{aligned} r(x_{f_{bu_i}(l)}, x_{f_b(i)}) &= (x_{f_{bu_l}(i)}, x_{f_b(l)}) \\ &= (x_{f_{bu_l}(j)}, x_{f_b(l)}) \\ &= r(x_{f_{bu_j}(l)}, x_{f_b(j)}). \end{aligned}$$

So $f_b(j) = f_b(i)$. However, since $\deg(b) = m-1$, we know that f_b is a permutation. Hence $j = i$, a contradiction. The claim implies that if $c = u_i^m$ then the set $\{f_c(j) \mid 1 \leq j \leq n,\ i \neq j\}$ has $n-1$ elements.

We are now able to deal with the second case, that is, when $a = u_i^m$. By the previous, the set $\{f_a(j) \mid 1 \leq j \leq n,\ i \neq j\}$ has $n-1$ elements. So we define $f_a(i)$ as the unique element of $\{1, \ldots, n\}$ that is not contained in the latter set. Hence $f_a \in \mathrm{Sym}_n$.

We now prove that (8.13) also holds in the only remaining case, that is, when $a = bu_i$ and $b = u_i^{m-1}$. Since r and all maps f_d with $d \in \mathrm{FaM}_n$ (and $\deg(d) \leq m$) are bijective, there exist k, l so that

$$r(x_{f_{bu_k}(l)}, x_{f_b(k)}) = (x_{f_{bu_i}(i)}, x_{f_b(i)}).$$

We need to show that $k = l = i$. Suppose the contrary. Since $b = u_i^{m-1}$, we then have that $bu_k \neq u_l^m$ and $bu_l \neq u_k^l$. Therefore, we already know that

$$r(x_{f_{bu_k(l)}}, x_{f_b(k)}) = (x_{f_{bu_l}(k)}, x_{f_b(l)}).$$

Hence $f_b(i) = f_b(l)$ and $f_{bu_i}(i) = f_{bu_l}(k)$. So $i = l$ and $f_{bu_i}(i) = f_{bu_i}(k)$. Consequently we also obtain that $i = k$, a contradiction. This concludes the proof of the second case. Therefore, we have constructed, for every $a \in \mathrm{FaM}_n$, a permutation $f_a \in \mathrm{Sym}_n$ so that (8.6) is satisfied.

We now define a bijective mapping $v : \mathrm{FaM}_n \longrightarrow S$ so that $v(1) = 1$ and $v(u_i) = x_{f(i)}$ for $1 \leq i \leq n$. Because the defining relations for S are homogeneous of degree 2, the monoid S is $\mathbb{N}$-graded. Let S_m denote the subset of S consisting of the elements of degree m. So S is the disjoint union of all S_m. Hence to define v it is sufficient to define for each $m \geq 0$ a bijective mapping $v_m : U_m \longrightarrow S_m$, where U_m denotes the set of elements of degree m in FaM_n. Clearly, $U_0 = \{1\}$ and $U_1 = \{u_1, \ldots, u_n\}$. For $m = 0$, put $v_0(1) = v(1) = 1$. For $m = 1$, put $v_1(u_i) = v(u_i) = x_{f(i)}$. So, let $m \geq 2$. Put $U = \{u_1, \ldots, u_n\}$. Consider the map

$$\widetilde{v_m} : U^m \longrightarrow X^m$$

defined by mapping $(u_{i_1}, u_{i_2}, \ldots, u_{i_m})$ onto

$$(x_{f_{u_{i_2}\cdots u_{i_m}}(i_1)}, \ \ldots, \ x_{f_{u_{i_{m-1}} u_{i_m}}(i_{m-2})}, \ x_{f_{u_{i_m}}(i_{m-1})}, \ x_{f_1(i_m)}).$$

Note that this mapping is bijective, because all maps f_b (with $b \in \mathrm{FaM}_n$) are bijective.

Obviously, the symmetric group Sym_m defines a natural action on U^m and U^m/Sym_m can be identified with U_m. It is well known that the symmetric group Sym_m is generated by the transpositions $s_i = (i\ i+1)$ (with $1 \leq i \leq n-1$) and that the defining relations are $s_i^2 = 1$ and $s_i s_{i+1} s_i = s_{i+1} s_i s_{i+1}$ (the braid relations), for $1 \leq i \leq n-2$. Since the mappings r_i satisfy the braid relations, we also have an action of Sym_m on X^m. For $1 \leq i \leq m-1$, let $\tau_i : U^m \longrightarrow U^m$ denote the mapping that is defined by interchanging the i-th and $(i+1)$-th generators. Because of (8.6), we obtain a commutative diagram

$$\begin{array}{ccc} U^m & \stackrel{\widetilde{v}}{\longrightarrow} & X^m \\ \tau_i \downarrow & & \downarrow r_i \\ U^m & \stackrel{\widetilde{v}}{\longrightarrow} & X^m \end{array} .$$

Hence the actions of Sym_m on U^m and X^m are compatible. So $\widetilde{v_m}$ defines a bijection between the orbits of U^m/Sym_m and X^m/Sym_m. Because S_m can be identified with X^m/Sym_m, it follows that $\widetilde{v_m}$ induces a bijection v_m between U_m and S_m. Thus we obtain the desired bijective mapping $v : \operatorname{FaM}_n \longrightarrow S$.

Let $a \in \operatorname{FaM}_n$ with $\deg(a) = m \geq 1$. Write $a = u_{i_1} u_{i_2} \cdots u_{i_m}$. So $a = bu_{i_1}$ with $b = u_{i_2} \cdots u_{i_m}$. Then, it follows from the above construction that

$$\begin{aligned} v(a) &= x_{f_{u_{i_2} \cdots u_{i_m}}(i_1)} \cdots x_{f_{u_{i_{m-1}} u_{i_m}}(i_{m-2})}\, x_{f_{u_{i_m}}(i_{m-1})} x_{f_1(i_m)} \\ &= x_{f_b(i_1)}\, v(b). \end{aligned}$$

Consequently, for every $b \in \operatorname{FaM}_n$,

$$\{v(u_1 b), \ldots, v(u_n b)\} = \{x_1 v(b), \ldots, x_n v(b)\}.$$

This shows that v is a left I-structure and thus S is a monoid of (left) I-type.

To finish the proof we need to show that the left I-structure is uniquely defined. So, suppose w is a left I-structure with $w(u_i) = x_{f(i)}$. Because of (8.1), for each $b \in \operatorname{FaM}_n$, there exists $f_b \in \operatorname{Sym}_n$ so that $w(bu_i) = x_{f_b(i)} v(b)$ for $1 \leq i \leq n$. Note that then $f_1 = f$. Hence, for $1 \leq i, j \leq n$, $w(bu_i u_j) = x_{f_{bu_i}(j)} v(bu_i) = x_{f_{bu_i}(j)} x_{f_b(i)} v(b)$ and also $w(bu_i u_j) = w(bu_j u_i) = x_{f_{bu_j}(i)} v(bu_j) = x_{f_{bu_j}(i)} x_{f_b(j)} v(b)$. Because S is right cancellative, it thus follows, from Lemma 8.1.2, that $x_{f_{bu_i}(j)} x_{f_b(i)} = x_{f_{bu_j}(i)} x_{f_b(j)}$. So the equations (8.6) must hold. Hence, we have shown above that the mappings f_a are uniquely determined. But then the condition $w(bu_i) = x_{f_b(i)} v(b)$, for all $b \in \operatorname{FaM}_n$ and $1 \leq i \leq n$, proves that w is unique. This finishes the proof. $\square$

8.2 Structure of monoids of I-type

In this section, we will describe monoids of right I-type as submonoids of semidirect products $\mathrm{FaM}_n \rtimes \mathrm{Sym}_n$ so that the projection on the first component is bijective. It then will follow that a monoid is of left I-type if and only if it is of right I-type. Furthermore, these monoids are cancellative and have a group of quotients that is abelian-by-finite. We will show that this group is torsion free and solvable and that in many cases it is poly-(infinite cyclic). It also will be shown that monoids of I-type often can be decomposed as products of monoids of I-type but on less generators.

Before obtaining a characterization via semidirect products, we first need to prove some more technicalities. In order to state these, we consider the natural action of the symmetric group Sym_n of degree n on the free abelian group Fa_n of rank n with generating set $\{u_1, \ldots, u_n\}$. So the action of $\sigma \in \mathrm{Sym}_n$ on an arbitrary element $u_{i_1} \cdots u_{i_p} \in \mathrm{Fa}_n$ is defined as

$$\sigma(u_{i_1} \cdots u_{i_p}) = u_{\sigma(i_1)} \cdots u_{\sigma(i_p)}.$$

This action yields the semidirect product $\mathrm{Fa}_n \rtimes \mathrm{Sym}_n$, with product

$$(a, \sigma_1)\,(b, \sigma_2) = (a\sigma_1(b), \sigma_1 \circ \sigma_2),$$

where $a, b \in \mathrm{Fa}_n$ and $\sigma_1, \sigma_2 \in \mathrm{Sym}_n$. Clearly,

$$\mathrm{FaM}_n \rtimes \mathrm{Sym}_n = \{(a, \sigma) \mid a \in \mathrm{FaM}_n,\ \sigma \in \mathrm{Sym}_n\}$$

is a submonoid of $\mathrm{Fa}_n \rtimes \mathrm{Sym}_n$.

Lemma 8.2.1. *Let $S = \langle x_1, \ldots, x_n \rangle$ be a monoid of left I-type with left I-structure v. A mapping $w : \mathrm{FaM}_n \longrightarrow S$ is a left I-structure for S (that is, it satisfies (8.1)) if and only if $w = v \circ \sigma$ for some $\sigma \in \mathrm{Sym}_n$.*

Proof. It is easily verified that for any $\sigma \in \mathrm{Sym}_n$ the mapping $w = v \circ \sigma$ is a left I-structure of S.

Conversely, assume $w : \mathrm{FaM}_n \longrightarrow S$ is a left I-structure. Since $\{w(u_i) \mid 1 \leq i \leq n\} = \{v(u_i) \mid 1 \leq i \leq n\} = \{x_i \mid 1 \leq i \leq n\}$, there exists a permutation $\sigma \in \mathrm{Sym}_n$ such that $w(u_i) = (v \circ \sigma)(u_i)$ for $1 \leq i \leq n$. Because of Theorem 8.1.4, we know that a left I-structure is uniquely determined by the values it takes on $\{u_1, \ldots, u_n\}$. Hence $w = v \circ \sigma$. □

Lemma 8.2.2. *Let FaM_n be a free abelian monoid of rank n with generating set $\{u_1, \ldots, u_n\}$. Put $u = u_1 \cdots u_n$. Suppose $S = \langle x_1, \ldots, x_n \rangle$ is a monoid of left I-type with left I-structure v. Then there exists a mapping*

$$\psi : \mathrm{FaM}_n \longrightarrow \mathrm{Sym}_n : a \mapsto \psi(a)$$

so that for all $a, b \in \mathrm{FaM}_n$,

$$\psi(ab) = \psi(\psi(b)(a)) \circ \psi(b) \tag{8.14}$$

and

$$v(ab) = v(\psi(a)(b))\, v(a).$$

Furthermore, any mapping mapping $\psi : \mathrm{FaM}_n \longrightarrow \mathrm{Sym}_n$ satisfying (8.14) has a unique extension

$$\psi : \mathrm{Fa}_n \longrightarrow \mathrm{Sym}_n,$$

so that (8.14) is valid for any $a, b \in \mathrm{Fa}_n$, and so that, for any positive integer k that is a multiple of the order of the permutation $\psi(u)$,

$$\psi(au^{-k}) = \psi(a) \tag{8.15}$$

for all $a \in \mathrm{Fa}_n$. Moreover,

$$\psi(a)^{-1} = \psi(\psi(a)(a^{-1})),$$

for any $a \in \mathrm{Fa}_n$.

Proof. Let $a \in \mathrm{FaM}_n$. Consider the following maps

$$S \longrightarrow \{v(ba) \mid b \in \mathrm{FaM}_n\} \; : \; s \mapsto s\, v(a), \tag{8.16}$$

$$\mathrm{FaM}_n \longrightarrow \{v(ba) \mid b \in \mathrm{FaM}_n\} \; : \; b \mapsto v(ab). \tag{8.17}$$

Because of part (5) of Lemma 8.1.2, we know that both maps are bijective. Hence, there exists a bijection

$$w_a : \mathrm{FaM}_n \longrightarrow S$$

so that, for all $b \in \mathrm{FaM}_n$,

$$w_a(b) = s \quad \text{whenever } v(ab) = sv(a).$$

In particular,

$$w_a(b)v(a) = v(ab),$$

for all $a, b \in \mathrm{FaM}_n$.

Because of Lemma 8.1.2, S is right cancellative. Hence, it is clear that $w_a(1) = 1$. Because v is a left I-structure for S, the above yields that

$$\begin{aligned}
\{x_1 w_a(b)v(a), \ldots, x_n w_a(b)v(a)\} &= \{x_1 v(ab), \ldots, x_n v(ab)\} \\
&= \{v(u_1 ab), \ldots, v(u_n ab)\} \\
&= \{w_a(u_1 b)v(a), \ldots, w_a(u_n b)v(a)\}.
\end{aligned}$$

Hence, again because S is right cancellative, w_a is a left I-structure for S. Consequently, by Lemma 8.2.1, there exists a permutation $\psi(a) \in \mathrm{Sym}_n$ so that

$$w_a = v \circ \psi(a).$$

Since a is an arbitrary element in FaM_n, we have constructed the map

$$\psi : \mathrm{FaM}_n \longrightarrow \mathrm{Sym}_n : a \mapsto \psi(a)$$

so that, for all $a, b \in \mathrm{FaM}_n$,

$$v(ab) = v(\psi(a)(b))\, v(a). \tag{8.18}$$

It follows that for $a, b, c \in \mathrm{FaM}_n$

$$\begin{aligned} v(abc) &= v\left(c(ab)\right) \\ &= v\left(\psi(c)(ab)\right)\ v(c) \\ &= v\left(\psi(c)(a)\ \psi(c)(b)\right)\ v(c) \\ &= v\left(\psi\left(\psi(c)(b)\right)\left(\psi(c)(a)\right)\right)\ v\left(\psi(c)(b)\right)\ v(c) \end{aligned}$$

and

$$\begin{aligned} v(abc) &= v(a(bc)) \\ &= v\left(\psi(bc)(a)\right)\ v(bc) \\ &= v\left(\psi(bc)(a)\right)\ v\left(\psi(c)(b)\right)\ v(c). \end{aligned}$$

Again, because S is right cancellative and because v is bijective, we obtain that

$$\psi(\psi(c)(b))\, \psi(c)(a) = \psi(bc)(a).$$

Since this holds for all $a \in \mathrm{FaM}_n$, we get that

$$\psi(bc) = \psi(\psi(c)(b))\, \psi(c). \tag{8.19}$$

We now show that the mapping $\psi : \mathrm{FaM}_n \longrightarrow \mathrm{Sym}_n$ can be extended to a mapping $\mathrm{Fa}_n \longrightarrow \mathrm{Sym}_n$, satisfying the required equations (8.14) and (8.15). In order to show this, we first notice that $\sigma(u) = u$ for any $\sigma \in \mathrm{Sym}_n$, where $u = u_1 \cdots u_n$. It follows by induction that $(\psi(u))^k = \psi(u^k)$ for any positive k. Indeed, for any $k \geq 2$, (8.19) gives that $\psi(\psi(u)^{-1}(u^{k-1})u) = \psi(\psi(u)(\psi(u)^{-1}(u^{k-1}))\psi(u) = \psi(u^{k-1})\psi(u)$. Hence the induction hypothesis yields that

$$\begin{aligned}(\psi(u))^k &= (\psi(u))^{k-1}\ \psi(u)\\ &= \psi(u^{k-1})\ \psi(u)\\ &= \psi(\psi(u)^{-1}(u^{k-1})\ u)\\ &= \psi(u^{k-1}\ u)\\ &= \psi(u^k).\end{aligned}$$

Let $z = u^l$ where l is the order of the permutation $\psi(u)$. It follows that $\psi(z) = \psi(u)^l = 1$ and thus $\psi(z^k) = 1$ for any nonnegative integer k. Clearly, every element of Fa_n can be written as $z^{-k}a$ with $a \in \mathrm{FaM}_n$ and k a nonnegative integer.

If $az^{-k_1} = bz^{-k_2}$, then $az^{k_2} = bz^{k_1}$. Because of (8.19), it is clear that $\psi(az^{k_2}) = \psi(\psi(z^{k_2})(a)) \circ \psi(z^{k_2}) = \psi(a)$. Similarly $\psi(bz^{k_1}) = \psi(b)$ and thus $\psi(a) = \psi(b)$. Hence, the map

$$\mathrm{Fa}_n \longrightarrow \mathrm{Sym}_n : az^{-k} \mapsto \psi(a)$$

is well defined. This is an extension of the mapping $\psi : \mathrm{FaM}_n \to \mathrm{Sym}_n$. Abusing notation, we also denote it by ψ. It is readily verified that (8.19) remains valid for this mapping.

To finish the proof we note that for any $a \in \mathrm{Fa}_n$, $1 = \psi(1) = \psi(aa^{-1}) = \psi(\psi(a)(a^{-1}))\psi(a)$. Hence $\psi(a)^{-1} = \psi(\psi(a)(a^{-1}))$, as desired. □

The following result describes monoids of I-type as submonoids of semidirect products.

Theorem 8.2.3. *A monoid S is of left I-type if and only if S is anti-isomorphic with a submonoid T of a semidirect product $\mathrm{FaM}_n \rtimes \mathrm{Sym}_n$ so that the natural projection $T \longrightarrow \mathrm{FaM}_n$ on the first component is a bijective map, that is,*

$$S \cong \{(a, \psi(a)^{-1})^{-1} \mid a \in \mathrm{FaM}_n\} \subseteq \mathrm{Fa}_n \rtimes \mathrm{Sym}_n,$$

for some map $\psi : \mathrm{FaM}_n \longrightarrow \mathrm{Sym}_n$.

Proof. Let FaM_n be the free abelian monoid of rank n with generating set $\{u_1, \ldots, u_n\}$. We first prove the necessity of the conditions. So, let $S = \langle x_1, \ldots, x_n\rangle$ be a monoid of left I-type with left I-structure v. Because of Lemma 8.2.2, there exists a map $\psi : \mathrm{Fa}_n \longrightarrow \mathrm{Sym}_n$ so that $\psi(ab) = \psi(\psi(b)(a))\,\psi(b)$ for all $a, b \in \mathrm{Fa}_n$. Furthermore, if $a, b \in \mathrm{FaM}_n$, then $v(ab) = v(\psi(a)(b))\ v(a)$ and hence

$$v(a)\, v(b) = v(b\, (\psi(b))^{-1}(a)).$$

Let FaM_n^{-1} be the free abelian monoid of rank n with generating set $\{u_1^{-1}, \ldots, u_n^{-1}\}$. Consider the mapping

$$f : S \longrightarrow \mathrm{FaM}_n^{-1} \rtimes \mathrm{Sym}_n$$

defined by

$$f\,(v(a)) = (\psi(a)(a^{-1}), \psi(a)) = (a, \psi(a)^{-1})^{-1}.$$

Obviously f is an injective map and $f(1) = (1, 1)$, the identity of the semidirect product. For any $a, b \in \mathrm{FaM}_n$, we obtain, from the above formula for $v(a)v(b)$ and from the above formula for the image under ψ of a product, that

$$\begin{aligned}
f(v(a)\, v(b)) &= f\left(v\left(b\, (\psi(b))^{-1}(a)\right)\right) \\
&= (\psi(b\, (\psi(b))^{-1}(a))(b^{-1}\, (\psi(b))^{-1}(a^{-1})),\; \psi(b\, (\psi(b))^{-1}(a))) \\
&= \left((\psi(a) \circ \psi(b))\left(b^{-1}(\psi(b))^{-1}(a^{-1})\right),\; \psi(a)\, \psi(b)\right) \\
&= \left(\psi(a)(\psi(b)(b^{-1})\, a^{-1}),\; \psi(a)\psi(b)\right) \\
&= \left(\psi(a)(a^{-1})\, \psi(a)(\psi(b)(b^{-1})),\; \psi(a)\psi(b)\right) \\
&= f(v(a))\, f(v(b)).
\end{aligned}$$

So f is a monoid monomorphism and $f(S) = \{(a, \psi(a)^{-1})^{-1} \mid a \in \mathrm{FaM}_n\}$ is anti-isomorphic with the monoid $T = \{(a, \psi(a)^{-1}) \mid a \in \mathrm{FaM}_n\}$. Clearly T is a submonoid of $\mathrm{FaM}_n \rtimes \mathrm{Sym}_n$ so that the natural projection $p : \mathrm{FaM}_n \rtimes \mathrm{Sym}_n \longrightarrow \mathrm{FaM}_n$ induces a bijection of T onto FaM_n. This proves the necessity of the conditions.

Conversely, let T be a submonoid of $\mathrm{FaM}_n \rtimes \mathrm{Sym}_n$ such that the natural projection $p : T \longrightarrow \mathrm{FaM}_n$ is bijective. Hence, there exists a mapping $\psi : \mathrm{FaM}_n \longrightarrow \mathrm{Sym}_n$ so that the elements of T can be written uniquely as $(a, \psi(a))$ with $a \in \mathrm{FaM}_n$. Since T is a monoid, we get that $\psi\,(a\,(\psi(a))(b)) = \psi(a)\, \psi(b)$ and thus $\psi(ab) = \psi(a)\, \psi\left(\psi(a)^{-1}(b)\right)$ for all $a, b \in \mathrm{FaM}_n$. Consider the mapping

$$v : \mathrm{FaM}_n \longrightarrow T : a \mapsto (a, \psi(a)).$$

For any $a \in \mathrm{FaM}_n$ and any integer i with $1 \leq i \leq n$, we get that

$$\begin{aligned}
v(au_i) &= (au_i, \psi(au_i)) \\
&= \left(au_i, \psi(a)\psi\left(\psi(a)^{-1}(u_i)\right)\right) \\
&= (a, \psi(a))\, \left(\psi(a)^{-1}(u_i), \psi\left(\psi(a)^{-1}(u_i)\right)\right) \\
&= v(a)\, v\left(\psi(a)^{-1}(u_i)\right).
\end{aligned}$$

Hence, $\{v(au_i) \mid 1 \le i \le n\} = \{v(a)\, v(u_i) \mid 1 \le i \le n\}$. Clearly $v(1) = 1$. Thus the map v defines a right I-structure on T and, therefore, T is a monoid of right I-type. If now S is a monoid so that $f : S \longrightarrow T$ is an anti-isomorphism, then the map

$$w : \mathrm{FaM}_n \longrightarrow S : a \mapsto f(v(a))$$

defines a left I-structure on S. Hence S is a monoid of left I-type. This finishes the proof. □

Corollary 8.2.4. *A monoid S is of left I-type if and only if it is of right I-type. In particular, these monoids are left and right non-degenerate.*

Proof. Because a monoid T is of right I-type if and only if T^{opp} is of left I-type, it is sufficient to show that a monoid of right I-type also is of left I-type. So let S be a monoid of right I-type. Because of Theorem 8.2.3, we may assume that $S = \{(a, \psi(a)^{-1}) \mid a \in \mathrm{FaM}_n\}$, a submonoid of the semidirect product $\mathrm{FaM}_n \rtimes \mathrm{Sym}_n$. Since S is a monoid, it follows that $\psi(a)^{-1}\psi(b)^{-1} = \psi(a\psi(a)^{-1}(b))^{-1}$ for any $a, b \in \mathrm{FaM}_n$. Equivalently, $\psi(\psi(a)(b))\psi(a) = \psi(ab)$ for any $a, b \in \mathrm{FaM}_n$. Because of Lemma 8.2.2, we may extend ψ to a mapping $\mathrm{Fa}_n \longrightarrow \mathrm{Sym}_n$, satisfying the same property. In particular, we have that $\psi(a)^{-1} = \psi(\psi(a)(a^{-1}))$ for any $a \in \mathrm{Fa}_n$.

Clearly, for any $a \in \mathrm{FaM}_n$, we have that

$$\left(a, \psi(a)^{-1}\right)^{-1} = \left(\psi(a)(a^{-1}), \psi(a)\right).$$

Hence we may consider the monoid homomorphism

$$S^{opp} \longrightarrow \mathrm{FaM}_n^{-1} \rtimes \mathrm{Sym}_n$$

defined by

$$\left(a, \psi(a)^{-1}\right) \mapsto \left(a, \psi(a)^{-1}\right)^{-1} = \left(\psi(a)(a^{-1}), \psi(a)\right).$$

Let $p : S^{opp} \to \mathrm{FaM}_n$ be the natural projection. So, $p((a, \psi(a)^{-1}) = \psi(a)(a^{-1})$. We claim that p is bijective. To prove it is injective, suppose $\psi(a)(a^{-1}) = \psi(b)(b^{-1})$ for some $a, b \in \mathrm{FaM}_n$. Then, by the above, $\psi(a)^{-1} = \psi(\psi(a)(a^{-1})) = \psi(\psi(b)(b^{-1})) = \psi(b)^{-1}$. Hence, it follows that $(\psi(a)(a^{-1}), \psi(a)) = (\psi(b)(b^{-1}), \psi(b))$ and thus $(a, \psi(a)^{-1}) = (b, \psi(b)^{-1})$, as desired. Therefore, also the map $\mathrm{FaM}_n \longrightarrow \mathrm{FaM}_n^{-1}$ defined by mapping a onto $\psi(a)(a^{-1})$ is injective. Since it preserves the lengths of elements (the length of a in the generators $u_1, \ldots, u_n$ of FaM_n is equal to the length of $\phi(a)^{-1}(a^{-1})$ in the generators $u_1^{-1}, \ldots, u_n^{-1}$ of FaM_n^{-1}), it must be also surjective. Hence p is surjective, and thus indeed bijective. Since S is anti-isomorphic with S^{opp}, Theorem 8.2.3 therefore implies that S indeed is of left I-type.

That S is left and right non-degenerate now follows from Theorem 8.1.4 and from its dual version. □

Because of Corollary 8.2.4, a monoid S of left or right I-type is simply called a monoid of I-type. By Theorem 8.2.3, S may be identified with a submonoid of a semidirect product $\mathrm{FaM}_n \rtimes \mathrm{Sym}_n$ so that the projection onto its first component is bijective. Hence

$$S = \{(a, \psi(a)^{-1}) \mid a \in \mathrm{FaM}_n\},$$

and, for all $a, b \in \mathrm{FaM}_n$,

$$\psi(ab) = \psi(\psi(b)(a))\, \psi(b). \tag{8.20}$$

Let Fa_n be the free abelian monoid of rank n and generating set $\{u_1, \ldots, u_n\}$ and $\mathrm{FaM}_n = \langle u_1, \ldots, u_n \rangle$. Put $u = u_1 \cdots u_n$ and $z = u^k$, where k is the order of permutation $\psi(u)$. Because of Lemma 8.2.2 there exists a mapping

$$\psi : \mathrm{Fa}_n \longrightarrow \mathrm{Sym}_n$$

so that (8.20) is satisfied for all $a, b \in \mathrm{Fa}_n$, and such that

$$\begin{aligned} \psi(az^{-l}) &= \psi(a), \\ \psi(z^l) &= 1, \end{aligned}$$

for all $a \in \mathrm{Fa}_n$ and any positive integer l.

With these notations we can formulate the following result.

Corollary 8.2.5. *Let $S = \{(a, \psi(a)^{-1}) \mid a \in \mathrm{FaM}_n\}$ be a monoid of I-type. The following properties hold.*

1. *$s = (u, \psi(u)^{-1})$ is a normal element of S, that is, $Ss = sS$.*

2. *$(z, 1)$ is a central element in S.*

3. *S has a (two-sided) group of quotients $G = \{(z^{-l}, 1)(a, \psi(a)^{-1}) \mid a \in \mathrm{FaM}_n,\ l \in \mathbb{N}\}$.*

4. *$G = \{(a, \psi(a)^{-1}) \mid a \in \mathrm{Fa}_n\}$.*

5. $G_\psi = \{(a,1) \mid a \in \mathrm{Fa}_n$ *and* $\psi(a) = 1\}$ *is a normal subgroup of* G *of finite index* l*, where* l *is a positive integer that is a divisor of* $n!$.
6. $\psi(u_i^l) = 1$*, for every* $1 \leq i \leq n$*, that is,* $(u_i^l, 1) \in G_\psi$.
7. $A = \langle (u_1^l, 1), \ldots, (u_n^l, 1)\rangle$ *is a free abelian submonoid of* S *so that*

$$S = \bigcup_{b \in B} (b, \psi(b)^{-1})A,$$

where $B = \{u_1^{m_1} \cdots u_n^{m_n} \mid 0 \leq m_i < l\}$*. Furthermore,* $sA = As$ *for all* $s \in S$.

Proof. First, again notice that $\psi(b)(u) = u$ for any $b \in \mathrm{FaM}_n$. Hence, for any $a \in \mathrm{FaM}_n$, we get that

$$\begin{aligned}(u, \psi(u)^{-1})(a, \psi(a)^{-1}) &= (u\psi(u)^{-1}(a), \psi(u)^{-1}\psi(a)^{-1}) \\ &= (b, \psi(b)^{-1})(u, \psi(u)^{-1}),\end{aligned}$$

where $b = \psi(u)^{-1}(a)$. Therefore, $(u, \psi(u)^{-1})S = S(u, \psi(u)^{-1})$. So $(u, \psi(u)^{-1}$ is a normal element of S, and this proves part (1).

Let $(a, \psi(a)^{-1}) \in S$. Since $\psi(a)^{-1}(z) = z$, we obtain that

$$(a, \psi(a)^{-1})(z,1) = (a\psi(a)^{-1}(z), \psi(a)^{-1}) = (az, \psi(a)^{-1}) = (z,1)(a, \psi(a)^{-1}).$$

Hence $(z, 1)$ is central in S. This proves part (2).

For $a \in \mathrm{FaM}_n$, it is easily seen that there exists $b \in \mathrm{FaM}_n$ so that $a\psi(a)^{-1}(b) = z^l$ for some $l \in \mathbb{N}$. Hence, $(a, \psi(a)^{-1})(b, \psi(b)^{-1}) = (z^l, 1)$. It follows that $G = \{s(z^{-l}, 1) \mid s \in S,\ l \in \mathbb{N}\} = \{s(z,1)^{-l} \mid s \in S,\ l \in \mathbb{N}\}$ is a group that contains S. So G is the group of quotients of S. Furthermore, $(a, \psi(a)^{-1})(z^{-l}, 1) = (a\psi(a)^{-1}(z^{-l}), \psi(a)^{-1}) = (az^{-l}, \psi(az^{-l})^{-1})$. Hence parts (3) and (4) have been shown.

Clearly, $G_\psi = \{(a,1) \mid a \in \mathrm{Fa}_n,\ \psi(a) = 1\}$ is the kernel of the group homomorphism $G \longrightarrow \mathrm{Sym}_n$ that maps $(a, \psi(a)^{-1})$ onto $\psi(a)^{-1}$. It follows that G_ψ is a normal subgroup of G of finite index l, with $l = |G/G_\psi|$ a divisor of $n!$. Let $A_\psi = \{a \in \mathrm{Fa}_n \mid \psi(a) = 1\}$, a subgroup of A. The natural bijective map $\mathrm{Fa}_n \longrightarrow G$ induces a bijection between Fa_n / A_ψ and G/G_ψ. Hence $|\mathrm{Fa}_n / A_\psi| = l$ and thus parts (5) and (6) follow.

To finish the proof, let $(a, \psi(a)^{-1}) \in S$ and let B be the set described as in the statement of (7). Clearly $a = cb$, for some $b \in B$ and $c \in \langle u_1^l, \ldots, u_n^l\rangle$. Recall that $\psi(u_i^l) = 1$, and thus it easily follows from (8.20) that $\psi(c) = 1$. Hence $\psi(a) = \psi(bc) = \psi(\psi(c)(b))\psi(c) = \psi(b)$. Therefore,

$(a, \psi(a)^{-1}) = (c, 1)(b, \psi(b)^{-1})$, and it follows that $S = \bigcup_{b \in B} A(b, \psi(b)^{-1})$. Since $\langle u_1^l, \ldots, u_n^l \rangle$ is invariant under Sym_n, it also readily is verified that $sA = As$ for all $s \in S$. This finishes the proof. □

Of course the group of quotients G of a monoid S of I-type is defined by the same generators and relations as S. These groups have been investigated by Etingof, Guralnick, Schedler and Soloviev in [36, 37], where they are called structural groups. We simply call G a group of I-type. By the previous result it follows that a group G is of I-type if and only if G is isomorphic with a subgroup of $\mathrm{Fa}_n \rtimes \mathrm{Sym}_n$, so that the natural projection onto the first component is a bijective map.

In order to prove that groups of I-type are solvable, we recall the following result (see [68, Hauptsatz VI.4.3]) due to Kegel and Wielandt.

Proposition 8.2.6. *Let $G_1, G_2, \ldots, G_n$ be subgroups of a finite group G. Suppose $G_iG_j = G_jG_i$ for all $1 \leq i, j \leq n$. If each group G_i is nilpotent then G is solvable.*

Also recall [20] that a group G is called a Bieberbach group if G is finitely generated torsion free and abelian-by-finite.

Corollary 8.2.7. *A group of I-type is solvable and Bieberbach.*

Proof. Let $G = \{(a, \psi(a)^{-1}) \mid a \in \mathrm{Fa}_n\}$ be a group of I-type. Because of Corollary 8.2.5, the group G is finitely generated and abelian-by-finite. To prove that G is a Bieberbach group, it remains to show that G is torsion free. In order to prove this we will introduce a fixed point free action of G on $\mathbb{R}^n$.

For this, first consider the left action of G on Fa_n defined by

$$g \cdot b = a\psi(a)^{-1}(b),$$

where $a, b \in \mathrm{Fa}_n$ and $g = (a, \psi(a)^{-1})$. So $g \cdot b = p\left((a, \psi(a)^{-1})(b, \psi(b)^{-1})\right)$, where p is the natural projection of G onto Fa_n. Put

$$\Phi : \mathbb{Z}^n \longrightarrow \mathrm{Fa}_n : (a_1, \ldots, a_n) \mapsto u_1^{a_1} \cdots u_n^{a_n},$$

the natural bijective map of $\mathbb{Z}^n$ onto the free abelian group Fa_n of rank n with generating set $\{u_1, \ldots, u_n\}$. The action of the group G on Fa_n induces an action $*$ of G on $\mathbb{Z}^n$ defined by

$$g * b = \Phi^{-1}(g \cdot \Phi(b)),$$

where $b \in \mathbb{Z}^n$ and $g \in G$. Let $(b_1,\ldots,b_n) \in \mathbb{Z}^n$ and $g = (a, \psi(a)^{-1}) \in G$. Then,

$$g * (b_1,\ldots,b_n) = \left(b_{\psi(a)^{-1}(1)} + a_1,\ldots,b_{\psi(a)^{-1}(i)} + a_i,\ldots,b_{\psi(a)^{-1}(n)} + a_n\right). \tag{8.21}$$

Of course this action can be extended naturally to an action on $\mathbb{R}^n$. This yields a left action of G on $\mathbb{R}^n$ via Euclidean transformations

$$f = \eta + t,$$

where η can be considered as a permutation matrix and t is a translation by a vector $(t_1,\ldots,t_n) \in \mathbb{Z}^n$. This action is fixed point free. Indeed, suppose that $(a, \psi(a)^{-1}) * b = b$, for some $b = (b_1,\ldots,b_n) \in \mathbb{R}^n$. So $a_i + b_{\psi(a)^{-1}(i)} = b_i$. Write $b_i = z_i + p_i$, with $z_i \in \mathbb{Z}$ and $p_i \in [0,1)$. Then it follows that $a_i + z_{\psi(a)^{-1}(i)} = z_i$ and thus $(a, \psi(a)^{-1})z = z$, where $z = (z_1,\ldots,z_n) \in \mathbb{Z}^n$. So z is a fixed point of $(a, \psi(a)^{-1})$. But, in the group G, this means that $(a, \psi(a)^{-1})(z, \psi(z)^{-1}) = (z, \psi(z)^{-1})$. So $(a, \psi(a)^{-1}) = 1$, as desired.

Suppose now that G is not torsion free. Then there exists $1 \neq g \in G$ so that $g^p = 1$ for some prime p. Let the action of g on $\mathbb{R}^n$ be determined by a mapping $\eta + t$ as described above. So η becomes a direct sum of permutation matrices η_i, each of which is either the identity matrix or determined by a cycle of length p. Therefore, $f = f_1 \oplus \cdots \oplus f_m$, where $f_i = \eta_i + v_i$ and v_i is a translation by a vector in $\mathbb{Z}^p$. So $f^p = 1$ is equivalent with all $f_i^p = 1$. Since $1 \neq g$, there exists i so that $f_i \neq 1$, and thus f_i is determined by a cycle of length p, say $(p, p-1,\ldots,2,1)$. Write $v_i = (w_{i,1},\ldots,w_{i,p}) \in \mathbb{Z}^p$. Because $f_i^p = 1$, it follows that $\sum_{j=1}^{p} w_{i,j} = 0$, and hence $(0, -w_{i,1}, -w_{i,1} - w_{i,2},\ldots,-w_{i,1} - w_{i,2} - \cdots - w_{i,p-1})$ is a fixed point of f_i. So, f has a fixed point, a contradiction. This finishes the proof of G being torsion free.

Next, we show that G is solvable. From Corollary 8.2.5, we know that $G_\psi = \{(a,1) \mid a \in \mathrm{Fa}_n \text{ with } \psi(a) = 1\}$ is a normal subgroup of G, and Fa_n / A_ψ is a finite abelian group, where $A_\psi = \{b \in \mathrm{Fa}_n \mid \psi(b) = 1\}$. Note that $\psi(a)(b) \in A_\psi$ for any $a \in \mathrm{Fa}_n$ and $b \in A_\psi$. Indeed, for such elements, equation (8.20) yields that $\psi(a) = \psi(\psi(b)(a))\psi(b) = \psi(ba) = \psi(\psi(a)(b))\psi(a)$, and thus $\psi(\psi(a)(b)) = 1$, as desired. It follows that for each $a \in \mathrm{Fa}_n$, the mapping $\psi(a)^{-1}$ induces an automorphism on the group Fa_n / A_ψ. This way we obtain a semidirect product $(\mathrm{Fa}_n / A_\psi) \rtimes \psi(G)$ and the natural map $\Psi : G/G_\psi \longrightarrow (\mathrm{Fa}_n / A_\psi) \rtimes \psi(G)$ is a group monomorphism. Furthermore, the projection p onto the first component yields a bijective mapping from $\Psi(G/G_\psi)$ onto Fa_n / A_ψ.

Let $B_1,\ldots,B_m$ be the distinct Sylow subgroups of the finite abelian group Fa_n/A_ψ. So $\mathrm{Fa}_n / A_\psi = B_1 \cdots B_m$ and $|B_i| = p_i^{n_i}$, where each p_i

is a prime number. Clearly $C_i = p^{-1}(B_i)$ is a subgroup of $\Psi(G/G_\psi)$ of order $p_i^{n_i}$ and $C_iC_j = C_jC_i$. So $C_1 \cdots C_m$ is a subgroup of $\Psi(G/G_\psi)$ of order $p_1^{n_1} \cdots p_m^{n_m}$. Hence $\Psi(G/G_\psi) = C_1 \cdots C_n$, a product of finitely many nilpotent finite groups, and thus, by Proposition 8.2.6, the group $\Psi(G/G_\psi)$ is solvable. Consequently, G/G_ψ is solvable. Since G_ψ is abelian, also G is solvable. □

The following result states that monoids (respectively groups) of I-type often can be decomposed as products of monoids (respectively groups) of I-type on less generators. In Theorem 8.2.3, we have proved that if S is a monoid of I-type then we may write $S = \{(a, \psi(a)^{-1}) \mid a \in \mathrm{FaM}_n\} \subseteq \mathrm{FaM}_n \rtimes \mathrm{Sym}_n$, for some map $\psi : \mathrm{FaM}_n \longrightarrow \mathrm{Sym}_n$. Of course, we can then can write $S = \{(a, \varphi(a)) \mid a \in \mathrm{FaM}_n\}$, where $\varphi : \mathrm{FaM}_n \longrightarrow \mathrm{Sym}_n$ is the map defined by $\varphi(a) = \psi(a)^{-1}$. Sometimes it will be easier to work with the latter representation of S.

Proposition 8.2.8. *Let $\mathrm{FaM}_n = \langle u_1, \ldots, u_n \rangle$ be a free abelian monoid of rank n with generating set $U = \{u_1, \ldots, u_n\}$, and let $S = \{(a, \psi(a)) \mid a \in \mathrm{FaM}_n\}$ be a monoid of I-type. Let $\sim$ denote the relation on U defined by*

$$u_i \sim u_j \quad \text{if } u_i = \psi(a)(u_j)$$

for some $a \in \mathrm{FaM}_n$. Then $\sim$ is an equivalence relation. Let $C_1, \ldots, C_m$ denote its equivalence classes. The following properties hold.

1. *Each $S_{(i)} = \{(a, \psi(a)) \mid a \in \langle u_j \mid u_j \in C_i \rangle\}$ is monoid of I-type.*

2. *Every element of S has a unique presentation of the form $s_1 s_2 \cdots s_m$ with $s_i \in S_{(i)}$. So $S = S_{(1)} \cdots S_{(m)}$ and, furthermore, $S_{(i)}S_{(j)} = S_{(j)}S_{(i)}$.*

3. *The group of quotients $G = S^{-1}S = G_{(1)}G_{(2)} \cdots G_{(m)}$, where $G_{(i)} = S_{(i)}^{-1}S_{(i)} = \{s_i f_i^{-q} \mid s_i \in S_{(i)},\ q \in \mathbb{N}\}$, with $f_i = (c_i, \psi(c_i))$ and $c_i = \prod_{c \in C_i} c$, and each element of $g \in G$ can be written uniquely as $g_1 g_2 \cdots g_m$ with $g_i \in G_i$. Furthermore, $G_{(i)}G_{(j)} = G_{(j)}G_{(i)}$.*

Proof. First we show that $\sim$ is an equivalence relation. Obviously it is reflexive. To show it is transitive, assume $u_i \sim u_j$ and $u_j \sim u_k$. Then $u_i = \psi(a)(u_j)$ and $u_j = \psi(b)(u_k)$ for some $a, b \in \mathrm{FaM}_n$. Because $\psi(a)\psi(b) = \psi(c)$ for some $c \in \mathrm{FaM}_n$, we get that $u_i = \psi(a)\psi(b)(u_k) = \psi(c)(u_k)$ and thus $u_i \sim u_k$. Also $\psi(a)^{-1} = \psi(d)$ for some $d \in \mathrm{FaM}_n$. Hence $\sim$ is symmetric and thus indeed it is an equivalence relation on U.

It is clear that each $S_{(i)} = \{(a, \psi(a)) \mid a \in \langle u_j \mid u_j \in C_i \rangle\}$ is a submonoid of S so that the projection onto the free abelian monoid $\langle u_j \mid u_j \in C_i \rangle$ is a bijection. Theorem 8.2.3 and Corollary 8.2.4 thus yield that each $S_{(i)}$ is a monoid of I-type. This proves part (1).

To prove part (2), let $a \in \mathrm{FaM}_n$ and write $a = bc$ with $b \in \langle C_1 \rangle$ and $c \in \langle C_2, \ldots, C_m \rangle$. It follows that (again we use several times that the second component of an element of S is uniquely determined by its first component)

$$\begin{aligned}(a, \psi(a)) &= (b, \psi(b))(\psi(b)^{-1}(c), \psi(\psi(b)^{-1}(c))) \\ &= (c, \psi(c))(\psi(c)^{-1}(b), \psi(\psi(c)^{-1}(b)))\end{aligned}$$

with $\psi(b)^{-1}(c) \in \langle C_2, \ldots, C_m \rangle$ and $\psi(c)^{-1}(b) \in \langle C_1 \rangle$. By an induction argument it then follows that an element of S can be written in the form $s_1 s_2 \cdots s_m$ with each $s_i \in S_{(i)}$. Furthermore $S_{(i)} S_{(j)} = S_{(j)} S_{(i)}$.

Because of Corollary 8.2.5, we know that the quotient group of $S_{(i)}$ is $\{s_i f_i^{-q} \mid s_i \in S_{(i)},\ q \in \mathbb{N}\}$ with $f_i = (c_i, \psi(c_i))$ and $c_i = \prod_{c \in C_i} c$. Furthermore, f_i is a normal element of $S_{(i)}$. Since $\psi(a)(c_i) = c_i$ for any $a \in \mathrm{Fa}_n$, it also follows that f_i is normal in S. We get that $G_{(i)} G_{(j)} = G_{(j)} G_{(i)}$ for all i, j. Consequently, $G_{(1)} G_{(2)} \cdots G_{(m)}$ is a group, and thus it is the group of quotients G of S. Clearly $G_{(i)} \cap \big(G_{(1)} \cdots G_{(i-1)} G_{(i+1)} \cdots G_{(m)}\big) = \{1\}$. Hence each element of G can be written uniquely as $g_1 \cdots g_m$ with $g_i \in G_{(i)}$. $\square$

A monoid S of I-type is said to be decomposable if the relation $\sim$ (defined in Proposition 8.2.8) is not trivial, otherwise we call S indecomposable. Obviously, if S is generated by only two elements, then S is indecomposable if and only if S is not abelian. Example 8.1.3 shows that there exist monoids of I-type that are indecomposable. In Section 8.4 we will investigate this property in great detail. Also note that S is indecomposable if and only if the left action $G \longrightarrow \mathrm{Sym}(U)$ defined by mapping $(a, \psi(a))$ onto $\psi(a)$ is transitive.

We now show that a large class of groups of I-type is poly-(infinite cyclic). For this, we consider an equivalence relation introduced in [37]. This naturally will yield an epimorphic image that again is a monoid (respectively) group of I-type.

Lemma 8.2.9. *Let* $\mathrm{FaM}_n = \langle u_1, \ldots, u_n \rangle$ *be a free abelian monoid of rank* n *and let* $S = \{(a, \psi(a)) \mid a \in \mathrm{Fa}_n\}$ *be a monoid of* I*-type. The retract relation* $\approx$ *on the set of generators* $X = \{x_i = (u_i, \psi(u_i)) \mid 1 \leq i \leq n\}$ *of* S *is defined by*

$$x_i \approx x_j \quad \text{if } \psi(u_i) = \psi(u_j).$$

Then $\approx$ is an equivalence relation that is compatible with the defining relations of S (respectively of $G = SS^{-1}$). That is, if $x_i x_j = x_p x_q$ and $x_w x_v = x_k x_l$ with $x_i \approx x_w$ and $x_j \approx x_v$, then $x_p \approx x_k$ and $x_q \approx x_l$.

Proof. Clearly $\approx$ is an equivalence relation on X. Recall from Corollary 8.2.5 that the monoid S has a group of quotients $G = \{(a, \psi(a)) \mid a \in \mathrm{Fa}_n\}$, with $\mathrm{Fa}_n = \mathrm{gr}(u_1, \ldots, u_n)$.

It is sufficient to prove the following two statements.

1. If $x_i x_j = x_p x_q$, $x_w x_j = x_k x_l$ and $x_i \approx x_w$ then $p = k$ and $x_q \approx x_l$.

2. If $x_i x_j = x_p x_q$, $x_i x_v = x_k x_l$ and $x_j \approx x_v$ then $x_p \approx x_k$ and $q = l$.

To prove the first statement, assume $x_i x_j = x_p x_q$ and $x_w x_j = x_k x_l$ are two defining relations for S and $x_i \approx x_w$. Clearly,

$$\begin{aligned} x_i x_j &= (u_i, \psi(u_i))\,(u_j, \psi(u_j)) \\ &= (u_i \psi(u_i)(u_j), \psi(u_i)\psi(u_j)) \\ &= (\psi(u_i)(u_j), \psi(\psi(u_i)(u_j))) \\ &\quad\ (\psi(\psi(u_i)(u_j))^{-1}(u_i), \psi(\psi(\psi(u_i)(u_j)))^{-1}(u_j)) \\ &= x_{\psi(u_i)(u_j)} x_{\psi(\psi(u_i)(u_j))^{-1}(u_i)} \end{aligned}$$

and, similarly

$$x_w x_j = x_{\psi(u_w)(u_j)} x_{\psi(\psi(u_w)(u_j))^{-1}(u_w)}.$$

Because of Theorem 8.1.4, $x_i x_j$ and $x_w x_j$ appear in exactly one defining relation. Since, by assumption, $\psi(u_i) = \psi(u_w)$ we thus get that $u_p = \psi(u_i)(u_j) = \psi(u_w)(u_j) = u_k$. Hence

$$\begin{aligned} x_q &= (u_q, \psi(u_q)) = x_p^{-1} x_i x_j \\ &= (u_q, \psi(u_p)^{-1}\psi(u_i)\psi(u_j)) \\ x_l &= (u_l, \psi(u_l)) = x_k^{-1} x_w x_j \\ &= (u_l, \psi(u_k)^{-1}\psi(u_w)\psi(u_j)). \end{aligned}$$

Since $p = k$, we thus obtain that

$$\psi(u_q) = \psi(u_p)^{-1}\psi(u_i)\psi(u_j) = \psi(u_k)^{-1}\psi(u_w)\psi(u_j) = \psi(u_l).$$

Hence $x_q \approx x_l$, as desired.

Next we prove the second statement. So assume $x_i x_j = x_p x_q$ and $x_i x_v = x_k x_l$ are two defining relations for S and $x_j \approx x_v$. Then $u_p = \psi(u_i)(u_j)$,

$u_q = \psi(\psi(u_i)(u_j))^{-1}(u_i)$, $u_k = \psi(u_i)(u_v)$ and $u_l = \psi(\psi(u_i)(u_v))^{-1}(u_i)$. Because of Lemma 8.2.2, we know that

$$\psi(\psi(u_j)(u_i))\psi(u_j) = \psi(u_i u_j) = \psi(\psi(u_i)(u_j))\psi(u_i).$$

Hence, since by assumption $\psi(u_j) = \psi(u_v)$,

$$\begin{aligned} \psi(u_i u_j) &= \psi(\psi(u_v)(u_i))\psi(u_v) \\ &= \psi(u_v u_i) \\ &= \psi(\psi(u_i)(u_v))\psi(u_i). \end{aligned}$$

So $\psi(\psi(u_i)(u_j))\psi(u_i) = \psi(\psi(u_i)(u_v))\psi(u_i)$, and therefore $\psi(\psi(u_i)(u_j)) = \psi(\psi(u_i)(u_v))$. Hence $q = l$ and $x_p \approx x_k$, as desired. □

Proposition 8.2.10. *Let* $\mathrm{FaM}_n = \langle u_1, \ldots, u_n \rangle$ *be a free abelian monoid of rank* n. *Let* $S = \{(a, \psi(a)) \mid a \in \mathrm{Fa}_n\}$ *be a monoid of I-type with group of quotients* G *and let* $\approx$ *be the retract relation on the set of generators* $X = \{x_i = (u_i, \psi(u_i)) \mid 1 \leq i \leq n\}$ *of* S. *Then the monoid*

$$S/\!\approx\, = \langle \overline{x_i} \mid 1 \leq i \leq n,\ \overline{x_i}\,\overline{x_j} = \overline{x_p}\,\overline{x_q} \text{ if } x_i x_j = x_p z_q \rangle$$

is of I-type with group of quotients

$$G/\!\approx\, = \mathrm{gr}(\overline{x_i} \mid 1 \leq i \leq n,\ \overline{x_i}\,\overline{x_j} = \overline{x_p}\,\overline{x_q} \text{ if } x_i x_j = x_p z_q),$$

where $\overline{x_i}$ *denotes the equivalence class of* x_i *in* X.

Furthermore, the kernel of natural group homomorphism $G \longrightarrow G/\approx$ *is a torsion free and finitely generated abelian group and it is the normal closure of the group*

$$\mathrm{gr}(x_i x_j^{-1} \mid 1 \leq i, j \leq n,\ x_i \approx x_j).$$

Proof. Consider the monoid

$$S/\!\approx\, = \langle \overline{x_i} \mid 1 \leq i \leq n,\ \overline{x_i}\,\overline{x_j} = \overline{x_p}\,\overline{x_q} \text{ if } x_i x_j = x_p x_q \rangle$$

and the group

$$G/\!\approx\, = \mathrm{gr}(\overline{x_i} \mid 1 \leq i \leq n,\ \overline{x_i}\,\overline{x_j} = \overline{x_p}\,\overline{x_q} \text{ if } x_i x_j = x_p z_q).$$

Because of Lemma 8.2.9, the defining relations of $S/\!\approx$ (respectively of $G/\!\approx$) are independent of the choice of the representatives. Hence the monoid $S/\!\approx$ (respectively the group $G/\!\approx$) is obtained by identifying some of the generators x_i of S (respectively G) under the equivalence relation $\approx$. Let $\overline{X} = X/\approx$, a generating set for $S/\approx$, and let $m = |\overline{X}|$. It easily is seen that $S/\approx$ (respectively $G/\approx$) is defined by $\binom{m}{2}$ quadratic relations, and that the associated map $\overline{r} : \overline{X}^2 \longrightarrow \overline{X}^2$ is a solution of the quantum Yang-Baxter equation and is non-degenerate. Hence, by Theorem 8.1.4, $S/\approx$ is a monoid of I-type. Since G is the group of quotients of S it also follows that $G/\approx$ is the group of quotients of $S/\approx$.

Let $\varphi : G \longrightarrow G/\approx$ denote the natural group homomorphism. It easily is verified that the kernel of φ is the normal closure of the group generated by the elements

$$\begin{aligned} x_i x_j^{-1} &= (u_i, \psi(u_i))\,(u_j, \psi(u_j))^{-1} \\ &= (u_i, \psi(u_i))\,(\psi(u_j)^{-1}(u_j^{-1}), \psi(u_j)^{-1}) \\ &= (u_i\psi(u_i)(\psi(u_j)^{-1}(u_j^{-1})), 1) \end{aligned}$$

with $x_i \approx x_j$. It follows that $\ker(\varphi)$ is isomorphic with a subgroup of Fa_n, and thus it is a torsion free finitely generated abelian group. □

For convenience sake, we introduce the following terminology and notation (this was introduced in [37]).

Definition 8.2.11. Let S be a monoid of I-type and $\approx$ the retract relation $\approx$. If G is the group of quotients of S, then denote by $\mathrm{Ret}^1(G)$ the group $G/\approx$ (again a group of I-type). For any integer $m \geq 1$, we define $\mathrm{Ret}^{m+1}(G) = \mathrm{Ret}^1(\mathrm{Ret}^m(G))$. If there exists a positive integer m so that $\mathrm{Ret}^m(G)$ is a cyclic group, then G is called retractable. The smallest such positive integer m is called the retractable level of G (or of S).

Note that, in [37], groups of I-type that have retractable level m are called a multipermutation solution of level m.

Proposition 8.2.12. *A retractable group of I-type is poly-(infinite cyclic).*

Proof. Let G be a group of I-type of retractable level m. Because of Proposition 8.2.10, G has a normal subgroup N that is torsion free abelian and finitely generated and $G/N = \mathrm{Ret}^1(G)$. Furthermore, $\mathrm{Ret}^1(G)$ again is a group of I-type. If $m = 1$ then G/N is cyclic and thus G is poly-(infinite cyclic). If $m > 1$ then G/N is a group of I-type of retractable level $m - 1$. By induction it follows that G is poly-(infinite cyclic). □

Note that, in [37], it is mentioned that almost all groups of I-type on at most 8 generators are retractable and decomposable. Actually there are only two groups of I-type on four generators that are indecomposable and not retractable. We will show in Example 8.2.14 that one of these groups is not poly-(infinite cyclic). In order to prove this, we make use of another well known example (Example 8.2.13).

For some time it was an open problem whether all torsion free groups are unique product groups (see [129]). This would have given an easy proof for the fact that the group algebra of a torsion free group is a domain. Moreover, this also would prove that the units of such a group algebra are trivial, solving another problem of Kaplansky. However the following example, due to Promislow [132], shows that not all torsion free groups are unique product groups.

Example 8.2.13. Let P be the group

$$P = \mathrm{gr}(x, y \mid x^{-1}y^2x = y^{-2},\ y^{-1}x^2y = x^{-2}).$$

Then $A = \mathrm{gr}(x^2, y^2, (xy)^2)$ is a normal and free abelian subgroup of P of rank 3 and P/A is the four group of Klein. Furthermore, G is torsion free but it is not a unique product group. In particular P, is not poly-(infinite cyclic).

Earlier, in [135], Rips and Segev also gave an example of a group with such properties.

Example 8.2.14. The monoid

$$\begin{aligned} S &= \langle x_1, x_2, x_3, x_4 \mid x_1x_2 = x_3x_3,\ x_2x_1 = x_4x_4,\ x_1x_3 = x_2x_4, \\ &\qquad x_1x_4 = x_4x_2,\ x_2x_3 = x_3x_1,\ x_3x_2 = x_4x_1 \rangle \end{aligned}$$

is of I-type. Its group of quotients is not poly-(infinite cyclic), and thus also not retractable.

Proof. Let $X = \{x_1, x_2, x_3, x_4\}$. Since there are $\binom{4}{2}$ relations and every word x_ix_j with $1 \leq i, j \leq 4$ appears at most once in one of the relations, we obtain an associated bijective map $r : X^2 \longrightarrow X^2$. It is readily verified that the conditions of Theorem 8.1.4 are satisfied. Hence, by Corollary 8.2.4, S is a monoid of I-type. Because of Theorem 8.2.3, we may consider $S = \{(a, \psi(a)) \mid a \in \mathrm{FaM}_n\}$ as a submonoid of $\mathrm{Fa}_4 \rtimes \mathrm{Sym}_4$ by identifying x_i with (u_i, ψ_i), where $\psi_i = \psi(u_i)$ and $\mathrm{FaM}_4 = \langle u_1, u_2, u_3, u_4 \rangle$. The permutations ψ_i easily can be determined from the defining relations. It follows that

$$\psi_1 = (23),\ \psi_2 = (14),\ \psi_3 = (1243),\ \psi_4 = (1342).$$

For example, $x_1x_2 = x_3x_3$ yields

$$\begin{aligned}(u_1\psi_1(u_2), \psi_1\psi_2) &= (u_1, \psi_1)(u_2, \psi_3)\\ &= (u_3, \psi_3)(u_3, \psi_3)\\ &= (u_3\psi_3(u_3), \psi_3\psi_3).\end{aligned}$$

Hence, $u_1\psi_1(u_2) = u_3\psi_3(u_3)$, and thus $\psi_1(u_2) = u_3$ and $\psi_3(u_3) = u_1$.

Note that, as a submonoid of Sym_4, we have $\langle \psi_i \mid 1 \leq i \leq 4\rangle = D_8$, the dihedral group of order 8. Let $G = SS^{-1}$, the group of quotients of S. Since $\psi_i \neq \psi_j$ for $i \neq j$, we get that $\mathrm{Ret}^1(G) = G$. So the group G is not retractable. We now check it is not poly-(infinite cyclic). In order to prove this consider the set

$$N = \{(a, \psi(a)) \mid a \in \mathrm{Fa}_4,\ |a| = 0\},$$

where $|a|$ denotes the total degree of a in the generators $u_1, \dots, u_n$. Clearly N is a normal subgroup of G. Also,

$$\begin{aligned}N &= \{(a, \psi(a)) \mid a = (u_1u_4^{-1})^{\alpha_1}(u_2u_4^{-1})^{\alpha_2}(u_3u_4^{-1})^{\alpha_3},\ \alpha_i \in \mathbb{Z}\}\\ &= \mathrm{gr}(x_ix_j^{-1} \mid 1 \leq i,j \leq 4)\\ &= \mathrm{gr}(x_4x_1^{-1}, x_4x_2^{-1}, x_4x_3^{-1})\end{aligned}$$

and

$$G/N \cong \mathbb{Z}.$$

Because of Corollary 8.2.7, the group G is torsion free. Hence so is N. Furthermore

$$N \subseteq \mathrm{Fa}_3 \rtimes \mathrm{Aut}(\mathrm{Fa}_3)$$

and again the natural projection $N \longrightarrow \mathrm{Fa}_3$ is a bijection. Note that N is not necessarily a group of I-type as the automorphisms of $\mathrm{Aut}(\mathrm{Fa}_3)$ that appear in the second component of elements of N do not necessarily permute the basis elements of Fa_3.

Consider the elements

$$\begin{aligned}p = x_4x_1^{-1} &= (u_4u_3^{-1}, (13)(24)),\\ s = x_4x_2^{-1} &= (u_4u_1^{-1}, (12)(34)),\\ t = x_4x_3^{-1} &= (u_4u_2^{-1}, (14)(23)).\end{aligned}$$

Then the group N is generated by s, p, t and one verifies easily that $spt = 1$ and $s^{-1}p^2s = p^{-2}, p^{-1}s^2p = s^{-2}$. So N is a homomorphic image of the group

P defined in Example 8.2.13. Since N is torsion free and both groups N and P have Hirsch rank 3, it follows that $N \cong P$. Example 8.2.13 therefore yields that N is not a poly-(infinite cyclic) group. Consequently, G is not a poly-(infinite cyclic) group. $\square$

8.3 Binomial monoids are of I-type

In [46, 47], Gateva-Ivanova introduced the class of monoids of skew polynomial type. These are finitely generated monoids that are defined by quadratic square free homogeneous relations. Their semigroup algebras $K[S]$ over a field K are called binomial skew polynomial rings, and they form a restricted class of skew polynomial rings with quadratic relations considered in work by Artin and Schelter [7]. In this section, we prove that the monoids S are monoids of I-type. A large part of the proof is devoted to show that S satisfies the exterior and interior cyclic condition (see Definition 8.3.3).

We begin with the formal definition of the monoids under consideration. As in [78], we simply call them binomial monoids.

Definition 8.3.1. A monoid S is called binomial if it is generated by a finite set $X = \{x_1, \ldots, x_n\}$ and it is subject to defining relations of the type

$$x_j x_i = x_{i'} x_{j'}$$

that satisfy the following conditions.

(B1) $j > i$, $i' < j'$ and $i' < j$.

(B2) Every monomial $x_i x_j$ with $i \neq j$ occurs exactly once in a defining relation (and thus there are precisely $\binom{n}{2}$ relations and they all are square free).

(B3) If $x_1^{\alpha_1} \cdots x_n^{\alpha_n} = x_1^{\beta_1} \cdots x_n^{\beta_n}$ for some nonnegative integers α_i and β_i then $\alpha_i = \beta_i$ for $1 \leq i \leq n$.

Conditions (B1) and (B2) yield that $S = \{x_1^{\alpha_1} \cdots x_n^{\alpha_n} \mid \alpha_i \in \mathbb{N},\ 1 \leq i \leq n\}$. Condition (B3) then implies that every element of S has a unique normal form.

Note that, in the definition of a binomial monoid, we use the degree lexicographic ordering on $S = \{x_1^{\alpha_1} \cdots x_n^{\alpha_n} \mid \alpha_i \in \mathbb{N},\ 1 \leq i \leq n\}$ with $x_1 < x_2 < \cdots < x_n$. Of course we can also write all the elements of S uniquely in the form $x_n^{\alpha_n} \cdots x_1^{\alpha_1}$ with all $\alpha_i \geq 0$. Hence S is also a binomial monoid for the reversed degree lexicographic ordering.

We also note that the three listed conditions given in the definition can be interpreted in terms of Bergman's diamond lemma [9] and noncommutative Gröbner basis techniques [114, 115].

Let $S = \langle x_1, \ldots, x_n \rangle$ be a binomial monoid with defining relations $x_j x_i = x_{i'} x_{j'}$ with $j > i$, $i' < j'$ and $i' < j$. In the following lemma, we prove that $i \leq j'$ (later it will actually follow that $i < j'$). To do this, it is convenient to introduce some technicalities.

For two distinct elements v and w in the free monoid $\mathrm{FM}_n = \langle X \rangle$, with free generating set $X = \{x_1, \ldots, x_n\}$, write

$$v \longrightarrow w,$$

and call this a reduction, if there exist $v', w' \in \mathrm{FM}_n$ and i, j with $1 \leq i < j \leq n$, so that $v = v'(x_j x_i)w'$ and $w = v'(x_{i'} x_{j'})w'$ where $x_j x_i = x_{i'} x_{j'}$ is a defining relation for S. Note that for given i, j, with $1 \leq i < j \leq n$, there exists a unique reduction

$$v \longrightarrow x_i x_j,$$

with $v \in \mathrm{FM}_n$. The natural image of $u \in \mathrm{FM}_n$ in S we will also denote by u. Of course $u = v$ in S if $u, v \in \mathrm{FM}_n$ and $u \longrightarrow v$.

Lemma 8.3.2. *The following properties hold for a binomial monoid $S = \langle x_1, \ldots, x_n \rangle$.*

1. *Let r and t be integers so that $1 \leq r < t \leq n$. If $w_0 = x_r x_r x_t$ or $w_0 = x_r x_t x_t \in \langle X \rangle = \mathrm{FM}_n$, then there exists a unique sequence of elements $w_1, \ldots, w_k \in \mathrm{FM}_n$ ($k \leq 2n$) so that*

   (a) $w_k \longrightarrow w_{k-1} \longrightarrow \ldots \longrightarrow w_1 \longrightarrow w_0$,

   (b) no reduction $w \longrightarrow w_k$ is possible, with $w \in \mathrm{FM}_n$.

2. *In S we have that $x_k x_j x_i, x_i x_j x_k \neq x_r x_r x_t$ and $x_k x_j x_i, x_i x_j x_k \neq x_r x_t x_t$, for any i, j, k, r, t with $n \geq k > j > i \geq 1$ and $1 \leq r \leq t \leq n$.*

3. *Let $1 \leq i < j \leq n$. If $x_j x_i = x_{i'} x_{j'}$ is a defining relation for the binomial monoid S (so $i' < j'$), then $1 \leq i' < j' \leq n$, $i' < j$ and $i \leq j'$.*

Proof. (1) Because of the type of the defining relations, we note that if $x_p x_q x_r \in \mathrm{FM}_n$, with either $p \geq q$ or $q \geq r$, then there is at most one $w = x_{p_1} x_{q_1} x_{r_1} \in \mathrm{FM}_n$ so that $w \longrightarrow x_p x_q x_r$. Furthermore, it is clear that either $p_1 \geq q_1$ and $p_1 > p$, or $q_1 \geq r_1$ and $q_1 > q$. Hence we can repeat the process at most $2n$ times until we reach $w = x_{p_k}x_{q_k}x_{r_k} \in \mathrm{FM}_n$ with $p_k \geq q_k \geq r_k$.

The statement (1) now follows if we take a word $x_p x_q x_r$ with $p = q$ or $q = r$.

(2) Clearly, there are two different reductions starting from $w = x_k x_j x_i$:

$$w \longrightarrow x_{j'}x_{k'}x_i$$

with $x_k x_j = x_{j'}x_{k'}$, $j' < k'$ and $j' < k$, and

$$w \longrightarrow x_k x_{i''}x_{j''}$$

with $x_j x_i = x_{i''}x_{j''}$, $i'' < j''$ and $i'' < j$. Let now $w_0 = x_r x_r x_t$ or $w_0 = x_r x_t x_t$. It follows that $w \neq w_0$ in S, for otherwise we obtain two different sequences of reductions starting in w and ending w_0, and this would be in contradiction with part (1). Because of the normal form of the elements we also have that $x_i x_j x_k$ and w_0 are different as elements of S.

(3) From the defining relations, we know that $i' < j'$ and $i' < j$. In S we have $x_j x_i x_{j'} = x_{i'}x_{j'}x_{j'}$. Hence, since also $i < j$, we obtain from part (2) that $i \leq j'$. $\square$

As mentioned earlier, a crucial step in proving that binomial monoids are of I-type is to show that they satisfy the following condition.

Definition 8.3.3. A binomial monoid $S = \langle x_1, \ldots x_n \rangle$ is said to satisfy the interior cyclic condition for a pair j, i with $1 \leq i < j \leq n$ if there exist integers $i_1, i_2, \ldots, i_s, p$ with $1 \leq i < p \leq n$ (and $s < j$), $1 \leq i_k < j$ and $i_k < p$ for all k, so that

$$x_j x_i = x_{i_1} x_p,\ x_j x_{i_1} = x_{i_2} x_p,\ \ldots, x_j x_{i_s} = x_i x_p.$$

Similarly, one says that a pair j, i with $1 \leq i < j \leq n$ satisfies the exterior cyclic condition if there exist integers $j_1, j_2, \ldots, j_t, r$ with $1 \leq r < j$ (and $t < n - i$), $i < j_k \leq n$ and $r < j_k$ for all k, so that

$$x_j x_i = x_r x_{j_1},\ x_{j_1} x_i = x_r x_{j_2},\ \ldots, x_{j_t} x_i = x_r x_j.$$

Note that, in the definition of the respective cyclic conditions, one requires that the relations involved are of the type $x_j x_i = x_{i'} x_{j'}$ with $j > i$, $i' < j'$ and $i < j'$. The last inequality is an essential part of this definition.

We now prove a series of technical lemmas before showing that both the exterior and interior cyclic conditions hold for a binomial monoid.

Lemma 8.3.4. *The following properties hold for a binomial monoid* $S = \langle x_1, \ldots, x_n \rangle$.

1. *Let* $1 < j \leq n$ *and suppose* $x_j x_1 = x_r x_q$ *for some* r, q *with* $1 \leq r < q \leq n$ *(and thus* $r < j$*). Then there exists* p *with* $p > r$ *and so that* $x_p x_1 = x_r x_j$.

2. *The exterior cyclic condition is satisfied for all pairs* $j, 1$ *with* $1 < j \leq n$.

Proof. (1) Let $1 < j \leq n$. Suppose $x_j x_1 = x_r x_q$, for some r, q with $1 \leq r < q \leq n$, and thus also $r < j$. Let p, k, with $1 \leq k < p \leq n$, be so that $x_p x_k = x_r x_j$. In S we have $x_p x_k x_1 = x_r x_j x_1 = x_r x_r x_q$. Since $r < q$ and $p > k \geq 1$, Lemma 8.3.2 implies that $k = 1$. So $x_p x_1 = x_r x_j$, as desired.

(2) Let $1 < j \leq n$. Write $x_j x_1 = x_r x_q$ with $1 \leq r < q \leq n$, $r < j$. We claim that there exists a sequence of pairwise distinct integers $j = j_0, j_1, \ldots, j_s = q$ so that $x_{j_0} x_1 = x_r x_{j_s}$ and $x_{j_{i+1}} x_1 = x_r x_{j_i}$ for $0 \leq i < s$. Hence the pair $j, 1$ satisfies the exterior cyclic condition.

Indeed, if $q = j_0$ then we take $s = 0$. If $q \neq j_0$, then by part (1), there exists j_1 so that $x_{j_1} x_1 = x_r x_{j_0}$. Because every word $x_u x_v$ with $u \neq v$ appears exactly once in one of the defining relations, it is clear that j_1 and j_0 are distinct. If $j_1 = q$, then the desired exterior condition is satisfied again. If $j_1 \neq q$, then we can repeat this process. Suppose that we have constructed pairwise distinct integers $j_0, j_1, \ldots, j_t$, all also different from q, so that $x_{j_{i+1}} x_1 = x_r x_{j_i}$ for $0 \leq i < t$. Then, again by part (1), there exists j_{t+1} so that $x_{j_{t+1}} x_1 = x_r x_{j_t}$. If $j_{t+1} = q$, then the desired exterior condition holds again. In the other case, again because of the unique appearance of the words $x_u x_v$ with $u \neq v$, we also obtain that $j_{t+1} \neq j_i$ with $0 \leq i \leq t$. So the integers $j_0, j_1, \ldots, j_t, j_{t+1}$ are distinct. Since all these integers are bounded by n, this process must stop after say s steps, and thus $x_{j_s} = x_q$. This proves the claim. $\square$

Because of the previous lemma, the exterior cyclic condition holds for all pairs k, j with $k = 1$ and $j > k$. To prove this condition holds for all pairs k, j with $j > k$, we may proceed by induction and assume from now on the following condition.

(IH) k_0 *is an integer so that* $1 < k_0 \leq n$ *and the exterior cyclic condition holds for all pairs* j, k *with* $j > k$ *and* $k < k_0$.

We need to show that the exterior cyclic condition holds for all pairs k_0, j with $n \geq j > k_0$.

Lemma 8.3.5. *Let $S = \langle x_1, \ldots, x_n \rangle$ be a binomial monoid and suppose condition (IH) is satisfied. Then the following properties hold.*

1. *Let $i \leq k_0$ and $v_1, v_2 < i$. Suppose*

$$x_i x_{v_1} = x_{r_1} x_{m_1} \text{ and } x_i x_{v_2} = x_{r_2} x_{m_2}$$

for some r_1, r_2, m_1, m_2 with $r_1 < m_1$ and $r_2 < m_2$. Then $r_1 = r_2$ if and only if $v_1 = v_2$.

2. *If $a < i \leq k_0$, then there exists v, with $v < i$, such that*

$$x_i x_v = x_a x_m,$$

for some m, $m > a, v$.

Proof. (1) Suppose $x_i x_{v_1} = x_{r_1} x_{m_1}$ and $x_i x_{v_2} = x_{r_2} x_{m_2}$, for some positive integers $v_1, v_2, r_1, r_2, m_1, m_2$ with $r_1 < m_1$, $r_2 < m_2$, $v_1, v_2 < i \leq k_0$.

If $v_1 = v_2 = v$, then $x_i x_v = x_{r_1} x_{m_1} = x_{r_2} x_{m_2}$. Since S is a binomial monoid and $x_i x_v$ can appear at most once in a defining relation, we get that $r_1 = r_2$ (and $m_1 = m_2$).

Conversely, assume $r_1 = r_2 = r$. So $x_i x_{v_1} = x_r x_{m_1}$ and $x_i x_{v_2} = x_r x_{m_2}$. Note that $v_1, v_2 < k_0$. Hence, because of the assumption (IH), the exterior cyclic condition holds for the pairs i, v_1 and i, v_2. Therefore, $v_1 < m_1$ and $v_2 < m_2$, and we get integers a, b, with $v_1 < a \leq n$ and $v_2 < b \leq n$, so that $x_a x_{v_1} = x_r x_i$ and $x_b x_{v_2} = x_r x_i$. Hence, again because $x_r x_i$ can appear at most once in a defining relation, we get that $v_1 = v_2$.

(2) Let $i \leq k_0$. Because of the assumption (IH), the exterior cyclic condition holds for all pairs i, j with $i > j$. Hence, we get equalities

$$x_i x_{i-1} = x_{a_{i-1}} x_{m_{i-1}}, \ \ldots, \ x_i x_1 = x_{a_1} x_{m_1},$$

where $a_j < m_j$ and $j < m_j$ for all j with $1 \leq j \leq i-1$. Because of part (1), the integers $a_1, \ldots, a_{i-1}$ are pairwise distinct. Since all $a_j < i$, part (2) follows. □

Lemma 8.3.6. *Let $S = \langle x_1, \ldots, x_n \rangle$ be a binomial monoid and suppose condition (IH) is satisfied. Let $1 \leq i < j \leq n$ with $i \leq k_0$. If $x_j x_i = x_{i'} x_{j'}$ with $1 \leq i' < j' \leq n$, then $i < j'$.*

Proof. Because of Lemma 8.3.2, we know that $i \leq j'$. So we need to show that $i \neq j'$. Assume the contrary, that is, suppose $j' = i$. Then $x_j x_i = x_{i'} x_i$

with $i' < i$. Because of Lemma 8.3.5, there exists v, with $v < i$, so that $x_i x_v = x_{i'} x_m$, for some m with $m > i', v$. Hence in S we get

$$x_j x_i x_v = x_{i'} x_i x_v = x_{i'} x_{i'} x_m.$$

Since $j > i > v$ and $i' < m$ this yields a contradiction with Lemma 8.3.2. □

We now show that, under the assumption (IH), the interior cyclic condition holds for all pairs j, m_0 provided $1 \leq m_0 < j \leq n$ and $x_j x_{m_0} = x_{m_1} x_t$ for some m_1 with $m_1 < t \leq k_0$. Note that then because of Lemma 8.3.2 and Lemma 8.3.6, we get that $m_0 < t$.

Lemma 8.3.7. *Let $S = \langle x_1, \dots, x_n \rangle$ be a binomial monoid and suppose condition (IH) is satisfied. Let $t \leq k_0$. For any triple j, m_0, m_1 with $1 \leq m_0, m_1 < j$ and $m_0, m_1 < t$, the equality $x_j x_{m_1} = x_{m_0} x_t$ implies $x_j x_{m_0} = x_m x_t$ for some m with $m < t$.*

Proof. We prove this by decreasing induction on j. First assume $j = n$ and thus $x_n x_{m_1} = x_{m_0} x_t$. If $m_1 = m_0$, then the result follows. If, on the other hand, $m_1 \neq m_0$, then there exists a defining relation $x_p x_{m_2} = x_{m_1} x_t$ with $1 \leq m_2 < p \leq n$ and $m_1 < p$. Because of Lemma 8.3.2, we know that $m_2 \leq t \leq k_0$. Hence, Lemma 8.3.6 implies that $m_2 < t$. In S we also get that

$$x_n x_p x_{m_2} = x_n x_{m_1} x_t = x_{m_0} x_t x_t.$$

Since $m_0 < t$ and $n \geq p > m_2$, we obtain from Lemma 8.3.2 that $p = n$. Hence, we have shown that the equality $x_n x_{m_1} = x_{m_0} x_t$ implies that

$$x_n x_{m_2} = x_{m_1} x_t,$$

for some integer m_2 with $1 \leq m_2, m_1 < n$ and $m_2 < t$. Because $m_1 \neq m_0$ and since $x_n x_{m_2}$ occurs precisely once in the defining relations, we also obtain that $m_2 \neq m_1$. If $m_2 \neq m_0$, then this reasoning can be done again. Repeating this process sufficiently many times we get pairwise distinct integers $m_1, \dots, m_l = m_0$ so that $x_n x_{m_{i+1}} = x_{m_i} x_t$, for $0 \leq i < l$. Hence $x_n x_{m_0} = x_{m_{l-1}} x_n$. This finishes the proof for $j = n$.

Suppose now that the lemma holds for all $j > j_0$. Let m_0, m_1 be integers so that $1 \leq m_0, m_1 < j_0 \leq n$ and $m_0, m_1 < t$, and also

$$x_{j_0} x_{m_1} = x_{m_0} x_t. \tag{8.22}$$

We need to prove that $x_{j_0} x_{m_0} = x_m x_t$ for some m with $m < t$. Since $m_0, m_1 < t \leq k_0$ and because of the assumption (IH), we know that the exterior cyclic condition holds for the pair j, m_1. Therefore equation (8.22) implies that

$$x_t x_{m_1} = x_{m_0} x_v \tag{8.23}$$

for some v with $m_0, m_1 < v$. Because S is a binomial monoid and because of the assumption (IH), Lemma 8.3.2 and Lemma 8.3.6 yield a sequence of integer pairs

$$(j_1, m_1),\ (j_2, m_2),\ \ldots \tag{8.24}$$

where $1 \leq m_i < t$ and $m_i < j_{i-1}$ for all $i > 1$, so that

$$x_{j_1} x_{m_2} = x_{m_1} x_t,\ x_{j_2} x_{m_3} = x_{m_2} x_t,\ \ldots,\ x_{j_i} x_{m_{i+1}} = x_{m_i} x_t. \tag{8.25}$$

Note that $j_{i-1} \leq j_i$ for each $i \geq 1$. Suppose the contrary, that is, $j_{i-1} > j_i$ for some $i \geq 1$. Then $j_{i-1} > j_i > m_{i+1}$, and ,by (8.25), we have that

$$x_{j_{i-1}} x_{j_i} x_{m_{i+1}} = x_{j-1} x_{m_i} x_t = x_{m_{i-1}} x_t x_t.$$

Thus, for the triple $j_{i-1} > j_i > m_{i+1}$, we have

$$x_{j_{i-1}} x_{j_i} x_{m_{i+1}} = x_{m_{i-1}} x_t x_t.$$

Since $m_{i-1} < t$, this is in contradiction with Lemma 8.3.2.

Next we claim that $j_{i-1} = j_i$ for each $i \geq 1$. Indeed, assume the contrary. So from the above, $j_i > j_{i-1} \geq j_0$. Hence, the induction hypothesis (and therefore the statement of the lemma) holds for $j = j_i$. The equality $x_{j_i} x_{m_{i+1}} = x_{m_i} x_t$ thus implies that

$$x_{j_i} x_{m_i} = x_m x_t \tag{8.26}$$

for some m with $m < t$. Also, by the hypothesis, $t \leq k_0$. Thus $m_i < k_0$ and, by the assumption (IH), the exterior cyclic condition holds for the pair j_i, m_i. It then follows from (8.26) that

$$x_t x_{m_i} = x_m x_{t'} \tag{8.27}$$

for some t' with $t' > m$. Also, consider the equality

$$x_{j_{i-1}} x_{m_i} = x_{m_{i-1}} x_t. \tag{8.28}$$

Again the exterior cyclic condition yields

$$x_t x_{m_i} = x_{m_{i-1}} x_{t''}$$

for some $t'' > m_{i-1}$. Since $x_t x_{m_i}$ appears exactly once in one of the defining relations, (8.27) and (8.28) yield that $m = m_{i-1}$. Consequently, (8.26) and (8.28) imply that

$$x_{j_i} x_{m_i} = x_{m_{i-1}} x_t = x_{j_{i-1}} x_{m_i}.$$

However, this is in contradiction with Lemma 8.3.6.

Since all m_i are bounded by t, we get that $m_a = m_b$ for two nonnegative integers a and b with $a < b$. Let a be the smallest such integer. We claim that $a = 0$. Indeed, suppose the contrary, that is $0 < a < b \leq n$. Then equation (8.25) and the assumption $m_a = m_b$ yield that

$$x_{j_{b-1}} x_{m_a} = x_{j_{b-1}} x_{m_b} = x_{m_{b-1}} x_t.$$

Note that $m_a \leq k_0$. Hence, because of the assumption (IH) the exterior cyclic condition for the pair j_{b-1}, m_a yields

$$x_t x_{m_a} = x_{m_{b-1}} x_u \tag{8.29}$$

for some u with $u > m_{b-1}$. The exterior cyclic condition applied to the relation $x_{j_{a-1}} x_{m_a} = x_{m_{a-1}} x_t$ (note that by assumption $a \geq 1$) implies that

$$x_t x_{m_a} = x_{m_{a-1}} x_v, \tag{8.30}$$

for some v with $v > m_{a-1}$. Since $x_t x_{m_a}$ appears precisely once in one of the defining relations, it follows from (8.29) and (8.30) that $m_{b-1} = m_{a-1}$. This is in contradiction with the minimality of a. Hence, the claim is proved, and thus $a = 0$.

So $m_0 = m_b$ for some $b > 0$. Since $j_{b-1} = j_0$, we thus obtain from (8.25) that $x_{j_0} x_{m_0} = x_{m_b} x_t$. This finishes the proof. □

Lemma 8.3.8. *Let $S = \langle x_1, \ldots, x_n \rangle$ be a binomial monoid and suppose condition (IH) is satisfied. Let $k_0 < j \leq n$ and suppose $x_j x_{k_0} = x_r x_q$ for some r, q with $1 \leq r < q \leq n$. Then there exists j_1 with $j_1 > k_0$ and $x_{j_1} x_{k_0} = x_r x_j$.*

Proof. Because $r < j$, there exists a defining relation

$$x_{j_1} x_k = x_r x_j, \tag{8.31}$$

with $1 \leq k < j_1 \leq n$. We must prove that $k = k_0$.

Note that

$$x_{j_1} x_k x_{k_0} = x_r x_j x_{k_0} = x_r x_r x_q.$$

Because $j_1 > k$ and $r < q$, Lemma 8.3.2 implies that $k \leq k_0$.

We prove the result by contradiction. So assume $k \neq k_0$, that is, suppose $k < k_0$. Hence

$$x_{t_1} x_{k_1} = x_k x_{k_0}, \tag{8.32}$$

for some t_1, k_1 with $1 \leq k_1 < t_1 \leq n$ and, because of Lemma 8.3.2, $k_1 \leq k_0$. It follows from the assumption (IH) and Lemma 8.3.6 that $k_1 < k_0$. In FM_n consider the following reductions

$$\begin{aligned} x_{j_1}[x_{t_1} x_{k_1}] &\longrightarrow x_{j_1}[x_k x_{k_0}] = [x_{j_1} x_k] x_{k_0} \\ &\longrightarrow [x_r x_j] x_{k_0} = x_r [x_j x_{k_0}] \longrightarrow x_r [x_r x_q]. \end{aligned} \tag{8.33}$$

It thus follows from Lemma 8.3.2 that $j_1 \leq t_1$. We claim that $j_1 < t_1$. Suppose, on the contrary, that $t_1 = j_1$. Then (8.32) gives $x_{j_1} x_{k_1} = x_k x_{k_0}$. Hence, because of Lemma 8.3.7,

$$x_{j_1} x_k = x_m x_{k_0} \tag{8.34}$$

for some m with $m < k_0$. So, (8.31) and (8.34) yield that

$$x_{j_1} x_k = x_r x_j = x_m x_{k_0}.$$

Since $x_{j_1} x_k$ appears precisely once in a defining relation, we obtain that $j = k_0$, in contradiction with the assumption of the lemma.

Put $r_1 = k$, $j_0 = r$ and $j = t_0$. We now show that we can extend (8.33) to the left as follows (for any positive integer s):

$$w_{2s+2} \longrightarrow w_{2s+1} \longrightarrow w_{2s} \longrightarrow \ldots \longrightarrow w_3 \longrightarrow w_2 \longrightarrow w_1 \longrightarrow w_0, \tag{8.35}$$

where

$$w_0 = x_r x_r x_q, \quad w_1 = x_r x_j x_{k_0}, \quad w_2 = x_{j_1} x_k x_{k_0},$$

$$w_3 = x_{j_1} x_{t_1} x_{k_1}, \quad w_4 = x_{j_2} x_{r_2} x_{k_1}$$

and

$$j_2 > r_2 \text{ and } x_{j_2} x_{r_2} = x_{j_1} x_{t_1}.$$

Indeed, assume that we already have constructed the following elements and reductions in FM_n

$$w_{2s} \longrightarrow w_{2s-1} \longrightarrow \ldots \longrightarrow w_1 \longrightarrow w_0, \tag{8.36}$$

such that for $1 \leq i \leq s$ we have

$$\begin{aligned} w_{2i} = [x_{j_i} x_{r_i}] x_{k_{i-1}} &\longrightarrow w_{2i-1} = x_{j_{i-1}} [x_{t_{i-1}} x_{k_{i-1}}] \\ &\longrightarrow w_{2(i-1)} = x_{j_{i-1}} x_{r_{i-1}} x_{k_{i-2}} \end{aligned} \tag{8.37}$$

and the following relations hold

$$k_{i-1} < k_0,\ k_{i-1} < t_{i-1},\ r_i < k_{i-1},\ r_i < j_i \leq n,\ j_{i-1} < t_{i-1},\ j_i > j_{i-1}, \tag{8.38}$$

$$x_{t_{i-1}} x_{k_{i-1}} = x_{r_{i-1}} x_{k_{i-2}},\ x_{j_i} x_{r_i} = x_{j_{i-1}} x_{t_{i-1}}. \tag{8.39}$$

We shall now find w_{2s+1} and $w_{2(s+1)}$ so that (8.37), (8.38) and (8.39) hold for $i = s+1$. The inequality $r_s < k_{s-1}$ implies that there exist t_s and k_s, with $k_s < t_s \leq n$, such that

$$x_{t_s} x_{k_s} = x_{r_s} x_{k_{s-1}}. \tag{8.40}$$

Note that, because of Lemma 8.3.2, $k_s \leq k_{s-1}$. Let $w_{2s+1} = x_{j_s}[x_{t_s} x_{k_s}]$. So we have the reduction

$$w_{2s+1} = x_{j_s}[x_{t_s} x_{k_s}] \longrightarrow x_{j_s}[x_{r_s} x_{k_{s-1}}] = w_{2s}. \tag{8.41}$$

Hence we get that $x_{j_s} x_{t_s} x_{k_s} = x_r x_r x_q$ in S. We claim that $t_s \neq j_s$. Indeed, if, on the contrary, $t_s = j_s$, then it follows from (8.40) that

$$x_{j_s} x_{k_s} = x_{r_s} x_{k_{s-1}}. \tag{8.42}$$

Since $k_s \leq k_{s-1} \leq k_0$, the assumption (IH) and Lemma 8.3.6 imply that $k_s < k_{s-1}$. Hence, by Lemma 8.3.7 we obtain

$$x_{j_s} x_{r_s} = x_u x_{k_{s-1}} \tag{8.43}$$

for some u with $u < k_{s-1}$. But (8.39) and (8.38) imply

$$x_{j_s} x_{r_s} = x_{j_{s-1}} x_{t_{s-1}},\ k_{s-1} < t_{s-1}. \tag{8.44}$$

So (8.43) and (8.44) yield that $x_u x_{k_{s-1}} = x_{j_{s-1}} x_{t_{s-1}}$. Since both sides of the equation are in normal form, this gives a contradiction. Therefore, we have shown that $t_s \neq j_s$. Hence, as $t_s < k_s$ and $r < q$, (8.41) and Lemma 8.3.2 yield that $j_s < t_s$. Consequently, there exists an element $w_{2(s+1)}$ in FM_n so that

$$w_{2(s+1)} = [x_{j_{s+1}} x_{r_{s+1}}] x_{k_s} \longrightarrow [x_{j_s} x_{t_s}] x_{k_s} = w_{2s+1}$$

with

$$x_{j_{s+1}} x_{r_{s+1}} = x_{j_s} x_{t_s},\ j_{s+1} > r_{s+1} \text{ and } j_{s+1} > j_s.$$

We claim that $r_{s+1} \neq k_s$. Indeed, suppose the contrary. Then $x_{j_{s+1}} x_{k_s} x_{k_s} = x_{j_s} x_{t_s} x_{k_s} = x_r x_r x_q$. On the other hand, because of (8.40) and (8.43), we get that $x_{j_s} x_{t_s} x_{k_s} = x_{j_s} x_{r_s} x_{k_{s-1}} = x_u x_{k_{s-1}} x_{k_{s-1}}$. So we have shown that

$x_u x_{k_{s-1}} x_{k_{s-1}} = x_r x_r x_q$. However, this gives a contradiction as $u < k_{s-1}$ and $r < q$, and thus both sides of the equation are in normal form.

Now, because $x_{j_{s+1}} x_{r_{s+1}} x_{k_s} = x_r x_r x_q$, $r < q$, $j_{s+1} > r_{s+1}$, Lemma 8.3.2 implies that $r_{s+1} < k_s$. This shows the construction of an infinite sequence (8.35) with the required properties. However, the fact that $j_1 < j_2 < \cdots < j_s < \cdots \leq n$ yields a contradiction. Thus, the assumption $k < k_0$ is impossible and this finishes the proof. □

Proposition 8.3.9. *A binomial monoid satisfies the exterior and interior cyclic condition.*

Proof. Let $S = \langle x_1, \ldots, x_n \mid x_j x_i = x_{i'} x_{j'},\ j > i,\ i' < j',\ i' < j \rangle$ be a binomial monoid. We first show that the exterior cyclic condition holds for any pair of integers k, j with $j > k$. We prove this by induction on k. For $k = 1$, this follows from Lemma 8.3.4. So assume $k_0 > 1$ and the exterior cyclic condition holds for any pair j, k with $1 \leq k < j \leq n$ and $k < k_0$. We need to prove the exterior cyclic condition for a pair j, k_0 with $k_0 < j \leq n$. For this suppose $x_j x_{k_0} = x_r x_{j_1}$ with $r < j_1$. Because of Lemma 8.3.8, there exist pairwise distinct integers $j^{(0)} = j,\ j^{(1)}, \ldots, j^{(t)} = j_1$, with $t < n - k_0$, and so that

$$x_{j^{(i+1)}} x_{k_0} = x_r x_{j^{(i)}} \quad \text{and} \quad x_j x_{k_0} = x_r x_{j^{(t)}},$$

for $0 \leq i < t$. It follows that

$$x_j x_{k_0} = x_r x_{j^{(t)}},\ \ x_{j^{(t)}} x_{k_0} = x_r x_{j^{(t-1)}},\ \ \ldots, x_{j^{(1)}} x_{k_0} = x_r x_j.$$

This proves that the exterior cyclic condition indeed holds for the pair k_0, j. So S satisfies the exterior cyclic condition.

The proof made use of the ordering $x_1 < x_2 < \cdots < x_n$. Using the reversed lexicographic ordering $x_n < x_{n-1} < \cdots < x_1$ we also obtain the exterior cyclic condition with respect to this order. In other words, if $1 \leq j_1 < r \leq n$ then there exist j, k so that $1 \leq j < k \leq n$ and there exist integers j_i so that

$$x_j x_k = x_r x_{j_1},\ \ x_{j_1} x_k = x_r x_{j_2},\ \ \ldots,\ x_{j_t} x_k = x_r x_j,$$

with $j_i > r$, $k > j_i$ for each $1 \leq i \leq t$. Hence

$$x_r x_{j_1} = x_j x_k,\ \ x_r x_j = x_{j_t} x_k,\ \ x_r x_{j_t} = x_{j_{t-1}} x_k,\ \ \ldots,\ x_r x_{j_2} = x_{j_1} x_k.$$

So the interior cyclic condition holds for any pair r, j_1 with $r > j_1$. □

We now show that all binomial monoids are of I-type.

Theorem 8.3.10. *Binomial monoids are monoids of I-type.*

Proof. Let $S = \langle x_1, \ldots, x_n \mid x_j x_i = x_{i'} x_{j'} \rangle$ be a binomial monoid. So the defining relations satisfy the conditions (B1), (B2) and (B3) of Definition 8.3.1. In particular, there are $\binom{n}{2}$ defining relations of the type $x_j x_i = x_{i'} x_{j'}$ with $j > i$, $i' < j'$, $i' < j$ and $i < j'$ (we make use of Lemma 8.3.6 to get the last inequality). Furthermore, every monomial $x_i x_j$ with $i \neq j$ appears exactly once in a defining relation.

Let $X = \{x_1, \ldots, x_n\}$. As in (8.2), consider the associated bijective map

$$r : X^2 \longrightarrow X^2$$

defined by

$$r(x_i, x_i) = (x_i, x_i), \quad r(x_j, x_i) = (x_k, x_l) \text{ and } r(x_k, x_l) = (x_j, x_i)$$

if $x_j x_i = x_k x_l$ is a defining relation (so $j > i$). Clearly $r^2 = \mathrm{id}_{X^2}$. Hence, because of Theorem 8.1.4 and Corollary 8.2.4, to prove the result it is sufficient to show that the following properties hold.

(IT.2) r is a set theoretic solution of the Yang-Baxter equation.

(IT.3) For $i, j \in \{1, \ldots, n\}$ there exist unique $k, l \in \{1, \ldots, n\}$ such that $r(x_k, x_i) = (x_l, x_j)$.

Recall (8.5) that the mapping $r_i : X^m \longrightarrow X^m$ is defined by $r_i = \mathrm{id}_{X^{i-1}} \times r \times \mathrm{id}_{X^{m-i-1}}$. The diagonal of X^m we denote by $\mathcal{D}_m$.

Assume $r(x_i, x_j) = (x_k, x_l)$. Because of Proposition 8.3.9, the monoid S satisfies the interior and exterior cyclic condition. Hence $r(x_i, x_k) = (x_m, x_l)$ for some m. It follows that $r_1 r_2(x_i, x_i, x_j) = (x_m, x_l, x_l)$. Consequently,

$$r_1 r_2(\mathcal{D}_2 \times X) = X \times \mathcal{D}_2. \tag{8.45}$$

We first prove condition (IT.3). Consider the map

$$\varphi : X^2 \longrightarrow X^2 : (z, t) \mapsto (t, y),$$

where y is such that $r(z, t) = (x, y)$ (note that since r is a bijective map, the element y is uniquely determined by z and t). Suppose $\varphi(z, t) = (t, y)$ and $r(z, t) = (x, y)$. Then, $(x, y, y) \in X \times \mathcal{D}_2$ and thus $r_2 r_1(x, y, y) \in \mathcal{D}_2 \times X$. Hence $r_2 r_1(x, y, y) = r_2(z, t, y) \in \mathcal{D}_2 \times X$. Consequently, there exists $x_1 \in X$ so that $r_2(z, t, y) = (z, z, x_1)$. Therefore, $r(t, y) = (z, x_1)$ and thus z is uniquely determined by (t, y). It follows that φ is injective, and thus also bijective. Hence, for $t', y' \in X$ we get that $\{\varphi(z', t') \mid z' \in X\} \subseteq \{(t', x) \mid x \in X\}$. Since both sets have the same cardinality, one gets that $\varphi(z', t') = (t', y')$ for some $z' \in X$, that is , $r(z', t') = (x', y')$ for some $x' \in X$. Moreover, such pair (z', x') is unique. Indeed, for suppose also $z_1, x_1 \in X$ are such that $r(z_1, t') = (x_1, y')$. Then $\varphi(z', t') = (t', y') = \varphi(z_1, t')$. Hence $z' = z_1$ and thus also $(x_1, y') = r(z_1, t') = r(z', t') = (x', y')$. So $(z', x') = (z_1, x_1)$, as desired. Thus we have shown that condition (IT.3) holds.

It remains show that condition (IT.2) holds. This is done by a careful examination of the equivalence classes on X^3 for the equivalence relation $\sim$ defined by $w_1 \sim w_2$ if and only if $w_1 = w_2$ in S, or equivalently, $w_2 = r_{i_1} r_{i_2} \cdots r_{i_p} w_1$ for some $p, i_1, \ldots, i_p$.

Recall that every equivalence class contains exactly one element of the form $(x_{a_1}, x_{a_2}, x_{a_3})$ with $1 \leq a_1 \leq a_2 \leq a_3 \leq n$. Let D_∞ denote the infinite dihedral group $\mathrm{gr}(r_1, r_2 \mid r_1^2 = r_2^2 = e)$.

Clearly, D_∞ acts on X^3 and the $\sim$ equivalence classes correspond to the D_∞-orbits. Let $\mathcal{O}$ be such an orbit. We show that $\mathcal{O}$ has either 1, 3 or at least 6 elements. To prove this we consider three mutually exclusive cases. First, $\mathcal{O} \cap \mathcal{D}_3 \neq \emptyset$. Then, since the defining relations are square free, $|\mathcal{O}| = 1$. Second, $(\mathcal{O} \cap ((\mathcal{D}_2 \times X) \cup (X \times \mathcal{D}_2))) \setminus \mathcal{D}_3 \neq \emptyset$. Then it follows from (8.45) that $|\mathcal{O}| = 3$. Indeed, if, for example, $(x, x, y) \in \mathcal{O}$ (with $x \neq y$), then $\mathcal{O} = \{(x, x, y), (x, x', y'), (x'', y', y')\}$, where $xy = x'y'$ and $xx' = x''y'$. Third, $\mathcal{O} \cap ((\mathcal{D}_2 \times X \cup X \times \mathcal{D}_2)) = \emptyset$. Then, $\mathcal{O} = \{w, r_1 w, r_2 r_1 w, \ldots\}$ for some $w \in X^3$. Thus a generic member of $\mathcal{O}$ is of the form $(r_2 r_1)^a w$ or $r_1 (r_2 r_1)^a w$, with $a \in \mathbb{N}$. It is readily verified that $|\{w, r_1(w), r_2(w)\}| = 3$. Since $\mathcal{O}$ is finite there must be some repetition in the listed presentations of the elements of the set. If $(r_2 r_1)^a w = r_1 (r_2 r_1)^b w$, for some $a, b \in \mathbb{N}$, then define

$$w_1 = \begin{cases} r_1 (r_2 r_1)^{\lfloor \frac{a+b}{2} \rfloor} w & \text{if } a + b \text{ is odd} \\ (r_2 r_1)^{\lfloor \frac{a+b}{2} \rfloor} w & \text{if } a + b \text{ is even.} \end{cases}$$

Thus $r_1 w_1 = w_1$ or $r_2 w_2 = w_2$ (depending on whether $a + b$ is odd or even), whence w is equivalent with w_1 and $w_1 \in \mathcal{D}_2 \times X \cup X \times \mathcal{D}_2$, contradicting the hypotheses.

Hence there exists a positive integer p so that $(r_2 r_1)^p w = w$. Let p be the smallest such number. Then

$$\mathcal{O} = \{w, (r_2 r_1) w, \ldots, (r_2 r_1)^{p-1} w, r_1 w, r_1 (r_2 r_1) w, \ldots, r_1 (r_2 r_1)^{p-1} w\}. \tag{8.46}$$

In particular, $|\mathcal{O}| = 2p$ is even. Because of the above, we also know that $2p > 3$. We claim that $p \neq 2$. Suppose the contrary. Then, with $w = x_a x_b x_c$

(thus with $a < b < c$), we obtain that the elements of the orbit $\mathcal{O}$ satisfy the following commutative diagram:

$$\begin{array}{ccc} x_a x_b x_c & \xrightarrow{r_2} & x_a x_d x_e \\ r_1 \downarrow & & \downarrow r_1 \\ x_f x_g x_c & \xrightarrow{r_2} & x_f x_h x_e \end{array}$$

This implies that S has the following defining relations:

$$x_b x_c = x_d x_e \tag{8.47}$$

$$x_a x_b = x_f x_g \tag{8.48}$$

$$x_a x_d = x_f x_h \tag{8.49}$$

$$x_g x_c = x_h x_e \tag{8.50}$$

The relations (8.47) and (8.50) yield that $\varphi(x_b, x_c) = (x_c, x_e)$ and $\varphi(x_g, x_c) = (x_c, x_e)$. Since φ is bijective, we thus obtain that $x_b = x_g$. Hence, (8.48) becomes $x_a x_b = x_f x_b$. However, because of Lemma 8.3.6, this implies that $x_a = x_b = x_f = x_g$, a contradiction. So $|\mathcal{O}| = 2p$ with $p \geq 3$. In particular $|\mathcal{O}| \geq 6$.

Thus, we find that there are n orbits with a single element, $n(n-1)$ orbits with 3 elements and $n(n-1)(n-2)/6$ orbits with at least 6 elements. From the equality

$$|X^3| = n^3 = 1 \cdot n + 3 \cdot n(n-1) + 6 \cdot \frac{n(n-1)(n-2)}{6},$$

we deduce that the orbits of the third type contain exactly 6 elements.

Now it is easily shown that r is a set theoretic solution of the Yang-Baxter equation, that is, condition (IT.2) holds. Indeed, if the orbit of w has 6 elements, then, from (8.46), it follows that $(r_2 r_1)^3 w = w$. If the orbit has 3 elements, then $(r_2 r_1)^3 w = w$ follows directly from (8.45). Finally if the orbit has only 1 element, then $r_1 w = r_2 w = w$. Hence, we always have that $(r_2 r_1)^3(w) = w$ and thus $r_2 r_1 r_2(w) = r_1 r_2 r_1(w)$, as desired. □

8.4 Decomposable monoids of I-type

In this section, we prove Rump's result which says that all non-cyclic monoids of I-type are decomposable provided that they have square free defining relations. As an application, it is then shown that monoids of I-type that are defined by square free relations are binomial monoids, a result due to

Gateva-Ivanova. We finish with describing the three and four generated binomial monoids and the monoids of I-type generated by a prime number of elements.

We begin with recalling some terminology and notation. Let $\mathrm{FaM}_n = \langle u_1, \ldots, u_n \rangle$ be a free abelian monoid of rank n, and let $S = \{(a, \psi(a) \mid a \in \mathrm{FaM}_n\}$ be a monoid of I-type. So $\psi(a\psi(a)(b)) = \psi(a)\,\psi(b)$, for all $a, b \in \mathrm{FaM}_n$, and S is a submonoid of the natural semidirect product $\mathrm{FaM}_n \rtimes \mathrm{Sym}_n$. In Section 8.2, we have shown that $S = \langle x_1, \ldots, x_n \rangle$ with $x_i = (u_i, \psi(u_i))$ and that S has a group of quotients $G = SS^{-1} = \{(a, \psi(a)) \mid a \in \mathrm{Fa}_n\}$. Also, for every $a \in \mathrm{Fa}_n$, there exists $b \in \mathrm{FaM}_n$ so that $\psi(a) = \psi(b)$. Further, recall that S (or equivalently G) is said to be indecomposable if the left action $G \longrightarrow \mathrm{Sym}_n$, defined by mapping $(a, \psi(a))$ onto $\psi(a)$, is transitive. Equivalently, with the notation of Proposition 8.2.8, for any $1 \le i, j \le n$, there exists $a \in \mathrm{FaM}_n$ so that $\psi(a)(u_i) = u_j$. The latter is denoted by $u_i \sim u_j$.

Suppose now that the defining relations of S are square free, that is, if $x_i^2 = x_k x_l$ then $k = l = i$. Let $v = \psi(u_i)(u_i)$. Since

$$\begin{aligned} x_i^2 &= (u_i, \psi(u_i))\,(u_i, \psi(u_i)) \\ &= (u_i v, \psi(u_i)\psi(u_i)) \\ &= (v, \psi(v))\,(\psi(v)^{-1}(u_i), \psi(\psi(v)^{-1}(u_i))), \end{aligned}$$

we get that S is square free if and only if

$$\psi(u_i)(u_i) = u_i$$

for all i with $1 \le i \le n$. Hence, in this case, $x_i^w = (u_i^w, 1)$ with $w = n!$. Consequently, if $\psi(a)(u_i) = u_j$, for some $a \in \mathrm{FaM}_n$ then $x_i^w = (u_i^w, 1)$ and $(a, \psi(a)) x_i^w (a, \psi(a))^{-1} = (\psi(a)(u_i^w), 1) = x_j^w$. Therefore, $u_i \sim u_j$ if and only if there exists $s \in S$ so that $s x_i^w s^{-1} = x_j^w$.

Lemma 8.4.1. *Let $S = \{(a, \psi(a)) \mid a \in \mathrm{FaM}_n\} = \langle x_1, \ldots, x_n \rangle$ be a monoid of I-type, where $x_i = (u_i, \psi(u_i))$ and $\mathrm{FaM}_n = \langle u_1, \ldots, u_n \rangle$. If $\psi(u_i)(u_i) = u_i$ for all $i = 1, \ldots, n$, then*

$$\psi(\langle u_1^k, \ldots, u_n^k \rangle) = \langle \psi(u_1)^k, \ldots, \psi(u_n)^k \rangle,$$

for all integers k.

Proof. Put $\psi_i = \psi(u_i)$ for $1 \le i \le n$. Let $\sigma, \tau \in \psi(\langle u_1^k, \ldots, u_n^k \rangle)$. Then there exist $u, v \in \mathrm{FaM}_n$ such that $\psi(u^k) = \sigma$ and $\psi(v^k) = \tau$. Hence,

$\sigma\tau = \psi(u^k)\psi(v^k) = \psi(u^k\ \psi(u^k)(v^k))$. It follows that $\sigma\tau \in \psi(\langle u_1^k, \ldots, u_n^k\rangle)$, and thus the latter is a subgroup of Sym_n.

Because, by assumption, $\psi_i(u_i) = u_i$ for all i, it is easily verified that $\psi_i^k = \psi(u_i^k) \in \psi(\langle u_1^k, \ldots, u_n^k\rangle)$. Therefore, $\langle \psi_1^k, \ldots, \psi_n^k\rangle \subseteq \psi(\langle u_1^k, \ldots, u_n^k\rangle)$.

To prove the result it is now sufficient to show that

$$\psi(\langle u_1^k, \ldots, u_j^k\rangle) \subseteq \langle \psi_1^k, \ldots, \psi_n^k\rangle,$$

for all $1 \leq j \leq n$. We prove this by induction on j. Since $\psi(u_i^k) = \psi_i^k$, the case $j = 1$ is obvious. Suppose that $1 < j \leq n$ and that $\psi(\langle u_1^k, \ldots, u_{j-1}^k\rangle) \subseteq \langle \psi_1^k, \ldots, \psi_n^k\rangle$. Let $u \in \langle u_1^k, \ldots, u_{j-1}^k\rangle$ and let $u_i = \psi(u)^{-1}(u_j)$. Then, $\psi(uu_j^{km}) = \psi(u)\ \psi(\psi(u)^{-1}(u_j^{km})) = \psi(u)\ \psi(u_i^{km}) = \psi(u)\ \psi_i^{km}$. By the induction hypothesis, $\psi(u) \in \langle \psi_1^k, \ldots, \psi_n^k\rangle$. So, we get that $\psi(\langle u_1^k, \ldots, u_j^k\rangle) \subseteq \langle \psi_1^k, \ldots, \psi_n^k\rangle$. Therefore $\psi(\langle u_1^k, \ldots, u_n^k\rangle) = \langle \psi_1^k, \ldots, \psi_n^k\rangle$, as desired. □

We need another technical lemma. For the proof we recall the notion of primitivity. Let G be a transitive permutation group G on a set A. A nonempty subset B of A is said to be a subset of imprimitivity if $Bg = Bh$ or $Bg \cap Bh = \emptyset$, for any $g, h \in G$. If $B = A$ or $|B| = 1$, then B always is a set of imprimitivity. We call these the trivial subsets of imprimitivity. The group G is said to be imprimitive if it has a nontrivial subset of imprimitivity. Otherwise G is said to be primitive.

Lemma 8.4.2. *Suppose H is a transitive group of permutations of the set $\{1, \ldots, n\}$. If H is nilpotent and $H = \langle \bigcup_{i=1}^n \mathrm{St}(i)\rangle$, where $\mathrm{St}(i)$ denotes the stabilizer of i in H, then H is trivial.*

Proof. We prove the lemma by induction on n. The case $n = 1$ is clear. So let $n > 1$. Suppose first that H is a primitive permutation group. We claim that H is then abelian, that is $\mathrm{Z}(H) = H$. Suppose the contrary. Then, because H is nilpotent, let h be an element in the second center of H but $h \notin \mathrm{Z}(H)$. Consider the subgroup $N = \mathrm{gr}(\mathrm{Z}(H), h)$. Clearly, N is abelian and normal in H. Proposition 4.4 in [127] says that a nontrivial normal subgroup of a primitive permutation group is transitive. Hence it follows that N is transitive. Proposition 3.2 in [127] states that in a transitive abelian group all stabilizers are trivial. Applying this statement to both N and $\mathrm{Z}(H)$, it follows that $|N| = n = |\,\mathrm{Z}(H)|$. However, this contradicts with $\mathrm{Z}(H)$ being properly contained in N. This proves the claim, and thus H indeed is abelian. Hence, again using Proposition 3.2 in [127], the assumption $H = \langle \bigcup_{i=1}^n \mathrm{St}(i)\rangle$ implies that $H = \{1\}$, as desired.

Second, assume that H is imprimitive and let $B \subseteq \{1, \ldots, n\}$ be a nontrivial subset of imprimitivity. So $1 < |B| < n$. Then H acts by permutations on the set $A_1 = \{Bh \mid h \in H\}$ and it inherits the hypothesis with respect to this action. Since $1 < |A_1| < n$, the induction hypothesis implies that $H = \{1\}$. □

Theorem 8.4.3. *Any monoid of I-type with square free defining relations is decomposable if and only if it is not cyclic.*

Proof. Let $S = \{(a, \psi(a) \mid a \in \mathrm{FaM}_n\}$ be a monoid of I-type with $\mathrm{FaM}_n = \langle u_1, \ldots, u_n \rangle$. Obviously, S is indecomposable if it is a cyclic monoid. Conversely, assume S is not cyclic and its defining relations are square free. Because of Corollary 8.2.7, the group $T = \{\psi(a) \mid a \in \mathrm{Fa}_n\}$ is a solvable subgroup of Sym_n.

Let $G_\psi = \{(a, 1) \mid a \in \mathrm{Fa}_n, \psi(a) = 1\}$, $A_\psi = \{a \in \mathrm{Fa}_n \mid \psi(a) = 1\}$ and $A = \mathrm{Fa}_n / A_\psi$. So, the map $G/G_\psi \longrightarrow T$ defined by $(a, \psi(a))G_\psi \mapsto \psi(a)$ is a group isomorphism. Define $\overline{\psi}\colon \mathrm{Fa}_n / A_\psi \longrightarrow T$ by $\overline{\psi}(uA_\psi) = \psi(u)$ for all $u \in \mathrm{Fa}_n$. This mapping is well defined, because if $u, v \in \mathrm{Fa}_n$ are such that $uv^{-1} \in A_\psi$, then $\psi(u) = \psi((uv^{-1})v) = \psi(uv^{-1})\, \psi(\psi(uv^{-1})^{-1}(v)) = \psi(v)$.

The mapping $\overline{\psi}$ is injective, and thus bijective. Indeed, suppose $a, b \in \mathrm{Fa}_n$ are such that $\overline{\psi}(aA_\psi) = \overline{\psi}(bA_\psi)$, that is, $\psi(a) = \psi(b)$. Because $\psi(\psi(b)^{-1}(b^{-1})) = \psi(b)^{-1}$, we get that $\psi(ab^{-1}) = \psi(a)\, \psi(\psi(a)^{-1}(b^{-1})) = \psi(b)\, \psi(\psi(b)^{-1}(b^{-1})) = \psi(b)\psi(b)^{-1} = 1$. So, $ab^{-1} \in A_\psi$ and thus $aA_\psi = bA_\psi$.

We now prove the result by contradiction. So, suppose that G is decomposable, that is, the action of G, and thus of T on $\{u_1, \ldots, u_n\}$ is transitive. Since $n > 1$, it is clear that T is nontrivial. Hence, T contains a maximal intransitive normal subgroup N_1. Let N_2 be a normal subgroup of T such that $N_1 \subset N_2$ and N_2/N_1 is a minimal nontrivial normal subgroup of T/N_1. Since T is solvable, there exists a prime p such that N_2/N_1 is a p-subgroup of T/N_1. In particular, $|A| = |T| = p^m k$, with p and k coprime and $m \geq 1$.

Let $A^k = \{a^k \mid a \in A\}$ and let $T_k = \overline{\psi}(A^k)$. Then $|A^k| = |T_k| = p^m$. By Lemma 8.4.1, $T_k = \langle \psi_1^k, \ldots, \psi_n^k \rangle$, where $\psi_i = \psi(u_i)$. Because $|T_k| = p^m$ and $(p, k) = 1$, we have that T_k is a Sylow p-subgroup of T. Thus $(T_k N_1)/N_1$ is a Sylow p-subgroup of T/N_1. Since N_2/N_1 is a normal subgroup of T/N_1, we have that $N_2/N_1 \subseteq (T_k N_1)/N_1$.

Of course the action of T on $\{u_1, \ldots, u_n\}$ induces an action of any subgroup of T on this set. Let $O_1, \ldots, O_q$ be the different orbits of N_1. Note that $q > 1$ since N_1 acts intransitive. By the maximal condition on N_1 we know that the subgroup N_2 of T acts transitive. Hence the natural induced action of N_2/N_1 on $\{O_1, \ldots, O_q\}$ is transitive as well. Thus $(T_k N_1)/N_1$ also acts transitively on $\{O_1, \ldots, O_q\}$.

Because the sets $O_1, \ldots, O_q$ form a partition of $\{u_1, \ldots, u_n\}$, we have that, for $1 \leq j \leq n$, there exists i so that $u_j \in O_i$. The square free assumption on the relations yields that then $\psi_j^k(u_j) = u_j \in O_i$. Hence, $\psi_j^k N_1(O_i) = O_i$ and thus $\psi_j^k N_1 \in \mathrm{St}\,(O_i)$, where $\mathrm{St}\,(O_i)$ denotes the stabilizer of O_i in $(T_k N_1)/N_1$. So we have shown that

$$(T_k N_1)/N_1 = \langle \bigcup_{i=1}^{q} \mathrm{St}\,(O_i) \rangle.$$

Since $(T_k N_1)/N_1$ is nilpotent, acts transitively on $\{O_1, \ldots, O_q\}$ and because it is generated by the stabilizers of the O_i, it follows from Lemma 8.4.2 that $(T_k N_1)/N_1$ is trivial. But this yields a contradiction as $q > 1$, and the action is transitive. $\square$

Example 8.1.3 shows that Theorem 8.4.3 cannot be extended to monoids of I-type for which the defining relations are not square free.

In Theorem 8.3.10, we have proved that binomial monoids are of I-type with square free defining relations. We now prove the converse.

Corollary 8.4.4. *A monoid of I-type with square free defining relations is a binomial monoid.*

Proof. Let S be a monoid of I-type with square free defining relations. So S has a finite generating set X, say of order n, and the associated bijective map $r : X^2 \longrightarrow X^2$ satisfies conditions (IT.1), (IT.2) and (IT.3) stated in Theorem 8.1.4.

We prove the result by induction on n. If $n = 1$, then S is the infinite cyclic monoid and thus S is binomial. So assume $n > 1$. Because of Theorem 8.4.3, we know then that S is decomposable. In particular, because of Proposition 8.2.8, $X = Y \cup Z$, the disjoint union of two nonempty sets Y and Z so that both submonoids $\langle Y \rangle$ and $\langle Z \rangle$ are monoids of I-type and $YZ = ZY$ (that is, r induces a bijective map $Z \times Y \longrightarrow Y \times Z$). It follows from the induction hypothesis that the elements of Y (respectively Z) can be ordered, say $y_1 < \cdots < y_k$ (respectively $z_1 < \cdots < z_m$), so that conditions (B1), (B2) and (B3) of Definition 8.3.1 are satisfied, and thus $\langle Y \rangle$ (respectively $\langle Z \rangle$) is a binomial monoid. On $X = \{y_1, \ldots, y_k, z_1, \ldots, z_m\}$ we put the order $y_1 < \ldots < y_k < z_1 < \ldots < z_m$.

It remains to show that, with respect to this chosen order, conditions (B1), (B2) and (B3) are satisfied for X. That (B2) holds follows from

Theorem 8.1.4. Then, since S is a monoid of I-type, the number of elements of a particular degree, say k, is the same as the number of elements of degree k in a free abelian monoid FaM_n. Hence, (B3) follows.

So, it remains to show that (B1) holds. For this suppose $r(x, y) = (x', y')$ with $x > y$. We need to prove that $x' < y'$ and $x' < x$. Because of the induction hypothesis, this clearly is satisfied if either $x, y \in Y$ (and thus also $x', y' \in Y$), or $x, y \in Z$ (and thus also $x', y' \in Z$). So we need to consider the remaining case that $x \in Z$ and $y \in Y$. Since $ZY = YZ$, it follows that $x' \in Y$ and $y' \in Z$. Hence, by the choice of the ordering, we get that $x' < y'$ and $x' < x$, as desired. □

Of course it is easy to describe all binomial monoids on three generators in a direct manner. However, this also follows immediately from Proposition 8.2.8 and Corollary 8.4.4. Indeed, such a monoid has to be decomposed as a product of such monoids on respectively one and two generators. Since the latter again can be decomposed, necessarily as the product of two cyclic monoids, we obtain that, up to isomorphism, there are only two binomial monoids on three generators.

Example 8.4.5. The binomial monoids generated by three elements (and not less) are

$$\begin{aligned} B^{3,1} &= \langle x_1, x_2, x_3 \mid x_3x_1 = x_1x_3,\ x_3x_2 = x_2x_3,\ x_2x_1 = x_1x_2\rangle, \\ B^{3,2} &= \langle x_1, x_2, x_3 \mid x_3x_1 = x_2x_3,\ x_3x_2 = x_1x_3,\ x_2x_1 = x_1x_2\rangle. \end{aligned}$$

Nonabelian binomial monoids generated by four elements decompose into products of binomial monoids generated by at most three elements. Hence, it is readily verified that there are precisely five such monoids.

Example 8.4.6. The binomial monoids generated by four elements (and not less) are

$$\begin{aligned} B^{4,1} &= \langle x_1, x_2, x_3, x_4 \mid x_4x_1 = x_1x_4,\ x_4x_2 = x_2x_4,\ x_4x_3 = x_3x_4, \\ &\qquad x_3x_1 = x_1x_3,\ x_3x_2 = x_2x_3,\ x_2x_1 = x_1x_2\rangle, \\ B^{4,2} &= \langle x_1, x_2, x_3, x_4 \mid x_4x_1 = x_2x_4,\ x_4x_2 = x_1x_4,\ x_4x_3 = x_3x_4, \\ &\qquad x_3x_1 = x_1x_3,\ x_3x_2 = x_2x_3,\ x_2x_1 = x_1x_2\rangle = \langle x_1, x_2, x_3\rangle\langle x_4\rangle, \\ B^{4,3} &= \langle x_1, x_2, x_3, x_4 \mid x_4x_1 = x_2x_4,\ x_4x_2 = x_3x_4,\ x_4x_3 = x_1x_4, \\ &\qquad x_3x_1 = x_1x_3,\ x_3x_2 = x_2x_3,\ x_2x_1 = x_1x_2\rangle = \langle x_1, x_2, x_3\rangle\langle x_4\rangle, \end{aligned}$$

$$\begin{aligned}
B^{4,4} &= \langle x_1, x_2, x_3, x_4 \mid x_4x_1 = x_2x_4,\ x_4x_2 = x_1x_4,\ x_4x_3 = x_3x_4, \\
&\quad x_3x_1 = x_2x_3,\ x_3x_2 = x_1x_3,\ x_2x_1 = x_1x_2\rangle = \langle x_1, x_2, x_3\rangle\langle x_4\rangle, \\
B^{4,5} &= \langle x_1, x_2, x_3, x_4 \mid x_2x_1 = x_3x_4,\ x_2x_3 = x_1x_4,\ x_2x_4 = x_4x_2, \\
&\quad x_1x_2 = x_4x_3,\ x_1x_3 = x_3x_1,\ x_3x_2 = x_4x_1\rangle \\
&= \langle x_1, x_3\rangle\ \langle x_2, x_4\rangle.
\end{aligned}$$

We finish this section describing the indecomposable monoids of I-type generated by a prime number of elements.

Corollary 8.4.7. *Let p be a prime number and let $S = \{(a, \psi(a)) \mid a \in \mathrm{FaM}_p\}$ be a monoid of I-type. If S is indecomposable, then $S \cong \{(a, \sigma^{|\deg(a)|}) \mid a \in \mathrm{FaM}_p\}$, where σ is the p-cycle $(1\ 2\ \ldots\ p)$.*

Proof. Let $G = SS^{-1}$ be the group of quotients of S. So $G = \{(a, \psi(a)) \mid a \in \mathrm{Fa}_p\}$, a group of I-type. Assume S is indecomposable, that is, the group $T = \{\psi(a) \mid a \in \mathrm{FaM}_p\}$ acts transitively on the basis $\{u_1, \ldots, u_p\}$ of FaM_p. From Corollary 8.2.5 we know that $G_\psi = \{(a, 1) \mid a \in \mathrm{Fa}_p,\ \psi(a) = 1\} = A_\psi \times \{1\}$ is a normal subgroup of G, with $A_\psi = \{a \in \mathrm{Fa}_p \mid \psi(a) = 1\}$. Clearly, $G/G_\psi \longrightarrow T$ defined by mapping $(a, \psi(a))G_\psi$ onto $\psi(a)$ is a group isomorphism and $|\,\mathrm{Fa}_p / A_\psi| = |G/G_\psi| = |T|$.

For each $1 \leq i \leq p$, let $T_i = \{\psi(a) \mid a \in \mathrm{Fa}_p,\ \psi(a)(u_i) = u_i\}$, the stabilizer of u_i in T. Because the action is transitive, $[T : T_i] = p$. Furthermore, because T_i is a subgroup of Sym_p and since p is prime, it follows that the order of any element of T_i is a divisor of $(p-1)!$ and thus relatively prime with p. Hence, $|T_i| = n$ with $(n, p) = 1$. Consequently, $|T| = np$. Hence Fa_p / A_ψ is an abelian group of order np. Write $\mathrm{Fa}_p / A_\psi = \mathbb{Z}_p \rtimes A_0'$, where A_0' consists of the elements of Fa_p / A_ψ that are of order relatively prime to p. Let A_0 be its inverse image in Fa_p. Clearly A_0 is invariant under the action of T and thus $B = \{(a_0, \psi(a_0)) \mid a_0 \in A_0\}$ is a subgroup of G of index p. Because, for every $1 \leq i \leq p$, the group T_i consists of elements of order relatively prime with p, it follows that $T_i \subseteq \psi(A_0)$. Since both groups are of index p in T, this yields that $T_i = \psi(A_0)$. Hence $\psi(A_0) = \bigcap_{i=1}^p T_i = \{1\}$. Thus $T_i = \{1\}$ and $T \cong \mathbb{Z}_p$.

Because of Lemma 8.2.2, we know that if $a \in \mathrm{FaM}_p$ with $\deg(a) > 1$, then $\psi(a) = \psi(b)\psi(c)$ for some $b, c \in \mathrm{FaM}_p$, with $1 \leq \deg(b), \deg(c) < \deg(a)$, and $\deg(b) + \deg(c) = \deg(a)$. It follows that $\psi(u_x)$ has order p for some $x \in \{1, \ldots, p\}$. Without loss of generality, we may assume that $x = 1$ and $\psi(u_1) = \sigma = (1\,2\ldots p)$. Write $\psi(u_p) = \sigma^i$ for some i with $0 \leq i \leq p-1$. Clearly

$$(u_p, \psi(u_p))\,(u_1, \sigma)^{-1} = (u_p, \sigma^i)\,(u_p^{-1}, \sigma^{-1}) = (u_p u_{p+i}^{-1}, \sigma^{i-1}),$$

where we agree that $u_{p+i} = u_i$ if $i \neq 0$. Furthermore,

$$(u_p u_{p+i}^{-1}, \sigma^{i-1})^p = (\gamma, 1)$$

with

$$\begin{aligned}
\gamma &= u_p u_{p+i}^{-1}\, \sigma^{i-1}(u_p u_{p+i}^{-1}) \cdots \sigma^{(i-1)(p-1)}(u_p u_{p+i}^{-1}) \\
&= (u_p \sigma^{i-1}(u_p) \cdots \sigma^{(i-1)(p-1)}(u_p))\, (u_{p+i} \sigma^{i-1}(u_{p+i}) \cdots \sigma^{(i-1)(p-1)}(u_{p+i}))^{-1} \\
&= \left(\prod_{j=0}^{p-1} \sigma^{(i-1)j}(u_p) \right) \left(\prod_{j=0}^{p-1} \sigma^{(i-1)j}(u_{p+i}) \right)^{-1} .
\end{aligned}$$

If $i \neq 1$, then σ^{i-1} is a generator of the group $T = \langle \sigma \rangle$ and it follows that $u_p u_{p-1} \cdots u_1 = \prod_{j=0}^{p-1} \sigma^{(i-1)j}(u_p) = \prod_{j=0}^{p-1} \sigma^{(i-1)j}(u_{p+i})$ and thus $\gamma = 1$. But then $(u_p u_{p+i}^{-1}, \sigma^{i-1})$ is a nontrivial periodic element in SS^{-1}. However, this yields a contradiction as, by Corollary 8.2.7, the group SS^{-1} is torsion free. Thus we have shown that $i = 1$ and thus $\psi(u_p) = \psi(u_1) = \sigma$. Using the same argument it follows that $\psi(u_p) = \psi(u_{p-1}) = \ldots = \psi(u_1)$. Again by Lemma 8.2.2, we thus get that $\psi(a) = \sigma^{\deg(a)}$, and the result follows. □

8.5 Algebras of monoids of I-type

In this section, we investigate the algebraic structure of the semigroup algebra $K[S]$ of a monoid $S = \{(a, \psi(a)) \mid a \in \mathrm{FaM}_n\}$ of I-type. First, it is proved that such algebras are Noetherian domains that satisfy a polynomial identity and $\mathrm{GK}(K[S]) = \mathrm{clKdim}(K[S]) = n$. Next, the minimal prime ideals of the monoid S are described. It turns out that all these ideals are principal and generated by a normal element. As an application, we obtain that both the monoid S and the algebra $K[S]$ are maximal orders.

Some of the properties follow immediately from the results in the previous sections.

Theorem 8.5.1. *Let $S = \{(a, \psi(a)) \mid a \in \mathrm{FaM}_n\}$ be a monoid of I-type and let K be field. Then the semigroup algebra $K[S]$ is a Noetherian domain that satisfies a polynomial identity and* $\mathrm{clKdim}(K[S]) = \mathrm{GK}(K[S]) = n$.

Proof. Because of Corollary 8.2.7, the monoid S has a a group of quotients G that is Bieberbach. Hence, by Theorem 3.3.4, the group algebra $K[G]$ and thus also $K[S]$ is a domain. Furthermore, because of Corollary 8.2.5 and Proposition 8.2.8, the group G contains a free abelian subgroup A of rank n, so that A is normal in G and $A \cap S$ is finitely generated. So, by Theorem 3.1.9 and Theorem 4.4.7, the algebra $K[S]$ is Noetherian and satisfies a polynomial identity. Since $\mathrm{GK}(K[S]) = n$, we also obtain from Theorem 3.5.6 that $\mathrm{clKdim}(K[S]) = \mathrm{GK}(K[S]) = n$. $\square$

In order to prove that $K[S]$ also is a maximal order, we first consider the corresponding property for the semigroup S.

Theorem 8.5.2. *Let $S = \{(a, \psi(a)) \mid a \in \mathrm{FaM}_n\}$ be a monoid of I-type and let $C_1, \dots, C_m$ be the equivalence classes for the relation $\sim$ on the generating set $U = \{u_1, \dots, u_m\}$ of FaM_n (we use notations as in Proposition 8.2.8). For each i with $1 \leq i \leq n$, let $c_i = \prod_{c \in C_i} c$ and $f_i = (c_i, \psi(c_i))$. Then the minimal prime ideals of S are the ideals $Sf_1 = f_1 S, \dots, Sf_m = f_m S$. Furthermore, $\langle f_1, \dots, f_m \rangle$ is a free abelian monoid of rank m and each $f_i^{n!}$ is central in S.*

Put $x_i = (u_i, \psi(u_i))$, for $1 \leq i \leq n$. If, furthermore, S is a binomial monoid then each f_i is a product, in some order, of all elements in $X_i = \{x_j \mid u_j \in C_i\}$. In particular, $f_1 \cdots f_m = (u_1 \cdots u_n, \psi(u_1 \cdots u_n))$ is the product of all generators $x_1, \dots, x_n$ in some order.

Proof. Let $w = n!$. Because of Corollary 8.2.5, we know that $\psi(a)^w = 1$ for all $a \in \mathrm{FaM}_n$. Clearly, $\psi(a)(c_i) = c_i$ for all $a \in \mathrm{FaM}_n$. It follows that each f_i is a normal element of S. Furthermore, $f_i^w = (c_i^w, 1)$ is central in S and $\langle f_1, \dots, f_m \rangle$ is an abelian monoid. Because of the unique presentation of the elements of S as stated in Proposition 8.2.8, this monoid is free abelian of rank m.

Again, because of Proposition 8.2.8, we know that the group of quotients of S is $G = \{sf^{-q} \mid s \in S,\ q \in \mathbb{N}\} = \{(a, \psi(a)) \mid a \in \mathrm{Fa}_n\}$, where f is the normal element $(u_1 \cdots u_n,\ \psi(u_1 \cdots u_n))$. Hence, every (one-sided) ideal of S contains a power of f. Since $f = f_1 \cdots f_m$ and all f_i are normal elements of S, we get that each prime ideal of S contains at least one of the elements $f_1, \dots, f_m$.

Let P be a minimal prime ideal of S and let K be a field. Assume i is such that $f_i \in P$. Because of Corollary 8.2.7, G is a torsion free polycyclic-by-finite group. Hence, from Proposition 4.5.8, we get that $K[P]$ is a height one prime ideal of $K[S]$. Let $A = \langle v_1, \dots, v_n \rangle$ where $v_j = (u_j^w, 1)$. Note that the group generated by the elements $u_1^w, \dots, u_n^w$ is invariant under all automorphisms $\psi(a)$ with $a \in \mathrm{Fa}_n$. Hence, AA^{-1} is a normal subgroup of finite index in G. Consequently, $K[S]$ has a natural G/AA^{-1}-gradation and its component of degree e (the identity of G/AA^{-1}) is the polynomial algebra $K[A] = K[v_1, \dots, v_n]$. It then follows from Theorem 3.4.8 that

$P\cap K[A]=Q_1\cap\cdots\cap Q_r$, where $Q_1,\ldots,Q_r$ are all the height one primes of $K[A]$ that are minimal over $P\cap K[A]$. Since $f_i^w=(c_i^w,1)=\prod_{c\in C_i}(c^w,1)=v_{j_1}\cdots v_{j_l}$, with all factors distinct, and because $f_i^w\in P\cap K[A]$, we get that each ideal Q_j is generated by one of the elements $v_{j_1},\ldots,v_{j_l}$. Consequently, $K[A]\cap P=K[A]z'$ and $A\cap P=Az'$ with $z'=(z,1)$ and z a product of distinct generators u_j^w. Obviously, $f_i^w\in Az'$.

We now show that $z'=f_i^w$. Let $1\le j\le n$ and let k be such that $u_k^{-1}=\psi(u_j)^{-1}(u_j^{-1})$. Then consider the element

$$\begin{aligned}
s &= v_kx_j^{-1}z'z_j\\
&= (u_k^w,1)\,(u_j,\psi(u_j))^{-1}(z,1)(u_j,\psi(u_j))\\
&= (u_k^w,1)(\psi(u_j)^{-1}(u_j^{-1}),\psi(u_j)^{-1})(z,1)(u_j,\psi(u_j))\\
&= (u_k^{w-1}\psi(u_j)^{-1}(z),\psi(u_j)^{-1})(u_j,\psi(u_j))\\
&= (u_k^w\psi(u_j)^{-1}(z),1).
\end{aligned}$$

Since

$$\begin{aligned}
v_kx_j^{-1} &= (u_k^w,1)\,(u_j,\psi(u_j))^{-1}\\
&= (u_k^w\psi(u_j)^{-1}(u_j^{-1}),\psi(u_j)^{-1})\\
&= (u_k^{w-1},\psi(u_j)^{-1})\in S,
\end{aligned}$$

we get that $s\in P\cap A=Az'$. Hence, comparing degrees it follows that

$$u_k^w\psi(u_j)^{-1}(z)=u_l^wz,$$

for some l. It follows that if u_j^w is a factor of z then u_k^{2w} is a factor of u_l^wz and thus $k=l$. Consequently, $\psi(u_j)^{-1}(z)=z$. On the other hand, if u_j^w is not a factor of z then consider the following product

$$\begin{aligned}
s' &= x_jz'x_j^{-1}v_j\\
&= (u_j,\psi(u_j))\,(z,1)(u_j,\psi(u_j))^{-1}(u_j^w,1)\\
&= (u_j,\psi(u_j))(z,1)(\psi(u_j)^{-1}(u_j^{-1}),\psi(u_j)^{-1})(u_j^w,1)\\
&= (u_j\psi(u_j)(z)u_j^{-1}u_j^w,1)\\
&= (\psi(u_j)(z)u_j^w,1).
\end{aligned}$$

Since

$$\begin{aligned} x_j^{-1}v_j &= (u_j, \psi(u_j))^{-1}(u_j^w, 1) \\ &= (\psi(u_j)^{-1}(u_j^{-1}), \psi(u_j)^{-1})(u_j^w, 1) \\ &= (u_k^{-1}, \psi(u_j)^{-1})(u_j^w, 1) \\ &= (u_k^{-1}u_k^w, \psi(u_j)^{-1}) \\ &= (u_k^{w-1}, \psi(u_j^{-1})) \in S, \end{aligned}$$

it follows that $s' \in P \cap A = Az'$ and thus

$$u_j^w \psi(u_j)(z) = u_q^w z,$$

for some q. Since, by assumption, u_j^w is not a factor of z, it follows that $j = q$ and thus again $\psi(u_j)(z) = z$. So we have shown that $\psi(u_j)(z) = z$ for all j. Lemma 8.2.2 therefore yields that $\psi(a)(z) = z$ for all $a \in A$. On the other hand, we know that $f_i^w \in Az'$, and thus z divides c_i^w. So, let u_j^w be a factor of z. It follows that, for any $a \in \mathrm{FaM}_n$, $\psi(a)(u_j^w)$ also is a factor of z. Hence c_i^w is a factor of z. Since we already know that z is a factor of c_i^w, it thus follows that $z = c_i^w$, and thus $z' = f_i^w$, as desired.

Therefore, we have shown that $P \cap A = Af_i^w$. Suppose now that $s_1 s_2 \cdots s_m \in P$ with each $s_j \in S_{(j)} = \langle (c, \psi(c)) \mid c \in C_j \rangle$ (see Proposition 8.2.8). Recall that each $S_{(j)}$ also is a monoid of I-type and thus, from the above, each right ideal of $S_{(j)}$ contains a central element $f_j^{n_j}$ for some nonzero positive integer n_j that is a multiple of w. In particular, for each $j \neq i$, there exists $d_j \in S_{(j)}$ so that $s_j d_j = f_j^{n_j}$ for some positive integer n_j. Since each $f_j^{n_j}$ is central, it then easily follows that $s_i d \in P$ for some element $d \in \langle f_1^w, \ldots, f_{i-1}^w, f_{i+1}^w, \ldots, f_m^w \rangle$, with $d \in \mathrm{Z}(S)$. Since $A \cap P = Af_i^w$ and because $\langle f_1^w, \ldots, f_m^w \rangle$ is a free abelian monoid of rank m, it is clear that $d \notin P$. So $s_i \in P$. Hence we have shown that $P \subseteq S_{(1)} \cdots S_{(i-1)}(P \cap S_{(i)})S_{(i+1)} \cdots S_{(m)}$. The reverse inclusion is obvious. Consequently,

$$P = S_{(1)} \cdots S_{(i-1)}(P \cap S_{(i)})S_{(i+1)} \cdots S_{(m)}.$$

Next let $(a, \psi(a)) \in P \cap S_{(i)}$. Write $a = \prod_{b \in C_i} b^{n_b}$ with each $n_b \geq 0$. For each $b \in C_i$, let m_b be the minimal nonnegative integer so that $m_b + n_b \in w\mathbb{Z}$. Let $d = \prod_{b \in C_i} b^{m_b} \in \mathrm{FaM}_n$. Then

$$(a, \psi(a))(\psi(a)^{-1}(d), \psi(\psi(a)^{-1}(d))) = (ad, \psi(ad)) \in A \cap P = Af_i^w.$$

Hence $m_b+n_b \geq w$ for any $b \in C_i$. Since $m_b = 0$ if $n_b = 0$, it follows that all n_b must be strictly positive. Therefore $(a, \psi(a)) \in f_i S_{(i)}$. Hence $P = f_i S$.

So, we have shown that if P is a minimal prime ideal of S, then $f_i \in P$, for some $1 \leq i \leq m$ and then $P = Sf_i = f_i S$. To finish the proof, it is sufficient to show that, for each $1 \leq i \leq m$, the element f_i belongs to a minimal prime of S. Because of Theorem 8.5.1, the algebra $K[S]$ is a Noetherian domain. As f_i is a normal element of $K[S]$, it then follows from the principal ideal theorem (Theorem 3.2.4) that $f_i \in Q$ for some height one prime ideal Q of $K[S]$. From Proposition 4.5.8 it follows that $Q \cap S$ is a minimal prime ideal of S. As $f_i \in Q \cap S$, the required property follows.

Finally, assume that, furthermore, $S = \langle x_1, \ldots, x_n \rangle$ is a binomial monoid. We prove by induction on n that each f_i is a product of distinct generators $x_j = (u_j, \psi(u_j))$, and $f_1 \cdots f_m$ is the product of all generators $x_1, \ldots, x_n$ in some order. If $n = 1$, then S is the infinite cyclic group and the claim is obvious. So assume $n > 1$. Then, because of Theorem 8.4.3, $m > 1$. Hence, because of Proposition 8.2.8 and Corollary 8.4.4, $S = S_{(1)} \cdots S_{(m)}$, with each $S_{(i)}$ a binomial monoid with generating set $X_i = \{(u_j, \psi(u_j)) \mid u_j \in C_i\}$ and $\{x_1, \ldots, x_n\} = \bigcup_{i=1}^{m} X_i$, a disjoint union. By the induction hypothesis, $(c_i, \psi(c_i)) = f_i$ is the product, in some order, of all $x_j \in X_i$. Hence, the result follows. □

Note that in the proof of Example 8.1.3, we have shown that $S = \langle x, y \mid x^2 = y^2 \rangle = \{(a, \psi(a)) \mid a \in \mathrm{FaM}_2 = \langle u_1, u_2 \rangle\}$ with $\psi(a)$ the permutation that interchanges u_1 and u_2 if a has even length, otherwise $\psi(a) = 1$. It follows from Theorem 8.5.2 that $S(u_1 u_2, 1) = S(u_1, \psi(u_1))^2 = S(u_2, \psi(u_2))^2$ is the unique minimal prime ideal of S (compare also with Example 4.5.11). So, the final statement of the theorem does not necessarily hold for monoids of I-type that are not binomial.

We formulate an immediate consequence of the previous result.

Corollary 8.5.3. *A monoid of I-type is indecomposable if and only if it has only one minimal prime ideal.*

From Theorem 8.5.2, we know that the minimal primes of a monoid S of I-type are generated by a normal element. Because of Proposition 8.2.8, we also know that the group of quotients SS^{-1} of S is obtained by inverting all these normal elements. We now prove that if such monoid also satisfies the ascending chain condition on ideals, then it is a maximal order. In particular, monoids of I-type are maximal orders. But first we prove the following lemma.

Lemma 8.5.4. *Let S be a monoid which is an order (in its group of quotients $G = SS^{-1}$) and suppose S satisfies the ascending chain condition on two-sided ideals. Any ideal J of S maximal for the condition $S \subset (J :_l J)$ is a prime ideal of S. Furthermore, S is a maximal order if and only if $(P :_l P) = (P :_r P) = S$ for all $P \in \mathrm{Spec}(S)$.*

Proof. Let J be an ideal of S that is maximal for the condition $S \subset (J :_l J)$. Clearly, $J \neq S$. Suppose J is not a prime ideal. Hence, there exist ideals J_1 and J_2 of S properly containing J and such that $J_1 J_2 \subseteq J$. Clearly, $J_1 \subseteq (J :_l J_2)$ and $(J :_l J_2)J_2 \subseteq J \subseteq J_2$. So $(J :_l J_2) \subseteq (J_2 :_l J_2)$. By the maximality condition, we thus obtain that $(J :_l J_2) \subseteq S$. So $(J :_l J_2)$ is an ideal of S containing J_1 and such that $(J :_l J_2)J_2 \subseteq J$. Hence, we may assume that $J_1 = (J :_l J_2)$. Now, since $J_1 J_2 \subseteq J$, we obviously get that $(J :_l J)J_1 J_2 \subseteq J$ and thus $(J :_l J)J_1 \subseteq (J :_l J_2) = J_1$. Therefore, $(J :_l J) \subseteq (J_1 :_l J_1)$. Again because of the maximality condition, it thus follows that $(J :_l J) \subseteq S$, a contradiction. So, J is a prime ideal of S. This proves the first part of the statement.

Suppose I is an ideal of S so that $S \subset (I :_l I)$. By the ascending chain condition on two-sided ideals of S and because of the first part of the lemma, there then exists a prime ideal P such that $S \subset (P :_l P)$. The second part of the statement is now clear. □

Lemma 8.5.5. *Assume S is a submonoid of a group G and S satisfies the ascending chain condition on two sided ideals. Let T be a submonoid of S generated by elements $p \in \mathrm{N}(S)$ so that pS is a prime ideal of S. If $G = ST^{-1}$, then S is a maximal order.*

Proof. Let $Z = \{p \in \mathrm{N}(S) \mid pS \text{ is a prime ideal }\}$. Since $G = ST^{-1}$, it follows that every ideal of S intersects $T \subseteq \langle Z \rangle$. Hence, every prime ideal intersects Z.

Let $p \in Z$. We show that pS is a minimal prime ideal of S. Suppose the contrary. Then there exists a prime Q of S with $Q \subset pS$. By the above, there exists $q \in Z \cap Q$. So $qS \subset pS$. Write $q = ps$ for some $s \in S$. Since qS is prime and p is normal, it follows that $s \in Sq = qS$ and thus $s = tq$ for some $t \in S$. Then $q = ps = ptq$ and so $1 = pt$, a contradiction.

Suppose S is not maximal order. Because of Lemma 8.5.4 (and by symmetry), we may assume that there exists a prime ideal P of S and $g \in G \setminus S$ such that $gP \subseteq P$; moreover, this P can be chosen to be maximal for the existence of such an element $g \in G\setminus S$. Write $g = z^{-1}s$, $z = p_1^{a_1} \cdots p_n^{a_n}$ with all $p_i \in Z$ and $a_i > 0$. Cancelling some p_i, if necessary, we may assume that $s \notin p_i S$ for every i. Clearly, $sP \subseteq zP \subseteq p_1 S$. Because, by assumption,

p_1S is a prime ideal, we get $P \subseteq p_1S$ (as $s \notin p_1S$) and thus $P \subseteq p_1^{-1}P \subseteq S$. So

$$s'(p_1^{-1}P) = p_1^{-1}sP \subseteq p_1^{-1}zP = z'(p_1^{-1}P),$$

where $s', z' \in S$ are p_1-conjugates of s and z respectively, and $g' = (z')^{-1}s' \notin S$ (because, otherwise, $g \in S$, as p_1 is normal). So, $(z')^{-1}s'(p_1^{-1}P) \subseteq p_1^{-1}P$ and $p_1^{-1}P = Sp_1^{-1}P$ is an ideal of S. The maximality condition now implies that $P = p_1^{-1}P$. Thus $p_1P = P$. Let $q \in Z$ be such that $q \in P$. Then $q = p_1x$ for some $x \in P$. Thus $qS \subseteq p_1S$ and as these ideals are both minimal prime, it follows that $qS = p_1S$. Hence, $x \in \mathrm{U}(S) \cap P$, a contradiction. $\square$

We now prove that semigroup algebras of monoids of I-type are indeed maximal orders.

Theorem 8.5.6. *Let K be a field and let $S = \{(a, \psi(a)) \mid a \in \mathrm{FaM}_n\}$ be a monoid of I-type and $C_1, \ldots, C_m$ the equivalence classes for the relation $\sim$ on the generating set $U = \{u_1, \ldots, u_m\}$ of FaM_n (with notations as as in Proposition 8.2.8) . For each i with $1 \leq i \leq n$, let $f_i = (c_i, \psi(c_i))$ with $c_i = \prod_{c \in C_i} c$. Then $K[S]$ is a maximal order and the ideals $K[S]f_i$, with $1 \leq i \leq m$, form a complete set of height one prime ideals of $K[S]$ that intersect S nontrivially.*

In particular, S is a maximal order and the subsemigroup $\mathrm{N}(S)$ consisting of normalizing elements of S is the free abelian monoid $\langle f_1, \ldots, f_m \rangle$ of rank m.

Furthermore, if $P_i = Sf_i$ then $S_{P_i} = S\langle f_j \mid 1 \leq j \leq n,\ j \neq i \rangle^{-1} = (\prod_{j, j \neq i} G_{(j)}) \rtimes S_{(i)}$, with $G_{(j)}$ the group of quotients of $S_{(j)}$, and $\mathrm{Spec}^0(S_{P_i}) = \{f_i S_{P_i}\}$.

Proof. It follows from Theorem 8.5.1, Theorem 8.5.2 and Lemma 8.5.5 that S is a maximal order with minimal prime ideals $Sf_1, \ldots, Sf_m$ and each $f_i^{n!}$ is central in S.

Because of Theorem 7.2.7, in order to prove that $K[S]$ is a maximal order, it is now sufficient to show that, for each minimal prime ideal $P = Sf_i$ of S, the monoid S_P has only one minimal prime ideal. From Section 7.1, recall that $S_P \subseteq S_{[P]} = \{g \in G \mid Jg \subseteq S$ for some ideal $J \not\subseteq P\}$. Because S is a maximal order, it follows that $S_{[P]} = \{g \in G \mid Ig \subseteq S$ for some divisorial ideal I of S with $I \not\subseteq P\}$. Because the minimal prime ideals form a basis for the divisor group of S, it follows that $S_{[P]}$ is obtained from S by localizing at the Ore set $\langle f_1, \ldots, f_{i-1}, f_{i+1}, \ldots, f_m \rangle$. Hence, $S_{[P]}$ is the localization with respect to the central set $\langle f_1^{n!}, \ldots, f_{i-1}^{n!}, f_{i+1}^{n!}, \ldots, f_m^{n!} \rangle$. For $j \neq i$, it is clear that $f_j^{n!} \notin P$ and thus $f_j^{-1} \in S_P$. Consequently, we get that $S_P = S_{[P]} = S\langle f_j \mid 1 \leq j \leq n,\ j \neq i\rangle^{-1} = (\prod_{j,j\neq i} G_{(j)}) \rtimes S_{(i)}$, with $G_{(j)}$ the group of quotients of $S_{(j)}$. Because S_P is a central localization of S, it also follows that the minimal prime ideals of S_P are of the form $S_P Q$ with Q a minimal prime ideal of S that does not intersect the set $\langle f_1^{n!}, \ldots, f_{i-1}^{n!}, f_{i+1}^{n!}, \ldots, f_m^{n!}\rangle$. Hence S_P has only one minimal prime ideal, namely $S_P P = S_P f_i$. This proves that $K[S]$ is a maximal order.

From Proposition 4.5.8, we get that the ideals $K[S]f_i$, with $1 \leq i \leq m$, form a complete set of height one prime ideals of $K[S]$ that intersect S nontrivially.

Finally, if $s \in \mathrm{N}(S)$, then Ss is an invertible ideal of S. Since the minimal primes are generated by a normal element (and thus the minimal prime ideals are invertible ideals), it follows that $Ss = Sf_1^{n_1} \cdots Sf_m^{n_m}$ for some nonnegative integers n_i. Because all units of S are trivial, it follows that $s = f_1^{n_1} \cdots f_m^{n_m}$. Hence, $\mathrm{N}(S) = \langle f_1, \ldots, f_m\rangle$, a free abelian monoid by Theorem 8.5.2. □

If S is a monoid of I-type with group of quotients G, then Brown's result (Theorem 3.6.4) yields that $K[G]$ is a Noetherian maximal order and all its height one primes are principal and generated by a normal element. Theorem 8.5.6 gives an analogue for the height one primes of $K[S]$ intersecting S nontrivially. As explained in Section 4.5, the primes of height one of $K[S]$ that do not intersect S are all the ideals of the form $Q \cap K[S]$ where Q is a prime of height one in $K[G]$.

8.6 Comments and problems

Monoids S of I-type were introduced by Gateva-Ivanova and Van den Bergh in [53]; they are the monoid analogue of an algebra situation [150]. In [37], Etingof, Schedler and Soloviev investigated groups of I-type.

The characterization via the Yang-Baxter equation obtained in Section 8.1 comes from [53]. For more results on solutions of the Yang-Baxter equation, we refer the reader to [36, 37, 48, 49, 50, 53, 63, 105, 146, 154]. That groups of I-type are solvable and Bieberbach comes from [53, 37]. The other results in Section 8.2 are taken from [78, 87]. That binomial monoids satisfy the interior and exterior cyclic condition is a result of Gateva-Ivanova [46, 47]. That such monoids are binomial is due to Gateva-Ivanova and Van den Bergh [53]. Rump's decomposability result in Section 8.4 comes from [138]. That decomposable monoids of I-type are binomial is a result due to Gateva-Ivanova [50] (see also [52, 48]). The description of indecomposable monoids of I-type generated by a prime number of elements is due to Etingof, Schedler and Soloviev [37].

In [37], Etingof, Schedler and Soloviev verified Rump's result, using computer computations, for monoids of I-type generated by n elements with $n \leq 8$ elements. Gateva-Ivanova proved this also for $n = pq$ with p, q prime in an earlier version of [50]. Jespers and Okniński in [87] proved the result if the group $\{\psi(a) \mid a \in \mathrm{FaM}_n\}$ is nilpotent.

In [37, 36, 138] monoids of I-type with an infinite generating set are also considered. Some of the results remain valid. However, the decomposability does not have an extension to the infinitely generated case. A counterexample is given in [138].

Monoids of I-type are related to some of the work of Dehornoy [32]. Namely, it may be verified that the defining presentation of every monoid S of I-type is complete in the sense of Dehornoy, or equivalently, the so called (strong) cube condition holds (see [32, Proposition 4.4]). In order to see this, we use $u, v, w \in X$ as in Picture 3.1 in [32]. Then, having relations $uv_1 = wu_0, vu_1 = wv_0, u_0v_2 = v_0u_2$ with $u_i, v_i \in X$ (the three given faces of the cube), and knowing that there is a relation $u_1u_2 = u''w''$ for some $u'', w'' \in X$ and a relation $vu'' = zv''$ for some $z, v'' \in X$, one checks easily that the Yang-Baxter relation $r_1r_2r_1(u, v_1, v_2) = r_2r_1r_2(u, v_1, v_2)$ yields $z = u$ and also $v_1v_2 = v''w''$. Therefore, the cube condition follows. Because of the results proved in [32], this yields another proof of the fact that S is cancellative and has a torsion free group G of quotients, but also implies some other properties of G, for example that G has a solvable word problem.

In Section 8.5, it is proved that the semigroup algebra $K[S]$ of a monoid S of I-type is a Noetherian domain that is a maximal order and satisfies a polynomial identity. This is a result due to Gateva-Ivanova and Van den Bergh [53]. The results on prime ideals in this section come from [78, 87].

In [53], Gateva-Ivanova and Van den Bergh proved that several other homological properties hold for the semigroup algebra of a monoid S of I-type. In particular, $K[S]$ has finite global dimension. So it turns out that these algebras have a very rich algebraic structure that resembles that of a polynomial algebra in finitely many commuting elements. As an application, one obtains then another proof for the fact that $K[S]$ is a domain. For this, recall that a semiprime Noetherian ring is a domain, provided that all finitely generated left R-modules have finite projective dimension and all finitely generated projective left R-modules are stably free (see for example [130]). That $K[S]$ is a maximal order can then also be obtained as a consequence of the following result by Stafford and Zhang [147]: if R is a positively graded, connected, Noetherian algebra satisfying a polynomial identity and of finite global dimension, then R is a domain and a maximal order.

The reasons and tools for dealing with the mentioned properties come from the study of homological properties of Sklyanin algebras by Tate and Van den Bergh [150] and from the work of Gateva-Ivanova on skew polynomial rings with binomial relations [46, 47]. Some other related types of algebras defined by quadratic relations have been investigated, for example in [7, 99]. The reader is also referred to a forthcoming book [131] of Polishchuk and Positselski in which other recent developments in the study of algebras defined by quadratic relations are studied. A central notion in this work is that of Koszul algebra.

Problems

1. Let G be a group of I-type and K a field. Is the group of units of $K[G]$ trivial, that is, is every unit of $K[G]$ of the type kg with $0 \neq k \in K$ and $g \in G$? See the comment before Example 8.2.13.

2. Characterize the groups of I-type that are unique product groups. Also see the comment before Example 8.2.13.

3. Describe the indecomposable monoids of I-type. In particular if $S = \{(a, \psi(a)) \mid a \in \mathrm{FaM}_n\}$ is such a monoid, then describe the group $T = \{\psi(a) \mid a \in \mathrm{FaM}_n\}$. See Proposition 8.2.8: the indecomposable binomial monoids are the building blocks of monoids of I-type.

4. Let S be a monoid of I-type and let K be a field. Is every height one prime ideal P of $K[S]$ generated by a normal element of $K[S]$? Equivalently, is $K[S]$ a unique factorization ring in the sense of Chatters and Jordan? See Theorem 8.5.6: if $P \cap S \neq \emptyset$ then P is principal. See also Theorem 3.6.4: every height one prime of $K[SS^{-1}]$ is principal.

Chapter 9

Monoids of skew type

In this chapter, we investigate finitely presented monoids S, say with generating set $\{x_1, \ldots, x_n\}$, that have square free quadratic homogeneous defining relations, that is, the relations are of the form $x_i x_j = x_k x_l$, where $i \neq j$ and $k \neq l$. If S satisfies the ascending chain condition on right ideals, then it turns out that there have to be at least $\binom{n}{2}$ such relations. These monoids will be called monoids of skew type. In Section 9.2, we study monoids of skew type that satisfy the cyclic condition. The latter is a generalization of the interior and exterior cyclic conditions (see Section 8.3) of which we have shown that they play a crucial role in the investigations on binomial monoids. We prove that the cyclic condition implies that the monoid is left and right non-degenerate and we investigate these semigroups in Section 9.3. It is proved that right non-degenerate monoids of skew type satisfy the ascending chain condition on right ideals. An important tool in the proof is the ideal chain obtained from left and right divisibility by generators. It also turns out that such monoids contain an ideal that is a cancellative semigroup. This ideal will play a crucial role in the investigations. In Section 9.4, it is then proved that the algebras of non-degenerate monoids of skew type are examples of Noetherian algebras finite Gelfand-Kirillov dimension that satisfy a polynomial identity. In Section 9.5, we give a description of the least cancellative congruence on a non-degenerate monoid S of skew type and we describe the prime radical of the algebra $K[S]$.

9.1 Definition

Let S be a finitely presented monoid generated by the set $\{x_1, \ldots, x_n\}$ and with defining relations of the form $x_i x_j = x_k x_l$, where $i \neq j$ and $k \neq l$.

Assume that S satisfies the ascending chain condition on right ideals. We claim that then at least $\binom{n}{2}$ relations are needed. Indeed, for $i \neq j$, consider the following ascending chain of right ideals

$$x_j x_i S \subseteq x_j x_i S \cup x_j^2 x_i S \subseteq \cdots$$

Hence, for some $0 < m < n$, there exists $s \in S$ such that

$$x_j^n x_i = x_j^m x_i s.$$

Since, by assumption, the relations are square free, this implies that the word $x_j x_i$ must appear in one of the defining relations. Hence, we have shown that each word $x_j x_i$, with $i \neq j$, appears in one of the relations. Because there are $2\binom{n}{2}$ such words, we indeed need at least $\binom{n}{2}$ relations.

The following example shows that in general the existence of $\binom{n}{2}$ relations is not sufficient for $K[S]$ to be Noetherian.

Example 9.1.1. Let $S = \langle x_1, x_2, x_3 \rangle$ be the monoid defined by the relations

$$x_2 x_1 = x_3 x_1, x_1 x_2 = x_3 x_2, x_1 x_3 = x_2 x_3.$$

Then S is a left cancellative monoid that does not satisfy the ascending chain condition on left ideals. However, S satisfies the ascending chain condition on right ideals and it is not right cancellative.

Furthermore, for any field K, the semigroup algebra $K[S]$ is neither right nor left Noetherian and $\mathrm{GK}(K[S]) = 2$.

Proof. For any integer $n \geq 2$, let $a_n = x_1 x_2^{n+1} - x_1^n x_2^2 \in K[S]$. We claim that $a_n K[S]$ is the K-vector space spanned by the set $\{a_n x_2^j | j \geq 0\}$. Indeed, first note that

$$x_1 x_2 x_1 = x_1 x_3 x_1 = x_2 x_3 x_1 = x_2 x_2 x_1.$$

Next,

$$\begin{aligned} x_1 x_2^2 x_1 &= x_1 x_2 x_3 x_1 = x_1 x_1 x_3 x_1 = x_1 x_1 x_2 x_1 = x_1 x_3 x_2 x_1 \\ &= x_2 x_3 x_2 x_1 = x_2 x_1 x_2 x_1 = x_2 x_1 x_3 x_1 = x_2 x_2 x_3 x_1 = x_2^3 x_1. \end{aligned}$$

An induction argument yields that, for every integer $a \geq 1$,

$$x_1 x_2^a x_1 = x_2^{a+1} x_1.$$

We have shown this equality for $a = 1$ and $a = 2$. That it also holds for $a > 2$ follows from

$$x_1 x_2^a x_1 = x_1 x_2^{a-1} x_3 x_1 = x_1 x_2^{a-2} x_1 x_3 x_1 = x_2^{a-1} x_1 x_3 x_1 = x_2^{a+1} x_1.$$

Hence, again by induction, it follows that, for $n \geq 0$,

$$x_1^n x_2^2 x_1 = x_2^{n+2} x_1.$$

It follows that, for $n \geq 2$,

$$a_n x_1 = x_1 x_2^{n+1} x_1 - x_1^n x_2^2 x_1 = x_2^{n+2} x_1 - x_2^{n+2} x_1 = 0$$

and

$$a_n x_3 = (x_1 x_2^n - x_1^n x_2) x_2 x_3 = (x_1 x_2^n - x_1^n x_2) x_1 x_3 = 0.$$

So the claim follows.

Because the generator x_2 only appears in the defining relation $x_1 x_2 = x_3 x_2$, it easily is seen that each element $x_1^k x_2^q$ only can be rewritten as syx_2^q, for some $s \in S$ and $y \in \{x_1, x_3\}$. Hence, if $k < n$ then $x_1 x_2^n \notin S x_1 x_2^k$. So, S does not satisfy the ascending chain condition on left ideals and thus $K[S]$ is not left Noetherian. It also easily follows that, for $n \geq 3$, there do not exist $\lambda_j \in K$ so that

$$a_n = \sum_{j=2}^{n-1} \lambda_j (x_1 x_2^{j+1} - x_1^j x_2^2) x_2^{n-j}.$$

Therefore, $a_n \notin \sum_{j=2}^{n-1} a_j K[S]$ for every n. So, indeed, $K[S]$ is not right Noetherian. That S satisfies the ascending chain condition on right ideals will follow from Proposition 9.3.9.

Obviously, the relations imply that S is not right cancellative. However, S is left cancellative. Indeed, suppose $s, t_1, t_2 \in S$ are such that $st_1 = st_2$. Because the relations are of the form $x_k x_i = x_l x_i$, it then follows that $t_1 = x_i t_1'$ and $t_2 = x_i t_2'$, for some $t_1', t_2' \in S$ and some i. We can now repeat the argument on the equality $(sx_i)t_1' = (sx_i)t_2'$. An induction on the length of t_1 (note that S has a natural length function) then implies that $t_1 = t_2$.

The relations imply that if $|\{i, j, k\}| = 3$, then $x_i^m x_k^n = x_j^m x_k^n$. Hence an element of S can be written in the form $x_i^m x_j^n$, with $m, n \geq 0$. As all words $x_1^m x_2^n$ uniquely determine the ordered pair (m, n), it follows that $\mathrm{GK}(K[S]) = 2$. □

Definition 9.1.2. A monoid S is said to be of skew type if it has a presentation of the form

$$S = \langle x_1, x_2, \ldots, x_n \mid R \rangle,$$

where R is a finite set consisting of $\binom{n}{2}$ defining relations that are of the form $x_i x_j = x_k x_l$, with $i \neq j$, $k \neq l$, $(i, j) \neq (k, l)$, and every word $x_i x_j$

(with $i \neq j$) appears in exactly one relation. For simplicity reasons, we will often denote the generating set $\{x_1, \ldots, x_n\}$ as X and S as $\langle X; R\rangle$. Clearly, such a monoid has a well defined length function. The length of $s \in S$ we denote by $|s|$. So $|s| = k$ if $s = x_{i_1} \cdots x_{i_k}$ with $1 \leq i_j \leq n$ for $1 \leq j \leq k$.

As explained in Section 8.1, if S is a monoid of skew type with defining generating set $X = \{x_1, \ldots, x_n\}$, then we obtain an associated bijective mapping

$$r : X \times X \longrightarrow X \times X$$

defined by

$$r(x_i, x_j) = (x_k, x_l), \;\; r(x_k, x_l) = (x_i, x_j), \;\; r(x_i, x_i) = (x_i, x_i),$$

if $x_i x_j = x_k x_l$ is a defining relation for S. Clearly r^2 is the identity mapping. For each $x \in X$, we obtain mappings f_x and g_x on X so that

$$r(x_i, x_j) = (f_{x_i}(x_j), g_{x_j}(x_i)).$$

Recall that S is said to be left non-degenerate (respectively right non-degenerate) if each g_x (respectively each f_x) is bijective. If both conditions are satisfied then S is simply said to be non-degenerate. Note that if S is left non-degenerate, then there are no relations of the type $xy = zy$ with $x \neq z$, and, for every $x, y \in X$, there exists a relation of the type $ux = vy$ for some $u, v \in X$.

In Example 9.1.1 an example is given of a monoid of skew type that is right non-degenerate but not left non-degenerate. This example also shows that right non-degenerate monoids of skew type do not necessarily have the ascending chain condition on left ideals. The following example shows that a monoid of skew type that is neither left nor right non-degenerate even can have a free monoid of rank two as an epimorphic image. Obviously, such a monoid lacks the ascending chain condition on left an right ideals.

Example 9.1.3. The free monoid of rank two is an epimorphic image of the monoid of skew type

$$\langle x_1, x_2, x_3 \mid x_1x_2 = x_1x_3, \; x_2x_1 = x_3x_1, \; x_3x_2 = x_2x_3\rangle.$$

Proof. Let ρ be the least congruence on S containing the pair (x_2, x_3). Then S/ρ is a free monoid of rank two. □

9.2 Monoids of skew type and the cyclic condition

Because of Lemma 8.1.2 and Corollary 8.4.4, monoids of I-type are binomial if and only if they are of skew type. Furthermore, Theorem 8.1.4 and Corollary 8.2.4 show that these monoids also are non-degenerate. A crucial property in the proof of these results is that binomial monoids satisfy the interior and exterior cyclic condition (Proposition 8.3.9). We therefore consider a generalization of this condition (we do not impose any assumption on the ordering of the generating set X).

Definition 9.2.1. A monoid S generated by a finite set X is said to satisfy the cyclic condition, simply denoted (CC), if, for every pair $x, y \in X$, there exist elements $x = x_1, x_2, \ldots, x_k, y' \in X$ such that

$$yx = x_2y', \; yx_2 = x_3y', \; \ldots, \; yx_k = xy'.$$

If S is a monoid of skew type that satisfies the cyclic condition, then it follows that for distinct x and y in X there exist $x', y' \in X$ such that $yx' = xy'$. Hence, S is right non-degenerate. We will now show that the cyclic condition on S is symmetric, and thus S is left and right non-degenerate. To prove this we first show two technical lemmas.

Lemma 9.2.2. *Let $S = \langle X; R \rangle$ be a monoid of skew type. Assume that, if $yx = x_2y'$ is one of the defining relations for S (and thus $y \neq x$) then there also is a defining relation of the form $yx_2 = x_3y'$ for some $x_3 \in X$. Then, S satisfies the cyclic condition and the elements $x_2, x_3, \ldots, x_k$ in condition (CC) can be chosen distinct.*

Proof. First notice that if $yx = x_2y'$ is a defining relation then $y \neq x_2$. Indeed, the assumption implies that there exists a defining relation $yx_2 = x_3y'$ for some $x_3, y' \in X$; hence $y \neq x_2$ because S is of skew type. So, we may apply the assumption several times to get

$$yx = x_2y', \; yx_2 = x_3y', \; \ldots, \; yx_k = x_{k+1}y',$$

where $y \neq x_i \in X$ for every i, $x_{k+1} = x_r$ for some $r \leq k$ (with $x_1 = x$). Let r be minimal integer with this property. If $r \geq 2$ then $yx_{r-1} = x_ry'$ and $yx_k = x_{k+1}y' = x_ry'$ imply that $yx_{r-1} = yx_k$. Because S is right non-degenerate, it follows that $x_{r-1} = x_k$, contradicting minimality of r. Consequently, $r = 1$ and we get $x = x_{k+1}$, as desired. The last assertion is an immediate consequence of the proof. $\square$

Lemma 9.2.3. *Let $S = \langle X; R\rangle$ be a monoid of skew type that satisfies the cyclic condition. If $ax_1 = x_2b$, for some $a, b, x_1, x_2 \in X$, then the following condition hold.*

1. *There exist $c, x_3, x_0 \in X$ such that: (a) $ax_2 = x_3b$, (b) $ax_0 = x_1b$, and (c) $cx_1 = x_2a$.*

2. *If $ax_1 = x_2b$, $ax_2 = x_3b$ and $cx_1 = x_2a$, then $cx_2 = x_3a$.*

Proof. Condition (CC) applied to $ax_1 = x_2b$ implies the existence of $x_0, x_3 \in X$ so that $ax_0 = x_1b$ and $ax_2 = x_3b$. Then, applying condition (CC) to $x_2b = ax_1$ yields $x_2a = cx_1$, for some $c \in X$. This proves the first part of the statement.

For the second part, assume that $ax_1 = x_2b$, $ax_2 = x_3b$ and $cx_1 = x_2a$. Then, applying the first part to $ax_2 = x_3b$ yields that $tx_2 = x_3a$, for some $t \in X$. Applying the first part to the latter equation, we obtain that $ts = x_2a$, for some $s \in X$. Since also $cx_1 = x_2a$ and because S is a monoid of skew type, it follows that $t = c$. Hence $cx_2 = x_3a$, as desired. □

Proposition 9.2.4. *Let $S = \langle X; R\rangle$ be a monoid of skew type. Assume S satisfies the cyclic condition. Then the full cyclic condition (FC) holds in S, that is, for any distinct elements x and y in X there exist distinct elements $x = x_1, x_2, \ldots, x_k$ and distinct $y = y_1, y_2, \ldots, y_p$ in X such that*

$$y_1x_1 = x_2y_2,\ y_1x_2 = x_3y_2,\ \ldots,\ y_1x_k = x_1y_2,$$

$$y_2x_1 = x_2y_3,\ y_2x_2 = x_3y_3,\ \ldots,\ y_2x_k = x_1y_3,$$

$$\vdots$$

$$y_px_1 = x_2y_1,\ y_px_2 = x_3y_1,\ \ldots,\ y_px_k = x_1y_1.$$

This is called a cycle of type $k \times p$.

Proof. Let x and y be distinct elements in X. From the cyclic condition (CC) and Lemma 9.2.2, it follows that there exist distinct $x = x_1, \ldots, x_k \in X$ so that

$$yx_1 = x_2z,\ yx_2 = x_3z,\ \ldots,\ yx_k = x_1z, \tag{9.1}$$

for some $z \in X$. By the first part of Lemma 9.2.3, there exists $y^{(1)} \in X$ so that

$$y^{(1)}x_1 = x_2y. \tag{9.2}$$

Hence $yx_1 = x_2z$, $yx_2 = x_3z$ and $y^{(1)}x_1 = x_2y$ (if $k = 1$ then we put $x_2 = x_3 = x_1$ and if $k = 2$ then we put $x_3 = x_1$). By the second part of Lemma 9.2.3, we get $y^{(1)}x_2 = x_3y$. It follows, by an induction procedure, that $y^{(1)}$ is compatible with the condition (9.1), that is,

$$y^{(1)}x_1 = x_2y,\ y^{(1)}x_2 = x_3y,\ \ldots,\ y^{(1)}x_k = x_1y.$$

Applying the same procedure to (9.2), we obtain $y^{(2)} \in X$ such that

$$y^{(2)}x_1 = x_2y^{(1)},\ y^{(2)}x_2 = x_3y^{(1)},\ \ldots,\ y^{(2)}x_k = x_1y^{(1)}.$$

The cyclic condition (CC) and Lemma 9.2.2 applied to x_2y imply that, after finitely many such steps, we obtain distinct $y^{(1)}, \ldots, y^{(p-1)} \in X$ such that $x_2y^{(i)} = y^{(i+1)}x_1$, for $i = 1, \ldots, p-2$ and $x_2y^{(p-1)} = yx_1$. Since $yx_1 = x_2z$, we get $y^{(p-1)} = z$. Also, by the above procedure,

$$y^{(i+1)}x_1 = x_2y^{(i)},\ y^{(i+1)}x_2 = x_3y^{(i)},\ \ldots,\ y^{(i+1)}x_k = x_1y^{(i)},$$

for $i = 1, \ldots, p-2$. The lemma now follows by putting $y_p = y^{(1)}, y_{p-1} = y^{(2)}, y_{p-2} = y^{(3)}, \ldots, y_2 = z = y^{(p-1)}$, $y_1 = y$. □

Corollary 9.2.5. *A monoid of skew type that satisfies the cyclic condition is non-degenerate.*

Proof. Let $S = \langle X; R \rangle$ be a monoid of skew type that satisfies the cyclic condition. Because of Proposition 9.2.4, the monoid S satisfies the full cyclic condition (FC). Hence, for distinct $x, y \in X$, there exist $x', x'', y', y'' \in X$ so that $xy' = yx'$ and $y''x = x''y$. Consequently, the mappings f_x and g_x are surjective, and thus, as X is finite, they are bijective. This shows that S, indeed, is non-degenerate. □

Of course, there are many examples of non-degenerate monoids of skew type that do not satisfy the cyclic condition. One such example is the monoid S generated by x_1, x_2, x_3, x_4 and that is subject to the relations $x_2x_1 = x_1x_3$, $x_3x_1 = x_2x_4$, $x_4x_1 = x_1x_2$, $x_3x_2 = x_1x_4$, $x_4x_2 = x_2x_3$, $x_4x_3 = x_3x_4$.

The relevance of the cyclic condition is once more illustrated in the following corollary that relates all classes of monoids of skew type so far considered, provided that the associated map $r : X \longrightarrow X$ is a set theoretic solution of the Yang-Baxter equation.

Corollary 9.2.6. *Let $S = \langle X; R \rangle$ be a monoid of skew type. Assume the associated map $r : X \longrightarrow X$ is a set theoretic solution of the Yang-Baxter equation. Then the following conditions are equivalent.*

1. *S is a monoid of I-type.*
2. *S is a binomial monoid.*
3. *S satisfies the cyclic condition.*
4. *S is right non-degenerate.*
5. *S is left non-degenerate.*

Proof. Because S is defined by square free relations, Theorem 8.3.10 and Corollary 8.4.4 yield the equivalence of (1) and (2). That (2) implies (3) follows from Proposition 8.3.9. However this can also be shown as a direct consequence of the assumptions that S is a monoid of skew type so that the associated map $r : X \longrightarrow X$ is a set theoretic solution of the Yang-Baxter equation. To prove this, let $x_i \neq x_j$. Then, there exists a defining relation $x_i x_j = x_k x_l$, with $k \neq l$ and $i \neq k$. Hence, there also is a relation $x_i x_k = x_{k'} x_{l'}$, for some k', l'. Because of Lemma 9.2.2, it is now sufficient to show that $l = l'$. Clearly, the two equalities $x_i x_j = x_k x_l$ and $x_i x_k = x_{k'} x_{l'}$ imply that

$$r_2 r_1 r_2(x_i, x_i, x_j) = r_2 r_1(x_i, x_k, x_l) = r_2(x_{k'}, x_{l'}, x_l).$$

Since the relations are square free, we also have

$$r_1 r_2 r_1(x_i, x_i, x_j) = r_1 r_2(x_i, x_i, x_j) = r_1(x_i, x_k, x_l) = (x_{k'}, x_{l'}, x_l).$$

Since r is a set theoretic solution of the Yang-Baxter equation, it thus follows that

$$r_2(x_{k'}, x_{l'}, x_l) = (x_{k'}, x_{l'}, x_l).$$

Again, because the relations are square free, it follows that $l = l'$, as desired.

That (3) implies (4) and (5) follows at once from Corollary 9.2.5. Finally, both (4) and (5) imply (1), because of Theorem 8.1.4 and Corollary 8.2.4. □

9.3 Non-degenerate monoids of skew type

In this section, we will prove that monoids of skew type that are right non-degenerate satisfy the ascending chain condition on right ideals. In order to prove this, we first show that S has an ideal chain determined via the left and right divisibility by generators.

Let $X = \{x_1, \ldots, x_n\}$ and let $S = \langle X; R \rangle$ be a monoid of skew type with associated bijective map $r : X^2 \longrightarrow X^2$. For convenience sake, we will often abuse notation by identifying the elements in X^m with words of length m in the free monoid on X.

For any $m \geq 2$ and for any $i = 1, \ldots, m-1$, we define a mapping $g_i : X^m \longrightarrow X^m$ by putting, for $y_1, \ldots, y_m \in X$,

$$g_i(y_1 \cdots y_m) = y_1 \cdots y_{i-1} \overline{y}_i \overline{y}_{i+1} y_{i+2} \cdots y_m$$

if $r(y_i, y_{i+1}) = (\overline{y}_i, \overline{y}_{i+1})$. Note that if $y_i = y_{i+1}$ then $\overline{y}_i = \overline{y}_{i+1}$. Then, consider the mapping $g : X^m \longrightarrow X^m$ defined by

$$g(y_1 \cdots y_m) = g_{m-1} \cdots g_2 g_1 (y_1 \cdots y_m).$$

Next, define, for each $y_1 \in X$, the mapping

$$f_{y_1} : X^{m-1} \longrightarrow X^{m-1}$$

by putting

$$f_{y_1}(y_2 \cdots y_m) = s_1 \cdots s_{m-1} \text{ if } g(y_1 \cdots y_m) = s_1 \cdots s_m,$$

with each $s_i \in X$. Note that, in case $m = 2$, the notation of the mapping f_{y_1} corresponds with that used in the definition of non-degenerate monoids.

For simplicity reasons (and if unambiguous), we will often use the same notation for an element of X^m and its natural image in S. It is then clear that if $y_1 \in X$, $f_{y_1}(y_2 \cdots y_m) = s_1 s_2 \cdots s_{m-1}$ and $g(y_1 \cdots y_m) = s_1 \cdots s_m$ then $y_1 y_2 \cdots y_m = f_{y_1}(y_2 \cdots y_m) s_m$ in S.

Lemma 9.3.1. *Assume that $S = \langle X; R \rangle$ is a right non-degenerate monoid of skew type. If $y_1 \in X$ and $m \geq 2$, then $f_{y_1} : X^{m-1} \longrightarrow X^{m-1}$ is a one-to-one mapping.*

Proof. We prove this by induction on m. The case $m = 2$ is clear because, by assumption, S is right non-degenerate. So assume $m > 2$. Suppose

$$g(y_1 \cdots y_m) = s_1 \cdots s_m,$$

with all $y_i, s_i \in X$. It is sufficient to show that $s_1 \cdots s_{m-1}$ and y_1 determine $y_2 \cdots y_m$. By definition we have that

$$g_1(y_1 \cdots y_m) = s_1 h(y_1 y_2) y_3 \cdots y_m,$$

with $r(y_1, y_2) = (s_1, h(y_1 y_2))$. Moreover,

$$s_1 g(h(y_1 y_2) y_3 \cdots y_m) = g(y_1 \cdots y_m) = s_1 \cdots s_m$$

and thus

$$g(h(y_1y_2)y_3\cdots y_m) = s_2\cdots s_m.$$

By the induction hypothesis, it then follows that $s_2\cdots s_{m-1}$ and $h(y_1y_2)$ determine $y_3\cdots y_m$. Since S is right non-degenerate, y_1 and s_1 determine $h(y_1y_2)$ and y_2. Hence y_1 and $f_{y_1}(y_2\cdots y_m) = s_1\cdots s_{m-1}$ determine $y_2\cdots y_m$, as desired. □

We introduce another technical condition.

Definition 9.3.2. A finitely generated monoid $M = \langle m_1,\ldots,m_n\rangle$ is said to satisfy the over-jumping property (with respect to the given generating set $m_1,\ldots,m_n$) if for every $a \in M$ and $1 \leq i \leq n$ there exist $k \geq 1$ and $w \in M$ such that $aw = m_i^k a$.

The over-jumping property is formally stronger than the condition mentioned in the following lemma, which holds for all monoids that satisfy the ascending chain condition on right ideals.

Lemma 9.3.3. *If a finitely generated monoid $M = \langle m_1,\ldots,m_n\rangle$ satisfies the ascending chain condition on right ideals, then, for every $a \in M$ and every i, with $1 \leq i \leq n$, there exist positive integers p, q and $w \in M$ so that $m_i^p aw = m_i^{p+q}a$.*

Proof. This is proved in a similar fashion as in the proof of Lemma 4.1.5. Let $a \in M$ and $1 \leq i \leq n$. Consider the following chain of right ideals of M,

$$m_iaM \subseteq m_iaM \cup m_i^2aM \subseteq m_iaM \cup m_i^2aM \cup m_i^3aM \subseteq \cdots$$

Since, by assumption, M satisfies the ascending chain condition on right ideals, there exist positive integers q, p so that $m_i^{p+q}a \in m_i^p aM$. Hence $m_i^{p+q}a = m_i^p aw$ for some $w \in M$. □

Proposition 9.3.4. *Assume that $S = \langle X; R\rangle$ is a right non-degenerate monoid of skew type. Then S satisfies the over-jumping property.*

Proof. Let $X = \{x_1,\ldots,x_n\}$. Fix $y_1 \in X = \{x_1,\ldots,x_n\}$. Because of Lemma 9.3.1, we know that, for each $m \geq 2$, the mapping $f = f_{y_1} : X^{m-1} \longrightarrow X^{m-1}$ is bijective. Hence f^r is the identity map, for some $r \leq (|X|^{m-1})! = (n^{m-1})!$. In particular, for any $y_2,\ldots,y_m \in X$, we have that

$$f^r(y_2\cdots y_m) = y_2\cdots y_m.$$

Hence, in S (see the comment before Lemma 9.3.1), we get that

$$y_1y_2\cdots y_m = f_{y_1}(y_2\cdots y_m)s_m,$$

for some $s_m \in X$. Since $f_{y_1}(y_2\cdots y_m)$ is a product of $m-1$ generators, we thus also get that

$$y_1^2y_2\cdots y_m = y_1f_{y_1}(y_2\cdots y_m)s_m = f_{y_1}(f_{y_1}(y_2\cdots y_m))s_{m+1}s_m,$$

for some $s_{m+1} \in X$. Proceeding this way, we obtain that

$$\begin{aligned} y_1^r y_2\cdots y_m &= f_{y_1}^r(y_2\cdots y_m)s_{m+r-1}\cdots s_{m+1}s_m \\ &= y_2\cdots y_m s_{m+r-1}\cdots s_{m+1}s_m, \end{aligned}$$

for some $s_i \in X$ with $m \leq i \leq m+r-1$. This means that, in S, we have

$$y_1^r y_2\cdots y_m = y_2\cdots y_m w,$$

for some $w \in S$.

We have thus shown that, for every $m \geq 1$, there exists $r \geq 1$ ($r \leq (n^{m-1})!$) such that if $a \in S$, $|a| < m$ and $1 \leq i \leq n$, then $aw = x_i^r a$ for some $w \in S$. Hence, the result follows. □

Lemma 9.3.5. *Assume that $S = \langle X; R\rangle$ is a right non-degenerate monoid of skew type. Then, for every $x, y \in S$, there exist $t, w \in S$ such that $|w| = |y|$ and $xw = yt$. In particular, S satisfies the right Ore condition.*

Proof. We prove this by induction on the length of x. First, assume $|x| = 1$, that is, $x \in X$. Suppose $|y| = m$. By Lemma 9.3.1, the mapping $f_x : X^m \longrightarrow X^m$ is bijective. Thus, there exists $w \in$ with $|w| = |y|$, so that $f_x(w) = y$ and $xw = f_x(w)t = yt$ for some $t \in X$, as desired. Second, suppose $|x| = m > 1$ and assume that the assertion holds for all $x \in S$ of length strictly less than m. Write $x = z_1\cdots z_m$, with each $z_i \in X$. By the induction hypothesis, $z_1\cdots z_{m-1}u = yw$ for some $u, w \in S$ with $|u| = |y|$. We also know that $z_m v = us$, for some $v, s \in S$ such that $|v| = |u|$. Consequently,

$$xv = z_1\cdots z_{m-1}z_m v = z_1\cdots z_{m-1}us = yws$$

and $|v| = |y|$, again as desired. □

The sets introduced in the following definition will play a crucial role in the investigations.

Definition 9.3.6. Let $S = \langle X; R\rangle$ be a monoid of skew type. For a subset Y of $X = \{x_1, \dots, x_n\}$ define

$$S_Y = \bigcap_{y \in Y} yS, \quad S'_Y = \bigcap_{y \in Y} Sy$$

and

$$D_Y = \{s \in S_Y \mid \text{ if } s = xt \text{ for some } x \in X \text{ and } t \in S, \text{ then } x \in Y\},$$

$$D'_Y = \{s \in S_Y \mid \text{ if } s = tx \text{ for some } x \in X \text{ and } t \in S, \text{ then } x \in Y\}.$$

Because of Lemma 9.3.5, each such set S_Y is nonempty. However, it may happen that $S_Y = S_Z$ for different subsets Y and Z of X. As we shall see in examples (Proposition 10.3.1), it even can occur that $D_Y = \emptyset$.

Theorem 9.3.7. *Let $S = \langle X; R\rangle$ be a monoid of skew type with $X = \{x_1, \dots, x_n\}$. If S is right non-degenerate, then the following properties hold.*

1. *For each integer i, with $1 \le i \le n$, the set $S_i = \bigcup_{Y, |Y|=i} S_Y$ is an ideal of S. Furthermore,*

$$S_X = S_n \subseteq S_{n-1} \subseteq \cdots \subseteq S_1 \subseteq S.$$

2. $S_1 \setminus S_2 = \bigcup_{x \in X} \prec x \succ$.

3. *S is the union of sets of the form $\{y_1^{a_1} \cdots y_k^{a_k} : a_i \ge 0\}$, where $y_1, \dots, y_k \in X$ and $k \le n$.*

In particular, $\mathrm{GK}(K[S]) \le n$ *for any field K.*

Proof. Let Y be a subset of X. If $x \in X$, then let Z be the largest subset of X such that $xS_Y \subseteq S_Z$. Since X is right non-degenerate, it follows that $|Z| \ge |Y|$. Moreover, if $x \notin Y$, then $|Z| > |Y|$. Consequently, each $S_j = \bigcup_{Y, |Y|=j} S_Y$ is an ideal of S and, clearly,

$$S_X = S_n \subseteq S_{n-1} \subseteq \cdots \subseteq S_1 \subseteq S.$$

Put $S_{n+1} = \emptyset$. Obviously, if $j = |Y|$ then $D_Y = S_Y \setminus S_{j+1}$. So

$$S_j \setminus S_{j+1} = \bigcup_{Z,\, |Z|=j} D_Z, \tag{9.3}$$

a disjoint union.

We claim that if y belongs to a nonempty subset Y of X, with $|Y| = j$, then

$$D_Y \subseteq \prec y \succ (S \setminus S_j). \tag{9.4}$$

First, assume that $j = 1$ and thus $Y = \{y\}$. Suppose the claim is false. So, let $w \in D_Y \setminus \langle y \rangle$. Then $w = y^k xt$, for some $k \geq 1$, $x \in X$, $t \in S$, and $y^k x \in D_Y \setminus \langle y \rangle$. Since S is right non-degenerate and $y \neq x$, there exist distinct elements $u_1, \ldots, u_r = x \in X \setminus \{y\}$ and elements $w_1, \ldots, w_r \in X$ so that $yx = u_1 w_1$, $yu_1 = u_2 w_2$, $\ldots$, $yu_{r-1} = u_r w_r$. Therefore, $y^q x \in u_1 S \cup \cdots \cup u_r S$ for every $q \geq 1$. In particular, $y^k x \in u_i S$ for some i. But, as $y^k x \in D_Y$, this implies that $u_i \in Y = \{y\}$, a contradiction. Hence, we have shown that $D_Y = \prec y \succ$, for $Y = \{y\}$. Second, assume $j > 1$ and let $s \in D_Y$. Let $r \geq 1$ be the maximal integer so that $s = y^r t$, for some $t \in S$. We need to show that $t \notin S_j$. Suppose the contrary, that is, assume $t \in S_Z$ for some $Z \subseteq X$ with $|Z| = |Y|$. If $y \notin Z$ then, since S is right non-degenerate, $yt \in x_i S$ for at least $|Y| + 1$ different indices i. So $yt \in S_{j+1}$. Because S_{j+1} is an ideal of S, this implies that $s \in S_{j+1}$, a contradiction. Consequently, we have $y \in Z$. But then $t \in yS$, which contradicts the maximality of r. Hence, we have shown that $t \notin S_j$, as desired.

The equalities (9.3) and the inclusions (9.4) imply, by induction on the length of elements of S, that S is the (finite) union of sets of the form $\{y_1^{a_1} \cdots y_k^{a_k} \mid a_i \geq 0\}$, where $y_1, \ldots, y_k \in X$ and $k \leq n$.

It remains to show that $\mathrm{GK}(K[S]) \leq n$ for any field K. Clearly, $S \setminus S_2 = \bigcup_{i=1}^n \langle x_i \rangle$. Hence, there are $nm + 1$ elements of S that are words of length at most m and that belong to $S \setminus S_2$. Proceeding by induction on j, assume that the number of elements of $S \setminus S_j$ that are words of length at most m is bounded by a polynomial of degree $j - 1$ in m. Let $Y \subseteq X$ and assume $|Y| = j$. Since $D_Y \subseteq \prec y \succ (S \setminus S_j)$ for $y \in Y$, it is readily verified that the number of elements of D_Y that are words of length at most m are bounded by a polynomial of degree j. As $S_j \setminus S_{j+1}$ is a finite union of such D_Y, the same is true of the elements of the set $S_j \setminus S_{j+1}$. This proves the inductive claim. It follows that the growth of S is bounded by a polynomial of degree not exceeding n, so that $\mathrm{GK}(K[S]) \leq n$. □

The left-right symmetric dual of the set S_i will be denoted by S'_i. So, for each i with $1 \le i \le n$,

$$S'_i = \bigcup_{Y,\, |Y|=i} S'_Y.$$

Of course, if S is a monoid of skew type which is left non-degenerate then, as in Theorem 9.3.7, we obtain that each S'_i is an ideal of S and

$$S'_X = S'_n \subseteq S'_{n-1} \subseteq \cdots \subseteq S_1 \subseteq S.$$

Clearly, $S_1 = S'_1 = S \setminus \{1\}$.

We prove one more crucial technical lemma.

Lemma 9.3.8. *Let $S = \langle X; R\rangle$ be a right non-degenerate monoid of skew type. Let Y be a subset of X. Assume $|Y| = i - 1$ and Z is a subset of Y. If $b \in D_Z$ and $|b| = k$, then*

$$(S_{i-1})^k \cap D_Y \subseteq bS.$$

Furthermore,

$$(S_{i-1})^{k+1} \cap D_Y \subseteq bS_{i-1} \text{ and } (S_{i-1} \cap S'_{i-1})^{k+1} \cap D_Y \subseteq b(S_{i-1} \cap S'_{i-1}).$$

Proof. If $k = 1$, then the statements are obvious. So, assume $k \ge 2$. Write $b = y_k \cdots y_1$, with each $y_j \in X$. Let $q \ge k$ and $a = a_q \cdots a_1 \in D_Y$, with each $a_j \in S_{i-1}$. Since S_i is an ideal of S and $a \in D_Y \subseteq S_{i-1} \setminus S_i$, it is clear that each $a_j \in S_{i-1} \setminus S_i$. Because $b \in D_Z$, with $Z \subseteq Y$, we get $y_k \in Y$, and therefore $a_q \in y_k S$. So $a_q = y_k b_k$ for some $b_k \in S$. Then $a_q a_{q-1} = y_k c_k$, where $c_k = b_k a_{q-1}$. Because $a_q a_{q-1} \in S_{i-1} \setminus S_i$ it is clear that $c_k \in S_{i-1} \setminus S_i$.

Suppose we already have shown that

$$a_q \cdots a_{q-r} = y_k \cdots y_{k-r+1} c_{k-r+1}, \tag{9.5}$$

for some $r \ge 1$ (with $r < k$) and $c_{k-r+1} \in S_{i-1} \setminus S_i$. We claim that $c_{k-r+1} \in y_{k-r} S$. Let $W \subseteq X$ be so that $|W| = i - 1$ and $c_{k-r+1} \in D_W$. Consider the set

$$U = \{x \in X \mid y_k \cdots y_{k-r+1} x \in D_V, \text{ for some } V \subseteq Y\}.$$

Because S is right non-degenerate, an induction argument on r yields that $|U| \le |Y|$. Since the left hand side of equation (9.5) is an initial segment of $a \in D_Y$, it follows that $a_q \cdots a_{q-r} \in D_Y$. Because $c_{k-r+1} \in D_W$, we thus obtain that $W \subseteq U$. So $W = U$. Since $y_k \cdots y_{k-r+1} y_{k-r}$ is an initial segment of $b \in D_Z$ and $Z \subseteq Y$, we also get that $y_{k-r} \in U = W$. Hence $c_{k-r+1} \in y_{k-r}S$. This proves the claim.

Now, write $c_{k-r+1} = y_{k-r}b_{k-r}$ for some $b_{k-r} \in S$. So,

$$a_q \cdots a_{q-r} a_{q-r-1} \quad = \quad y_k \cdots y_{k-r+1} y_{k-r} b_{k-r} a_{q-r-1}.$$

Define $c_{k-r} = b_{k-r}a_{q-r-1}$. Then $c_{k-r} \in S_{i-1} \setminus S_i$.

We have shown that, for any $q \geq k$, $a_q \cdots a_{q-k+1} \in y_k \cdots y_1 S = bS$. If $q = k$, then the first assertion of the lemma follows. On the other hand, if $q = k+1$, then we obtain $a = a_{k+1} \cdots a_1 \in bSa_1 \subseteq bS_{i-1}$. Hence, the second and third assertion follow. □

Proposition 9.3.9. *A right non-degenerate monoid of skew type satisfies the ascending chain condition on right ideals.*

Proof. Let S be a right non-degenerate monoid of skew type. Let $S_n \subseteq S_{n-1} \subseteq \cdots \subseteq S_1 \subseteq S$ be the ideal chain defined in Theorem 9.3.7. Put $S_{n+1} = \emptyset$. We again agree that $S/S_{n+1} = S$. Clearly $S/S_1 = \{1\}$ satisfies the ascending chain condition on right ideals. Hence, to prove the result, it is sufficient to show that if S/S_i (with $1 \leq i \leq n$) satisfies the ascending chain condition on right ideals, then so does S/S_{i+1}.

So, let $1 \leq i \leq n$ and suppose S/S_i satisfies the ascending chain condition on right ideals. From Theorem 9.3.7, we know that $S = \{z_1^{a_1} z_2^{a_2} \cdots z_m^{a_m} : a_j \geq 0\}$, for some $m \geq 1$ and $z_1, \ldots, z_m \in X$ (not all z_j are necessarily different). We claim that S_i/S_{i+1} is finitely generated as a right ideal of S/S_{i+1}. For $1 \leq k \leq m$, put $C_k = \{z_k^{a_k} \cdots z_m^{a_m} | a_j \geq 0\}$. To prove the claim, it is sufficient to show, by induction on $m-k$, that the right ideal of S/S_{i+1} generated by $C_k \cap (S_i \setminus S_{i+1})$ is finitely generated. If $m - k = 0$, then this is clear. So, assume $1 \leq k < m$. Let $B = \{b \in C_{k+1} | z_k^a b \in S_i$ for some $a > 0\}$. If $y \in (C_k \setminus C_{k+1}) \cap S_i$ then $y \in z_k^a B$ for some positive integer a. Let a be maximal with this property and write $y = z_k^a t$ with $t \in B$. If $t \in S_Z$ with $|Z| = i$ then, by the maximality of a, we get that $z_k \notin Z$. But, as S is right non-degenerate, we then obtain that $z_k^a t \in x_j S$ for at least $i+1$ different indices j. So $y = z_k^a t \in S_{i+1}$. On the other hand, if $t \in S \setminus S_i$ then $y \in z_k^a (B \cap (S \setminus S_i))$ for some a. Because, by assumption, S/S_i satisfies the ascending chain condition on right ideals, we also have that $(B \cap (S \setminus S_i))S \subseteq b_1 S \cup \cdots \cup b_r S$ for some $b_j \in B \cap (S \setminus S_i)$. Since S_i is an ideal of S, it follows that

$$
\begin{aligned}
&((C_k \setminus C_{k+1}) \cap S_i)\, S \,\cup\, S_{i+1} \\
&\subseteq \bigcup_{t\geq N}\bigcup_{j=1}^{r} z_k^t b_j S \,\cup\, \bigcup_{j=1}^{N-1}(z_k^j B\cap S_i)S \,\cup\, S_{i+1} \\
&\subseteq (C_k \cap S_i)\, S \,\cup\, S_{i+1},
\end{aligned}
$$

where N is chosen so that $z_k^N b_j \in S_i$ for $j = 1, \ldots, r$. Hence

$$
\begin{aligned}
&(C_k \cap S_i)\, S \cup S_{i+1} \\
&= \bigcup_{t\geq N}\bigcup_{j=1}^{r} z_k^t b_j S \,\cup\, \bigcup_{j=1}^{N-1}(z_k^j B\cap S_i)S \,\cup\, (C_{k+1}\cap S_i)S \,\cup\, S_{i+1}.
\end{aligned} \tag{9.6}
$$

By the inductive hypothesis, $C_{k+1} \cap S_i$ generates a finitely generated ideal modulo S_{i+1}. Clearly, for $1 \leq j \leq N-1$, we have that $z_k^j B \cap S_i = z_k^j (B \cap S_i) \cup (z_k^j (B \setminus S_i) \cap S_i)$. Since S_i is an ideal, we get from the definition of C_{k+1} that $C_{k+1} \cap S_i = B \cap S_i$. Hence, $B \cap S_i$, and thus also $z_k^j (B \cap S_i)$ generates a finitely generated right ideal modulo S_{i+1}. On the other hand, $(z_k^j (B \setminus S_i) \cap S_i) S = z_k^j (\bigcup_b bS)$, where the union runs through all $b \in (B \setminus S_i)$ so that $z_k^j b \in S_i$. Because, by assumption, S/S_i satisfies the ascending chain condition, we get that this union $\bigcup_b bS$, and thus also $z_k^j (B \setminus S_i) \cap S_i$, is a finitely generated right ideal of S. Hence we have shown that the right ideal $\bigcup_{j=1}^{N-1} (z_k^j B \cap S_i) S$ is finitely generated modulo S_{i+1}.

Next we show that the double union, in (9.6), also is a finitely generated right ideal of S. Because of Proposition 9.3.4, we know that S has the over-jumping property. Consequently, for every j, there exist $w_j \in S$ and a positive integer q_j such that

$$b_j w_j = z_k^{q_j} b_j.$$

Hence $z_k^{q_j N} b_j \subseteq b_j S$ and thus $z_k^{p+q_j N} b_j \subseteq z_k^p b_j S$, for every $p \geq 0$. It follows that the right ideal $\bigcup_{t\geq N} z_k^t b_j S = \bigcup_{t=N}^{N+q_j} z_k^t b_j S$ is finitely generated, as claimed.

The above shows that $C_k \cap S_i$ generates a finitely generated right ideal modulo S_{i+1}. So, we proved the claim that S_i/S_{i+1} is a finitely generated right ideal of S/S_{i+1}, say

$$S_i \;=\; s_1 S \cup \cdots \cup s_q S \cup S_{i+1}, \tag{9.7}$$

for some $s_1, \ldots, s_q \in S_i$.

We are now in a position to prove that S/S_{i+1} satisfies the ascending chain condition under the assumption that S/S_i satisfies this property; and then the result follows. We show this by contradiction. So, suppose there is an infinite sequence $a_1, a_2, \ldots$ in $S \setminus S_{i+1}$ yielding a properly increasing set of right ideals

$$a_1S \subset a_1S \cup a_2S \subset \cdots \subset a_1S \cup \cdots \cup a_kS \subset \cdots.$$

Since S is the union of finitely many sets D_Y, with $Y \subseteq X$, we may assume that all $a_j \in D_Y$ for some Y. As, by assumption, S/S_i has the ascending chain condition on right ideals, it follows that $D_Y \subseteq S_i \setminus S_{i+1}$ and thus $|Y| = i$. Let $p = |a_1|$. From Lemma 9.3.8, we obtain that $a_j \notin S_i^p$. It then easily follows from (9.7) that the sequence (a_j) has a subsequence contained in $s_{k_1} \cdots s_{k_l}(S \setminus S_i)$, for some $l < p$ and some k_j. However, this leads to a proper ascending chain of right ideals in S/S_i, and this on its turn is in contradiction with the fact that S/S_i satisfies the ascending chain condition on right ideals. $\square$

We finish this section with another characterization of monoids of I-type within the class of monoids of skew type that are right non-degenerate.

Theorem 9.3.10. *Let $S = \langle x_1, x_2, \ldots, x_n; R\rangle$ be a right non-degenerate monoid of skew type. For each i, with $1 \leq i \leq n$, denote by $\sigma_i \in \operatorname{Sym}_n$ the permutation defined by $f_{x_i}(x_j) = x_{\sigma_i(j)}$. The following conditions are equivalent.*

1. *S is of I-type.*
2. *$\sigma_i \circ \sigma_{\sigma_i^{-1}(j)} = \sigma_j \circ \sigma_{\sigma_j^{-1}(i)}$ for all i, j.*
3. *For every defining relation $x_ix_j = x_kx_l$ of S, we have $\sigma_i \circ \sigma_j = \sigma_k \circ \sigma_l$.*

Proof. As before, we denote by $r_k \colon X^3 \longrightarrow X^3$, for $k = 1, 2$, the mappings $r_1 = r \times \mathrm{id}_X$ and $r_2 = \mathrm{id}_X \times r$. Let $1 \leq i, j \leq n$ with $i \neq j$. By definition of σ_i, we have that $r(x_i, x_j) = (x_{\sigma_i(j)}, x_m)$, for some m with $1 \leq m \leq n$. Because $r^2 = \mathrm{id}_{X^2}$, we then have that $r(x_{\sigma_i(j)}, x_m) = (x_i, x_j)$. So, by definition of $\sigma_{\sigma_i(j)}$, we get that $\sigma_{\sigma_i(j)}(m) = i$ and thus $m = \sigma_{\sigma_i(j)}^{-1}(i)$. So we have shown that

$$r(x_i, x_j) = (x_{\sigma_i(j)}, x_{\sigma_{\sigma_i(j)}^{-1}(i)}).$$

Consequently, for $1 \leq i, j, k \leq n$,

$$\begin{aligned}(r_1 \circ r_2 \circ r_1)(x_i, x_j, x_k) &= (r_1 \circ r_2)(x_{\sigma_i(j)}, x_{\sigma_{\sigma_i(j)}^{-1}(i)}, x_k) \qquad (9.8)\\ &= r_1(x_{\sigma_i(j)}, x_{\sigma_{\sigma_{\sigma_i(j)}^{-1}(i)}(k)}, x_{\sigma_{\sigma_{\sigma_{\sigma_i(j)}^{-1}(i)}(k)}^{-1}(\sigma_{\sigma_i(j)}^{-1}(i))})\end{aligned}$$

and

$$
\begin{aligned}
(r_2 \circ r_1 \circ r_2)(x_i, x_j, x_k) &= (r_2 \circ r_1)(x_i, x_{\sigma_j(k)}, x_{\sigma^{-1}_{\sigma_j(k)}(j)}) \\
&= r_2(x_{\sigma_i(\sigma_j(k))}, x_{\sigma^{-1}_{\sigma_i(\sigma_j(k))}(i)}, x_{\sigma^{-1}_{\sigma_j(k)}(j)}). \quad (9.9)
\end{aligned}
$$

From Theorem 8.1.4 and Corollary 8.2.4, recall that S is of I-type if and only if r yields a set theoretic solution of the Yang-Baxter equation, that is, $r_1 \circ r_2 \circ r_1 = r_2 \circ r_1 \circ r_2$. Therefore, if S is of I-type, then by (9.8) and (9.9), we have

$$\sigma_{\sigma_i(j)}(\sigma_{\sigma^{-1}_{\sigma_i(j)}(i)}(k)) = \sigma_i(\sigma_j(k)),$$

for all i, j, k. Thus

$$\sigma_{\sigma_i(j)} \circ \sigma_{\sigma^{-1}_{\sigma_i(j)}(i)} = \sigma_i \circ \sigma_j, \qquad (9.10)$$

for all i, j. Put $j' = \sigma_i(j)$. Then we can write (9.10) as

$$\sigma_{j'} \circ \sigma_{\sigma^{-1}_{j'}(i)} = \sigma_i \circ \sigma_{\sigma^{-1}_i(j')},$$

for all i, j'. Therefore (2) is a consequence of (1).

Suppose that

$$\sigma_i \circ \sigma_{\sigma^{-1}_i(j)} = \sigma_j \circ \sigma_{\sigma^{-1}_j(i)},$$

for all i, j. We will prove that r yields a set theoretic solution of the Yang-Baxter equation and thus S is of I-type. Because of (9.8) and (9.9), it is sufficient to prove the following equalities.

(a) $\sigma_{\sigma_i(j)}(\sigma_{\sigma^{-1}_{\sigma_i(j)}(i)}(k)) = \sigma_i(\sigma_j(k))$.

(b) $\sigma^{-1}_{\sigma_{\sigma_i(j)}(\sigma_{\sigma^{-1}_{\sigma_i(j)}(i)}(k))}(\sigma_i(j)) = \sigma_{\sigma^{-1}_{\sigma_i(\sigma_j(k))}(i)}(\sigma^{-1}_{\sigma_j(k)}(j))$.

(c) $\sigma^{-1}_{\sigma_{\sigma^{-1}_{\sigma_i(j)}(i)}(k)}(\sigma^{-1}_{\sigma_i(j)}(i)) = \sigma^{-1}_{\sigma_{\sigma^{-1}_{\sigma_i(\sigma_j(k))}(i)}(\sigma^{-1}_{\sigma_j(k)}(j))}(\sigma^{-1}_{\sigma_i(\sigma_j(k))}(i))$.

The equality (a) follows from

$$\sigma_{j'} \circ \sigma_{\sigma^{-1}_{j'}(i)} = \sigma_i \circ \sigma_{\sigma^{-1}_i(j')},$$

with $j' = \sigma_i(j)$. By (a), the equality (b) is equivalent to

$$\sigma^{-1}_{\sigma_i(\sigma_j(k))}(\sigma_i(j)) = \sigma_{\sigma^{-1}_{\sigma_i(\sigma_j(k))}(i)}(\sigma^{-1}_{\sigma_j(k)}(j)), \qquad (9.11)$$

and the latter follows from our assumption

$$\sigma_l^{-1} \circ \sigma_i = \sigma_{\sigma_l^{-1}(i)} \circ \sigma^{-1}_{\sigma_i^{-1}(l)},$$

with $l = \sigma_i(\sigma_j(k))$. By (a), the equality (c) is equivalent to

$$\sigma^{-1}_{\sigma^{-1}_{\sigma_i(j)}(\sigma_i(\sigma_j(k)))}(\sigma^{-1}_{\sigma_i(j)}(i)) = \sigma^{-1}_{\sigma_{\sigma_i(\sigma_j(k))}^{-1}(i)^{(\sigma^{-1}_{\sigma_j(k)}(j))}}(\sigma^{-1}_{\sigma_i(\sigma_j(k))}(i)).$$

The latter follows from our assumption

$$\sigma^{-1}_{\sigma_{j'}^{-1}(l)} \circ \sigma_{j'}^{-1} = \sigma^{-1}_{\sigma_l^{-1}(j')} \circ \sigma_l^{-1},$$

with $l = \sigma_i(\sigma_j(k))$ and $j' = \sigma_i(j)$. Hence r yields a set theoretic solution of the Yang-Baxter equation and (2) implies (1).

Finally, notice that $x_i x_p = x_j x_q$ if and only if $\sigma_i(p) = j$ and $\sigma_j(q) = i$. The latter is equivalent to $\sigma_i^{-1}(j) = p$ and $\sigma_j^{-1}(i) = q$. Hence, saying that $\sigma_i \sigma_p = \sigma_j \sigma_q$ whenever $x_i x_p = x_j x_q$ is equivalent to saying that $\sigma_i \sigma_{\sigma_i^{-1}(j)} = \sigma_j \sigma_{\sigma_j^{-1}(i)}$. So conditions (2) and (3) are equivalent. This completes the proof of the theorem. □

9.4 Algebras of non-degenerate monoids of skew type

In this section, we will prove that semigroup algebras of monoids S of skew type that are non-degenerate are Noetherian algebras that have finite Gelfand-Kirillov dimension. Hence, because of Theorem 5.1.6, these algebras also satisfy a polynomial identity. To prove this, we will rely on the fact that we already know that S satisfies the ascending chain condition on one-sided ideals (Proposition 9.3.9) and we will make use of Theorem 5.3.7. In particular, we will show that S has an ideal chain with factors that are either nilpotent or a uniform subsemigroup of a Brandt semigroup.

The assumption that S is left and right non-degenerate is essential. Indeed, if S is the monoid considered in Example 9.1.1, then S is a right non-degenerate monoid of skew type that is not left non-degenerate. Furthermore S satisfies the ascending chain condition on right ideals but its semigroup algebra is neither left nor right Noetherian.

Lemma 9.4.1. *Let $S = \langle X; R\rangle$ be a monoid of skew type that is non-degenerate. If $|X| = n$ then, for all $1 \leq k \leq n$,*

$$S_k^k \subseteq S'_k \text{ and } (S'_k)^k \subseteq S_k.$$

In particular, $I_k = S_k \cap S'_k$ *is an ideal of* S *such that* S_k/I_k *is nilpotent.*

Proof. Because S is non-degenerate, Theorem 9.3.7 yields that both S_k and S'_k are ideals of S, for $1 \leq k \leq n$. Since $S_1 = S'_1$, the result is obvious for $k = 1$. So assume that $k > 1$. Let $1 \neq a \in S$ and $b \in S_k$. Let $Y \subseteq X$ be maximal such that $a \in S'_Y$. If $|Y| < k$, then there exists $x \in X$ such that $b \in xS$ and $x \notin Y$. Hence, we may write $b = xc$ for some $c \in S$. Obviously $ax \in Sx$. But, as S is left non-degenerate, $x \notin Y$ and $a \in S'_Y$, the element ax also is contained in $|Y|$ different left ideals of the form Sw, with $x \neq w \in X$. Therefore $ax \in S'_Z$, for some $Z \subseteq X$ with $|Z| > |Y|$. Hence $ab = axc \in S'_{|Z|}$. By induction, it easily follows that $S_k^k \subseteq S'_k$, as desired. As $S_k^k \subseteq S_k \cap S'_k = I_k$, also the last statement follows. □

Theorem 9.4.2. *Let* K *be a field. If* $S = \langle X; R\rangle$ *is a non-degenerate monoid of skew type, then* $K[S]$ *is a Noetherian algebra that satisfies a polynomial identity. In particular,* $K[S]$ *embeds in a matrix ring over a field.*

Furthermore, for some positive integer N*, the ideal* S_X^N *is a cancellative subsemigroup of* S*.*

Proof. From Theorem 9.3.7, we know that $\mathrm{GK}(K[S])$ is finite. Because of Proposition 9.3.9, we also know that S satisfies the ascending chain condition on one-sided ideals. In view of Theorem 5.3.7 and its dual, to prove that $K[S]$ is Noetherian, it is sufficient to show that S has an ideal chain with each factor either nilpotent or a uniform subsemigroup of a Brandt semigroup.

Let $|X| = n$. Put $S_{n+1} = S'_{n+1} = \emptyset$ and again adopt the convention $S/\emptyset = S$. By induction on i, we will prove that S/S_i has an ideal chain of the desired type. The case $i = n+1$ then yields the result. We know, from Theorem 9.3.7, that $S \setminus S_2$ is the disjoint union of all $D_{\{x_i\}} = \prec x_i \succ$ and $\{1\}$. So S/S_2 has an ideal chain with abelian 0-cancellative factors. Hence, it has a chain of the type desired. So, now assume that, for some $i \geq 3$, the semigroup S/S_{i-1} has an ideal chain with factors either nilpotent or a uniform subsemigroup of a Brandt semigroup. We need to prove that this property also holds for S/S_i.

Because S satisfies the ascending chain condition on ideals, there exists an ideal J' of S so that $S_i \subseteq J'$ and J'/S_i is the maximal nil ideal of S/S_i. Put $I = S_{i-1} \cap S'_{i-1}$ and $J = J' \cap I$. Note that, because of Lemma 9.4.1, $(S'_i)^i \subseteq S_i \subseteq J$. Hence $S'_i \subseteq J'$. So, in S, we get the ideal chain

$$S_i \cap I \subseteq (S_i \cup S'_i) \cap I \subseteq J \subseteq I \subseteq S_{i-1}.$$

Again, because of Lemma 9.4.1, the first and the last Rees factors are nilpotent. Also, because of Theorem 2.4.10, J/S_i is nilpotent. The main point of the rest of the proof is to show that $J \subseteq I$ can be refined to an ideal chain of S with the required properties.

For subsets Y and W of X, put $I_{YW} = D_Y \cap D'_W$ and $I_Y = I_{YY}$. If $|Y| = i-1$, $D'_Y \neq \emptyset$ and $x \in X \setminus Y$, then $D'_Y x$ is contained in the left ideals $Sy_i x$, with $Y = \{y_1, \ldots, y_{i-1}\} \subseteq X$. Since S is left non-degenerate, it follows that $y_i x = x'_i y'_i$ with $x, y_i \in X$, $y'_i \neq x$, and $y'_i \neq y'_j$ if $i \neq j$. So, $D'_Y x \subseteq (Sy'_1 \cap \cdots \cap Sy'_{i-1} \cap Sx)$ and thus $D'_Y x \subseteq S'_i$. It then easily follows that $D'_Y D_Z \subseteq S'_i$, for every $Y, Z \subseteq X$ of cardinality $i-1$ with $Z \neq Y$, provided that $D_Y \neq \emptyset$ and $D'_Z \neq \emptyset$. Hence we get

$$I \setminus (S_i \cup S'_i) = (S_{i-1} \cap S'_{i-1}) \setminus (S_i \cup S'_i) \quad = \quad \bigcup_{Y,Z} I_{YZ}, \tag{9.12}$$

a disjoint union, and

$$I_{YZ}\, I_{Y'Z'} \quad \subseteq \quad \begin{cases} S_i \cup S'_i & \text{if } Z \neq Y', \\ I_{YZ'} \cup S_i \cup S'_i & \text{if } Z = Y'. \end{cases} \tag{9.13}$$

Since $S_i \cup S'_i \subseteq J'$, it then easily follows that each $D_Y \cup J'$ (respectively $D_Z \cup J'$) is a right ideal (respectively left ideal) of S. Hence $(D_Y \cap I) \cup J$ (respectively $(D_Z \cap I) \cup J$) is a right ideal (respectively left ideal) of S.

Let $Y \subseteq X$ with $|Y| = i-1$. We consider, separately, the following two mutually exclusive cases.

Case 1 : Either $I_Y = \emptyset$ or there exist $b \in D'_Y$ and $x \in D_Y$ such that $bx \in J$.

Case 2 : $I_Y \neq \emptyset$ and $D'_Y D_Y \subseteq (S_{i-1} \cap S'_{i-1}) \setminus J$, so, in particular, I_Y is a subsemigroup of $(S_{i-1} \cap S'_{i-1}) \setminus J$.

In Case 1 we claim that $D_Y \cap I$ and $D'_Y \cap I$ are contained in J. Because of (9.13), it readily is verified that $(D_Y \cap I)^2 = (\bigcup_{Z,|Z|=i-1} I_{YZ} \cap I)^2 \subseteq (I_{YY}^2 \cup S_i \cup S'_i) \cap I \subseteq (I_Y \cup S_i \cup S'_i) \cap I$. So, if $I_Y = \emptyset$ we get that $(D_Y \cap I)^2 \subseteq J$. Similarly $(D'_Y \cap I)^2 \subseteq J$. Because both $(D_Y \cap I) \cup J$ and $(D'_Y \cap I) \cup J$ are one-sided ideals, it follows that $D_Y \cap I$, $D'_Y \cap I \subseteq J$, as desired.

So assume that $I_Y \neq \emptyset$ and that there exist $b \in D'_Y$ and $x \in D_Y$ with $bx \in J$. Let q be the maximum of the lengths of b and x. Then, by Lemma 9.3.8 and its right-left dual, we get that

$$I_Y^{2q} \subseteq (D'_Y)^q D_Y^q \subseteq (Sb)(xS) \cup S_i \cup S'_i \subseteq J'.$$

So, I_Y is nilpotent modulo J'. Because of (9.13), we thus also get that the left ideal $(D'_Y \cap I) \cup J$ is nil modulo J. Consequently, $D'_Y \cap I \subseteq J$. Similarly, $D_Y \cap I \subseteq J$. This proves the claim in Case 1.

Note that it follows that if $|Y| = i-1$ and $D_Y \cap I \not\subseteq J$ or $D'_Y \cap I \not\subseteq J$ then Y satisfies the conditions listed in Case 2.

In Case 2, we will show that $I^r \cap I_Y$ is a cancellative semigroup for some $r \geq 1$. In order to prove this we introduce some notation and develop some machinery. For $a, b \in S$, we write

$$a \tau b \quad \text{if } az = bz \notin J \text{ for some } z \in I = S'_{i-1} \cap S_{i-1}.$$

Since S is a monoid of skew type, it is clear that if $a\tau b$ then a and b have the same length. Hence, for a given $a \in S$, there are only finitely many $b \in S$ so that $a\tau b$.

Let $\mathcal{A}$ be the set of all elements $d \in I$ such that every proper initial segment of d is not in I. In other words, $\mathcal{A}$ is the (unique) minimal set of generators of I as a right ideal of S. Because of the ascending chain condition on right ideals in S, we know that $\mathcal{A}$ is a finite set. Let $\{a_j, b_j\}$, with $a_j \in \mathcal{A}$ and $a_j \neq b_j$, be all the distinct pairs such that $a_j \tau b_j$, where $1 \leq j \leq q$.

Let $a\tau b$ for some $a \in I, b \in S$; so $az = bz \notin J$ for some $z \in I$. Let $Y, Z \subseteq X$ be such that $a \in D'_Y$ and $b \in D'_Z$. Because of (9.13), we obtain that $z \in D_Y$. Let $t \in Z$ and thus $b = t't$ for some $t' \in S$. Because S_i is an ideal, and because $bz \notin J$ and thus $bz \notin S_i$, it is clear that $tz \notin S_i$. As S is a right non-degenerate monoid of skew type, this implies that $t \in Y$. Hence, we have shown that $Z \subseteq Y$. In particular, $a, b \in Sx$ for every $x \in Z$. Thus, there exists an element $s \in S$ so that $a = a's$ and $b = b's$, for some $a' \in I, b' \in S$, and such that if $a' = a''x, b = b''x$ for $x \in X, a'', b'' \in S$, then $a'' \notin I$. Note that $a'sz = az = bz = b'sz$. Because $a'(sz) = az \notin J$ and $z \in I$, we have that $a'\tau b'$. The definition of $\mathcal{A}$ and the argument above applied to $a' \in I$ and $b' \in S$ also yield that $a' \in \mathcal{A}$.

So, we have shown that for every $a \in I$ and $b \in S$, with $a\tau b$ and $a \neq b$, there exists $s \in S$ so that $a = a_j s$ and $b = b_j s$. For each j, with $1 \leq j \leq q$, let $z_j \in I$ be such that $a_j z_j = b_j z_j \notin J$. Let $N = \max\{|z_j| \mid j = 1, \ldots, q\}$.

For a subset Y of X that satisfies Case 2, we now make the following claim. If $a, b \in D'_Y \cap I$, $a\tau b$ and $a \neq b$ then

$$at = bt \notin J \quad \text{for every } t \in I^N \cap I_Y. \tag{9.14}$$

In order to prove this, write $a = a_j s$ and $b = b_j s$, for some j and some $s \in S$. Since $a_j \in I \setminus J$, there exist $W, Z \subseteq X$, each of cardinality $i-1$, so that $a_j \in I_{WZ}$. As $a \in D'_Y \setminus J$, we thus get that $a \in I_{WY}$. Moreover, (9.13) and $a_j z_j = b_j z_j \notin J$ yield that $z_j \in I_{ZV}$ for some $V \subseteq X$. Now $a_j s = a \in D'_Y$ implies that $aI_Y = a_j s I_Y$ and, because Y satisfies Case 2, the former does not intersect J. In particular, $sI_Y \subseteq I_{ZY}$. Let $t \in I^N \cap I_Y$. Then, since $N \geq |z_j|$, we get, using Lemma 9.3.8, that $st \in s(I^N \cap I_Y) \subseteq D_Z \cap I^{|z_j|} \subseteq z_j S$. So $st = z_j u$ for some $u \in S$, and thus $at = a_j st = a_j z_j u$ and $at \in I_{WY} I_Y \subseteq D'_Y D_Y$. As Y satisfies Case 2, it thus follows that $at \notin J$. Also $bt = b_j st = b_j z_j u = a_j z_j u = at$. So $at = bt$. This proves the claim.

For every Y that satisfies Case 2, choose $c_Y \in I^N \cap I_Y$. Write $r = \max\{|c_Y| + 1\}$ and let $T = I^r$. Then $T/(J \cap T) = I^r/(J \cap I^r)$ and

$$T/(J \cap T) = \bigcup_{Y,W} T_{YW} \cup \{\theta\},$$

(a disjoint union) where

$$T_{YW} = (I_{YW} \cap T) \setminus J$$

and Y, W run through the set of $i-1$-element subsets of X. Because of (9.13),

$$T_{YZ}\, T_{Y'Z'} \subseteq \begin{cases} J \cap T & \text{if } Z \neq Y', \\ T_{YZ'} \cup (J \cap T) & \text{if } Z = Y'. \end{cases} \tag{9.15}$$

Put $T_Y = T \cap I_Y$. If $a \in T_{YZ}$, then $a \in (I^r \cap D'_Z) \setminus J$ and thus, by the comment made after the proof of Case 1, Z satisfies the conditions given in Case 2. If, furthermore, $b \in T_{ZW}$ then $b \in (I^r \cap D_Z) \setminus J$. Hence $ab \in (D'_Z D_Z \cap I^r) \setminus J$. So, because of (9.15), $ab \in T_{YW}$. Consequently,

$$\text{if } T_{YZ} \neq \emptyset \text{ and } T_{ZW} \neq \emptyset \text{ then } T_{YZ}\, T_{ZW} \subseteq T_{YW}, \tag{9.16}$$

in particular $T_{YW} \neq \emptyset$.

Let A be a maximal subsemigroup of $T/(J \cap T)$ of the form

$$A = \bigcup_{Y,W \in \mathcal{P}} T_{YW} \cup \{\theta\},$$

where $\mathcal{P}$ is a set of $(i-1)$-element subsets of X such that every T_{YW} is not empty. Let $Y \in \mathcal{P}$. Suppose that $T_{YW} \neq \emptyset$, for some $W \notin \mathcal{P}$ of cardinality $i-1$. Then $\emptyset \neq T_{ZY} T_{YW} \subseteq T_{ZW}$, for every $Z \in \mathcal{P}$. Clearly $T_W \neq \emptyset$ because, again by the comment made above, W satisfies Case 2. Furthermore, if $T_{WY'} \neq \emptyset$ for some $Y' \in \mathcal{P}$, then $T_{YW} T_{WY'} \subseteq T_{YY'}$. But this easily implies that $\bigcup_{Y,W\in\mathcal{P}\cup\{W\}} T_{YW} \cup \{\theta\}$ is a semigroup with each $T_{YW} \neq \emptyset$, in contradiction with the maximality of $\mathcal{P}$. It follows that

$$B = \bigcup_{Y\in\mathcal{P},\, V\notin\mathcal{P}} T_{YV} \cup \{\theta\}$$

is a right ideal of $T/(J\cap T)$. Now, also note that if $b \in B$ and $\theta \neq bs \in A$, for some $s \in S/J$, then $\theta \neq bsx \in A$ for some $x \in A$. However, $sx \in T/(J\cap T)$ implies that $bsx \in B$. Hence $\theta \neq bsx \in A \cap B$, a contradiction. This shows that B is a right ideal of S/J. From (9.15), it follows that B is nilpotent and this contradicts with the definition of J. Consequently,

$$T_{YV} = \emptyset \text{ if } V \notin \mathcal{P},\ |V| = i-1,\ Y \in \mathcal{P},$$

and similarly

$$T_{VY} = \emptyset \text{ if } V \notin \mathcal{P},\ |V| = i-1,\ Y \in \mathcal{P}.$$

Hence one obtains a decomposition

$$T/(J\cap T) = A_1 \cup \cdots \cup A_k,$$

where each A_i is a subsemigroup of the type as A above. This union is 0-disjoint, each A_i is an ideal of S/J and

$$A_iA_j \;=\; \theta \text{ if } i \neq j. \tag{9.17}$$

Fix some $A = A_i$, say $i = 1$. Let Y be such that $T_Y \subseteq A$. We now prove that if $a, b \in T \cap D'_Y$ satisfy $az = bz \notin J$ for some $z \in T$, then $a = b$. In order to prove this, we first note that it follows from (9.15) that $z \in D_Y \setminus J$. Hence, (9.17) yields that $z \in A$. Because of (9.16), $z(T\cap D'_Y) \cap T_Y \neq \emptyset$. So, we may assume that $z \in T_Y$. Now consider the elements $c = c_Y \in I^N \cap I_Y$, introduced above. Since $a \in I^r \cap D'_Y$ and $r \geq |c|+1$, the dual of Lemma 9.3.8 yields that $a = a'c$ and $b = b'c$ for some $a', b' \in I$. Because $c \in I_Y$ and $a', b' \in I$, the inclusions stated in (9.13) also imply that $a', b' \in D'_Y$. Consequently, $a'cz = b'cz \notin J$ and $a'\tau b'$. Hence, by (9.14), $a = a'c = b'c = b$, as desired. Since the previous is valid for any Y with $T_Y \subseteq A$, we have shown that in A the requirement $az = bz \neq \theta$ implies that $a = b$. By a symmetric argument, we thus obtain the following property

$$\text{if } a, b, z \in A, \text{ and } az = bz \neq \theta \text{ or } za = zb \neq \theta \text{ then } a = b. \tag{9.18}$$

In particular, A is weakly 0-cancellative and if Y satisfies Case 2 then $T_Y = I_Y \cap T$ is a cancellative semigroup.

Let $\mathcal{P} = \{Y_1, \ldots, Y_m\}$. Recall that each Y_i satisfies the conditions listed in Case 2 and thus I_{Y_i} is a subsemigroup of $I \setminus J$. Hence it follows from Lemma 9.3.8 that any two right ideals of each I_{Y_i} intersect nontrivially. Using (9.15), one then obtains that the same holds for right ideals in the semigroup T_{Y_i}. A symmetric argument yields that any two left ideals of T_{Y_i} also intersect nontrivially. Consequently, each T_{Y_i} is a cancellative right and left Ore semigroup. It follows that the weakly 0-cancellative semigroup A has a combinatorial Brandt congruence with classes the sets $T_{Y_iY_j}$ and $\{\theta\}$. Furthermore the non-null classes are cancellative and satisfy the Ore conditions. Theorem 2.1.6, therefore, yields that A has a Brandt semigroup of quotients with a maximal subgroup equal to the group of quotients of T_{Y_i}.

Hence $A = A_1$, and thus each A_i is a uniform subsemigroup of a Brandt semigroup. Clearly, A_1 is isomorphic with $(T/(J \cap T))/(A_2 \cup \cdots \cup A_k)$. Hence it follows that $T/(J \cap T)$ has an ideal chain (the ideals in the chain are determined by the A_i) whose factors are uniform subsemigroups of Brandt semigroups. Consider the ideal chain

$$S_i \subseteq S_i \cup (J \cap T) \subseteq S_i \cup T \subseteq S_{i-1} \subseteq S.$$

We know that $S_i \cup (J \cap T)$ is nilpotent modulo S_i and S_{i-1} is nilpotent modulo $S_i \cup T$. The factor $(S_i \cup T)/(S_i \cup (J \cap T))$ is naturally identified with $T/(J \cap T)$, because $S_i \cap T \subseteq J$. It follows that S/S_i has an ideal chain with each factor either nilpotent or a uniform subsemigroup of a Brandt semigroup. This completes the inductive step, and thus we have shown that $K[S]$ is Noetherian. Because of Theorem 5.1.6, it also follows that $K[S]$ satisfies a polynomial identity. Theorem 3.5.2 then yields that $K[S]$ embeds in a matrix ring over a field.

Finally, applying the above Case 2 to the set $I_X = S_X = D_X \cap D'_X = S_n \cap S'_n$, we obtain that, for some positive integer N, the set S_X^N is a cancellative semigroup. Hence the second part of the statement of the result follows. □

In the last paragraph of the proof we have shown that each A_i is an order in a Brandt semigroup. For this we made use of Theorem 2.1.6. One could have used instead classical localization techniques at the semigroup algebra level.

Lemma 9.4.1 says that, for a monoid S of skew type, there exists a positive integer m so that $S_i^m \subseteq S'_i$ for each i. We now show that for monoids of I-type we get more.

Proposition 9.4.3. *Let $S = \langle x_1, \ldots, x_n \rangle$ be a monoid of I-type, then $S_i = S'_i$, for every integer i with $1 \leq i \leq n$. Furthermore, the ideals S_i are semiprime.*

Proof. Because of the results in Section 8.2, we know that $S = \{(a, \psi(a)) \mid a \in \mathrm{FaM}_n\} \subseteq \mathrm{FaM}_n \rtimes \mathrm{Sym}_n$, with $\psi : \mathrm{FaM}_n \longrightarrow \mathrm{Sym}_n$. Let $\{u_1, \ldots, u_n\}$ be a generating set for the free abelian monoid $A = \mathrm{FaM}_n$ of rank n. Thus, we may assume that $x_i = (u_i, \psi(u_i))$. Let $s = (a, \psi(a)) \in S$. It readily is verified that $s \in x_i S$ if and only if $a \in u_i A$.

We also know (see the proof of Corollary 8.2.4) that the mapping

$$S^{opp} \longrightarrow \mathrm{FaM}_n^{-1} \rtimes \mathrm{Sym}_n$$

defined by mapping $(a, \psi(a))$ onto $(\psi(a)^{-1}(a^{-1}), \psi(a)^{-1}) = (a, \psi(a))^{-1}$ is an embedding so that the natural projection $S^{opp} \longrightarrow \mathrm{FaM}_n^{-1}$ is bijective. Let $B = \mathrm{FaM}_n^{-1}$, the free abelian monoid of rank n with generating set $\{u_1^{-1}, \ldots, u_n^{-1}\}$. For $1 \leq i \leq n$, put $y_i = x_i^{-1} = (\psi(u_i)^{-1}(u_i^{-1}), \psi(u_i)^{-1})$. So $S^{opp} \cong S^{-1} = \langle y_1, \ldots, y_n \rangle$.

Clearly $s = (a, \psi(a)) \in S x_i$ if and only if $s \in x_i S^{opp}$, or equivalently $s^{-1} = (\psi(a)^{-1}(a^{-1}), \psi(a)^{-1}) \in y_i S^{-1}$. By the above remark, the latter means that $\psi(a)^{-1}(a^{-1}) \in \psi(u_i^{-1})(u_i^{-1}) B$. Of course, this can be rewritten as $a \in \psi(a)\left(\psi(u_i^{-1})(u_i)\right) A$. Using again the above remark, we can formulate this condition as $s \in (u(i), \psi(u(i)) S$ with $u(i) = \psi(a)\left(\psi(u_i^{-1})(u_i)\right)$. Since $x_i^{-1} = (\psi(u_i^{-1})(u_i^{-1}), \psi(u_i)^{-1})$, we get that $\psi(u_i^{-1})(u_i^{-1}) \neq \psi(u_j^{-1})(u_j^{-1})$ for $i \neq j$. Hence, it follows that $u(i) \neq u(j)$ for $i \neq j$. Consequently, we have shown that if $s \in S_i'$ then $s \in S_i$. By symmetry, the reverse inclusion follows. This proves the first part of the result.

To prove the second part, put $w = n!$ and let $(a, \psi(a)) \in S \setminus S_i$. So $(a, \psi(a))$ is left divisible precisely by the k generators $(u_{i_j}, \psi(u_{i_j}))$, with $1 \leq j \leq k$, $k < i$ and each $m_j > 0$. Write $a = u_{i_1}^{m_1} \cdots u_{i_k}^{m_k} \in \mathrm{FaM}_n$. Let $b_k = \phi(a)^{-1}(u_k^{m_k(w-1)})$. Then

$$(a, \psi(a))(b_k, \psi(b_k)) = (u_{i_1}^{m_1} \cdots u_{i_{k-1}}^{m_{k-1}} u_{i_k}^{m_k w}, \psi(u_{i_1}^{m_1} \cdots u_{i_{k-1}}^{m_{k-1}} u_{i_k}^{m_k w})).$$

Continuing this process we get elements $b_k, \ldots, b_1 \in \mathrm{FaM}_n$ so that

$$(a, \psi(a))(b_k, \psi(b_k)) \cdots (b_1, \psi(b_1)) = (u_{i_1}^{m_1 w} \cdots u_{i_k}^{m_k w}, 1).$$

Let $(b, \psi(b)) = (b_k, \psi(b_k)) \cdots (b_1, \psi(b_1))$. Then

$$\begin{aligned} t &= (a, \psi(a))(b, \psi(b))(a, \psi(a)) \\ &= (u_{i_1}^{m_1(w+1)} \cdots u_{i_k}^{m_k(w+1)}, \psi(a)) \\ &\in (a, \psi(a)) S (a, \psi(a)). \end{aligned}$$

It follows that t is left divisible precisely by the elements $(u_{i_j}, \psi(u_{i_j}))$ with $1 \leq j \leq k$, and thus $t \in S \setminus S_i$. Consequently, S_i is a semiprime ideal of S. $\square$

The following example shows that a monoid S of skew type that does not satisfy the cyclic condition but is non-degenerate can have $S_i \neq S'_i$, for some i. Indeed, let S be the monoid generated by x_1, x_2, x_3, x_4 that is subject to the following relations $x_4x_1 = x_2x_4$, $x_4x_2 = x_3x_1$, $x_4x_3 = x_1x_2$, $x_3x_2 = x_1x_4$, $x_3x_4 = x_2x_3$ and $x_2x_1 = x_1x_3$. Clearly, S is non-degenerate, but $x_2x_1x_1 = x_1x_3x_1 = x_1x_4x_2 = x_3x_2x_2 \in S_3 \setminus S'_3$.

We finish this section with giving a much shorter proof of Theorem 9.4.2 for monoids of skew type that satisfy the cyclic condition. It turns out that such monoids have a large abelian submonoid.

Proposition 9.4.4. *Let $S = \langle X; R\rangle$ be a monoid of skew type that satisfies the cyclic condition. Then, for some positive integer p (a divisor of $(n-1)!$) and some finite subset C of S, the monoid $A = \langle x^p \mid x \in X\rangle$ is abelian, $cA = Ac$ for every $c \in C$, and*

$$S = \bigcup_{c \in C} cA.$$

In particular, $K[S]$ is a finite left and right module over the finitely generated commutative subalgebra $K[A]$, and thus $K[S]$ is a Noetherian algebra that satisfies a polynomial identity.

Proof. Let x and y_1 be distinct elements in X. Because S satisfies the cyclic condition, there exist $t, y_2, \ldots, y_s \in X$ so that

$$xy_1 = y_2t, \; xy_2 = y_3t, \; \ldots, xy_s = y_1t. \tag{9.19}$$

Hence, $x^s y_i = y_i t^s$ for all $i = 1, \ldots, s$. Thus, for every distinct $x, y \in X$, there exists $t \in X$ such that $x^r y = yt^r$ (where r is the least common multiple of the numbers s needed in all (9.19). Clearly r can be taken to be a divisor of $(n-1)!$. Corollary 9.2.5 yields that S is non-degenerate and thus $y \in X$ acts as a bijection on X by mapping x to t if $xy = y't$. By the above, we obtain that y also acts as a bijection on the set $\{x^r \mid x \in X\}$. Hence, there is a multiple p of r (and again a divisor of $(n-1)!$), such that $x^p y^p = y^p x^p$ for all $x, y \in X$.

Because of Theorem 9.3.7, we know that S is the union of sets of the form $\{x_1^{a_1} \cdots x_k^{a_k} \mid a_i \geq 0\}$, where $x_1, \ldots, x_k \in X$ and $k \leq n = |X|$. It thus follows that $S = CA$, where $C = \{x_1^{a_1} \cdots x_n^{a_n} \mid 0 \leq a_i < r, x_1, \ldots, x_n \in X\}$. Moreover, because each c acts as a bijection on the set of generators of A, we also have that $cA = Ac$. □

9.5 The cancellative congruence and the prime radical

In this section, we give a description of the prime radical of a semigroup algebra $K[S]$ of a monoid S of skew type. Recall that ρ_{can} denotes the least cancellative congruence on a semigroup S. So it is the intersection of all congruences $\sim$ on S such that $S/\sim$ is cancellative. We begin with describing these congruences.

Proposition 9.5.1. *Let $S = \langle X; R\rangle$ be a non-degenerate monoid of skew type. Let N be a positive integer so that the ideal $I = S_X^N$ is a cancellative semigroup (Theorem 9.4.2). Then, the least right cancellative congruence on S coincides with the least cancellative congruence ρ_{can} on S.*

Furthermore, for $s, t \in S$, the following conditions are equivalent.

1. $s \, \rho_{\text{can}} \, t$.
2. $sx = tx$ *for some* $x \in I$.
3. $xs = xt$ *for some* $x \in I$.
4. $I(s - t) = 0$.
5. $(s - t)I = 0$.

In particular, ρ_{can} is a homogeneous congruence, that is, if $s \, \rho_{\text{can}} \, t$ then s and t have the same degree.

Moreover, for any field K, the ideal of $K[S]$ determined by ρ_{can} is of the form

$$I(\rho_{\text{can}}) = \sum_{i=1}^{k} (s_i - t_i)K[S] = \sum_{i=1}^{k} K[S](s_i - t_i),$$

for some $k \geq 1$ and some $s_i, t_i \in S$ with $s_i \, \rho_{\text{can}} \, t_i$.

Proof. Note that, because of Lemma 9.3.5, the semigroup S satisfies the right and left Ore conditions, that is, $sS \cap tS \neq \emptyset$ and $Ss \cap St \neq \emptyset$ for every $s, t \in S$.

Consider on S the relation $\sim$ defined by $s \sim t$ if $sx = tx$ for some $x \in S$. Clearly $\sim$ is a reflexive and symmetric relation. We now show that it also is transitive. So suppose $s \sim t$ and $t \sim v$. Then $sx = tx$ and $ty = vy$ for some $x, y \in S$. Because of the right Ore condition, there exist $u, w \in S$ such that $xu = yw$. Thus $sxu = txu = tyw = vyw = vxu$ and so $s \sim v$. Hence $\sim$ is transitive and thus an equivalence relation. Obviously it is a left congruence. We next show it also is a right congruence. So suppose $s \sim t$ and let $x \in S$ be so that $sx = tx$. Let $z \in S$. Then, again by the right Ore condition, $za = xb$ for some $a, b \in S$. Hence,

$$sza = sxb = txb = tza,$$

and thus $sz \sim tz$, as desired. So we have shown that $\sim$ is a congruence on S. It is clear that it is the least congruence on S such that $S/\sim$ is right cancellative. Using a symmetric argument, it follows that two elements $s, t \in S$ are related for the least left cancellative congruence if and only if there exists $x \in S$ so that $xs = xt$.

Let $I = S_X^N$ be the ideal of S as defined in the statement of the result. Since it is a cancellative semigroup and because of the Ore condition on S, we obtain that I has a group of quotients. Hence, by Lemma 2.5.1, we can consider S as a subsemigroup of the semigroup $\widehat{S} = (S \setminus I) \cup II^{-1}$. Let $e = e^2 \in II^{-1}$, the identity of the group II^{-1}. For any $s, t \in S$ and $x \in I$, we obtain that $(s-t)x = 0$ implies $(s-t)xI = 0$, and thus $(s-t)e = (s-t)I = 0$. Since e is the identity of an ideal in $\widehat{S}$, the idempotent e is central in $\widehat{S}$. This yields that $e(s-t) = 0$ and therefore $I(s-t) = x(s-t) = 0$. It follows that conditions (2), (3), (4) and (5) in the statement of the result are equivalent. By the first part of the proof, condition (2) (respectively (3)) says that s and t are related for the least left cancellative congruence (respectively right cancellative congruence). It follows that the least right cancellative congruence coincides with the least left cancellative congruence on S, and thus equals ρ_{can}.

Let K be a field. Recall that the ideal $I(\rho_{\text{can}})$ is by definition the kernel of the natural epimorphism from $K[S]$ to $K[S/\rho_{\text{can}}]$ and thus $I(\rho_{\text{can}}) = \sum_{s,t \in S,\, s\rho_{\text{can}} t} K(s-t)$. Because of Theorem 9.4.2, we know that $K[S]$ is Noetherian. Hence the ideal $K[S]$ is finitely generated as a right and left ideal. The last part of the statement hence follows. $\square$

We now show that the congruence ρ_{can} is fundamental in the description of the prime radical $\mathcal{B}(K[S])$ of $K[S]$.

Proposition 9.5.2. *Let K be a field. If S is a non-degenerate monoid of skew type, then the following properties hold.*

1. *$I(\rho_{\text{can}}) = \text{ann}_{K[S]}(S_X^N) = \text{la}_{K[S]}(S_X^N) = \text{ra}_{K[S]}(S_X^N)$, for some $N \geq 1$ so that S_X^N is cancellative.*

2. *$I(\rho_{\text{can}}) \subseteq P$, for any $P \in \text{Spec}^0(K[S])$ with $P \cap S = \emptyset$.*

3. *$S_X \subseteq P$, for any $P \in \operatorname{Spec}^0(K[S])$ with $P \cap S \neq \emptyset$.*

4. *There exists at least one minimal prime P of $K[S]$ so that $P \cap S = \emptyset$.*

If, furthermore, $\operatorname{char}(K) = 0$ *then*

$$\begin{aligned} \mathcal{B}(K[S]) &= I(\rho_{\mathrm{can}}) \cap \bigcap_{P \in \operatorname{Spec}^0(K[S]), P \cap S \neq \emptyset} P \\ &= I(\rho_{\mathrm{can}}) \cap (\bigcap_{P \in \operatorname{Spec}^0(K[S]), S_X \subseteq P} P) \end{aligned}$$

Proof. Because of Proposition 9.5.1, we obtain at once that $I(\rho_{\mathrm{can}})$ is contained in $\operatorname{ann}_{K[S]}(S_X^N)$. Conversely, assume that $\alpha \in K[S]$ is such that $\alpha I = 0$ with $I = S_X^N$. Let $s \in I$ and write $\alpha = \alpha_1 + \cdots + \alpha_m$ with $|\operatorname{supp}(\alpha_i)s| = 1$ and $\operatorname{supp}(\alpha_i)s \neq \operatorname{supp}(\alpha_j)s$ for $i \neq j$. It follows that $\alpha_i s = 0$ for all i. So the augmentation of α_i is zero. Because $xs = ys$ for all $x, y \in \operatorname{supp}(\alpha_i)$, the definition of ρ_{can} yields that $x \rho_{\mathrm{can}} y$ and thus each $\alpha_i \in I(\rho_{\mathrm{can}})$. Similarly, $I\alpha = 0$ implies that $\alpha \in I(\rho_{\mathrm{can}})$. Hence we have shown that $I(\rho_{\mathrm{can}}) = \operatorname{la}(I) = \operatorname{ra}(I)$.

Because of Theorem 3.2.8, the algebra $K[S/\rho_{\mathrm{can}}]$ is semiprime if $\operatorname{char}(K) = 0$. In the latter case, we get that $\mathcal{B}(K[S]) \subseteq I(\rho_{\mathrm{can}})$ and thus $I(\rho_{\mathrm{can}})$ is the intersection of all (the minimal) prime ideals of $K[S]$ that contain $I(\rho_{\mathrm{can}})$.

From Proposition 9.5.1, we know that $s \rho_{\mathrm{can}} t$ if and only if $(s-t)I = 0 = I(s-t)$. So $I(\rho_{\mathrm{can}})\, I = I\, I(\rho_{\mathrm{can}}) = 0$. Hence, if P is a prime ideal of $K[S]$ with $P \cap S = \emptyset$, then $I(\rho) \subseteq P$. If, on the other hand, P is a prime ideal with $P \cap S \neq \emptyset$, then there exists $b \in P \cap S_X$. So, by Lemma 9.3.8, we get that $S_X^k \subseteq bS \subseteq P$ for some positive integer k. Hence $S_X \subseteq P$.

To finish the proof, it remains to show that there is at least one minimal prime ideal P of $K[S]$ so that $P \cap S = \emptyset$. Suppose the contrary. Then, from the above, $S_X \subseteq \mathcal{B}(K[S])$. Hence S_X consists of nilpotent elements, a contradiction. $\square$

Let $S = \langle X; R \rangle$ be a non-degenerate monoid of skew type. The prime ideals of $K[S]$ that intersect S nontrivially can be described in terms of the sets D_Y with $Y \subseteq X$.

Proposition 9.5.3. *Let $S = \langle X; R \rangle$ be a non-degenerate monoid of skew type. If P is a prime ideal of $K[S]$ that intersects S nontrivially, then $P \cap S = \bigcup D_Y$, where the union runs through all subsets $Y \subseteq X$ such that $P \cap D_Y \neq \emptyset$. Moreover, if $\emptyset \neq D_Y \subseteq P$ and $Y \subseteq Z$, then $D_Z \subseteq P$.*

Proof. Let P be a prime ideal of $K[S]$ so that $P \cap S \neq \emptyset$. Proposition 9.5.2 yields that $S_X \subseteq P$.

We prove the first part of the result by contradiction. So, suppose that $P \cap D_Y \neq \emptyset$, but $D_Y \not\subseteq P$ for some subset Y of X. Choose such subset Y with maximal possible $i = |Y|$. So $P \cap S_{i+1}$ is the union of all D_Z nontrivially intersecting P. Let $a \in P \cap D_Y$. Let $Z \subseteq X$ be such that $Y \subseteq Z$ and $D_Z \neq \emptyset$. From Lemma 9.3.8, we obtain that

$$D_Z^{|a|} \cap D_Z \subseteq aS \subseteq P. \tag{9.20}$$

If $Z \neq Y$ (so $|Z| > i$) and $D_Z^{|a|} \cap D_Z \neq \emptyset$, then $D_Z^{|a|} \subseteq S_{i+1}$ and thus, by the maximality of i we obtain that $D_Z \subseteq P$. If on the other hand $Z \neq Y$ and $D_Z^{|a|} \cap D_Z = \emptyset$, then $D_Z^{|a|} \subseteq \bigcup_{|W|>|Z|, Y \subseteq W} D_W$. The given argument can now be applied to every W with $D_W \neq \emptyset$. Continuing this process, after a finite number of steps, we obtain that $\bigcup_{Z \neq Y, Y \subseteq Z} D_Z$ is nilpotent modulo P. Since this is a right ideal of S and because $P \cap S$ is a prime ideal of S, it must be contained in P. Applying (9.20) to $Z = Y$, we thus have proved that

$$D_Y^{|a|} \subseteq (D_Y^{|a|} \cap D_Y) \cup \bigcup_{Z \neq Y, Y \subseteq Z} D_Z \subseteq aS \cup P \subseteq P.$$

Since we already know that $J = \bigcup_{Y \subset Z} D_Z \subseteq P$, we get that the right ideal $D_Y \cup J$ is nilpotent modulo the prime ideal P. Hence $D_Y \subseteq P$, a contradiction. This completes the proof of the first assertion.

The second assertion follows from the proof above. □

9.6 Comments and problems

Semigroups of skew type were introduced in [51]. The results in this chapter are mainly taken from [51]. Corollary 9.2.6 comes from [87], Theorem 9.3.10 is a partial generalization of Proposition 2.2(c) in [37] and it is proved in [17], Proposition 9.4.3 was proven in [87], Proposition 9.5.3 comes from [86]. In [87] it is shown that for monoids S of I-type all semiprime ideals S can be described in this manner. However, for the proof, one has to notice that the statements of Lemma 9.3.1, Definition 9.3.2, Proposition 9.3.4, Lemma 9.3.5, Definition 9.3.6, Theorem 9.3.7 part 1 and Lemma 9.3.8 remain valid for monoids of skew type that are right non-degenerate, but are not necessarily square free.

In order to prove Theorem 9.4.2, it was shown that non-degenerate monoids of skew type are linear semigroups with an ideal chain that can be improved into an ideal chain $S_X^N = I_1 \subseteq I_2 \subseteq \cdots \subseteq I_t = S$ for some $N \geq 1$, so that S_X^N is cancellative and each factor I_j/I_{j-1} either is nilpotent or a uniform subsemigroup of a Brandt semigroup. Furthermore, if I_j/I_{j-1} is a uniform subsemigroup of a Brandt semigroup, then $I_j \setminus I_{j-1} \subseteq (S_i \cap S_i') \setminus S_{i+1}$ for some i, and if a, b are in a cancellative component of I_j/I_{j-1}, then $a^{|b|} \in bS \cap Sb$. As a linear semigroup, S also has an ideal chain as in Theorem 2.4.3. Of course it can be embedded in many different ways in a full matrix ring over a field. In [86], it is shown that S can be embedded in a full matrix ring in such a manner that the above mentioned ideal chain can be refined further so that the structure of matrix factors of that chain correspond with that of the full matrix ring (as a multiplicative semigroup) and that they are at different rank levels. In [86], an example is given of a non-degenerate monoid of skew type (that does not satisfy the cyclic condition) with S_X not cancellative.

In [118] Odesskii, gives a survey of elliptic algebras, these are Poincaré-Birkhoff-Witt (PBW) algebras that are somehow related to an elliptic curve. Recall that a PBW-algebra is an associative $\mathbb{Z}_{\geq 0}$-graded algebra $A = \bigoplus_k A_k$, generated by n elements of degree 1, with $\frac{n(n-1)}{2}$ quadratic relations such that the dimension of A_k ($k \geq 1$) is $\frac{n(n+1)\ldots(n+k-1)}{k!}$. Examples are considered of such algebras, depending on two continuous parameters that are flat deformations of the polynomial ring in n variables. Several properties of these algebras are described.

Problems

1. Let S be a non-degenerate monoid of skew type and let K be a field. When is $K[S]$ a prime (Noetherian) maximal order? Because of Theorem 9.4.2 the algebra $K[S]$ is Noetherian. In case S is a monoid of I-type we know that $K[S]$ is a prime Noetherian maximal order, see Theorem 8.5.6.

2. Let S be a monoid of skew type and let K be a field. Find necessary and sufficient conditions, in terms of the presentation of S, for $K[S]$ to have a finite Gelfand-Kirillov dimension. See Theorem 9.3.7 in case S is right non-degenerate.

3. Let $S = \langle X; R \rangle$ be a monoid of skew type that satisfies the cyclic condition. Is it true that $S_i = S_i'$ for every $1 \leq i \leq |X|$? See Proposition 9.4.3 in case S is of I-type.

4. Let $S = \langle X; R \rangle$ be a monoid of skew type that satisfies the cyclic condition. Is S_X a cancellative monoid? See Theorem 9.4.2.

5. Let $S = \langle X; R \rangle$ be a monoid of skew type that satisfies the cyclic condition. Is the group $(S/\rho_{\text{can}})(S/\rho_{\text{can}})^{-1}$ solvable? See Corollary 8.2.7 in case S is of I-type.

The last three questions already were posed in [86] (where several other related questions can be found).

Chapter 10

Examples

The theory developed in the preceding chapters will be illustrated on several examples of monoids of skew type and their algebras. Our calculations also show that several ingredients of the structural description of these combinatorially defined algebras can be effectively computed.

If S is a non-degenerate monoid of skew type, then we have seen in Theorem 9.3.7 that $\mathrm{GK}(K[S]) \leq n$. Because of Theorem 9.4.2 and Theorem 3.5.2, we also know that $\mathrm{GK}(K[S])$ is an integer. If, furthermore, S satisfies the cyclic condition and S is not a cyclic monoid, then in the first section of this chapter we show that $2 \leq \mathrm{GK}(K[S]) \leq n$. It also will be proved that the upper bound is reached precisely when S is of I-type, or equivalently S is cancellative. So, the Gelfand-Kirillov dimension reflects some properties of the algebra $K[S]$ and the semigroup S. In Section 10.2, all four generated non-degenerate monoids of skew type are described. In Section 10.3, an example is given of a four generated monoid of skew type that satisfies the cyclic condition and such that its Gelfand-Kirillov dimension is two. It is then shown that such examples can be constructed on arbitrarily many generators. In Section 10.4, examples are given of non-degenerate monoids S of skew type generated by 4^n elements (with n an arbitrary positive integer) so that $\mathrm{GK}(K[S]) = 1$ and $K[S]$ is semiprime. In order to state the relevant information, the statements of the examples are rather elaborate.

In the last Section 10.5, we give several examples of cancellative monoids S that satisfy the ascending chain condition on one sided ideals and that are maximal orders. These illustrate that the algebraic structure of the localizations S_P, with respect to a minimal prime ideal, is much more complicated than in the commutative setting.

10.1 Monoids of skew type and the Gelfand-Kirillov dimension

Let $S = \langle x_1, x_2, \dots, x_n; R\rangle$ be a monoid of skew type that satisfies the cyclic condition. In this section, we show that $2 \leq \mathrm{GK}(K[S]) \leq n$. Furthermore, the upper bound is reached precisely when S is a monoid of I-type. This, also, turns out to be equivalent with S being cancellative.

In Proposition 9.4.4, it is shown that S contains an abelian submonoid $A = \langle x_1^p, \dots, x_n^p\rangle$, for some positive integer p that is a divisor of $(n-1)!$. In the following lemma we go into more detail and show that A is generated by free abelian groups of rank 2.

Lemma 10.1.1. *Let $S = \langle x_1, \dots, x_n; R\rangle$ be a monoid of skew type that satisfies the cyclic condition, with $n > 1$. If $x_{i_1}x_{j_1} = x_{j_2}x_{i_2}$ is a defining relation of S (so $i_1 \neq j_1$), then there exist positive integers r, s so that $r+s \leq n$, $x_{i_1}^r x_{j_1} = x_{j_1}x_{i_2}^r$, $x_{i_1}x_{j_1}^s = x_{j_2}^s x_{i_1}$ and $\langle x_{i_1}^r, x_{j_1}^s\rangle$ is free abelian of rank 2.*

Proof. Let $F = \langle y_1, y_2, \dots, y_n\rangle$ be a free monoid of rank n and let $\pi\colon F \longrightarrow S$ be the natural epimorphism, that is, $\pi(y_i) = x_i$ for $i = 1, \dots, n$. Let $x \in S$. We say that a word $w \in F$ represents x if $\pi(w) = x$. Clearly, two words $w, w' \in F$ represent the same element x in S if and only if there exists a finite sequence of words $w = w_0, w_1, w_2, \dots, w_m = w'$ in F such that w_i is obtained from w_{i-1} (with $i = 2, \dots, m$) by substituting a subword $y_j y_k$ by $y_p y_q$, where $x_j x_k = x_p x_q$ is a defining relation of S. In this case, we say that w_i is obtained from w_{i-1} by an S-relation.

Suppose $i_1 \neq j_1$ and $x_{i_1}x_{j_1} = x_{j_2}x_{i_2}$ is a defining relation of S. Since S satisfies the cyclic condition, we obtain from Proposition 9.2.4 that there exist positive integers r, s and $r+s$ distinct integers $i_1, i_2, \dots, i_s, j_1, j_2, \dots, j_r \in \{1, 2, \dots, n\}$ such that

$$\begin{array}{llll} x_{i_1}x_{j_1} = x_{j_2}x_{i_2}, & x_{i_2}x_{j_1} = x_{j_2}x_{i_3}, & \dots, & x_{i_s}x_{j_1} = x_{j_2}x_{i_1}, \\ x_{i_1}x_{j_2} = x_{j_3}x_{i_2}, & x_{i_2}x_{j_2} = x_{j_3}x_{i_3}, & \dots, & x_{i_s}x_{j_2} = x_{j_3}x_{i_1}, \\ \vdots & \vdots & & \vdots \\ x_{i_1}x_{j_r} = x_{j_1}x_{i_2}, & x_{i_2}x_{j_r} = x_{j_1}x_{i_3}, & \dots, & x_{i_s}x_{j_r} = x_{j_1}x_{i_1}. \end{array}$$

The relations in the columns yield that

$$x_{i_1}^r x_{j_1} = x_{j_1}x_{i_2}^r,\ x_{i_2}^r x_{j_1} = x_{j_1}x_{i_3}^r,\ \dots,\ x_{i_s}^r x_{j_1} = x_{j_1}x_{i_1}^r.$$

Hence $x_{i_1}^r x_{j_1}^s = x_{j_1}^s x_{i_1}^r$, and thus the submonoid $\langle x_{i_1}^r, x_{j_1}^s\rangle$ is abelian. The relations in the first row also yield that

$$x_{i_1}x_{j_1}^s = x_{j_2}^s x_{i_1}.$$

Since the defining relations are square free, we note that the only words that represent $x_{i_1}^m$ and $x_{j_1}^m$ are $y_{i_1}^m$ and $y_{j_1}^m$ respectively.

Let p, q be positive integers and $x = x_{i_1}^{rp} x_{j_1}^{sq}$. We claim that any word $w \in F$ that represents x is of the form

$$w = y_{i_{l_1}}^{n_1} y_{j_{k_1}}^{m_1} y_{i_{l_2}}^{n_2} y_{j_{k_2}}^{m_2} \cdots y_{i_{l_{g-1}}}^{n_{g-1}} y_{j_{k_{g-1}}}^{m_{g-1}} y_{i_{l_g}}^{n_g}, \qquad (10.1)$$

where g is an integer greater than 1, n_1, n_g are nonnegative integers, $l_1 = l_g = 1$, and $n_2, n_3, \ldots, n_{g-1}$, $m_1, m_2, \ldots, m_{g-1}$ are positive integers such that the following properties hold.

(i) $n_1 + k_1 \equiv k_{g-1} - n_g \equiv 1 \pmod r$ and $l_{t+1} - l_t \equiv m_t \pmod s$, for all $1 \le t \le g-1$.

(ii) If $g > 2$, then $k_u - k_{u+1} \equiv n_{u+1} \pmod r$ for all $1 \le u \le g-2$.

(iii) $n_1 + n_2 + \cdots + n_g = rp$ and $m_1 + m_2 + \cdots + m_{g-1} = sq$.

Note that the word $y_{i_1}^{rp} y_{j_1}^{sq}$ represents x and satisfies conditions (i), (ii) and (iii). Therefore, in order to prove the claim, it is sufficient to see that given any word w of the form (10.1) that satisfies conditions (i), (ii) and (iii), all the words obtained from w by an S-relation also satisfy conditions (i), (ii) and (iii). Suppose that $g = 2$. In this case

$$w = y_{i_1}^{n_1} y_{j_{k_1}}^{m_1} y_{i_1}^{n_2},$$

with $m_1 = sq > 0$, $n_1 + n_2 = rp$ and $n_1 + k_1 \equiv k_1 - n_2 \equiv 1 \pmod r$. If $n_1 > 0$, then, using the S-relation $x_{i_1} x_{j_{k_1}} = x_{j_{k_1+1}} x_{i_2}$, we obtain the word

$$w' = y_{i_1}^{n_1-1} y_{j_{k_1+1}} y_{i_2} y_{j_{k_1}}^{m_1-1} y_{i_1}^{n_2},$$

where $k_1 + 1$ is taken modulo r in the set $\{1, \ldots, r\}$. It is easily seen that w' satisfies conditions (i), (ii) and (iii). If $n_2 > 0$, then we can obtain by an S-relation the word

$$w' = y_{i_1}^{n_1} y_{j_{k_1}}^{m_1-1} y_{i_s} y_{j_{k_1-1}} y_{i_1}^{n_2-1},$$

where $k_1 - 1$ is taken modulo r in the set $\{1, \ldots, r\}$, and again it is easily verified that w' satisfies the conditions (i), (ii) and (iii). Similarly, it is straightforward to prove that, if $g > 2$, then all the words w' obtained from w by an S-relation satisfy conditions (i), (ii) and (iii). Now condition (iii) implies that the monoid $\langle x_{i_1}^r, x_{j_1}^s \rangle$ is free abelian of rank 2. $\square$

Theorem 10.1.2. *Let $S = \langle x_1, x_2, \ldots, x_n; R\rangle$ be a non-cylic monoid of skew type that satisfies the cyclic condition. Let $A = \langle x_1^m, \ldots, x_n^m\rangle$, where $m = (n-1)!$, and let K be a field. Then, $2 \leq \mathrm{GK}(K[S]) = \mathrm{GK}(K[A]) = k \leq n$, where k is the maximal rank k of a free abelian submonoid of the form $\langle x_{i_1}^m, \ldots, x_{i_k}^m\rangle \subseteq S$.*

Proof. Because of Proposition 9.4.4, we know that $K[S]$ is finitely generated as a right (and left) module over the commutative subalgebra $K[A]$, where $A = \langle x_1^m, \ldots, x_n^m\rangle$ and $m = (n-1)!$. Hence $\mathrm{GK}(K[S]) = \mathrm{GK}(K[A])$. Because of Theorem 9.3.7 and Lemma 10.1.1, we also get that $2 \leq \mathrm{GK}(K[S]) \leq n$. The result then follows from Theorem 3.5.6. □

In the following result we show that the upper bound for the Gelfand-Kirillov dimension occurs precisely when the monoid is of I-type.

Theorem 10.1.3. *Let $S = \langle x_1, x_2, \ldots, x_n; R\rangle$ be a monoid of skew type and let K be a field. If S satisfies the cyclic condition, then the following conditions are equivalent.*

1. $\mathrm{GK}(K[S]) = n$.
2. *S is of I-type.*
3. *S is cancellative.*

Proof. First, we show that (1) implies (2). So, suppose that $\mathrm{GK}(K[S]) = n$. Let $m = (n-1)!$. We know, from Lemma 10.1.1, that $A = \langle x_1^m, \ldots, x_n^m\rangle$ is abelian. Moreover, by Theorem 3.5.6 and Theorem 10.1.2, $\mathrm{GK}(K[A]) = \mathrm{GK}(K[S]) = \mathrm{clKdim}(K[A]) = n = \mathrm{rk}(A)$. This implies that A is a free abelian monoid of rank n. Indeed, suppose the contrary. Then the natural epimorphism $K[y_1, \ldots, y_n] \longrightarrow K[A]$ has a nontrivial kernel. Hence $\mathrm{clKdim}(K[A]) < n$, a contradiction.

Because of Corollary 9.2.5, the monoid S is non-degenerate. Hence, as in Theorem 9.3.10, we consider the permutations $\sigma_i \in \mathrm{Sym}_n$ defined by $\sigma_i(j) = i$ if $j = i$, otherwise $\sigma_i(j) = k$ provided $x_i x_j = x_k x_l$ is a defining relation of S. So, by Lemma 10.1.1,

$$x_i x_j^m = x_{\sigma_i(j)}^m x_i, \tag{10.2}$$

for all i, j.

Suppose that $x_i x_j = x_k x_l$ is a defining relation of S. Then, for all $t \in \{1, \ldots, n\}$, we have, by (10.2),

$$x_i x_j x_t^m = x_i x_{\sigma_j(t)}^m x_j = x_{\sigma_i(\sigma_j(t))}^m x_i x_j.$$

Also we have,

$$x_i x_j x_t^m = x_k x_l x_t^m = x_k x_{\sigma_l(t)}^m x_l = x_{\sigma_k(\sigma_l(t))}^m x_k x_l.$$

Since

$$x_i x_j x_j^{m-1} x_i^{m-1} = x_i x_j^m x_i^{m-1} = x_k^m x_i x_i^{m-1} = x_k^m x_i^m,$$

multiplying the two previous equalities by $x_j^{m-1} x_i^{m-1}$ on the right, we get

$$x_{\sigma_i(\sigma_j(t))}^m x_k^m x_i^m = x_{\sigma_k(\sigma_l(t))}^m x_k^m x_i^m.$$

Since A is free abelian, this implies that

$$\sigma_i(\sigma_j(t)) = \sigma_k(\sigma_l(t)).$$

Hence, we have shown that if $x_i x_j = x_k x_l$ is a defining relation, then $\sigma_i \sigma_j = \sigma_k \sigma_l$. Theorem 9.3.10 therefore implies that S is of I-type.

That (2) implies (1) already is mentioned in Section 8.1. That (2) implies (3) follows at once from Corollary 8.2.5 .

We now show that (3) implies (2). So assume S is cancellative. Suppose that $x_i x_j = x_k x_l$ is a defining relation of S. Then for all $t \in \{1, \ldots, n\}$, as in the first part of the proof, we get

$$x_{\sigma_i(\sigma_j(t))}^m x_i x_j = x_{\sigma_k(\sigma_l(t))}^m x_k x_l.$$

Since S is cancellative, this implies that $x_{\sigma_i(\sigma_j(t))}^m = x_{\sigma_k(\sigma_l(t))}^m$. Because the defining relations are square free, it follows that $\sigma_i(\sigma_j(t)) = \sigma_k(\sigma_l(t))$. Theorem 9.3.10 then yields that S is of I-type. □

Corollary 10.1.4. *Let S be a monoid of skew type. Then S is of I-type if and only if S is cancellative and satisfies the cyclic condition.*

Proof. That the conditions are necessary follows from Theorem 8.1.4, Corollary 8.2.5 and Corollary 9.2.6. The sufficiency follows at once from Theorem 10.1.3. □

10.2 Four generated monoids of skew type

Let S be a non-degenerate monoid of skew type. Clearly, if S is generated by two elements, then S is the free abelian monoid of rank two. If S is generated by three elements then it is easy to verify that S is one of the two binomial monoids $B^{3,1}$ and $B^{3,2}$ stated in Example 8.4.5. In this section, we describe all four generated monoids of skew type that are non-degenerate. First we state those that satisfy the cyclic condition. Apart from the five binomial monoids given in Example 8.4.6, there are three other monoids. We say that a monoid of skew type $S = \langle x_1, \ldots, x_n; R\rangle$ has a normal generator if some x_i is a normal element in S, that is, $Sx_i = x_iS$.

Proposition 10.2.1. *Let $S = \langle x_1, x_2, x_3, x_4; R\rangle$ be a non-degenerate monoid of skew type. If S has a normal generator, then S satisfies the cyclic condition and is isomorphic to one of the following binomial monoids,*

$$\begin{aligned}
B^{4,1} &= \langle x_1, x_2, x_3, x_4 \mid x_4x_1 = x_1x_4,\ x_4x_2 = x_2x_4,\ x_4x_3 = x_3x_4,\\
&\qquad x_3x_1 = x_1x_3,\ x_3x_2 = x_2x_3,\ x_2x_1 = x_1x_2\rangle,\\
B^{4,2} &= \langle x_1, x_2, x_3, x_4 \mid x_4x_1 = x_2x_4,\ x_4x_2 = x_1x_4,\ x_4x_3 = x_3x_4,\\
&\qquad x_3x_1 = x_1x_3,\ x_3x_2 = x_2x_3,\ x_2x_1 = x_1x_2\rangle,\\
B^{4,3} &= \langle x_1, x_2, x_3, x_4 \mid x_4x_1 = x_2x_4,\ x_4x_2 = x_3x_4,\ x_4x_3 = x_1x_4,\\
&\qquad x_3x_1 = x_1x_3,\ x_3x_2 = x_2x_3,\ x_2x_1 = x_1x_2\rangle,\\
B^{4,4} &= \langle x_1, x_2, x_3, x_4 \mid x_4x_1 = x_2x_4,\ x_4x_2 = x_1x_4,\ x_4x_3 = x_3x_4,\\
&\qquad x_3x_1 = x_2x_3,\ x_3x_2 = x_1x_3,\ x_2x_1 = x_1x_2\rangle,
\end{aligned}$$

or to one of the following non-cancellative monoids,

$$\begin{aligned}
C^{4,2} &= \langle x_1, x_2, x_3, x_4 \mid x_4x_1 = x_3x_4,\ x_4x_2 = x_2x_4,\ x_4x_3 = x_1x_4,\\
&\qquad x_3x_1 = x_2x_3,\ x_3x_2 = x_1x_3,\ x_2x_1 = x_1x_2\rangle,\\
C^{4,3} &= \langle x_1, x_2, x_3, x_4 \mid x_4x_1 = x_2x_4,\ x_4x_2 = x_3x_4,\ x_4x_3 = x_1x_4,\\
&\qquad x_3x_1 = x_2x_3,\ x_3x_2 = x_1x_3,\ x_2x_1 = x_1x_2\rangle.
\end{aligned}$$

If S satisfies the cyclic condition but S does not have a normal generator, then it is isomorphic to the binomial monoid

$$\begin{aligned}
B^{4,5} &= \langle x_1, x_2, x_3, x_4 \mid x_4x_1 = x_2x_3,\ x_4x_2 = x_1x_3,\ x_4x_3 = x_3x_4,\\
&\qquad x_3x_1 = x_2x_4,\ x_3x_2 = x_1x_4,\ x_2x_1 = x_1x_2\rangle\\
&= \langle x_1, x_2\rangle\, \langle x_3, x_4\rangle
\end{aligned}$$

or to the non-cancellative monoid

$$\begin{aligned}
C^{4,1} &= \langle x_1, x_2, x_3, x_4 \mid x_4x_1 = x_3x_4,\ x_4x_2 = x_2x_3,\ x_4x_3 = x_1x_4,\\
&\qquad x_3x_1 = x_1x_2,\ x_3x_2 = x_2x_4,\ x_2x_1 = x_1x_3\rangle.
\end{aligned}$$

Proof. First, assume S is non-degenerate and has a normal generator, say x_i. Without loss of generality, we may assume that $i = 4$. Since S is non-degenerate, we get that, for any $1 \leq i \leq 3$, there exists a unique $1 \leq j \leq 3$ such that $x_4 x_i = x_j x_4$. Hence x_4 determines a permutation f on $\{x_1, x_2, x_3\}$. Because S is of skew type it therefore follows that $\langle x_1, x_2, x_3 \rangle$ also is a non-degenerate monoid of skew type. Consequently $\langle x_1, x_2, x_3 \rangle$ is isomorphic to either $B^{3,1}$ or $B^{3,2}$ (the only nonabelian binomial monoid on three generators).

The permutation f either is the identity (that is, x_4 is central), a three cycle or a transposition.

Assume $\langle x_1, x_2, x_3 \rangle = B^{3,1}$. If f is the identity, then, of course, $S = B^{4,1}$. If f is a transposition, then, without loss of generality, we may assume that f interchanges x_1 and x_2. Hence $x_4 x_1 = x_2 x_4$, $x_4 x_2 = x_2 x_4$ and thus $x_4 x_3 = x_3 x_4$; so $S = B^{4,2}$. If f is a three cycle, then we may assume that $f = (x_1\ x_2\ x_3)$. Thus $x_4 x_1 = x_2 x_4$, $x_4 x_2 = x_3 x_4$ and $x_4 x_3 = x_1 x_4$ and therefore $S = B^{4,3}$.

Next assume that $\langle x_1, x_2, x_3 \rangle = \langle x_1, x_2, x_3 \mid x_3 x_1 = x_2 x_3,\ x_3 x_2 = x_1 x_3,\ x_2 x_1 = x_1 x_2 \rangle = B^{3,2}$. If $f = 1$, then S is isomorphic to $B^{4,2}$ (interchange x_3 and x_4 in the given presentation in the statement). If f is a transposition, then x_4 commutes with one generator of $B^{3,1}$. Hence there are three possible cases. First, x_4 commutes with x_1 and thus $S = \langle x_1, x_2, x_3, x_4 \mid x_4 x_1 = x_1 x_4,\ x_4 x_2 = x_3 x_4,\ x_4 x_3 = x_2 x_4,\ x_3 x_1 = x_2 x_3,\ x_3 x_2 = x_1 x_3,\ x_2 x_1 = x_1 x_2 \rangle$. Interchanging x_1 with x_2, we obtain that this semigroup is $C^{4,2}$. Second, x_4 commutes with x_2 and thus again $S = C^{4,2} = \langle x_1, x_2, x_3, x_4 \mid x_4 x_2 = x_2 x_4,\ x_4 x_1 = x_3 x_4,\ x_4 x_3 = x_1 x_4,\ x_3 x_1 = x_2 x_3,\ x_3 x_2 = x_1 x_3,\ x_2 x_1 = x_1 x_2 \rangle$. Note that $C^{4,2}$ is not cancellative as $x_1 x_3 x_4 = x_1 x_2 x_4$. Third, x_4 commutes with x_3 and then it follows that $S = B^{4,4}$.

If f is a three cycle, then there are two possibilities, either $S = C^{4,3} = \langle x_1, x_2, x_3, x_4 \mid x_4 x_1 = x_2 x_4,\ x_4 x_2 = x_3 x_4,\ x_4 x_3 = x_1 x_4,\ x_3 x_1 = x_2 x_3,\ x_3 x_2 = x_1 x_3,\ x_2 x_1 = x_1 x_2 \rangle$ or $S = \langle x_1, x_2, x_3, x_4 \mid x_4 x_1 = x_3 x_4,\ x_4 x_3 = x_2 x_4,\ x_4 x_2 = x_1 x_4,\ x_3 x_1 = x_2 x_3,\ x_3 x_2 = x_1 x_3,\ x_2 x_1 = x_1 x_2 \rangle$. Interchanging x_1 with x_2, we see that both these semigroups are isomorphic. Again we get that $x_1 x_3 x_4 = x_1 x_2 x_4$ and thus $C^{4,3}$ is not cancellative. From Proposition 8.3.9, we know that binomial monoids satisfy the cyclic condition. It also easily is verified that the monoids $C^{4,2}$ and $C^{4,3}$ satisfy this condition. Hence we have finished the proof of the first part of the proposition.

So, for the remainder of the proof assume that S satisfies the cyclic condition and does not have a normal generator. In particular, S is not abelian and hence, without loss of generality, we may assume that x_1 and x_2 do not commute. Thus either $x_1 x_2 = x_k x_1$, for $k \neq 1, 2$, or $x_1 x_2 \notin S x_1$. In the former case, without loss of generality, we may assume that $k = 3$.

First, assume that $x_1x_2 = x_3x_1$. Since x_1 is not normal, the cyclic condition implies that $x_1x_3 = x_2x_1$. Because S is left non-degenerate, it thus follows that $x_4x_1 \in Sx_4$. As x_1 is not normal, we get that $x_4x_1 \neq x_1x_4$. So either $x_4x_1 = x_2x_4$ or $x_4x_1 = x_3x_4$. The latter situation reduces to the former one if one interchanges x_2 with x_3. So, without loss of generality, we may assume that $x_4x_1 = x_2x_4$. Again, since x_4 is not normal and S satisfies the cyclic condition, we get that $x_4x_2 = x_1x_4$. Since S is right non-degenerate, we then obtain that $x_4x_3 \in x_3S$. Again, since x_4 is not normal, and thus $x_4x_3 \neq x_3x_4$, we must have $x_4x_3 = x_3x_2$. Since S is of skew type, the only possible relation containing x_3x_4 is therefore $x_3x_4 = x_2x_3$. Interchanging x_2 with x_3 we thus get that $S \cong C^{4,1}$. Note that in $C^{4,1}$ we have that $x_1x_3x_4 = x_2x_3x_4$ and therefore $C^{4,1}$ is not cancellative.

Second, assume that $x_1x_2 \notin Sx_1$. Note that, because of the left non-degeneracy, we also know that $x_1x_2 \neq x_kx_2$ for $k \neq 1$. So, without loss of generality, we may assume that $x_1x_2 = x_kx_3$ for some k. Of course $k \neq 3$. The right non-degeneracy yields that $k \neq 1$. If $k = 2$, then $x_2x_3 = x_1x_2$ and we are in the previous situation; so $S = C^{4,1}$. Hence, we may assume that $x_1x_2 = x_4x_3$. The cyclic condition then implies that $x_1x_4 = x_lx_3$ with $l = 2$ or 1. However the latter is impossible, because of the right non-degeneracy. So $x_1x_4 = x_2x_3$. The cyclic condition applied to the equality $x_4x_3 = x_1x_2$ yields that $x_4x_1 = x_3x_2$. Similarly $x_2x_3 = x_1x_4$ yields $x_2x_1 = x_3x_4$. Again, using the non-degeneracy, one finds out that the remaining relations are uniquely determined. Namely, $S = \langle x_1, x_2, x_3, x_4 \mid x_4x_1 = x_3x_2,\ x_4x_2 = x_2x_4,\ x_4x_3 = x_1x_2,\ x_3x_1 = x_1x_3,\ x_2x_4 = x_4x_2,\ x_2x_1 = x_3x_4\rangle$. Interchanging x_2 with x_3 we get that $S \cong B^{4,5}$. $\square$

To complete the list of all four generated monoids of skew type that are non-degenerate we now present those that do not satisfy the cyclic condition.

Proposition 10.2.2. *Let $S = \langle x_1, x_2, x_3, x_4; R\rangle$ be a non-degenerate monoid of skew type. If S does not satisfy the cyclic condition, then S is isomorphic to one of the following monoids.*

$$\begin{aligned}
D^{4,1} &= \langle x_1, x_2, x_3, x_4 \mid x_2x_1 = x_3x_4,\ x_2x_3 = x_4x_1,\ x_2x_4 = x_1x_3, \\
&\qquad x_1x_2 = x_4x_3,\ x_1x_4 = x_3x_2,\ x_3x_1 = x_4x_2\rangle, \\
D^{4,2} &= \langle x_1, x_2, x_3, x_4 \mid x_2x_1 = x_1x_3,\ x_2x_3 = x_3x_2,\ x_2x_4 = x_4x_3, \\
&\qquad x_1x_2 = x_3x_4,\ x_1x_4 = x_4x_1,\ x_3x_1 = x_4x_2\rangle, \\
D^{4,3} &= \langle x_1, x_2, x_3, x_4 \mid x_2x_1 = x_1x_3,\ x_2x_3 = x_3x_4,\ x_2x_4 = x_4x_2, \\
&\qquad x_1x_2 = x_4x_3,\ x_1x_4 = x_3x_1,\ x_3x_2 = x_4x_1\rangle,
\end{aligned}$$

$$\begin{aligned} D^{4,4} &= \langle x_1,x_2,x_3,x_4 \mid x_2x_1 = x_1x_3,\ x_2x_3 = x_4x_2,\ x_2x_4 = x_3x_1, \\ &\quad x_1x_2 = x_3x_4,\ x_1x_4 = x_4x_3,\ x_3x_2 = x_4x_1\rangle, \\ D^{4,5} &= \langle x_1,x_2,x_3,x_4 \mid x_2x_1 = x_1x_3,\ x_2x_3 = x_3x_4,\ x_2x_4 = x_4x_1, \\ &\quad x_1x_2 = x_4x_3,\ x_1x_4 = x_3x_2,\ x_3x_1 = x_4x_2\rangle. \end{aligned}$$

All these monoids are pairwise non-isomorphic. Furthermore, the permutation $(2\ 3)$ *defines an anti-isomorphism between* $D^{4,4}$ *and* $D^{4,5}$.

Proof. Let $S = \langle x_1, x_2, x_3, x_4; R\rangle$ be a non-degenerate monoid of skew type that does not satisfy the cyclic condition. Because of Proposition 10.2.1, we know that S does not have a normal generator. In particular, S is nonabelian.

Suppose that exactly three different generators are in one of the defining relations of S. Because S is non-degenerate, without loss of generality, we may assume that this relation is $x_2x_1 = x_1x_3$. Again, by the non-degeneracy, the word x_3x_1 then only can occur in one of the following possible defining relations: (1) $x_3x_1 = x_1x_2$, (2) $x_3x_1 = x_1x_4$, (3) $x_3x_1 = x_2x_4$, or (4) $x_3x_1 = x_4x_2$. We deal with each of these cases separately.

(1) $x_3x_1 = x_1x_2$. Since x_1 is not normal in S, it follows that $x_4x_1 \neq x_1x_4$. Further, since x_4 is not normal in S, one easily verifies that $x_2x_3 \neq x_3x_2$. Again, because S is non-degenerate, we thus have two possibilities, either $x_4x_1 = x_2x_4$ or $x_4x_1 = x_3x_4$.

Suppose that $x_4x_1 = x_2x_4$. As $x_2x_3 \neq x_3x_2$ and because S is non-degenerate, we get that $x_2x_3 = x_3x_4$, $x_3x_2 = x_4x_3$ and $x_1x_4 = x_4x_2$. Hence S satisfies the cyclic condition, a contradiction. So $x_4x_1 = x_3x_4$. Because $x_2x_3 \neq x_3x_2$ and, again since S is non-degenerate, we obtain that $x_3x_2 = x_2x_4$, $x_2x_3 = x_4x_2$ and $x_1x_4 = x_4x_3$. Consequently, $S = C^{4,1}$ and thus S satisfies the cyclic condition, a contradiction.

(2) $x_3x_1 = x_1x_4$. Since x_1 is not normal in S, we get that $x_4x_1 \neq x_1x_2$. Because S is non-degenerate, $x_1x_2 = x_4x_3$, $x_2x_3 = x_3x_4$ and $x_2x_4 = x_4x_2$. Hence, we have $S = D^{4,3}$. Note that $D^{4,3}$ does not satisfy the cyclic condition, because $x_1x_2 = x_4x_3$ and $x_1x_4 = x_3x_1 \notin Sx_3$.

(3) $x_3x_1 = x_2x_4$. Since S is non-degenerate, we have two possibilities: either $x_4x_1 = x_1x_2$ or $x_4x_1 = x_3x_2$.

If $x_4x_1 = x_3x_2$ then $x_2x_3 = x_4x_2$, $x_3x_4 = x_1x_2$ and $x_1x_4 = x_4x_3$. Thus $S = D^{4,4}$. Note that $D^{4,4}$ does not satisfy the cyclic condition, because $x_2x_3 = x_4x_2$ and $x_2x_4 = x_3x_1 \notin Sx_3$.

If $x_4x_1 = x_1x_2$ then $x_1x_4 = x_3x_2$, $x_3x_4 = x_4x_3$ and $x_2x_3 = x_4x_2$. Thus

$$\begin{aligned} S &= \langle x_1,x_2,x_3,x_4 \mid x_2x_1 = x_1x_3,\ x_2x_3 = x_4x_2,\ x_2x_4 = x_3x_1, \\ &\quad x_1x_2 = x_4x_1,\ x_1x_4 = x_3x_2,\ x_3x_4 = x_4x_3\rangle. \end{aligned}$$

The mapping $f\colon D^{4,3} \longrightarrow S$ determined by $f(x_1) = x_1$, $f(x_2) = x_4$, $f(x_3) = x_2$ and $f(x_4) = x_3$ is a monoid isomorphism.

(4) $x_3x_1 = x_4x_2$. Since S is non-degenerate, we have 3 possibilities: $x_1x_4 = x_4x_1$, $x_1x_4 = x_3x_2$ or $x_1x_4 = x_4x_3$.

If $x_1x_4 = x_4x_1$ then $x_1x_2 = x_3x_4$, $x_2x_4 = x_4x_3$ and $x_2x_3 = x_3x_2$ and thus $S = D^{4,2}$.

If $x_1x_4 = x_3x_2$ then $x_3x_4 = x_2x_3$, $x_2x_4 = x_4x_1$ and $x_1x_2 = x_4x_3$ and thus $S = D^{4,5}$.

If $x_1x_4 = x_4x_3$ then $x_4x_1 = x_2x_4$, $x_3x_4 = x_1x_2$ and $x_3x_2 = x_2x_3$ and

$$\begin{aligned} S &= \langle x_1, x_2, x_3, x_4 \mid x_2x_1 = x_1x_3,\ x_2x_3 = x_3x_2,\ x_2x_4 = x_4x_1, \\ &\qquad x_1x_2 = x_3x_4,\ x_1x_4 = x_4x_3,\ x_3x_1 = x_4x_2\rangle. \end{aligned}$$

Then the mapping $g\colon D^{4,3} \longrightarrow S$, determined by $g(x_1) = x_4$, $g(x_2) = x_2$, $g(x_3) = x_1$ and $g(x_4) = x_3$, is a monoid isomorphism.

Note that there is an anti-isomorphism $h\colon D^{4,4} \longrightarrow D^{4,5}$ determined by $h(x_1) = x_1$, $h(x_2) = x_3$, $h(x_3) = x_2$ and $h(x_4) = x_4$. For any automorphism or anti-automorphism σ of $D^{4,4}$, we have $\sigma(x_3) = x_3$, because x_3 is the unique generator that appears in all the defining relations of $D^{4,4}$. It is easy to check that the permutation $(x_1\ x_2\ x_4)$ induces an automorphism of $D^{4,4}$. Since the permutation $(x_1\ x_2)$ does not induce an anti-automorphism of $D^{4,4}$, it easily follows that $D^{4,4}$ has no anti-automorphism. Hence $D^{4,4}$ is not isomorphic to $D^{4,5}$.

Since $D^{4,3}$ has exactly one pair of commuting generators, $D^{4,4}$ and $D^{4,5}$ have no pair of commuting generators and $D^{4,2}$ has two pairs of commuting generators, we have that $D^{4,2}$, $D^{4,3}$, $D^{4,4}$, $D^{4,5}$ are pairwise non-isomorphic.

Finally, suppose that in none of the defining relations of S there are exactly three generators (note that such a monoid is not isomorphic to any of the monoids $D^{4,2}$, $D^{4,3}$, $D^{4,4}$, $D^{4,5}$). Without loss of generality, since S is nonabelian, we may assume that $x_2x_1 = x_3x_4$ is one of the defining relations of S.

Since S is non-degenerate, there are two possibilities: either $x_3x_1 = x_1x_3$ or $x_3x_1 = x_4x_2$.

If $x_3x_1 = x_1x_3$, then $x_4x_1 = x_3x_2$, $x_4x_2 = x_2x_4$ and $x_1x_2 = x_4x_3$. Consequently $S = B^{4,5}$ and thus S satisfies the cyclic condition, a contradiction. So $x_3x_1 = x_4x_2$. Since S is non-degenerate, we have $x_4x_1 = x_2x_3$, $x_4x_3 = x_1x_2$ and $x_1x_3 = x_2x_4$. Thus $S = D^{4,1}$.

To finish the proof, note that $D^{4,1}$ does not satisfy the cyclic condition, because $x_2x_1 = x_3x_4$ and $x_2x_3 = x_4x_1$. □

10.3 Examples of Gelfand-Kirillov dimension 2

Let $S = \langle x_1, x_2, x_3, x_4 \rangle$ be a non-cyclic monoid of skew type that satisfies the cyclic condition. Because of Theorem 10.1.2, we know that $2 \leq \mathrm{GK}(K[S]) \leq 4$, for any field K. If S is one of the binomial monoids $B^{4,i}$ (with $1 \leq i \leq 5$) listed in Proposition 10.2.1, then we know from Theorem 8.3.10 that S is a monoid of I-type and thus $\mathrm{GK}(K[S]) = 4$ (see Chapter 8). We now show that $K[C^{4,2}] = 3$. To prove this, note that in this monoid, $x_1x_3x_4 = x_1x_4x_1 = x_4x_3x_1 = x_4x_2x_3 = x_2x_4x_3 = x_2x_1x_4 = x_1x_2x_4$ and $x_2x_3x_4 = x_3x_1x_4 = x_3x_4x_3 = x_4x_1x_3 = x_4x_3x_2 = x_1x_4x_2 = x_1x_2x_4$. Thus $x_1x_2x_4 = x_1x_3x_4 = x_2x_3x_4$. These equalities and the defining relations for $C^{4,2}$ easily imply that $x_1^a x_2^b x_3^c x_4 = x_1^{a+b+c-1} x_2 x_4$, in case that at least two of the exponents a, b, c are nonzero. It follows that

$$C^{4,2} = \langle x_1, x_2, x_3 \rangle \cup \{x_1^a x_4^b, x_2^a x_4^b, x_3^a x_4^b, x_1^a x_2 x_4^b,\ a, b \geq 0\}.$$

The submonoid $\langle x_1, x_2, x_3 \rangle$ is binomial and thus, since it is of I-type by Theorem 8.3.10, $\mathrm{GK}(K[\langle x_1, x_2, x_3 \rangle]) = 3$. Hence $\mathrm{GK}(K[C^{4,2}]) = 3$.

In this section, we will show that $\mathrm{GK}(K[C^{4,1}]) = 2$ and that for integers n and j, with $n \geq 4$ and $2 \leq j \leq n$, there exist monoids M of skew type, generated by n elements, that satisfy the cyclic condition and with $\mathrm{GK}(K[M]) = j$ for any field K. In order to prove this and to describe the algebraic structure of $C^{4,1}$ and $K[C^{4,1}]$, we study the ideal chain discussed in Chapter 9. Recall that

$$\begin{aligned} C^{4,1} \quad = \quad & \langle x_1, x_2, x_3, x_4 \mid x_4x_1 = x_3x_4,\ x_4x_2 = x_2x_3,\ x_4x_3 = x_1x_4, \\ & x_3x_1 = x_1x_2,\ x_3x_2 = x_2x_4,\ x_2x_1 = x_1x_3 \rangle. \end{aligned}$$

It follows that

$$C^{4,1} = \{x_1^{a_1} x_2^{a_2} x_3^{a_3} x_4^{a_4} \mid a_i \geq 0,\ 1 \leq i \leq 4\}$$

and

$$\begin{array}{llll} x_1x_2^2 = x_3^2x_1, & x_2x_1^2 = x_1^2x_2, & x_3x_1^2 = x_1^2x_3, & x_4x_1^2 = x_3^2x_4, \\ x_1x_3^2 = x_2^2x_1, & x_2x_3^2 = x_4^2x_2, & x_3x_2^2 = x_2^2x_3, & x_4x_2^2 = x_2^2x_4, \\ x_1x_4^2 = x_4^2x_1, & x_2x_4^2 = x_3^2x_2, & x_3x_4^2 = x_4^2x_3, & x_4x_3^2 = x_1^2x_4. \end{array}$$

It is readily verified that the monoid

$$A = \langle x_1^2,\ x_2^2,\ x_3^2,\ x_4^2 \rangle.$$

is abelian and that

$$As = sA,$$

for all $s \in C^{4,1}$. Moreover,

$$C^{4,1} = \bigcup_{f\in F} fA = \bigcup_{f\in F} Af,$$

where

$$F = \{x_1^{a_1}x_2^{a_2}x_3^{a_3}x_4^{a_4} \mid 0 \le a_i \le 1,\ 1 \le i \le 4\}.$$

Also note that the permutation $(x_1\ x_2\ x_4)$ extends to an automorphism σ of $C^{4,1}$.

For $1 \le i \le 4$, we denote the ideal of $C^{4,1}$ consisting of the elements that are left (respectively right) divisible by at least i generators (see Theorem 9.3.7) simply by C_i (respectively C_i'). Put $a = x_1x_2x_4$, $b = x_1x_3x_4$ and $c = x_1x_2x_3$. One gets the following equalities.

$$\begin{aligned} a &= x_1x_2x_4 = x_1x_3x_2 = x_2x_1x_2 = x_2x_3x_1 \\ &= x_4x_2x_1 = x_4x_1x_3 = x_3x_4x_3 = x_3x_1x_4, \end{aligned}$$

$$\begin{aligned} b &= x_1x_4x_1 = x_1x_3x_4 = x_2x_1x_4 = x_2x_4x_3 \\ &= x_3x_2x_3 = x_3x_4x_2 = x_4x_1x_2 = x_4x_3x_1, \end{aligned}$$

$$\begin{aligned} c &= x_1x_4x_2 = x_4x_3x_2 = x_4x_2x_4 = x_2x_3x_4 \\ &= x_2x_4x_1 = x_3x_2x_1 = x_3x_1x_3 = x_1x_2x_3. \end{aligned}$$

Hence $a, b, c \in C_4 \cap C_4'$.

Proposition 10.3.1. *Let*

$$\begin{aligned} C^{4,1} = \langle x_1, x_2, x_3, x_4 \mid\ & x_4x_1 = x_3x_4,\ x_4x_2 = x_2x_3,\ x_4x_3 = x_1x_4, \\ & x_3x_1 = x_1x_2,\ x_3x_2 = x_2x_4,\ x_2x_1 = x_1x_3\rangle. \end{aligned}$$

The following properties hold.

(1) $C_i = C_i'$, *for all* $1 \le i \le 4$.

(2) $C_1 \setminus C_2 = \bigcup_{1\le i\le 4} \prec x_i \succ$.

(3) $C_2 \setminus C_3 = \{x_i^n x_j^m \mid n, m > 0,\ 1 \le i < j \le 4\} = \{x_i^n x_j^m \mid n, m > 0,\ 1 \le i, j \le 4,\ i \ne j\} = B \cup \sigma(B) \cup \sigma^2(B)$, *where* $B = \{x_1^n x_2^m,\ x_1^n x_3^m \mid n, m > 0\}$. *Further,* $B\sigma(B) \subseteq C_4$ *and, modulo* C_4 *(using matrix notation),*

$$B = \begin{bmatrix} B_{11} & B_{12} \\ B_{21} & B_{22} \end{bmatrix},$$

where

(a) $B_{11} = \{x_1^{2n}x_2^m \mid n, m > 0\} \subseteq \langle x_1^2, x_2\rangle$, *a free abelian monoid of rank* 2,

(b) $B_{12} = \{x_1^{2n}x_2^m \mid n \geq 0, m > 0\}x_1 = x_1\{x_1^{2n}x_3^m \mid n \geq 0, m > 0\}$,

(c) $B_{21} = \{x_1^{2n}x_3^m \mid m > 0, n \geq 0\}x_1 = x_1\{x_1^{2n}x_2^m \mid m > 0, n \geq 0\}$,

(d) $B_{22} = \{x_1^{2n}x_3^m \mid n, m > 0\} \subseteq \langle x_1^2, x_3\rangle$, *a free abelian monoid of rank* 2.

(4) $C_3 \setminus C_4 = \emptyset$.

(5) C_4 *is a cancellative semigroup and*

$$C_4 = C^{4,1}a \cup C^{4,1}b \cup C^{4,1}c = aC^{4,1} \cup bC^{4,1} \cup cC^{4,1},$$

where $a = x_1x_2x_4$, $b = x_1x_3x_4$ *and* $c = x_1x_2x_3$. *The group of quotients of* $C^{4,1}$ *is*

$$C_4C_4^{-1} = \mathrm{gr}(a^{-2}(a^2x_1^2)) \rtimes D_{14},$$

a semidirect product of an infinite cyclic group with the dihedral group of order 14. *Furthermore,*

$$C_4C_4^{-1} = \mathrm{gr}(\beta,\, \alpha, a) = \mathrm{gr}(\alpha, \beta^{-1}a) \cong \mathrm{gr}(\alpha) \rtimes \mathbb{Z},$$

where $\alpha = a^{-2}x_1x_2x_3^2x_4^2$, $\beta = a^{-2}(a^2x_1^2)$ *and* $\alpha^7 = 1$, $a^2 = \beta^3$, $a\alpha = \alpha^{-1}a$, $\alpha\beta = \beta\alpha$ *and* $a\beta = \beta a$. *The torsion subgroup* $(C_4C_4^{-1})^+$ *of this group is the cyclic group of order* 7 *generated by the element* α.

(6) $C^{4,1}$ *is periodic modulo* A, *that is, for any* $s \in C^{4,1}$ *there exists a positive integer* n *so that* $s^n \in A$. *Also*

$$C^{4,1} = \bigcup_{f\in F} fA,$$

with

$$F = \{1, x_1, x_2, x_3, x_4, x_1x_2, x_1x_3, x_1x_4, x_2x_3, x_2x_4, x_3x_4, a, b, c\},$$

and

$$fA \cap f'A = \emptyset \text{ for } f \neq f'.$$

(7) *The least cancellative congruence* ρ_{can} *on* $C^{4,1}$ *is generated by the pairs* (x_1^2, x_2^2), (x_2^2, x_3^2) *and* (x_3^2, x_4^2). *Further,* $\mathrm{gr}(S/\rho_{can}) = C_4C_4^{-1}$.

(8) *A is a separative semigroup, that is, if* $a, b \in A$, *then* $a^2 = ab = b^2$ *implies* $a = b$.

For a field K *the semigroup algebra* $K[C^{4,1}]$ *satisfies the following properties.*

(9) $\mathcal{B}(K[C^{4,1}]) = \begin{cases} \{0\} & \textit{if } \operatorname{char}(K) \neq 7, \\ I(\eta) & \textit{if } \operatorname{char}(K) = 7, \end{cases}$

where η *is defined as follows:* $s\eta t$ *if either* $s = t$, *or* $s, t \in C_4$ *and* $s - t \in \omega(C_4C_4^{-1}, (C_4C_4^{-1})^+)$, *the augmentation ideal determined by* $(C_4C_4^{-1})^+$ *in* $C_4C_4^{-1}$, *that is, the kernel of the natural epimorphism from* $K[C_4C_4^{-1}]$ *to* $K[C_4C_4^{-1}/(C_4C_4^{-1})^+]$.

(10) *The minimal prime ideals of* $K[C^{4,1}]$ *are*

$$K[C^{4,1}]x_4K[C^{4,1}],\ K[C^{4,1}]x_1K[C^{4,1}],\ K[C^{4,1}]x_2K[C^{4,1}]$$

and also the primes P *containing* $I(\rho_{can})$ *so that*

$$P/I(\rho_{can}) = Q \cap K[C^{4,1}/\rho_{can}],$$

where Q *is a minimal prime of the group algebra*

$$K[(C^{4,1}/\rho_{can})(C^{4,1}/\rho_{can})^{-1}] \cong K[C_4C_4^{-1}].$$

In particular, if $\operatorname{char}(K) = 7$, *then there is only one prime of the latter type.*

(11) $K[C^{4,1}]$ *is embedded in the direct product*

$$K[C_4C_4^{-1}] \oplus \mathrm{M}_2(K[X, Y, X^{-1}, Y^{-1}])^3 \oplus K[X, X^{-1}]^4 \oplus K.$$

(12) $\mathrm{GK}(K[C^{4,1}]) = 2 = \mathrm{cl}(K[C^{4,1}])$.

Proof. The proof will show that the dual sets C_i' have the same description as given in parts (2),(3),(4) and (5). Hence part (1) follows.

Since $C^{4,1}$ is non-degenerate, it is clear that $x_i^a x_j \in C_2$ for any $a > 0$ and $i \neq j$. So part (2) follows.

We now prove part (3). First note that

$$C^{4,1}/C^{4,1}x_4C^{4,1} \cong B^{3,2}/B^{3,2}x_2x_3B^{3,2} = B^{3,2}/B^{3,2}x_2x_3,$$

where $B^{3,2}$ is the binomial monoid on 3 generators satisfying the relations $x_2x_1 = x_1x_3$, $x_3x_1 = x_1x_2$, $x_3x_2 = x_2x_3$. Each nonzero element of

$B^{3,2}/B^{3,2}x_2x_3$ can be presented uniquely either as $x_1^{a_1}x_2^{a_2}$ or as $x_1^{a_1}x_3^{a_2}$, with $a_1, a_2 \geq 0$. So, in $C^{4,1}$, we obtain that $x_1^{a_1}x_2^{a_2} \notin C^{4,1}x_4C^{4,1}$ and $x_1^{a_1}x_3^{a_2} \notin C^{4,1}x_4C^{4,1}$. Also, the set B_3, that is the elements in $B^{3,2}$ left divisible by the three generators, consists of the elements with $x_1x_2x_3$ as a subword. So $B_3 \subseteq B^{3,2}x_2x_3$. It follows thus that $B_{11} \cap C_3 = \emptyset$. Since, for any $b \in B$, there exist elements $x, y \in \{x_1, x_2, x_3, x_4, 1\}$ so that $xby \in B_{11}$, we get that $B \cap C_3 = \emptyset$. Because C_3 is invariant under the map σ, we thus have proved that $(B \cup \sigma(B) \cup \sigma^2(B)) \cap C_3 = \emptyset$ and therefore $(B \cup \sigma(B) \cup \sigma^2(B)) \subseteq (C_2 \setminus C_3)$. Note that, from the defining relations and equalities mentioned before the statement of the result, it is readily verified that

$$\begin{aligned} B \cup \sigma(B) \cup \sigma^2(B) &= \{x_i^{a_1}x_j^{a_2} \mid a_1, a_2 > 0,\ i < j\} \\ &= \{x_i^{a_1}x_j^{a_2} \mid a_1, a_2 > 0,\ i \neq j\}. \end{aligned}$$

To prove that $C_3 \setminus C_4 = \emptyset$ and that the set $C_2 \setminus C_3$ is as described in the statement of the proposition, it is sufficient to show that

$$(B \cup \sigma(B) \cup \sigma^2(B))\,\{x_1, x_2, x_3, x_4\} \subseteq (B \cup \sigma(B) \cup \sigma^2(B)) \cup C_4.$$

Since C_4 is invariant under σ, it is sufficient to show that, for every i,

$$Bx_i \subseteq B \cup C_4.$$

So, let $z \in B$. Assume $z \in B_{11}$ and thus $z = x_2^{a_1}x_1^{2a_2} = x_1^{2a_2}x_2^{a_1}$, for some $a_1, a_2 > 0$. Then, because $a, b, c \in C_4$,

$$\begin{aligned} zx_1 &= x_2^{a_1}x_1^{2a_2+1} \in B, \\ zx_2 &= x_2^{a_1}x_1^{2a_2}x_2 = x_2^{a_1+1}x_1^{2a_2} \in B, \\ zx_3 &= x_2^{a_1}x_1^{2a_2}x_3 = x_2^{a_1-1}ax_1^{2a_2-1} \in C_4, \\ zx_4 &= x_2^{a_1}x_1^{2a_2}x_4 = x_2^{a_1}x_4x_3^{2a_2} = x_2^{a_1-1}bx_3^{2a_2-1} \in C_4. \end{aligned}$$

Assume $z \in B_{12}$ and thus $z = x_2^{a_1}x_1^{2a_2+1}$, for some $a_1 > 0$ and $a_2 \geq 0$. Then,

$$\begin{aligned} zx_1 &= x_2^{a_1}x_1^{2a_2+2} \in B, \\ zx_2 &= x_2^{a_1}x_1^{2a_2+1}x_2 = x_2^{a_1-1}ax_1^{2a_2} \in C_4, \\ zx_3 &= x_2^{a_1}x_1^{2a_2+1}x_3 = x_2^{a_1-1}x_2x_1x_3x_1^{2a_2} = x_2^{a_1-1}x_2x_2x_1x_1^{2a_2} \in B, \\ zx_4 &= x_2^{a_1}x_1^{2a_2+1}x_4 = x_2^{a_1-1}x_2x_1x_4x_3^{2a_2} = x_2^{a_1-1}bx_3^{2a_2} \in C_4. \end{aligned}$$

Assume $z \in B_{21}$ and thus $z = x_3^{a_1}x_1^{2a_2+1}$, for some $a_1 > 0$ and $a_2 \geq 0$. Then,

$$\begin{aligned} zx_1 &= x_3^{a_1}x_1^{2a_2+2} \in B, \\ zx_2 &= x_3^{a_1}x_1^{2a_2+1}x_2 = x_3^{a_1-1}x_3x_1x_2x_1^{2a_2} = x_3^{a_1-1}x_3x_3x_1x_1^{2a_2} \in B, \\ zx_3 &= x_3^{a_1}x_1^{2a_2+1}x_3 = x_3^{a_1}x_2x_1x_1^{2a_2} = x_3^{a_1-1}cx_1^{2a_2} \in C_4, \\ zx_4 &= x_3^{a_1}x_1^{2a_2+1}x_4 = x_3^{a_1-1}ax_3^{2a_2} \in C_4. \end{aligned}$$

Assume $z \in B_{22}$ and thus $z = x_3^{a_1} x_1^{2a_2}$, for some $a_1, a_2 > 0$. Then,

$$\begin{aligned}
zx_1 &= x_3^{a_1} x_1^{2a_2+1} \in B, \\
zx_2 &= x_3^{a_1} x_1^{2a_2} x_2 = x_3^{a_1} x_2 x_1^{2a_2} = x_3^{a_1-1} c x_1^{2a_2-1} \in C_4, \\
zx_3 &= x_3^{a_1+1} x_1^{2a_2} \in B, \\
zx_4 &= x_3^{a_1} x_1^{2a_2} x_4 = x_3^{a_1} x_4 x_3^{2a_2} = x_3^{a_1-1} a x_3^{2a_2} \in C_4.
\end{aligned}$$

This finishes the proof of the description of the sets $C_2 \setminus C_3$ and $C_3 \setminus C_4$.

Since $x_1^2 x_2 = x_2 x_1^2$ and because an element $x_1^{2a_1} x_2^{a_2} \in B_{11}$ is uniquely determined by a_1 and a_2, we get that B_{11} is a free abelian semigroup. It also easily is verified that $B\sigma(B), B\sigma^2(B) \subseteq C_3$ and that, modulo C_4, the semigroup B is as described in the statement of the result. This finishes the proof of statements (3) and (4).

We now prove (5). First, we show that the set C_4 is as described in the statement. So, let $s \in C_4$. Since this element is left divisible by x_1, we get that $s = x_1^{a_1} t$, for some $a_1 > 0$ and $t \notin x_1 C^{4,1}$. Because $s \notin C_2 \setminus C_4$, we have that $t \in C_2$. So $t \in \{x_1x_2,\ x_2x_4,\ x_3x_4\}C^{4,1}$. Consequently $s \in x_1^{a_1-1} a C^{4,1} \cup x_1^{a_1-1} b C^{4,1} \cup x_1^{a_1-1} c C^{4,1}$.

We notice that

$$\begin{aligned}
x_1 a &= x_1(x_4 x_1 x_3) = (x_1 x_4 x_1) x_3 = b x_3, \\
x_1 b &= x_1(x_2 x_4 x_3) = (x_1 x_2 x_4) x_3 = a x_3, \\
x_1 c &= x_1(x_2 x_3 x_4) = (x_1 x_2 x_3) x_4 = c x_4.
\end{aligned}$$

Next we claim that for any $v \in \{a, b, c\}$,

$$vx_1^2 = x_1^2 v = x_2^2 v = v x_2^2 = x_3^2 v = v x_3^2 = x_4^2 v = v x_4^2. \qquad (10.3)$$

This fact, together with the previous one, implies that indeed $C_4 = aC^{4,1} \cup bC^{4,1} \cup cC^{4,1}$. By symmetry, we also get $C_4 = C^{4,1}a \cup C^{4,1}b \cup C^{4,1}c$. The claim for $v = a$ follows from the following identities.

$$\begin{aligned}
x_1^2 a &= x_1 x_1 x_4 x_1 x_3 = x_1 b x_3 = x_1 x_2 x_4 x_3 x_3 = a x_3^2, \\
x_2^2 a &= x_2 x_2 x_4 x_1 x_3 = x_2 c x_3 = x_2 x_3 x_2 x_1 x_3 = x_2 x_3 x_1 x_3 x_3 \\
&= a x_3 x_3 = x_1 x_3 x_2 x_3 x_3 = x_1 b x_3 = x_1 x_1 x_4 x_1 x_3 = x_1^2 a, \\
x_3^2 a &= x_3 x_3 x_4 x_2 x_1 = x_3 b x_1 = x_3 x_4 x_3 x_1 = a x_1^2, \\
x_1^2 a &= x_1 x_1 x_4 x_2 x_1 = x_1 c x_1 = x_1 x_2 x_4 x_1 = a x_1^2, \\
a x_1^2 &= x_1 x_2 x_4 x_1^2 = x_1 x_2 x_3^2 x_4 = x_1 x_4^2 x_2 x_4 = x_4^2 x_1 x_2 x_4 = x_4^2 a,
\end{aligned}$$

$$
\begin{aligned}
ax_2^2 &= x_1x_2x_4x_2^2 = x_1x_2x_2^2x_4 = x_1x_2^2x_2x_4 = x_3^2x_1x_2x_4 = x_3^2a, \\
ax_3^2 &= x_1x_2x_4x_3^2 = x_1x_2x_1^2x_4 = x_1x_1^2x_2x_4 = x_1^2x_1x_2x_4 = x_1^2a, \\
ax_4^2 &= x_1x_2x_4x_4^2 = x_1x_2x_4^2x_4 = x_1x_3^2x_2x_4 = x_2^2x_1x_2x_4 = x_2^2a.
\end{aligned}
$$

Applying the map σ to the above equalities, we also get that the claim holds for $v = c = \sigma(a)$ and $b = \sigma(c)$.

Since the generators normalize A and because $x_1x_2x_3x_4 = x_1^2x_4x_3$, it easily follows that

$$C^{4,1} = \bigcup_{f \in F} fA,$$

with

$$F = \{1, x_1, x_2, x_3, x_4, x_1x_2, x_1x_3, x_1x_4, x_2x_3, x_2x_4, x_3x_4, a, b, c\}.$$

Second, we deal with the structure of the semigroup C_4. For this, consider the following monoid

$$
\begin{aligned}
\overline{\overline{C}} &= \langle x_1, x_2, x_3, x_4 \mid x_4x_1 = x_3x_4,\ x_4x_2 = x_2x_3,\ x_4x_3 = x_1x_4, \\
&\qquad x_3x_1 = x_1x_2,\ x_3x_2 = x_2x_4,\ x_2x_1 = x_1x_3,\ x_1^2 = x_2^2 = x_3^2 = x_4^2 = 1\rangle.
\end{aligned}
$$

Clearly $\overline{\overline{C}}$ is a group. Since $C^{4,1} = \bigcup_{f \in F} fA$ and $|F| = 14$, it follows that $\overline{\overline{C}}$ has at most 14 elements. In the dihedral group

$$D_{14} = \mathrm{gr}(g, h \mid g^7 = 1,\ h^2 = 1,\ hg = g^6h),$$

the elements $x_1 = g,\ x_2 = gh,\ x_3 = hg,\ x_4 = g^3h$ satisfy the relations mentioned in the presentation of $\overline{\overline{C}}$. So we get that

$$\overline{\overline{C}} \cong D_{14}$$

and $|F| = 14$. In particular, part (6) of the statement follows.

Also consider the following monoid

$$
\begin{aligned}
\overline{C} &= \langle x_1, x_2, x_3, x_4 \mid x_4x_1 = x_3x_4,\ x_4x_2 = x_2x_3,\ x_4x_3 = x_1x_4, \\
&\qquad x_3x_1 = x_1x_2,\ x_3x_2 = x_2x_4,\ x_2x_1 = x_1x_3,\ x_1^2 = x_2^2 = x_3^2 = x_4^2\rangle.
\end{aligned}
$$

Identifying F with its natural image in $\overline{C}$, we get

$$\overline{C} = \bigcup_{f \in F} \langle x_1^2 \rangle f = \bigcup_{f \in F} f \langle x_1^2 \rangle,$$

and for distinct $f, f' \in F$ we have $\langle x_1^2 \rangle f \cap \langle x_1^2 \rangle f' = \emptyset$. We claim that $\overline{C}$ is a cancellative monoid. In order to show this, suppose that $f_1 x_1^{2k} z = f_2 x_1^{2l} z \in \overline{C}$, for some $z \in \overline{C}$. Because of the previous, $f_1 = f_2$. Comparing lengths (notice that $\overline{C}$ has a length function because the defining relations are homogeneous of degree 2) we obtain that $k = l$. This proves that $\overline{C}$ is right cancellative. Similarly the left cancellation law is shown.

The relations (10.3), together with the fact that $\overline{C}$ is cancellative, imply that indeed the relation ρ_{can} on $C^{4,1}$ defined by the pairs (x_1^2, x_2^2), (x_2^2, x_3^2) and (x_3^2, x_4^2) is the least cancellative congruence and thus $\overline{C} = C^{4,1}/\rho_{can}$.

Note that since C_4 is an ideal and $C_4 = aC4, 1 \cup bC^{4,1} \cup cC^{4,1} = C^{4,1}a \cup C^{4,1}b \cup C^{4,1}c$, it follows at once from the relations (10.3) and the above, that

$$I(\rho_{can})\{a, b, c\} = \{0\} = \{a, b, c\} I(\rho_{can}), \tag{10.4}$$

and thus also

$$I(\rho_{can}) C_4 = \{0\} = C_4 I(\rho_{can}).$$

Our next claim is that, for any $w \in C_4$ with $|w| \geq 4$, there exist distinct $u, v \in \{a, b, c\}$ and distinct $u', v' \in \{a, b, c\}$ so that

$$w \in uC^{4,1} \cap vC^{4,1} \cap C^{4,1}u' \cap C^{4,1}v'. \tag{10.5}$$

Indeed, let $w \in C_4$. Because $C_4 = aC^{4,1} \cup bC^{4,1} \cup cC^{4,1}$ and because σ permutes a, b and c, we may assume without loss of generality that $w = aw'$, for some $w' \in C^{4,1}$. One part of the claim now follows from the following identities.

$$\begin{aligned}
ax_1 &= x_1x_2x_4x_1 = x_1x_2x_3x_4 = cx_4, \\
ax_2 &= x_1x_2x_4x_2 = x_1x_2x_2x_3 = x_3x_1x_2x_3 = x_3cx_3x_4x_2x_4 = bx_2, \\
ax_3 &= x_1x_2x_4x_3 = x_1b = x_1x_3x_4x_2 = bx_2, \\
ax_4 &= x_1x_2x_4x_4 = x_1x_3x_2x_4 = x_1x_3x_3x_2 = x_2x_1x_3x_2 = x_2a \\
&= x_2x_3x_4x_3 = cx_3.
\end{aligned}$$

Similarly one proves the other part of the claim.

Finally, we prove that C_4 is cancellative. We only prove the right cancellation law. So, assume $xz = yz$ for some $x, y, z \in C_4$. Notice that, because the relations are homogeneous, $|x| = |y|$. Of course from the non-degeneracy, we obtain that $x = y$ if $|x| = |y| = 2$. Next, assume $|x| = |y| = 3$ and $x \neq y$. Since $x, y \in C_4 = aC^{4,1} \cup bC^{4,1} \cup cC^{4,1}$, we get $x, y \in \{a, b, c\}$. Because of the mapping σ, there is no loss of generality assuming that $x = a$ and $y = b$.

So $az = bz$. Since $zC_4 \cap A \neq \emptyset$, we may also assume that $z \in A$. However, as $aA \cap bA = \emptyset$, this yields a contradiction. Next, assume $|x| = |y| \geq 4$. Because of (10.5), there then exists $d \in \{a, b, c\}$ so that $x = x'd$ and $y = y'd$, for some $x', y' \in C^{4,1}$. Hence $x'dz = y'dz$ and thus $x'\rho_{can}y'$. But then (10.4) implies that $(x' - y')d = 0$. So $x = y$, as desired. This indeed shows that C_4 is cancellative.

It is easily seen that C_4 satisfies the Ore condition and hence it has a group of quotients $C_4C_4^{-1}$. We now determine the structure of this group. For this first notice that

$$\begin{aligned} a^2 &= (x_1x_2x_4)^2 = x_1x_2x_4x_1x_2x_4 = x_1x_3x_2x_1x_2x_4 \\ &= x_1x_3ax_4 = x_1x_3x_1x_2x_4x_4 = x_1x_1x_2x_2x_4x_4 = x_1^2x_2^2x_4^2. \end{aligned}$$

Hence $c^2 = (\sigma(a))^2 = \sigma(a^2) = \sigma(x_1^2x_2^2x_4^2) = x_2^2x_4^2x_1^2 = a^2$. Thus also $b^2 = (\sigma^2(a))^2 = \sigma^2(a^2) = a^2$ and therefore $c^2 = b^2 = a^2$. So

$$a^2 = b^2 = c^2 \in A.$$

Now

$$c^2 = (x_1x_2x_3)^2 = x_1x_2x_3x_1x_2x_3 = x_1x_1x_3x_2x_2x_3 = x_1^2x_2^2x_3^2.$$

Again using the map σ we get that

$$a^2 = x_1^2x_2^2x_4^2 = x_1^2x_2^2x_3^2 = x_2^2x_3^2x_4^2 = x_1^2x_3^2x_4^2.$$

It follows that a^2 is central in $C^{4,1}$.

Consider the semigroup homomorphism

$$\psi_1 : C^{4,1} \longrightarrow C_4C_4^{-1} : s \mapsto a^{-2}(a^2s).$$

For $x, y \in C^{4,1}$, it is clear that $\psi_1(x) = \psi_1(y)$ if and only if $a^2x = a^2y$, or equivalently (because C_4 is cancellative) $ax = ay$, that is, (because of (10.4)) $x\rho_{can}y$. So ψ_1 factors through $C^{4,1}/\rho_{can}$. Hence we get an injective homomorphism

$$\psi_2 : C^{4,1}/\rho_{can} \longrightarrow C_4C_4^{-1}$$

and we can identify the group of quotients of $C^{4,1}/\rho_{can}$ with $C_4C_4^{-1}$. Consequently, we obtain a homomorphism

$$\psi_3 : C_4C_4^{-1} \longrightarrow \overline{\overline{C}}$$

so that

$$\psi_3 \circ \psi_2 = \psi,$$

where ψ is the natural epimorphism $C^{4,1}/\rho_{can} \longrightarrow \overline{\overline{C}}$. So, ψ_3 is onto as well. Because $a^2x_1^2 = a^2x_i^2$ for $1 \le i \le 4$ and a^2 is central in $C^{4,1}$, it follows that $a^{-2}(a^2x_1^2)$ is central in $C_4C_4^{-1}$. Further, note that $\psi_2(A) = \langle a^{-2}(a^2x_1^2)\rangle$ and the index of $\psi_3(\mathrm{gr}(x_1^2))$ in $\overline{\overline{C}}$ is at most 14. It follows that $C_4C_4^{-1}$ is an extension of the cyclic subgroup $\mathrm{gr}(a^{-2}(a^2x_1^2))$ by D_{14}. In particular, it is a finitely generated finite conjugacy group and thus its torsion elements form a finite subgroup, which we denote by $(C_4C_4^{-1})^+$. Also, note that $x_1x_2x_2^2x_4^2 \in C_4 \setminus A$ and it is easily verified that

$$(x_1x_2x_2^2x_4^2)^2 = a^{14}.$$

Hence

$$(a^{-2}x_1x_2x_2^2x_4^2)^7 = 1.$$

Let $\alpha = a^{-2}x_1x_2x_2^2x_4^2$. Since $\beta = a^{-2}(a^2x_1^2)$ has infinite order, we get that $(C_4C_4^{-1})^+$ is embedded in $\overline{\overline{C}} \cong D_{14}$. So, we obtain that the cyclic group $\mathrm{gr}(\alpha)$ is a normal subgroup of order 7 in $C_4C_4^{-1}$. From the above we know that β is central in $C_4C_4^{-1}$, and thus its image in $\overline{\overline{C}} \cong D_{14}$ is trivial. Also, $\beta^3 = a^2$. Therefore, $\psi_3(a)$ is an element of order 2. Hence, we get that

$$(C_4C_4^{-1})^+ = \mathrm{gr}(\alpha)$$

and

$$C_4C_4^{-1} = \mathrm{gr}(\alpha, \beta, a) = \mathrm{gr}(\alpha, \beta^{-1}a).$$

Consequently,

$$C_4C_4^{-1} \cong \mathrm{gr}(\alpha) \rtimes \mathbb{Z}.$$

Further, we note that

$$\alpha a \alpha \in x_1x_2ax_1x_2A$$

and

$$x_1x_2ax_1x_2 = x_1x_2x_1x_2x_4x_1x_2 = x_1^2x_3x_2x_4x_1x_2 \in x_2x_1x_2A.$$

So $\alpha a \alpha \in aA$. It thus follows from (10.3) that $\alpha a \alpha = ax_1^{2v}$ for some $v \in \mathbb{Z}$. As $|\alpha a \alpha| = |a|$, we obtain that $\alpha a \alpha = a$. Hence, in $C_4C_4^{-1}$, we get $a\alpha = \alpha^{-1}a$. It is now readily verified that $C_4C_4^{-1}$ has a presentation as mentioned in the statement of the proposition. So this completes the proof of (1) to (7). Part (8) follows once (9) has been proved. Indeed, (9) implies that $\mathcal{B}(L[C^{4,1}]) = \{0\}$ for any field L of zero characteristic. As $L[C^{4,1}]$ is a normalizing extension of $L[A]$, this implies that $\mathcal{B}(L[A]) = \{0\}$. Hence if $a, b \in A$, with $a^2 = ab = b^2$, then $a - b$ is a nilpotent element in $L[A]$ and thus $a - b = 0$. So A is separative, that is, (8) holds.

Let K be a field. We now prove the statements on the algebra $K[C^{4,1}]$. Because of the descriptions obtained on $C^{4,1}/C_2$ and C_2/C_3, it is readily verified that $K_0[C^{4,1}/C_2]$ and $K_0[C_2/C_3]$ are semiprime. So $\mathcal{B}(K[C^{4,1}]) = \mathcal{B}(K[C_4])$, since $C_3 = C_4$. If $\mathrm{char}(K) \neq 7$, then, because of Theorem 3.2.8, the prime radical $\mathcal{B}(K[C_4])$ of the algebra of the cancellative semigroup C_4 is trivial. On the other hand, if $\mathrm{char}(K) = 7$, then it follows from Theorem 3.2.8 and a very special case of a result of Dyment and Zalesskii (see [129, Theorem 8.4.16]) that $\mathcal{B}(K[C_4]) = \omega(C_4C_4^{-1}, (C_4C_4^{-1})^+) \cap K[C_4]$. Hence (9) follows.

From Proposition 9.5.2, we know that every minimal prime P of $K[C^{4,1}]$ contains either $I(\rho_{can})$ or C_4. In the former case,

$$P/I(\rho_{can}) = Q \cap K[C^{4,1}/\rho_{can}],$$

for some (unique) minimal prime Q of the group algebra

$$K[(C^{4,1}/\rho_{can})(C^{4,1}/\rho_{can})^{-1}] \cong K[C_4C_4^{-1}].$$

Note that, if $\mathrm{char}(K) = 7$, then, since $C_4C_4^{-1} \cong \mathrm{gr}(\alpha) \rtimes \mathbb{Z}$, it follows that there is only one minimal prime of this type. Suppose now that P contains C_4. Since B, $\sigma(B)$ and $\sigma^2(B)$ are orthogonal modulo C_4, it follows that P contains two of these sets. Assume, for example, that $\sigma(B), \sigma^2(B) \subseteq P$. As $Bx_4 \subseteq C_4 \subseteq P$, we get that $x_4 \in P$ or $B \subseteq P$. In the former case, we noticed in the beginning of the proof that

$$C^{4,1}/(x_4) \cong B^{3,2}/(x_2x_3).$$

From Theorem 8.5.2, we know that $x_2x_3B^{3,2}$ is a minimal prime ideal of $B^{3,2}$. Thus, by Theorem 8.5.6, we obtain that $K[C^{4,1}x_4C^{4,1}]$ is a prime ideal of $K[C^{4,1}]$. Consequently, so are the ideals $K[C^{4,1}x_1C^{4,1}] = \sigma(K[C^{4,1}x_4C^{4,1}])$ and $K[C^{4,1}x_2C^{4,1}] = \sigma^2(K[C^{4,1}x_4C^{4,1}])$. Note that none of these prime ideals contains $I(\rho_{can})$. If these primes are not minimal or if we have not covered all minimal primes yet, then there exists a minimal prime P so that $B, \sigma(B), \sigma^2(B) \subseteq P$. Hence, three of the generators x_1, x_2, x_3, x_4 belong to P, a contradiction with the assumptions on P. Hence we have shown that $K[C^{4,1}x_1C^{4,1}]$, $K[C^{4,1}x_2C^{4,1}]$ and $K[C^{4,1}x_4C^{4,1}]$ are indeed the minimal prime ideals of $K[C^{4,1}]$ that contain C_4. So we have proved (10).

Clearly, the descriptions of the sets $C_i \setminus C_{i+1}$ yield that free abelian subsemigroups of $C^{4,1}$ have rank at most 2 and that such a subsemigroup exists. Hence, because of Corollary 5.4.5, we obtain (12).

Finally we prove (11). Because of Lemma 2.5.1, the set $\widehat{C^{4,1}} = (C^{4,1} \setminus C_4) \cup (C_4C_4^{-1}) = (C^{4,1} \setminus C_4) \cup (C_4\,\mathrm{gr}(a^2))$ has a semigroup structure that extends that of $C^{4,1}$. Then, since $K[C_4C_4^{-1}]$ is an ideal of $K[\widehat{C^{4,1}}]$ with identity,

$$K[C^{4,1}] \subseteq K[\widehat{C^{4,1}}] \cong K_0[\widehat{C^{4,1}}/C_4C_4^{-1}] \oplus K[C_4C_4^{-1}].$$

Repeating a similar argument on $K_0[\widehat{C^{4,1}}/C_4C_4^{-1}]$ the result will follow. □

In contrast with binomial monoids, the semigroup $C = C^{4,1}$ is indecomposable (that is, it is not a product of monoids of skew type each on less generators), the divisibility layer $C_3 \setminus C_4$ is empty and $C^{4,1}$ contains a cancellative ideal T whose group of quotients is not torsion free.

Let n be an arbitrary integer, with $n \geq 4$. Let $T^{(n)}$ be the monoid of skew type generated by $x_1, \dots, x_n$ and with defining relations

$$\begin{array}{lll} x_1x_2 = x_3x_1, \quad \dots, & x_1x_{n-2} = x_{n-1}x_1, & x_1x_{n-1} = x_2x_1, \\ x_nx_1 = x_{n-1}x_n, & x_nx_{n-1} = x_1x_n, & \\ x_ix_{i+1} = x_{i+2}x_i, \quad \dots, & x_ix_{n-1} = x_nx_i, & x_ix_n = x_{i+1}x_i, \end{array}$$

for all $2 \leq i \leq n-2$. Note that $T^{(n)}$ satisfies the cyclic condition. We will show that these monoids yield algebras that have Gelfand-Kirillov dimension two. For this, we first prove three lemmas.

Lemma 10.3.2. *Let $\rho = \rho_{\text{can}}$ be the least cancellative congruence on $T^{(n)}$. If $n > 4$, then $x_2x_1x_2 = x_nx_1x_2$ and $x_1 \,\rho\, x_2 \,\rho \dots \rho\, x_n$.*

Proof. The defining relations yield that

$$\begin{aligned} x_2x_1x_2 &= x_1x_{n-1}x_2 = x_1x_2x_{n-2} \\ &= x_3x_1x_{n-2} = x_3x_{n-1}x_1 \\ &= x_nx_3x_1 \quad\;\; = x_nx_1x_2. \end{aligned}$$

Since $x_2x_1x_2 = x_nx_1x_2$, it follows that $x_2 \,\rho\, x_n$. Hence, the relations

$$x_2x_3 = x_4x_2, \quad \dots, \quad x_2x_{n-1} = x_nx_2,$$

imply that $x_2 \,\rho\, x_3 \,\rho \dots \rho\, x_n$. Since $x_nx_1 = x_{n-1}x_n$, we also get

$$x_1 \,\rho\, x_2 \,\rho \dots \rho\, x_n.$$

□

For simplicity, we denote by T_n' the subset of $T^{(n)}$ consisting of all elements that are right divisible by all generators of $T^{(n)}$. Since $T^{(n)}$ is left non-degenerate, the dual of Theorem 9.3.7 yields that T_n' is an ideal of $T^{(n)}$.

Lemma 10.3.3. *Consider* $z = x_2x_1x_2 \in T^{(n)}$. *Then* $z \in T'_n$.

Proof. For $n = 4$ we have

$$z = x_2x_1x_2 = x_2x_3x_1 = x_4x_2x_1 = x_4x_1x_3 = x_3x_4x_3 = x_3x_1x_4 \in T'_n.$$

Suppose that $n > 4$. By Lemma 10.3.2, $z = x_nx_1x_2$ and thus

$$\begin{aligned} z &= x_nx_1x_2 = x_{n-1}x_nx_2 = x_{n-1}x_2x_{n-1} = x_2x_{n-2}x_{n-1} \\ &= x_2x_nx_{n-2} = x_3x_2x_{n-2} = x_3x_{n-1}x_2 = x_nx_3x_2 = x_nx_2x_n \\ &= x_2x_{n-1}x_n = x_2x_nx_1 = x_3x_2x_1 = x_3x_1x_{n-1} = x_1x_2x_{n-1} \\ &= x_1x_nx_2 = x_nx_{n-1}x_2 = x_nx_2x_{n-2} = x_2x_{n-1}x_{n-2}. \end{aligned}$$

We claim that $z = x_2x_{i+1}x_i$, for all $n-2 \geq i \geq 3$. If $n - i = 2$, the claim follows from the above equalities. In particular, this settles the case where $n = 5$. Suppose that $n > 5$. We prove the claim by induction on $n - i$. Assume we know that $z = x_2x_{i+1}x_i$, for some i with $4 \leq i \leq n-2$. Then

$$\begin{aligned} z &= x_2x_{i+1}x_i = x_2x_ix_n = x_{i+1}x_2x_n = x_{i+1}x_3x_2 \\ &= x_3x_ix_2 = x_3x_2x_{i-1} = x_2x_nx_{i-1} = x_2x_{i-1}x_{n-1} \\ &= x_ix_2x_{n-1} = x_ix_nx_2 = x_{i+1}x_ix_2 = x_{i+1}x_2x_{i-1} = x_2x_ix_{i-1}, \end{aligned}$$

which proves the inductive claim. It follows that $z \in T^{(n)}x_i$, for all $3 \leq i \leq n-2$. Since $z = x_2x_1x_2 = x_2x_3x_1 = x_{n-1}x_2x_{n-1} = x_nx_2x_n$, we have that $z \in T'_n$. □

Lemma 10.3.4. *If* $n > 4$ *and* $m = (n-1)!$, *then* $x_k^{2m}x_j^{2m}x_i^{2m} \in T'_n$ *when* $1 \leq i < j < k \leq n$.

Proof. Recall, from Lemma 10.1.1, that the submonoid $A = \langle x_1^m, \ldots, x_n^m \rangle$ of $T^{(n)}$ is abelian. From the relations $x_1x_2 = x_3x_1, \ldots, x_1x_{n-2} = x_{n-1}x_1$, $x_1x_{n-1} = x_2x_1$, it follows that

$$x_1^jx_i = x_{j+i}x_1^j, \quad x_1^{n-2}x_i = x_ix_1^{n-2}, \tag{10.6}$$

when $2 \leq i \leq n-1$ and $1 \leq j < n-i$. The relations

$$x_nx_1 = x_{n-1}x_n, \quad x_nx_{n-1} = x_1x_n, \tag{10.7}$$

yield that

$$x_n^2x_1 = x_1x_n^2 \quad \text{and} \quad x_n^2x_{n-1} = x_{n-1}x_n^2. \tag{10.8}$$

For each i with $2 \leq i \leq n-2$, the relations

$$x_i x_{i+1} = x_{i+2} x_i, \quad \ldots, \quad x_i x_{n-1} = x_n x_i, \quad x_i x_n = x_{i+1} x_i,$$

imply that

$$x_i^{n-i} x_j = x_j x_i^{n-i}, \tag{10.9}$$

for all j with $i < j \leq n$.

We now deal with three separate cases to finish the proof.

Case 1: $1 < i < j < k \leq n$. It is easy to see that then

$$x_k x_j^{k-j-1} = x_j^{k-j-1} x_{j+1} \tag{10.10}$$

and

$$x_j x_i^{j-i+1} = x_i^{j-i+1} x_{n-1}. \tag{10.11}$$

Thus we get

$$\begin{aligned}
x_k^{2m} x_j^{2m} x_i^{2m} &= x_j^{k-j-1} x_{j+1}^{2m} x_j^{2m-k+j+1} x_i^{2m} \quad \text{(by (10.10))} \\
&= x_j^{k-j-1} x_{j+1}^{2m} x_j^{m} x_i^{2m} x_j^{m-k+j+1} \quad \text{(by (10.9))} \\
&= x_j^{k-j-1} x_j^{m} x_{j+1}^{2m} x_i^{2m} x_j^{m-k+j+1} \quad \text{(by (10.9))} \\
&= x_j^{k-j-1} x_i^{j-i+1} x_{n-1}^{m} x_n^{2m} x_i^{2m-j+i-1} x_j^{m-k+j+1} \quad \text{(by (10.11))} \\
&= x_j^{k-j-1} x_i^{j-i+1} x_n x_1^{m} x_n^{2m-1} x_i^{2m-j+i-1} x_j^{m-k+j+1} \quad \text{(by (10.7))} \\
&= x_j^{k-j-1} x_i^{j-i+1} x_n x_1^{m} x_n^{2} x_i^{m} x_n^{2m-3} x_i^{m-j+i-1} x_j^{m-k+j+1} \quad \text{(by (10.9))} \\
&= x_j^{k-j-1} x_i^{j-i+1} x_n^{3} x_1^{m} x_i^{m} x_n^{2m-3} x_i^{m-j+i-1} x_j^{m-k+j+1} \quad \text{(by (10.8))} \\
&= x_j^{k-j-1} x_i^{j-i+1} x_n^{3} x_1^{m-n+i} x_2^{m} x_1^{n-i} x_n^{2m-3} \\
&\qquad x_i^{m-j+i-1} x_j^{m-k+j+1} \quad (\text{since } x_2 x_1^{n-i} = x_1^{n-i} x_i) \\
&= x_j^{k-j-1} x_i^{j-i+1} x_n x_1^{m-n+i-1} x_n^{2} x_1 x_2^{m} x_1^{n-i} x_n^{2m-3} \\
&\qquad x_i^{m-j+i-1} x_j^{m-k+j+1} \quad \text{(by (10.8))} \\
&= (x_j^{k-j-1} x_i^{j-i+1} x_n x_1^{m-n+i-1} x_n) z (x_2^{m-1} x_1^{n-i} x_n^{2m-3} \\
&\qquad x_i^{m-j+i-1} x_j^{m-k+j+1}).
\end{aligned}$$

From Lemma 10.3.3, we know that $z \in T_n'$. Since T_n' is an ideal of $T^{(n)}$, it follows that $x_k^{2m} x_j^{2m} x_i^{2m} \in T_n'$.

Case 2: $1 = i < j < k \leq n$ and $j < n-1$. Then we get

$$\begin{aligned}
x_k^{2m} x_j^{2m} x_1^{2m} &= x_j^{k-j-1} x_{j+1}^{2m} x_j^{2m-k+j+1} x_1^{2m} \quad \text{(by (10.10))} \\
&= x_j^{k-j-1} x_{j+1}^{2m} x_j^{m} x_1^{2m} x_j^{m-k+j+1} \quad \text{(by (10.6))} \\
&= x_j^{k-j-1} x_j^{m} x_{j+1}^{2m} x_1^{2m} x_j^{m-k+j+1} \quad \text{(by (10.9))} \\
&= x_j^{k-j-1} x_j^{m} x_{j+1} x_1^{2m} x_{j+1}^{2m-1} x_j^{m-k+j+1} \quad \text{(by (10.6))} \\
&= x_j^{k-j-1} x_j^{m} x_1^{j-2} x_3 x_1^{2m-j+2} x_{j+1}^{2m-1} x_j^{m-k+j+1} \quad \text{(by (10.6))} \\
&= x_j^{k-j-1} x_1^{j-2} x_2^{m} x_3 x_1^{2m-j+2} x_{j+1}^{2m-1} x_j^{m-k+j+1} \quad \text{(by (10.6))} \\
&= (x_j^{k-j-1} x_1^{j-2} x_2^{m-1}) z (x_1^{2m-j+1} x_{j+1}^{2m-1} x_j^{m-k+j+1}).
\end{aligned}$$

Again because of Lemma 10.3.3, we thus obtain that $x_k^{2m} x_j^{2m} x_i^{2m} \in T_n'$.
Case 3: $i = 1$, $j = n-1$ and $k = n$. Then we get

$$\begin{aligned}
x_n^{2m} x_{n-1}^{2m} x_1^{2m} &= x_n^{2m} x_{n-1} x_1^{2m} x_{n-1}^{2m-1} \quad \text{(by (10.6))} \\
&= x_n^{2m} x_1^{n-3} x_2 x_1^{2m-n+3} x_{n-1}^{2m-1} \quad \text{(by (10.6))} \\
&= x_n^{2m-1} x_{n-1}^{n-4} x_n x_1 x_2 x_1^{2m-n+3} x_{n-1}^{2m-1} \quad \text{(by (10.7))} \\
&= (x_n^{2m-1} x_{n-1}^{n-4}) z (x_1^{2m-n+3} x_{n-1}^{2m-1}).
\end{aligned}$$

Lemma 10.3.3 thus yields that $x_k^{2m} x_j^{2m} x_i^{2m} \in T_n'$.

So we have shown that $x_k^{2m} x_j^{2m} x_i^{2m} \in T_n'$, for all $1 \leq i < j < k \leq n$. □

Theorem 10.3.5. *Let K be a field. Then $\mathrm{GK}(K[T^{(n)}]) = 2$ for all $n \geq 4$.*

Proof. For $n = 4$, the result follows from assertion (12) in Proposition 10.3.1, because $T^{(4)}$ coincides with the monoid $C^{4,1}$.

Suppose that $n > 4$. Again let $m = (n-1)!$. From Proposition 9.5.2, we know that $(T_n')^q I(\rho) = 0$ for some q, where $I(\rho)$ is the ideal of $K[T^{(n)}]$ determined by the least cancellative congruence ρ on $T^{(n)}$. In particular, by Lemma 10.3.2, $x_k^m - x_j^m \in I(\rho)$ for all k, j. Therefore, from Lemma 10.3.4, it follows that

$$x_k^{2mq} x_j^{2mq} x_i^{2mq} (x_k^m - x_j^m) = 0,$$

when $1 \leq i < j < k \leq n$. This implies that x_k^m, x_j^m and x_i^m do not generate a free abelian semigroup. Therefore, Theorem 10.1.2 implies that $\mathrm{GK}(K[T^{(n)}]) = 2$. □

Corollary 10.3.6. *Let K be a field. Assume n and j are integers so that $n \geq 4$ and $2 \leq j \leq n$. Then, there exists a monoid $M = \langle x_1, x_2, \ldots, x_n \rangle$ of skew type, satisfying the cyclic condition and such that $\mathrm{GK}(K[M]) = j$.*

Proof. If $j = n$, then the free abelian monoid of rank n satisfies the required conditions.

Suppose that $j = n-1$. Let A be a monoid of skew type with 4 generators that satisfies the cyclic condition such that, for any field K, $\mathrm{GK}(K[A]) = 3$ (as mentioned earlier $C^{4,2}$ is such a semigroup). Let $M = A \times \mathrm{FaM}_{n-4}$. Then it is easy to see that M is a monoid of skew type with n generators that satisfies the cyclic condition. Since $K[M]$ is the polynomial algebra over $K[A]$ in $n-4$ commuting variables, we obtain, from Proposition 3.5.5, that $\mathrm{GK}(K[M]) = 3 + (n-4) = n-1$.

Suppose that $j \leq n-2$. Let $M = T^{(n-j+2)} \times \mathrm{FaM}_{j-2}$. It is easy to see that M is a monoid of skew type with n generators that satisfies the cyclic condition. Since $K[M]$ is the polynomial algebra over $K[T^{(n-j+2)}]$ in $j-2$ commuting variables, it follows again from Proposition 3.5.5 that $\mathrm{GK}(K[M]) = \mathrm{GK}(K[T^{(n-j+2)}]) + (j-2)$. By Theorem 10.3.5, $\mathrm{GK}(K[M]) = j$. □

10.4 Non-degenerate monoids of skew type of Gelfand-Kirillov dimension one

In this section, we construct examples of non-degenerate monoids S of skew type generated by 4^n elements (with n an arbitrary positive integer) so that $\mathrm{GK}(K[S]) = 1$ and $K[S]$ is semiprime for any field K of characteristic zero. We begin with showing that the four generated non-degenerate monoid $D^{4,1}$ of skew type satisfies such properties (see Proposition 10.2.2). To prove this, we describe in detail its algebraic structure. In particular, it is shown that $D^{4,1}$ is a semilattice of cancellative semigroups. Also all minimal prime ideals of its algebra are described

Again, the investigations will make use of the chains obtained from the ideals $D_i^{4,1}$ (respectively $(D_i^{4,1})'$) consisting of the elements of $D^{4,1}$ that are left (respectively right) divisible by at least i generators (see Theorem 9.3.7). For simplicity we denote these ideals by D_i and D_i' respectively.

Theorem 10.4.1. *Let*

$$\begin{aligned} D^{4,1} \;=\; & \langle x_1, x_2, x_3, x_4 \mid x_2x_1 = x_3x_4,\; x_2x_3 = x_4x_1,\; x_2x_4 = x_1x_3, \\ & x_1x_2 = x_4x_3,\; x_1x_4 = x_3x_2,\; x_3x_1 = x_4x_2 \rangle. \end{aligned}$$

The following properties hold.

(1) $D^{4,1} \setminus D_2 = \langle x_1 \rangle \cup \langle x_2 \rangle \cup \langle x_3 \rangle \cup \langle x_4 \rangle$.

(2) $D_2 \setminus D_3 = \{x_i x_j \mid 1 \leq i, j \leq 4,\ i \neq j\}$.

(3) $D_3 \setminus D_4 = \{s \in D^{4,1} \mid |s| = 3,\ s \neq x_i^3 \text{ for } 1 \leq i \leq 4\}$ *and this set has 12 elements.*

(4) $D_4 = \{s \in D^{4,1} \mid |s| \geq 4,\ s \neq x_i^n \text{ for } 1 \leq i \leq 4\}$.

(5) $D_i = D_i'$ *for* $i = 1, 2, 3, 4$.

(6) D_2 *is a cancellative semigroup and* $D^{4,1}$ *is a semilattice of cancellative semigroups.*

(7) $A = \langle x_1^2, x_2^2, x_3^2, x_4^2 \rangle$ *is a central submonoid.*

(8) $sA = As$ *and* $s^2 \in A$ *for every* $s \in D^{4,1}$.

(9) For any field K, $K[D^{4,1}]$ *is a Noetherian algebra that satisfies a polynomial identity and* $\mathrm{GK}(K[D^{4,1}]) = 1 = \mathrm{cl}(K[D^{4,1}])$.

(10) If K *is a field of characteristic not two, then* $K[D^{4,1}]$ *contains the following minimal primes that intersect* $D^{4,1}$ *trivially.*

$$M_1 = (x_1 - x_2, x_3 - x_4, x_1 - x_3), \qquad M_2 = (x_1 - x_2, x_3 - x_4, x_1 + x_3),$$
$$M_3 = (x_1 + x_2, x_3 + x_4, x_1 - x_3), \qquad M_4 = (x_1 + x_2, x_3 + x_4, x_1 + x_3).$$

Furthermore

$$\begin{aligned} I_1 &= (x_3^2 - x_2^2, x_4^2 - x_2^2, x_1^2 - x_4^2, x_4x_3 - x_3x_4, x_2x_4 - x_4x_2, x_2x_3 - x_3x_2) \\ &= M_1 \cap M_2 \cap M_3 \cap M_4. \end{aligned}$$

The other minimal primes of $K[D^{4,1}]$ *intersecting* $D^{4,1}$ *trivially are: in case* K *does not contain a primitive fourth root of unity*

$$I_2 = (x_1^2 + x_4^2, x_4^2 - x_2^2, x_3^2 + x_2^2, x_4x_3 + x_3x_4, x_2x_4 + x_4x_2, x_2x_3 - x_3x_2),$$
$$I_3 = (x_3^2 - x_2^2, x_4^2 + x_2^2, x_1^2 - x_4^2, x_4x_3 - x_3x_4, x_2x_4 + x_4x_2, x_2x_3 + x_3x_2),$$
$$I_4 = (x_3^2 + x_2^2, x_4^2 + x_2^2, x_1^2 + x_4^2, x_4x_3 + x_3x_4, x_2x_4 - x_4x_2, x_2x_3 + x_3x_2),$$

and in case K *contains a primitive fourth root of unity* ξ

$$\begin{aligned} J_2 &= (x_4 - \xi x_1, x_3 + \xi x_2, x_1^2 + x_2^2, x_1x_2 + x_2x_1), \\ J_2' &= (x_4 + \xi x_1, x_3 - \xi x_2, x_1^2 + x_2^2, x_1x_2 + x_2x_1), \\ J_3 &= (x_1 - \xi x_2, x_4 - \xi x_3, x_3^2 - x_2^2, x_2x_3 + x_3x_2), \\ J_3' &= (x_1 + \xi x_2, x_4 + \xi x_3, x_3^2 - x_2^2, x_2x_3 + x_3x_2), \\ J_4 &= (x_1 - \xi x_3, x_4 + \xi x_2, x_3^2 + x_2^2, x_2x_3 + x_3x_2), \\ J_4' &= (x_1 + \xi x_3, x_4 - \xi x_2, x_3^2 + x_2^2, x_2x_3 + x_3x_2). \end{aligned}$$

Also $J_i \cap J_i' = I_i$ *for* $2 \leq i \leq 4$.

(11) *The minimal primes of* $K[D^{4,1}]$ *intersecting* $D^{4,1}$ *are*

$$K[D^{4,1}x_1D^{4,1} \cup D^{4,1}x_2D^{4,1} \cup D^{4,1}x_3D^{4,1}],$$
$$K[D^{4,1}x_1D^{4,1} \cup D^{4,1}x_2D^{4,1} \cup D^{4,1}x_4D^{4,1}],$$
$$K[D^{4,1}x_1D^{4,1} \cup D^{4,1}x_3D^{4,1} \cup D^{4,1}x_4D^{4,1}],$$
$$K[D^{4,1}x_2D^{4,1} \cup D^{4,1}x_3D^{4,1} \cup D^{4,1}x_4D^{4,1}].$$

(12) *If* K *is a field of characteristic zero, then* $K[D^{4,1}]$ *is semiprime.*

(13) *The minimal primes of* $K[A]$ *not intersecting* A *are*

$$Q_1 = (x_3^2 - x_2^2, x_4^2 - x_2^2, x_1^2 - x_4^2), \qquad Q_2 = (x_3^2 + x_2^2, x_4^2 - x_2^2, x_1^2 + x_4^2),$$
$$Q_3 = (x_3^2 - x_2^2, x_4^2 + x_2^2, x_1^2 - x_4^2), \qquad Q_4 = (x_3^2 + x_2^2, x_4^2 + x_2^2, x_1^2 + x_4^2),$$

considered as ideals in $K[A]$.

Proof. We note that all even permutations on $\{1, 2, 3, 4\}$ define an automorphism of the monoid $D^{4,1}$. So the alternating group Alt_4 acts naturally on the monoid $D^{4,1}$.

We begin with determining the ideals D_i, with $1 \leq i \leq 4$. Of course,

$$D_1 \setminus D_2 = \bigcup_{1 \leq i \leq 4} \prec x_i \succ.$$

If $i \neq j$, then clearly $x_i x_j \in D_2 \setminus D_3$. There are 60 words of length 3 that are not a power of a generator. These are the five words listed in the following equalities

$$x_1^2 x_2 \;=\; x_1 x_4 x_3 \;=\; x_3 x_2 x_3 \;=\; x_3 x_4 x_1 \;=\; x_2 x_1^2$$

and their images under the action of Alt_4

$$\begin{aligned}
x_1^2 x_3 &= x_1 x_2 x_4 = x_4 x_3 x_4 = x_4 x_2 x_1 = x_3 x_1^2,\\
x_1^2 x_4 &= x_1 x_3 x_2 = x_2 x_4 x_2 = x_2 x_3 x_1 = x_4 x_1^2,\\
x_2^2 x_1 &= x_2 x_3 x_4 = x_4 x_1 x_4 = x_4 x_3 x_2 = x_1 x_2^2,\\
x_2^2 x_4 &= x_2 x_1 x_3 = x_3 x_4 x_3 = x_3 x_1 x_2 = x_4 x_2^2,\\
x_2^2 x_3 &= x_2 x_4 x_1 = x_1 x_3 x_1 = x_1 x_4 x_2 = x_3 x_2^2,\\
x_3^2 x_4 &= x_3 x_2 x_1 = x_1 x_4 x_1 = x_1 x_2 x_3 = x_4 x_3^2,\\
x_3^2 x_1 &= x_3 x_4 x_2 = x_2 x_1 x_2 = x_2 x_4 x_3 = x_1 x_3^2,\\
x_3^2 x_2 &= x_3 x_1 x_4 = x_4 x_2 x_4 = x_4 x_1 x_3 = x_2 x_3^2,
\end{aligned}$$

$$\begin{array}{rcl} x_4^2x_3 &=& x_4x_1x_2 = x_2x_3x_2 = x_2x_1x_4 = x_3x_4^2, \\ x_4^2x_2 &=& x_4x_3x_1 = x_1x_2x_1 = x_1x_3x_4 = x_2x_4^2, \\ x_4^2x_1 &=& x_4x_2x_3 = x_3x_1x_3 = x_3x_2x_4 = x_1x_4^2. \end{array}$$

We see that all the elements listed above belong to $D_3 \setminus D_4$ and that they are equal to a product in which a square is involved. So

$$\{x_i^2x_j \mid i \neq j\} \subseteq D_3 \setminus D_4.$$

Note that all twelve elements $x_i^2x_j$ (with $i \neq j$) are different in $D^{4,1}$. From the previous equalities one deduces that

$$\{x_1^2,\ x_2^2,\ x_3^2,\ x_4^2\} \subseteq \mathrm{Z}(D^{4,1})$$

and

$$\begin{array}{c} \{s \in D^{4,1} \mid |s| = 3,\ s \neq x_i^3 \text{ for } 1 \leq i \leq 4\} \subseteq D_3 \setminus D_4, \\ \{s \in D^{4,1} \mid |s| = 3,\ s \neq x_i^3 \text{ for } 1 \leq i \leq 4\} \subseteq D_3' \setminus D_4'. \end{array}$$

We next show that the latter inclusions actually are equalities. We do this by showing that any element of length three (that is not a power of a generator), multiplied on the right by a generator, yields an element in $D_4 \cap D_4'$. Because the ideal $D_4 \cap D_4'$ is invariant under the action of Alt_4, it is sufficient to prove this for the element $x_1^2x_2$. Hence, we have to show that $x_1^2x_2x_k \in D_4$, for every k. That these elements also belong to D_4' is shown similarly. Since $x_1^2x_2$ is left divisible by x_1, x_2 and x_3, we need to show that right multiplication by a generator also yields left divisibility by x_4. That this is the case follows from the following equalities.

$$\begin{array}{rcl} x_1^2x_2x_1 &=& x_1(x_1x_2x_1) = (x_1x_4^2)x_2 = x_4^2x_1x_2, \\ x_1^2x_2x_2 &=& x_1(x_1x_2^2) = (x_1x_2^2)x_1 = x_4x_3x_2x_1, \\ x_1^2x_2x_3 &=& (x_1^2x_2)x_3 = x_2(x_1^2x_3) = (x_2x_3x_1)x_1 = x_4x_1^3, \\ x_1^2x_2x_4 &=& (x_1^2x_2)x_4 = x_2(x_1^2x_4) = (x_2x_1x_3)x_2 = x_4x_2^3. \end{array}$$

So it follows that indeed

$$D_4 = \{s \in D^{4,1} \mid |s| \geq 4,\ s \neq x_i^n \text{ for } 1 \leq i \leq 4\},$$

$$D_3 \setminus D_4 = \{s \in D^{4,1} \mid |s| = 3,\ s \neq x_i^3 \text{ for } 1 \leq i \leq 4\}$$

and also

$$D_2 \setminus D_3 = \{s \in D^{4,1} \mid |s| = 2,\ s \neq x_i^2\} = \{x_ix_j \mid i \neq j\}.$$

Consequently

$$D_2^2 \subseteq D_4. \qquad (10.12)$$

Moreover, $D_i = D_i'$ for $1 = 1, 2, 3, 4$. We list a few more useful equalities.

$$\begin{aligned} x_1^2x_2^2 &= x_2x_1x_1x_2 = x_3x_4x_4x_3 = x_3^2x_4^2, \\ x_1^2x_3^2 &= x_3x_1x_1x_3 = x_4x_2x_2x_4 = x_2^2x_4^2, \\ x_1^2x_4^2 &= x_4x_1x_1x_4 = x_2x_3x_3x_2 = x_2^2x_3^2. \end{aligned} \qquad (10.13)$$

It follows that

$$x_1^2x_2^2x_3^2 = x_3^2x_3^2x_4^2 = x_2^2x_2^2x_4^2 = x_1^2x_1^2x_4^2.$$

More generally,

$$x_i^4x_j^2 = x_k^2x_l^2x_i^2 = x_k^4x_j^2,$$

for every i, j, k, l such that $\{1, 2, 3, 4\} = \{i, j, k, l\}$. Hence, we obtain that

$$x_1^6x_2^2 = x_1^2(x_1^4x_2^2) = x_1^2(x_3^4x_2^2) = (x_1^2x_3^4)x_2^2 = (x_1^2x_2^4)x_2^2 = x_1^2x_2^6$$

and thus

$$x_1^6x_2^2 = x_1^2x_2^6.$$

Let

$$A = \langle x_1^2,\ x_2^2,\ x_3^2,\ x_4^2 \rangle,$$

a central submonoid of $D^{4,1}$. Put $A_i = A \cap D_i$. Note that all elements a in A that are not of the form x_i^{2n} can be written as a multiple of x_1^2. Hence $a = x_1^{2n}x_i^{2m}$, with $n, m > 0$ and $i = 2, 3$ or 4. Moreover, using $x_1^2x_i^4 = x_1^2x_4^4$, we get that a is equal to either $x_1^{2n}x_2^{2m}$, $x_1^{2n}x_2^{2m}x_3^2$ or $x_1^{2n}x_2^{2m}x_4^2$. Using $x_1^2x_2^2x_i^2 = x_1^4x_j^2$, for $\{i, j\} = \{3, 4\}$, one deduces that a is either $x_1^{2n}x_2^{2m}$, $x_1^{2n}x_3^2$ or $x_1^{2n}x_4^2$. Finally, using $x_1^6x_2 = x_1^2x_2^6$, we obtain that if $a \in \langle x_1^2, x_2^2 \rangle$ then a is either x_1^2, $x_1^{2n}x_2^2$ or $x_1^{2n}x_2^4$. Thus we get that

$$A = A' \cup A_4, \qquad (10.14)$$

with

$$A' = \langle x_1^2 \rangle \cup \langle x_2^2 \rangle \cup \langle x_3^2 \rangle \cup \langle x_4^2 \rangle \qquad (10.15)$$

and

$$A_4 = x_1^2x_2^2\langle x_1^2 \rangle \cup x_1^2x_2^4\langle x_1^2 \rangle \cup x_1^2x_3^2\langle x_1^2 \rangle \cup x_1^2x_4^2\langle x_1^2 \rangle. \qquad (10.16)$$

Also

$$D^{4,1} = \left(\bigcup_{i=1}^{4} Ax_i\right) \cup \left(\bigcup_{1\leq i,j\leq 4, i\neq j} Ax_ix_j\right) \cup A. \tag{10.17}$$

Note that

$$\begin{aligned} x_1x_2x_1x_2 = x_4x_3x_1x_2 = x_4x_4x_2x_2 \in A, \\ x_1x_3x_1x_3 = x_2x_4x_1x_3 = x_2x_2x_3x_3 \in A, \\ x_1x_4x_1x_4 = x_3x_2x_1x_4 = x_3x_3x_4x_4 \in A. \end{aligned}$$

Now, using the action of Alt_4 on $D^{4,1}$, it is easy to see that $s^2 \in A$, for every $s \in D^{4,1}$. Since A is a central submonoid of $D^{4,1}$, we have that $sA = As$ for every $s \in D^{4,1}$. Therefore, $K[D^{4,1}]$ is a finitely generated module over $K[A]$. This shows that $K[D^{4,1}]$ is a Noetherian algebra that satisfies a polynomial identity (of course, we already know this from Proposition 9.4.4). Furthermore, because of (10.14), (10.15), (10.16) and (10.17), we get

$$\mathrm{GK}(K[D^{4,1}]) = \mathrm{GK}(K[A]) = 1.$$

From Proposition 3.5.2, we know that

$$\mathrm{cl}(K[D^{4,1}]) = \mathrm{cl}(K[A]) = 1.$$

Because of Proposition 9.5.2, we also have that every prime ideal P of $K[D^{4,1}]$ that intersects $D^{4,1}$ nontrivially contains D_4, and thus, by (10.12), it contains D_2. Since $K[D^{4,1}]/K[D_2] \cong \bigoplus_{1\leq i\leq 4} K[x_i]$, we get that there are precisely four minimal primes intersecting $D^{4,1}$ nontrivially, and these are

$$\begin{aligned} &K[D^{4,1}x_1D^{4,1} \cup D^{4,1}x_2D^{4,1} \cup D^{4,1}x_3D^{4,1}], \\ &K[D^{4,1}x_1D^{4,1} \cup D^{4,1}x_2D^{4,1} \cup D^{4,1}x_4D^{4,1}], \\ &K[D^{4,1}x_1D^{4,1} \cup D^{4,1}x_3D^{4,1} \cup D^{4,1}x_4D^{4,1}], \\ &K[D^{4,1}x_2D^{4,1} \cup D^{4,1}x_3D^{4,1} \cup D^{4,1}x_4D^{4,1}]. \end{aligned}$$

Later, we will show that D_2 is cancellative. It then follows that $D^{4,1}$ is a semilattice of cancellative semigroups and thus we obtain, from Corollary 3.2.9, that $K[D^{4,1}]$ is semiprime if K has characteristic zero.

For the remainder of the proof we need a few more equalities.

$$\begin{aligned} x_2^4x_1 &= x_3^4x_1 = x_4^4x_1, \quad x_1^4x_2 = x_3^4x_2 = x_4^4x_2, \\ x_1^4x_3 &= x_2^4x_3 = x_4^4x_3, \quad x_1^4x_4 = x_2^4x_4 = x_3^4x_4. \end{aligned} \tag{10.18}$$

To prove these, we first notice that

$$\begin{aligned} x_1x_1x_1x_2 &= x_1x_1x_4x_3 = x_1x_3x_2x_3 = x_2x_4x_2x_3 \\ &= x_2x_3x_1x_3 = x_2x_3x_2x_4 = x_2x_1x_4x_4. \end{aligned}$$

Hence

$$x_1x_1x_1x_1x_2 = x_1x_2x_1x_4x_4 = x_4x_4x_2x_4x_4 = x_4x_4x_4x_4x_2.$$

And thus $x_1^4x_2 = x_4^4x_2$. Applying the action of Alt_4, all the other equalities listed in (10.18) follow.

From (10.16) we obtain that

$$A_4 = a_2\langle x_1^2\rangle \cup a_3\langle x_1^2\rangle \cup a_4\langle x_1^2\rangle \cup b\langle x_1^2\rangle = \{a_2, a_3, a_4, b\}\langle x_1^2\rangle,$$

where

$$a_i = x_1^2x_i^2 \quad \text{and} \quad b = x_1^2x_2^4,$$

with $i = 2, 3, 4$. It is readily verified that the alternating group Alt_4 acts transitively on the set $\{a_2, a_3, a_4\}$ and that b is invariant under this action. We now determine relations among the (ideal) generators of A_4. Clearly,

$$b^2 = (x_1^2x_2^4)^2 = x_1^2(x_1^2x_2^6)x_2^2 = x_1^2(x_1^6x_2^2)x_2^2 = bx_1^6$$

and

$$a_2^2 = a_3^2 = a_4^2 = bx_1^2.$$

Further, it is easily seen that

$$a_ia_j = a_kx_1^4 \quad \text{for} \quad \{i, j, k\} = \{2, 3, 4\}$$

and

$$a_ib = a_ix_1^6.$$

Let T be the subsemigroup of $\mathbb{Z} \times \mathbb{Z}_2 \times \mathbb{Z}_2$ formed by the elements of the type (m, x, y) such that $m \geq 2$ and if $x = y = 0$ then $m \geq 3$. Consider the map

$$\phi : T \longrightarrow A_4$$

defined by

$$\phi(m, x, y) = \begin{cases} a_2x_1^{2(m-2)} & \text{if } (x, y) = (1, 0), \\ a_3x_1^{2(m-2)} & \text{if } (x, y) = (0, 1), \\ a_4x_1^{2(m-2)} & \text{if } (x, y) = (1, 1), \\ bx_1^{2(m-3)} & \text{if } (x, y) = (0, 0). \end{cases}$$

Using the previous equalities, it easily is verified that ϕ is a semigroup homomorphism. Define the following ideals of $K[A]$

$$\begin{aligned} Q_1 &= (x_3^2 - x_2^2,\ x_4^2 - x_2^2,\ x_1^2 - x_4^2), \ Q_2 = (x_3^2 + x_2^2,\ x_4^2 - x_2^2,\ x_1^2 + x_4^2), \\ Q_3 &= (x_3^2 - x_2^2,\ x_4^2 + x_2^2,\ x_1^2 - x_4^2), \ Q_4 = (x_3^2 + x_2^2,\ x_4^2 + x_2^2,\ x_1^2 + x_4^2). \end{aligned}$$

Considering the homomorphic images of A modulo Q_2, Q_3 and Q_4, for any field K of characteristic not equal to two, it is easy to see that the sets $a_2\langle x_1^2\rangle$, $a_3\langle x_1^2\rangle$, $a_4\langle x_1^2\rangle$ and $b\langle x_1^2\rangle$ are pairwise disjoint. Since the restriction of ϕ to each of these sets is injective, we get that ϕ is injective. Hence, ϕ is an isomorphism of semigroups. Consequently, A_4 is cancellative and its group of quotients is isomorphic with $\mathbb{Z} \times \mathbb{Z}_2 \times \mathbb{Z}_2$.

Suppose now that K is a field of characteristic not two. Then, we get that

$$\begin{aligned} K[A_4A_4^{-1}] = \ & K[\mathbb{Z}](1 + a_1a_2^{-1})(1 + a_1a_3^{-1}) \oplus K[\mathbb{Z}](1 + a_1a_2^{-1})(1 - a_1a_3^{-1}) \\ & \oplus K[\mathbb{Z}](1 - a_1a_2^{-1})(1 + a_1a_3^{-1}) \oplus K[\mathbb{Z}](1 - a_1a_2^{-1})(1 - a_1a_3^{-1}). \end{aligned}$$

Hence, the minimal prime ideals of $K[A_4A_4^{-1}]$ are the annihilators of respectively $(1+a_1a_2^{-1})(1+a_1a_3^{-1})$, $(1+a_1a_2^{-1})(1-a_1a_3^{-1})$, $(1-a_1a_2^{-1})(1+a_1a_3^{-1})$ and $(1 - a_1a_2^{-1})(1 - a_1a_3^{-1})$. And thus the minimal primes of $K[A_4]$ not intersecting A_4 are the annihilators of $(a_2+a_1)(a_3+a_1)$, $(a_2+a_1)(a_3-a_1)$, $(a_2 - a_1)(a_3 + a_1)$ and $(a_2 - a_1)(a_3 - a_1)$.

We now can determine all minimal primes of $K[A]$ that do not intersect A. So let Q be such an ideal. Then $Q \cap K[A_4]$ is a prime of $K[A_4]$ and hence contains a minimal prime of $K[A_4]$, that is, it contains the annihilator in $K[A_4]$ of one of the elements $(a_2 + a_1)(a_3 + a_1)$, $(a_2 + a_1)(a_3 - a_1)$, $(a_2 - a_1)(a_3 + a_1)$ or $(a_2 - a_1)(a_3 - a_1)$. Hence, Q contains one of the following four sets $\{a_2 - a_1, a_3 - a_1\}$, $\{a_2 + a_1, a_3 - a_1\}$, $\{a_2 - a_1, a_3 + a_1\}$, $\{a_2 + a_1, a_3 + a_1\}$. In the first case we get $x_1^2(x_3^2 - x_2^2), x_1^2(x_4^2 - x_2^2) \in Q$. Since $Q \cap A = \emptyset$, we get that $x_3^2 - x_2^2 \in Q$ and $x_4^2 - x_2^2 \in Q$. The equality $x_1^2x_2^2 = x_3^2x_4^2$ implies that also $x_1^2 - x_4^2 \in Q$. So Q contains the ideal $Q_1 = (x_3^2 - x_2^2, x_4^2 - x_2^2, x_1^2 - x_4^2)$. Since $K[A]/Q_1 \cong K[x_1^2]$, a polynomial algebra, we get that Q_1 is a prime ideal of $K[A]$ and thus $Q = Q_1$. The other three cases are dealt with similarly. Hence, we obtain that the minimal primes of $K[A]$ not intersecting A are Q_1, Q_2, Q_3 and Q_4.

Next, we determine all minimal primes of $K[D^{4,1}]$ that do not intersect $D^{4,1}$. So, let P be such a prime. Since $K[D^{4,1}]$ is a natural $\mathbb{Z}$-graded ring (via the length of the elements), Theorem 3.4.2 implies that P is homogeneous. As $K[D^{4,1} \setminus \{1\}]$ is the unique $\mathbb{Z}$-homogeneous maximal ideal and because

$P \cap D^{4,1} = \emptyset$, it follows that P is strictly contained in $K[D^{4,1} \setminus \{1\}]$. Hence, P is not a maximal ideal. Since $K[A]$ is central, we get that $P \cap K[A]$ is a prime ideal of $K[A]$. Because of Proposition 3.2.7, $P \cap K[A]$ is not maximal in $K[A]$. As $K[A]$ is of prime dimension one, we obtain that $P \cap K[A]$ is a minimal prime ideal of $K[A]$. Consequently $P \cap K[A] = Q_i$, for some $1 \leq i \leq 4$.

We deal with the case $i = 2$. So, $P \cap K[A] = (x_3^2 + x_2^2, x_4^2 - x_2^2, x_1^2 + x_4^2)$. The relations between the words of length three in $D^{4,1}$ yield that, for $1 \leq m, n \leq 4$ with $m \neq n$, $x_1 x_m x_n = x_k x_p^2$ and $x_1 x_n x_m = x_k x_q^2$, for some $1 \leq k, p, q \leq 4$. For example, $x_1 x_4 x_3 + x_1 x_3 x_4 = x_1^2 x_2 + x_4^2 x_2 \in P$. Hence $x_1^2(x_4 x_3 + x_3 x_4) \in P$ and thus, because $P \cap D^{4,1} = \emptyset$, we obtain that $x_4 x_3 + x_3 x_4 \in P$. Similarly, $x_1(x_2 x_4 + x_4 x_2) = (x_1^2 + x_2^2) x_3 \in P$ implies that $x_2 x_4 + x_4 x_2 \in P$. Also, $x_1(x_2 x_3 - x_3 x_2) = (x_3^2 - x_1^2) x_4 \in P$ implies $x_2 x_3 - x_3 x_2 \in P$. So we have shown that

$$I_2 = (x_1^2 + x_4^2, x_4^2 - x_2^2, x_3^2 + x_2^2, x_4 x_3 + x_3 x_4, x_2 x_4 + x_4 x_2, x_2 x_3 - x_3 x_2) \subseteq P.$$

Denote by y_i the natural image of x_i in the algebra $K[D^{4,1}]/I_2$. Because of the defining relations of $D^{4,1}$, it follows that y_i and y_j either commute or anti-commute, and y_i^2 and y_j^2 are either equal or additive inverses. Consequently, $K[D^{4,1}]/I_2$ has a K-spanning set $\langle y_1 \rangle \{1, y_2, y_3, y_4\}$ and

$$K[D^{4,1}]/I_2 = K[y_1] + K[y_1]\, y_2 + K[y_1]\, y_3 + K[y_1]\, y_4,$$

a normalizing extension of $K[y_1]$. It then follows that also $K[D^{4,1}]/P$ is a normalizing extension of the natural image V of $K[y_1]$ in $K[D^{4,1}]/P$. Since $K[D^{4,1}]/P$ is of dimension one and Noetherian, we obtain from Proposition 3.2.7 that also the algebra V is of dimension one. Hence, $V \cong K[y_1]$, a polynomial algebra.

We need to consider two separate cases. First, assume that K contains a primitive fourth root of unity, say ξ. So $\xi^2 = -1$. Note that $y_1 y_4 = y_4 y_1$ and $y_1^2 = -y_4^2$ and thus we get that

$$(y_1 y_4 - \xi y_1^2)(y_1 y_4 + \xi y_1^2) = 0.$$

Because y_1^2 and $y_1 y_4$ are central, it then follows that either $x_1 x_4 - \xi x_1^2 \in P$ or $x_1 x_4 + \xi x_1^2 \in P$. As $P \cap D^{4,1} = \emptyset$ and y_1 is a normal element in $K[D^{4,1}]/I_2$, we obtain that either $x_4 - \xi x_1 \in P$ or $x_4 + \xi x_1 \in P$. Since $y_2 y_3 = y_1 y_4$ and $y_1^2 = y_3^2$, we thus also get that if $x_4 - \xi x_1 \in P$ then $x_2 - \xi x_3 \in P$. In case $x_4 + \xi x_1 \in P$, we obtain $x_2 + \xi x_3 \in P$. So we have shown that either

$$J_2 = (x_4 - \xi x_1, x_3 + \xi x_2, x_1^2 + x_2^2, x_1 x_2 + x_2 x_1) \subseteq P$$

or

$$J_2' = (x_4 + \xi x_1, x_3 - \xi x_2, x_1^2 + x_2^2, x_1x_2 + x_2x_1) \subseteq P.$$

Denote by z_i the natural image of x_i in $K[D^{4,1}]/J_2$. Clearly,

$$K[D^{4,1}]/J_2 = K[z_1] + K[z_1]z_2 = K[z_1^2] + K[z_1^2]z_1 + K[z_1^2]z_2 + K[z_1^2]z_1z_2.$$

Suppose $\alpha, \beta \in K[z_1]$ are such that $\alpha + \beta z_2 = 0$ and $\alpha \neq 0$. Since J_2 is a $\mathbb{Z}$-homogeneous ideal and $D^{4,1} \cap P = \emptyset$, it follows that $K[z_1]$ is a polynomial algebra and there exists a nonzero $k \in K$ such that $z_1^{n+1} = kz_1^n z_2$, for some $n \geq 0$. Because z_1^2 is central, we obtain that $x_1 - kx_2 \in P$. As $x_1x_2 + x_2x_1 \in P$, we also get $2kx_2^2 \in P$. Since K is a field of characteristic not two, this yields $x_2^2 \in P$, a contradiction. Therefore, $\{1, z_2\}$ is a free $K[z_1]$-basis for $K[D^{4,1}]/J_2$. In particular, $K[z_1^2]$ consists of regular elements in $K[D^{4,1}]/J_2$.

Localizing at the nonzero elements of $K[z_1^2]$, we obtain a twisted group algebra of the Klein four group over the rational function field $K(z_1^2)$. So, this algebra has a natural grading by the Klein four group and, as $\mathrm{char}(K) \neq 2$, we obtain from Theorem 3.4.5 that it is semiprime. Since it is four dimensional over a central subfield and because it is noncommutative, it must be a prime algebra. Hence $K[D^{4,1}]/J_2$ is a prime algebra as well, that is, J_2 is a prime ideal of $K[D^{4,1}]$. Similarly one proves that J_2' is a prime ideal.

Next, assume K does not contain a primitive fourth root of unity. We now show that the set $\{1, y_2, y_3, y_4\}$ is a free $K[y_1]$-basis for $K[D^{4,1}]/I_2$. In order to prove this, suppose $\alpha = \alpha_1 1 + \alpha_2 y_2 + \alpha_3 y_3 + \alpha_4 y_4 = 0$ with each $\alpha_i \in K[y_1]$. Then, using the fact that the generators either anti-commute or commute, we get that

$$y_1\alpha + \alpha y_1 = 2y_1(\alpha_1 + \alpha_4 y_4) = 0.$$

Write $\alpha_1 = \alpha_{11} + \alpha_{12}y_1$ and $\alpha_4 = \alpha_{41} + \alpha_{42}y_1$, with each $\alpha_{kl} \in K[y_1^2]$. Then, since $\mathrm{char}(K) \neq 2$, we get that

$$\beta = \alpha_{11}y_1 + \alpha_{42}y_1^2y_4 = -(\alpha_{12}y_1^2 + \alpha_{41}y_1y_4)$$

is central. Because y_1 and y_4 anti-commute with y_2, it follows that

$$\alpha_{11}y_1 + \alpha_{42}y_1^2y_4 = 0.$$

Consequently,

$$\alpha_{12}y_1^2 + \alpha_{41}y_1y_4^2 = 0.$$

Hence also

$$\alpha_{11}y_1^2 + \alpha_{42}y_1^2 y_1 y_4 = 0.$$

If $\alpha_{42} \neq 0$, then using the fact that I_2 is a $\mathbb{Z}$-homogeneous ideal and $y_1^2 y_1 y_4$ has degree four, we get that, for some nonzero $k \in K$ and some $n \geq 0$, $y_1^{2n} y_1^2 y_1 y_4 = k y_1^2 y_1^2 y_1^{2n}$. So $y_1^{2n} y_1^3 (y_4 - k y_1) = 0$. Since $D^{4,1} \cap P = \emptyset$, this easily implies that $x_4 - kx_1 \in P$. As $x_4^2 + x_1^2 \in P$, it follows that $(k^2+1)x_1^2 \in P$ and thus $k^2 + 1 \in P$. So $k^2 = -1$. But, since by assumption K does not contain a primitive fourth root of unity, this yields a contradiction. So, we obtain that $\alpha_{42} = 0$. It then also easily is verified that $\alpha_{11} = 0$. Now, the identity $\alpha_{12}y_1^2 + \alpha_{41}y_1 y_4 = 0$ implies that $\alpha_{12}y_1^2 y_1 + \alpha_{41}y_1^2 y_4 = 0$. So by the above, $\alpha_{12} = \alpha_{41} = 0$. Thus we have shown that $\alpha_1 = \alpha_4 = 0$. The identity $y_1\alpha_2 y_2 + y_1\alpha_3 y_3 = 0$ implies that $y_1\alpha_2 y_2^2 + y_1\alpha_3 y_3 y_2 = -y_1\alpha_2 y_1^2 + y_1^2\alpha_3 y_4 = 0$. Hence, by the previous, $\alpha_2 = \alpha_3 = 0$. This proves that indeed $\{1, y_2, y_3, y_4\}$ is a $K[y_1]$-basis for $K[D^{4,1}]/I_2$.

Consequently, the nonzero elements of $K[y_1]$ form an Ore set of regular elements and localizing at this set one easily verifies that

$$K[D^{4,1}]/I_2 \subseteq K(y_1) * [\mathbb{Z}_2 \times \mathbb{Z}_2],$$

a crossed product of the Klein four group $\mathbb{Z}_2 \times \mathbb{Z}_2$ over the rational function field $K(y_1)$. Since $\mathrm{char}(K) \neq 2$ and considering this crossed product as graded by the Klein four group, Theorem 3.4.5 implies that this algebra is semiprime. So the central algebra $K[y_1^2] + K[y_1^2]y_1 y_4$ also is semiprime. Thus $K(y_1^2) + K(y_1^2)y_1 y_4$ either is a field or a direct sum of two fields. It easily is verified that the latter occurs precisely when K contains a primitive fourth root of unity. Since, by the assumption, the latter is not the case, we get that $K(y_1) * [\mathbb{Z}_2 \times \mathbb{Z}_2]$ is a noncommutative semiprime algebra which is four dimensional over the central subfield $K(y_1^2) + K(y_1^2)y_1 y_4$. So the algebra is simple. In particular, $K[D^{4,1}]/I_2$ is prime and thus I_2 is a prime ideal.

So, we are left to deal with minimal primes P of $K[D^{4,1}]$ with $P \cap D^{4,1} = \emptyset$ and $P \cap K[A] = Q_i$, where $i = 1, 3$ or 4. The previous approach also can be applied to the case $i = 3$ and $i = 4$. For $i = 3$, this yields the following prime ideals: if K does not contain a primitive fourth root of unity

$$I_3 = (x_3^2 - x_2^2,\ x_4^2 + x_2^2,\ x_1^2 - x_4^2,\ x_4 x_3 - x_3 x_4,\ \ x_2 x_4 + x_4 x_2,\ x_2 x_3 + x_3 x_2)$$

and if K contains a primitive fourth root of unity, say ξ,

$$J_3 = (x_1 - \xi x_2,\ x_4 - \xi x_3,\ x_3^2 - x_2^2,\ x_2 x_3 + x_3 x_2)$$

or

$$J_3' = (x_1 + \xi x_2,\ x_4 + \xi x_3,\ x_3^2 - x_2^2,\ x_2 x_3 + x_3 x_2).$$

In the case $i = 4$, one obtains the following prime ideals: if K does not contain a primitive fourth root of unity

$$I_4 = (x_3^2 + x_2^2,\ x_4^2 + x_2^2, x_1^2 + x_4^2,\ x_4x_3 + x_3x_4,\ x_2x_4 - x_4x_2,\ x_2x_3 + x_3x_2);$$

otherwise

$$J_4 = (x_1 - \xi x_3,\ x_4 + \xi x_2,\ x_3^2 + x_2^2,\ x_2x_3 + x_3x_2)$$

or

$$J_4' = (x_1 + \xi x_3,\ x_4 - \xi x_2,\ x_3^2 + x_2^2,\ x_2x_3 + x_3x_2).$$

In the case $i = 1$ we have

$$I_1 = (x_3^2 - x_2^2,\ x_4^2 - x_2^2,\ x_1^2 - x_4^2,\ x_4x_3 - x_3x_4,\ x_2x_4 - x_4x_2,\ x_2x_3 - x_3x_2) \subseteq P.$$

Using the defining relations it is readily verified that $K[D^{4,1}]/I_1$ is commutative and thus it is contained in the commutative crossed product (thus a twisted group algebra) $K(y_1) * [\mathbb{Z}_2 \times \mathbb{Z}_2]$. So, this algebra is a direct product of at most four fields. Clearly, the following prime ideals

$$\begin{aligned} M_1 &= (x_1 - x_2, x_3 - x_4, x_1 - x_3), & M_2 &= (x_1 - x_2, x_3 - x_4, x_1 + x_3), \\ M_3 &= (x_1 + x_2, x_3 + x_4, x_1 - x_3), & M_4 &= (x_1 + x_2, x_3 + x_4, x_1 + x_3) \end{aligned}$$

contain I_1. Since the algebra $K[D^{4,1}]$ and the factor algebras $K[D^{4,1}]/M_i$ are of dimension one, we get that M_1/I_1, M_2/I_1, M_3/I_1 and M_4/I_1 are all the minimal prime ideals of $K[D^{4,1}]/I_1$. Hence $P = M_1$, M_2, M_3 or M_4 and $M_1 \cap M_2 \cap M_3 \cap M_4 = I_1$.

Finally, we claim that D_2 is cancellative (and thus, by Corollary 3.2.9, $K[D^{4,1}]$ is semiprime if K is of characteristic zero). In order to prove this, we may assume that $K = \mathbb{Q}$. In this case, for $2 \leq j \leq 4$, each I_j is a prime ideal of $K[D^{4,1}]$ and the image of $D^{4,1}$ in $K[D^{4,1}]/I_j$ is cancellative. Let $s, t, v \in D^{4,1}$ be such that $sv = tv$. Then $s - t \in I_2 \cap I_3 \cap I_4$. By examining the words of length not exceeding three, it is easy to see that $s = t$ if $|s| = |t| \leq 3$. So it is enough to consider $s, t \in D_2$ with $|s| = |t| > 3$. In this case, let $z \in D^{4,1}$ be of maximal length such that $s, t \in zD^{4,1}$. Clearly $|z| \geq 1$. Then, we may write $s = zs_1, t = zt_1$, for some $s_1, t_1 \in D^{4,1}$ such that, for every j, either $s_1 \notin x_jD^{4,1}$ or $t_1 \notin x_jD^{4,1}$. Then $uz \in \mathrm{Z}(D^{4,1})$, for some $u \in D^{4,1}$, and $(s_1 - t_1)uz = u(s - t) \in I_2 \cap I_3 \cap I_4$. Since I_2, I_3 and I_4 are primes not intersecting $D^{4,1}$, we get $s_1 - t_1 \in I_2 \cap I_3 \cap I_4$. Now, $s_1, t_1 \notin D_4$, so that either $|s_1| = |t_1| \leq 3$ or $s_1 = x_j^m, t_1 = x_k^m$ for some j, k and $m \geq 4$. In the former case, we know that $s_1 = t_1$ and hence $s = t$. In the latter case, from the relations (10.18), it follows that $zx_j^m, zx_k^m \in zx_i^4 D^{4,1}$ for some i. This contradicts the choice of z, and thus this completes the proof of cancellativity of D_2. □

As an application we prove that there exist non-degenerate monoids of skew type on many generators and of Gelfand-Kirillov dimension 1, 2, 3 or 4.

Proposition 10.4.2. *Let m be a positive integer and $k \in \{1,2,3,4\}$. There exist non-degenerate monoids S of skew type generated by 4^m elements so that* $\mathrm{GK}(K[S]) = k$, *for any field K.*

Proof. Let $N^{(1)}, N^{(2)}, N^{(3)}, N^{(4)}$ be non-degenerate monoids of skew type so that $N^{(i)} \cap N^{(j)} = \{1\}$ for every $i \neq j$. Let $B^{(i)}$ denote their respective sets of defining generators and suppose these generating sets have the same cardinality. Consider the mapping

$$f : B^{(1)} \cup B^{(2)} \cup B^{(3)} \cup B^{(4)} \longrightarrow \{x_1, x_2, x_3, x_4\},$$

defined by $f(b) = x_i$ if $b \in B^{(i)}$. Then, consider any non-degenerate monoid S of skew type with generating set $B^{(1)} \cup B^{(2)} \cup B^{(3)} \cup B^{(4)}$ (the disjoint union) so that each $N^{(i)}$ is a submonoid of S and so that the defining relations on S are such that f extends to a monoid epimorphism

$$f : S \longrightarrow D^{4,1}. \tag{10.19}$$

Such an S can be constructed since all $B^{(i)}$ have the same cardinality (see also the comments after the proof). So we have

$$B^{(i)} B^{(j)} = B^{(k)} B^{(l)},$$

if $x_i x_j = x_k x_l$ in $D^{4,1}$. Here, we denote by $B^{(i)} B^{(j)}$ the set consisting of the elements $b_i b_j$ with $b_i \in B^{(i)}$ and $b_j \in B^{(j)}$. This implies that every equality $x_{i_1} \cdots x_{i_p} = x_{j_1} \cdots x_{j_p}$ leads to an equality $B^{(i_1)} \cdots B^{(i_p)} = B^{(j_1)} \cdots B^{(j_p)}$. Because of (10.14), (10.15), (10.16) and (10.17) we know that

$$D^{4,1} = \left(\bigcup_{1 \leq i \leq 4} A x_i \right) \cup \left(\bigcup_{1 \leq i,j \leq 4, i \neq j} A x_i x_j \right) \cup A,$$

where

$$A = \left(\bigcup_{1 \leq i \leq 4} \langle x_i^2 \rangle \right) \cup \left(\bigcup_{2 \leq i \leq 4} x_1^2 x_i^2 \langle x_1^2 \rangle \right) \cup x_1^2 x_2^4 \langle x_1^2 \rangle.$$

It therefore follows that

$$S = \left(\bigcup_{1 \le i \le 4} VB^{(i)} \right) \cup \left(\bigcup_{1 \le i,j \le 4, i \ne j} VB^{(i)}B^{(j)} \right) \cup V,$$

where

$$V = \left(\bigcup_{1 \le i \le 4} \langle (B^{(i)})^2 \rangle \right) \cup \left(\bigcup_{2 \le i \le 4} (B^{(1)})^2 (B^{(i)})^2 \langle (B^{(1)})^2 \rangle \right)$$
$$\cup\, (B^{(1)})^2 (B^{(2)})^4 \, \langle (B^{(1)})^2 \rangle.$$

Hence, $K[S]$ is a finitely generated $K[V]$-module and consequently

$$\mathrm{GK}(K[S]) = \mathrm{GK}(K[V]) = \max\{\mathrm{GK}(K[\langle (B^{(i)})^2 \rangle]) \mid 1 \le i \le 4\}.$$

Thus also

$$\mathrm{GK}(K[S]) = \max\{\mathrm{GK}(K[N^{(i)}]) \mid 1 \le i \le 4\}.$$

From Section 10.3 and Theorem 10.4.1, we know that there exist examples of four generated non-degenerate monoids $N^{(i)}$ of skew type so that $\mathrm{GK}(K[N^{(i)}])$ is 1, 2, 3 or 4. Hence, the result follows by an induction argument. □

We now present a concrete and inductive realization for the mapping (10.19). As a consequence, examples of non-degenerate monoids S of skew type are constructed so that $\mathrm{GK}(K[S]) = 1$ and $K[S]$ is semiprime, for any field K of characteristic zero. The construction starts in a somewhat more general setting. Let $T = \langle t_1, \ldots, t_n \rangle$ and $U = \langle u_1, \ldots, u_m \rangle$ be non-degenerate monoids of skew type. Then, consider

$$S = \langle x_{ij},\ i = 1, \ldots, n,\ j = 1, \ldots, m\ ; R \rangle$$

where R consists of relations of the following type

$$x_{ij}x_{kl} = x_{pr}x_{qs}$$

provided

$$t_i t_k = t_p t_q \quad \text{and} \quad u_j u_l = u_r u_s.$$

We give an interpretation of the multiplication of generators. First, the generators are ordered as follows $x_{11}, \ldots, x_{1m}, x_{21}, \ldots, x_{2m}, \ldots$ and hence they are divided into n blocks, indicated by the first index. The second index indicates the position within the block. The blocks multiply according to the multiplication in T, and the product of elements of any two blocks is determined by the multiplication in U. It is easy to see that S is non-degenerate, since both T and U are non-degenerate. Moreover, if T and U satisfy the cyclic condition, then so does S.

Clearly S embeds into the direct product $U \times T$. The process now continues with an extra requirement. Suppose, furthermore, $U = \bigcup_{\gamma \in \Gamma} U_\gamma$ and $T = \bigcup_{\lambda \in \Lambda} T_\lambda$ are semilattices of cancellative semigroups U_γ, T_λ, respectively (because of Theorem 10.4.1 this is the case if $T = U = D^{4,1}$). By Theorem 9.4.2, $K[U_\gamma]$ and $K[T_\lambda]$ are algebras that satisfy a polynomial identity. Hence, by Lemma 2.1.3, both U_γ and T_λ have a group of quotients. Moreover $S = \bigcup_{\gamma,\lambda} S_{\gamma,\lambda}$ is a semilattice of the semigroups $S_{\gamma,\lambda} = S \cap (U_\gamma \times T_\lambda)$. Corollary 3.2.9 therefore yields that $K[S]$ is semiprime if $\mathrm{char}(K) = 0$. Hence, starting with $T = U = D^{4,1}$ and using induction as in the proof of Proposition 10.4.2, we get the following result.

Corollary 10.4.3. *Let m be a positive integer and let K be a field of characteristic zero. There exist non-degenerate monoids S of skew type generated by 4^m elements so that $\mathrm{GK}(K[S]) = 1$ and $K[S]$ is semiprime.*

10.5 Examples of maximal orders

Examples of commutative maximal orders $K[S]$ are easily obtained via Theorem 3.6.1. On the other hand, Theorem 8.5.6 shows that semigroup algebras of monoids of I-type yield an interesting class of noncommutative examples with height one primes principally generated by a normal element. So, important information is determined by the submonoid $\mathrm{N}(S)$ of normal elements. This was proved as an application of Theorem 7.2.7 that reduces the problem of when a semigroup algebra of certain cancellative monoids is a maximal order to the monoid S, and in particular to associated monoids S_P with only one minimal prime ideal.

The aim in this section is to give more examples of maximal orders and, in particular, of their localizations S_P. In first instance they illustrate the rather complicated nature of these local monoids. In the commutative setting, using classical arguments as in the commutative ring case, one gets from Theorem 7.2.5 that S_P is a discrete valuation semigroup, that is, it is isomorphic with a direct product of a cyclic monoid and a group. Examples show that the structure of S_P, in general, is much more complicated.

Furthermore, they also show that in general there is no strong link with $\mathrm{N}(S)$.

A natural question that remains is to find a description of monoids with only one minimal prime ideal and that are a maximal order that satisfies the ascending chain condition on one-sided ideals. As indicated by the above mentioned examples, one might even ask whether the minimal prime ideal is principal.

First, recall the construction of Example 4.5.11, see also Example 8.1.3, Theorem 8.5.6 and Theorem 8.5.2.

Example 10.5.1. The monoid $M = \langle x, y \mid x^2 = y^2 \rangle$ has trivial unit group and yields a PI Noetherian domain $K[M]$ that is a maximal order with $\mathrm{Spec}^0(M) = \{Mx^2\}$ and $\mathrm{N}(M) = \langle x^2 \rangle$.

In order to construct more examples, we first prove an auxiliary result on a semidirect product $S = S_1 \rtimes S_2$ of monoids S_1, S_2 determined by a homomorphism $\sigma : S_2 \longrightarrow \mathrm{Aut}(S_1)$. We assume that each S_i has a group of quotients G_i, so that σ extends to $G_2 \longrightarrow \mathrm{Aut}(G_1)$ and S embeds into the semidirect product $G_1 \rtimes G_2$. Moreover, assume that $\sigma(S_2)$ is finite.

Recall that an ideal I of S_1 is said to be σ-invariant if $\sigma(s_2)(I) = I$ for any $s_2 \in S_2$. Note that $s_2 s_1 s_2^{-1} = \sigma(s_2)(s_1)$ for $s_1 \in S_1$. For an ideal J of S, define $\lambda(J) = \{j \in S_1 \mid js_2 \in J \text{ for some } s_2 \in S_2\}$. We claim that $\lambda(J)$ is a σ-invariant ideal of S_1. In order to prove this, let $s_2 \in S_2$ and $j \in \lambda(J)$. Then $jt_2 \in J$ for some $t_2 \in S_2$. Hence $s_2 j s_2^{-1} s_2 t_2 \in s_2 J \subseteq J$. Therefore $\sigma(s_2)(j) \in \lambda(J)$. So $\sigma(s_2)(\lambda(J)) \subseteq \lambda(J)$. Since $\sigma(s_2)$ is an automorphism of finite order, we also get $\sigma(s_2)^{-1}(\lambda(J)) \subseteq \lambda(J)$. Hence $\lambda(J)$ is σ-invariant. Clearly $\lambda(J)$ is a left ideal of S_1. To prove that it also is a right ideal, let $s_1 \in S_1$. Then $js_1 = jt_2(t_2^{-1} s_1 t_2) \in JS_1 \subseteq J$ and therefore $js_1 \in \lambda(J)$, as desired.

A σ-invariant ideal Q of S_1 is said to be σ-prime if $IJ \subseteq Q$ implies $I \subseteq Q$ or $J \subseteq Q$, for any σ-invariant ideals I, J of S_1. Note that, in this case, QS is an ideal of S. Moreover, if I and J are ideals of S such that $IJ \subseteq QS$, then $\lambda(I)\, \lambda(J) \subseteq Q$. Indeed, let $i \in \lambda(I)$ and $j \in \lambda(J)$. Since $\lambda(J)$ is σ-invariant, there exist $s_2, s_3 \in S_2$ so that $is_2 \in I$ and $(s_2^{-1} j s_2) s_3 \in J$. Hence $(is_2)(s_2^{-1} j s_2 s_3) = (ij)(s_2 s_3) \in IJ \subseteq QS$ and thus $ij \in Q$, as claimed. Because Q is σ-prime it thus follows that $\lambda(I) \subseteq Q$ or $\lambda(J) \subseteq Q$. Consequently $I \subseteq QS$ or $J \subseteq QS$. It follows that QS is a prime ideal of S.

Lemma 10.5.2. *Let S_1 and S_2 be monoids satisfying the ascending chain condition on one-sided ideals, each contained in a finitely generated torsion free abelian-by-finite group. Suppose $\sigma : S_2 \longrightarrow \mathrm{Aut}(S_1)$ is a monoid* *homomorphism with $\sigma(S_2)$ finite. Let $S = S_1 \rtimes_\sigma S_2$ be the corresponding semidirect product. Then also S satisfies the ascending chain condition on one-sided ideals. If each S_i is a maximal order, then S is a maximal order in a torsion free abelian-by-finite group.*

Proof. Let G_1 and G_2 be the groups of quotients of S_1 and S_2, respectively. Let A_1 and A_2' be abelian normal subgroups of finite index in G_1 and G_2, respectively. Because $\sigma(S_2)$ is finite and σ extends to a homomorphism $G_2 \longrightarrow \mathrm{Aut}(G_1)$ (also denoted by σ), there exists a subgroup A_2 of A_2' which is normal and of finite index in G_2 and such that A_2 acts trivially on S_1. It is clear that the group of quotients $G = SS^{-1}$ of S equals

$$SS_1^{-1} \rtimes_\sigma S_2S_2^{-1} = G_1 \rtimes_\sigma G_2$$

and $A = A_1 \times A_2$ is a normal abelian subgroup of finite index in G. Note that $S \cap A = (S \cap A_1) \times (S \cap A_2) = (S_1 \cap A_1) \times (S_2 \cap A_2)$ and thus it is a finitely generated monoid, because so is each $S_i \cap A_i$ by Theorem 4.1.6 and Theorem 4.1.7. Hence $K[S]$ is Noetherian by Theorem 4.4.7.

Assume now that each S_i is a maximal order. To prove that S is a maximal order, let $g \in G$ and let J be an ideal of S such that $gJ \subseteq J$. Then

$$gJS_1^{-1} \subseteq JS_1^{-1}. \tag{10.20}$$

Now, for any field K, the localization $K[S]S_1^{-1}$ of $K[S]$ with respect to the Ore set S_1 is a Noetherian ring and thus, by Theorem 3.2.6, the ideal $K[J]$ of $K[S]$ extends to the ideal $JK[S]S_1^{-1}$ of $K[S]S_1^{-1}$. Note that $K[S]S_1^{-1} \subseteq K[G]$. It follows that JS_1^{-1} is an ideal of SS_1^{-1}. Now $G_1 = S_1S_1^{-1}$ is a subgroup of SS_1^{-1} which is normal in G. Since $SS_1^{-1}/S_1S_1^{-1} \cong S_2$ is a maximal order, it follows easily that SS_1^{-1} is a maximal order as well. Hence (10.20) implies that $g \in G_1 \rtimes S_2$. So, $g = s_1^{-1}t_1s_2$ for some $s_1, t_1 \in S_1$, $s_2 \in S_2$ and $s_1^{-1}t_1s_2J \subseteq J$. Let $n > 0$ be so that $s_2^n \in A_2$. Let $J_1 = \{j_1 \in S_1 \mid j_1j_2 \in J$ for some $j_2 \in S_2\}$. Then, as $s_1^{-1}t_1s_2^nJ = (s_1^{-1}t_1s_2)s_2^{n-1}J \subseteq J$ and s_2^n acts trivially on S_1, we get $s_1^{-1}t_1J_1 \subseteq J_1$. As J_1 is an ideal of the maximal order S_1, it follows that $s_1^{-1}t_1 \in S$. Thus $g \in S$, as desired. That $Jg \subseteq J$ implies $g \in S$ is proved by a symmetric argument. Hence, indeed, S is a maximal order. □

We now will give a concrete example illustrating Lemma 10.5.2. It also shows that the local monoids S_P can be effectively calculated, but their structure is more complicated than in the commutative case. In Example 4.5.13 we considered the following finitely presented monoid

$$D = \langle a_1, a_2, b_1, b_2 \mid a_1a_2 = b_1b_2,\ a_ia_j = a_ja_i,$$
$$a_ib_j = b_ja_i,\ b_ib_j = b_jb_i,\ i, j = 1, 2\rangle.$$

First, we prove that D is a maximal order using the fact that $D \cong B = \langle xz, xw, yz, yw\rangle$ in the free abelian group $\mathrm{F} = \mathrm{gr}(x, y, w, z)$ of rank four and applying the characterization of Krull monoids in Theorem 3.6.5. Note that $xw = (xz)(zw^{-1})^{-1}$, $yz = (xz)(xy^{-1})^{-1}$ and $yw = (xy^{-1})^{-1}(xz)(zw^{-1})^{-1}$ and thus $BB^{-1} = \mathrm{gr}(xy^{-1}, zw^{-1}, xz)$. Take $g = (xy^{-1})^i(zw^{-1})^j(xz)^k \in \mathrm{F}_+ = \langle x, y, w, z\rangle$. This means that $i \le 0$, $j \le 0$, $i + k \ge 0$ and $j + k \ge 0$. If $i + j + k \ge 0$ then $g = (xz)^{i+j+k}(xw)^{-j}(yz)^{-i} \in B$. If on the other hand $i + j + k < 0$ then $g = (xw)^{i+k}(yz)^{j+k}(yw)^{-i-j-k} \in B$. Hence it follows that $B = BB^{-1} \cap \mathrm{F}_+$ and thus B indeed is a maximal order.

So, D is an abelian maximal order satisfying the ascending chain condition on ideals, with group of quotients

$$A_1 = G_1 = DD^{-1} = \mathrm{gr}\{a_1, a_2, b_1\},$$

a free abelian group of rank 3 and

$$\mathrm{Spec}^0(D) = \{Q_1, Q_2, Q_3, Q_4\},$$

where $Q_1 = (\{a_1, b_1\})$, $Q_2 = (\{a_2, b_2\})$, $Q_3 = (\{a_1, b_2\})$ and $Q_4 = (\{a_2, b_1\})$, see Example 4.5.13. Here (X) stands for the ideal of D generated by a subset X. Also note that it is easily verified that $D = \{a_1^{\alpha_1}a_2^{\alpha_2}b_1^{\alpha_3} \mid \text{all } \alpha_i \ge 0\} \cup \{a_1^{\alpha_1}a_2^{\alpha_2}b_2^{\alpha_3} \mid \text{all } \alpha_i \ge 0\}$, a disjoint union of two free abelian monoids of rank three. So every element of D has a unique presentation in one of these two forms.

Let $A_2 = \mathrm{gr}\{(xy)^2, (yx)^2, x^2\} \subseteq G_2 = MM^{-1}$, where M is as in Example 10.5.1. Then $[G_2 : A_2] = 4$ and A_2 contains a subgroup of finite index that is free abelian of rank 2. Now, x and y act on D as follows:

$$a_1^x = a_2,\ b_1^x = b_2,\ a_2^x = a_1, b_2^x = b_1,$$

$$a_1^y = b_1,\ b_1^y = a_1,\ a_2^y = b_2,\ b_2^y = a_2.$$

Since the action of x and y preserves the relation $a_1a_2 = b_1b_2$, we can extend it to an action σ of M on D. Thus, by Lemma 10.5.2,

$$\begin{aligned} S = D \rtimes_\sigma M \ &= \ \langle a_1, a_2, b_1, b_2, x, y \mid a_1a_2 = b_1b_2,\ a_ia_j = a_ja_i,\ b_ib_j = b_jb_i, \\ &\quad a_ib_j = b_ja_i,\ x^2 = y^2,\ a_1^x = a_2,\ a_2^x = a_1,\ b_1^x = b_2, \\ &\quad b_2^x = b_1, a_1^y = b_1,\ b_1^y = a_1,\ a_2^y = b_2,\ b_2^y = a_2\rangle \end{aligned}$$

is a maximal order satisfying the ascending chain condition on one-sided ideals and $S \subseteq G = G_1 \rtimes G_2$.

Example 10.5.3. The monoid $S = D \rtimes_\sigma M$ is a maximal order satisfying the ascending chain condition on one-sided ideals with $\mathrm{Spec}^0(S) = \{P_1 = (\{a_1a_2, a_1b_2, a_2b_1\})S,\ P_2 = (\{a_1a_2, a_1b_1, a_2b_2\})S,\ P_3 = x^2S\}$ and $\mathrm{N}(S) = \langle a_1a_2, x^2\rangle$, a free abelian monoid of rank 2. Furthermore,

$$\begin{aligned} S_{P_1} &= U\langle a_1, a_2\rangle \rtimes (MM^{-1}) = \langle a_1, a_2\rangle\ \mathrm{U}(S_{P_1}),\\ S_{P_2} &= V\langle a_1, a_2\rangle \rtimes (MM^{-1}) = \langle a_1, a_2\rangle\ \mathrm{U}(S_{P_2}),\\ S_{P_3} &= DD^{-1} \rtimes M \end{aligned}$$

and

$$\mathrm{Spec}^0(S_{P_1}) = \{a_1a_2S_{P_1}\},\ \mathrm{Spec}^0(S_{P_2}) = \{a_1a_2S_{P_2}\},\ \mathrm{Spec}^0(S_{P_3}) = \{x^2S_{P_3}\},$$

for some subgroups U and V of MM^{-1}. For any field K, $K[S]$ is a Noetherian PI domain that is a maximal order.

Proof. By Lemma 10.5.2 S has a group of quotients $G = DD^{-1} \rtimes MM^{-1}$ which is torsion free and abelian-by-finite and, by Theorem 7.0.1, $K[S]$ is a Noetherian PI domain.

First, we show that $\mathrm{N}(S) = \langle a_1a_2, x^2\rangle$. By the defining relations we get $a_1a_2, x^2 \in \mathrm{Z}(S) \subseteq \mathrm{N}(S)$. Let $n_1 \in D$ and $n_2 \in M$ be such that $n_1n_2 \in \mathrm{N}(S)$. Clearly, $n_2 \in \mathrm{N}(M)$, and $\mathrm{N}(M) = \langle x^2\rangle$ by Example 10.5.1. Since M normalizes D, we get $\mathrm{N}(M) \subseteq \mathrm{N}(S)$. So also $n_1 \in \mathrm{N}(S)$. Hence, for every $s_2 \in M$ there exist $x_1 \in D$ and $x_2 \in M$ so that $s_2n_1 = n_1x_1x_2$. As $s_2n_1s_2^{-1} \in D$ and $n_1x_1 \in D$, it follows that $s_2 = x_2$ and therefore $x_1 = n_1^{-1}s_2n_1s_2^{-1} \in D$. Write $n_1 = a_1^{\alpha_1}a_2^{\alpha_2}b_1^{\beta_1}b_2^{\beta_2}$ with $\alpha_i, \beta_i \geq 0$. Taking $s_2 = x$ we then get

$$x_1 = (a_1^{-\alpha_1}a_2^{-\alpha_2}b_1^{-\beta_1}b_2^{-\beta_2})(a_2^{\alpha_1}a_1^{\alpha_2}b_2^{\beta_1}b_1^{\beta_2}) \in D.$$

This yields that $x_1^x = x_1^{-1} \in D^x = D$. Therefore $x_1 \in \mathrm{U}(D) = \{1\}$ and thus

$$\begin{aligned} a_1^{\alpha_2-\alpha_1}a_2^{-(\alpha_2-\alpha_1)} &= b_1^{\beta_1-\beta_2}b_2^{-(\beta_1-\beta_2)}\\ &= b_1^{\beta_1-\beta_2}\left(b_1(a_1a_2)^{-1}\right)^{\beta_1-\beta_2}\\ &= b_1^{2(\beta_1-\beta_2)}a_1^{-(\beta_1-\beta_2)}a_2^{-(\beta_1-\beta_2)}. \end{aligned}$$

As $\mathrm{gr}\{a_1, a_2, b_1\}$ is free abelian of rank 3, we get $\beta_1 - \beta_2 = 0$ and thus also $\alpha_2 - \alpha_1 = 0$. Hence

$$n_1 = (a_1a_2)^{\alpha_1}(b_1b_2)^{\beta_1} = (a_1a_2)^{\alpha_1+\beta_1}.$$

Thus indeed $\mathrm{N}(S) = \langle a_1a_2, x^2\rangle$.

Next we determine the minimal primes of S. So let $P \in \mathrm{Spec}^0(S)$. We have to consider two possibilities. First, assume $P \cap D = \emptyset$. Since $K[S]$ is Noetherian, Proposition 4.5.8 and Theorem 3.2.6 imply that primes not intersecting D are in a natural one-to-one correspondence with the primes of the localization $(DD^{-1}) \rtimes M$, and thus with the primes of M. Hence, it follows that $P \cap M \in \mathrm{Spec}^0(M) = \{x^2M\}$. Moreover $(P \cap M)S = D \rtimes (P \cap M) = x^2S$ is an ideal of S. Since $P \cap M$ is a prime ideal of M, it follows that $D \rtimes (P \cap M)$ is a prime ideal of $S = D \rtimes M$. Hence $(P \cap M)S \in \mathrm{Spec}(S)$. So it is the only minimal prime ideal of S not intersecting D.

Second, assume that $P \cap D \neq \emptyset$. Again, since $P \in \mathrm{Spec}^0(S)$, we get from Proposition 4.5.8 that $K[P] \in \mathrm{Spec}^1(K[S])$. Since $(xy)^2, (yx)^2, x^2$ act trivially on A_1, it follows that $A_1A_2 = A_1 \times A_2$, a direct product. Lemma 7.1.1 and Lemma 7.1.4 imply that $K[P] \cap K[A_1 \times A_2] = Q_1 \cap \cdots \cap Q_n$, where the Q_i are all the height one prime ideals of $K[S] \cap K[A_1 \times A_2]$ containing $K[P] \cap K[A_1 \times A_2]$ and all Q_i are G-conjugate. Because of Proposition 4.5.8 we know that each $Q_i = K[Q_i \cap (A_1 \times A_2)]$. Hence $P \cap (A_1 \times A_2)$ is an intersection of minimal primes of $S \cap (A_1 \times A_2) = (D \cap A_1) \times (M \cap A_2) = D \times (M \cap A_2)$, all of which are G-conjugate and all of which intersect D. Now note that if Q is a prime ideal of D then $Q \times (M \cap A_2)$ is a prime ideal of $D \times (M \cap A_2)$. Hence the minimal primes of $D \times (M \cap A_2)$ intersecting D are precisely the ideals of the form $R \times (M \cap A_2)$ with R a minimal prime in D. It follows that $P \cap D = P \cap (A_1 \times A_2) \cap D$ is an intersection of minimal primes of D, all of which are also G-conjugate. Hence $P \cap D$ contains an intersection of a full G-orbit of minimal primes of D. So $P \cap D$ contains a σ-prime ideal Q. Because P is a minimal prime, the remark preceding Lemma 10.5.2 implies that $P = QS = (P \cap D)S$. Because of the description of the minimal primes in D, we easily get that P is one of the following primes:

$$\begin{aligned} P_1 &= (Q_1 \cap Q_2)S = (\{a_1a_2, a_1b_2, a_2b_1\})S, \\ P_2 &= (Q_3 \cap Q_4)S = (\{a_1a_2, a_1b_1, a_2b_2\})S. \end{aligned}$$

Hence $\mathrm{Spec}^0(S) = \{P_1, P_2, P_3 = Sx^2S\}$.

Next, we determine S_{P_1} and its minimal primes. For this, note that the conjugacy class $C = \{a_1b_1, a_2b_2\} \not\subseteq P_1$. But multiplying each of $b_2a_2^{-1}$, $b_2^{-1}a_2$, $a_1^{-1}b_1$ and $a_1b_1^{-1}$ by C gives a subset of S. So

$$\{b_2a_2^{-1} = a_1b_1^{-1},\ b_2^{-1}a_2 = a_1^{-1}b_1\} \subseteq \mathrm{U}(S_{P_1}).$$

Also $x^2 \in \mathrm{U}(S_{P_1})$ as $C_{x^2} \not\subseteq P_1$ and $C_{x^2}x^{-2} = \{1\} \subseteq S$. It follows that $U = \mathrm{gr}(b_2a_2^{-1},\, b_2^{-1}a_2,\, a_1^{-1}b_1,\, a_1b_1^{-1}) \subseteq S_{P_1}$ and thus

$$U\langle a_1, a_2\rangle \rtimes MM^{-1} \subseteq S_{P_1} \subseteq DD^{-1} \rtimes MM^{-1} = G.$$

Hence

$$S_{P_1} \subseteq (S_{P_1} \cap DD^{-1}) \rtimes MM^{-1}.$$

We claim that $S_{P_1} \cap DD^{-1} = U\langle a_1, a_2\rangle$. Multiplying, if necessary, with an element of U, it is sufficient to show that if $g \in S_{P_1} \cap \mathrm{gr}(a_1, a_2)$, then $g \in \langle a_1, a_2\rangle$. So, let g be such an element and let C be a G-conjugacy class contained in S such that $C \not\subseteq P_1$ and $Cg = gC \subseteq S$. Since $P_1 = (\{a_1a_2, a_1b_2, a_2b_1\})S = (P_1 \cap D)S$, it follows that $gC_1 \subseteq D$ for some subset C_1 of D closed under the action M and $C_1 \not\subseteq (\{a_1a_2, a_1b_2, a_2b_1\})$. So C_1 contains one of the following conjugacy classes: $\{1\}$, $\{a_1^\alpha, b_1^\alpha, a_2^\alpha, b_2^\alpha\}$ or $\{(a_1b_1)^\alpha, (a_2b_2)^\alpha\}$ for some $\alpha > 0$. Hence, in all cases $g\{(a_1b_1)^\alpha, (a_2b_2)^\alpha\} \subseteq D$ for some $\alpha > 0$. Now write $g = a_1^{\alpha_1}a_2^{\alpha_2}$ with $\alpha_i \in \mathbb{Z}$. Then $a_1^{\alpha+\alpha_1}a_2^{\alpha_2}b_1^\alpha = g(a_1b_1)^\alpha, a_1^{\alpha_1}a_2^{\alpha+\alpha_2}b_2^\alpha = g(a_2b_2)^\alpha \in D$. Because of the earlier mentioned unique presentation of the elements of D, we get that $\alpha_1, \alpha_2 \geq 0$. So indeed $g \in \langle a_1, a_2\rangle$. It follows that

$$S_{P_1} = U\langle a_1, a_2\rangle \rtimes MM^{-1} = \mathrm{U}(S_{P_1})\langle a_1, a_2\rangle. \tag{10.21}$$

Moreover, applying the homomorphism $\chi : DD^{-1} \longrightarrow \mathrm{gr}(a_1, a_2)$ defined by $\chi(b_i) = \chi(a_i) = a_i, i = 1, 2$, we clearly get that $\mathrm{gr}(a_1, a_2) \cap U\langle a_1, a_2\rangle = \langle a_1, a_2\rangle$. It follows that $\mathrm{U}(S_{P_1}) \cap \mathrm{gr}(\langle a_1, a_2\rangle) = \{1\}$. By (10.21), one-sided ideals of S_{P_1} are generated by their intersection with the free abelian monoid $\langle a_1, a_2\rangle$. In particular, by Lemma 10.5.2, S_{P_1} is a maximal order satisfying the ascending chain condition on one-sided ideals. It is also clear that every ideal of $\langle a_1, a_2\rangle$ contains a power of a_1a_2. So, since a_1a_2 is central, every prime ideal of S_{P_1} contains a_1a_2. Suppose I, J are σ-invariant ideals of $U\langle a_1, a_2\rangle$ properly containing $T = a_1a_2U\langle a_1, a_2\rangle$ and such that $IJ \subseteq T$. Then $a_1^\alpha \in I$ or $a_2^\alpha \in I$ for some $\alpha \geq 1$. As I is invariant, we get $a_1^\alpha, a_2^\alpha \in I$ for some α. Since the same holds for J, this implies that IJ contains a power of a_1, contradicting $IJ \subseteq T$. Therefore T is σ-prime and the remark given before Lemma 10.5.2 implies that $a_1a_2S_{P_1} = TS_{P_1}$ is a prime ideal of S_{P_1}. Hence $\mathrm{Spec}^0(S_{P_1}) = \{a_1a_2S_{P_1}\}$.

Similarly, $S_{P_2} = V\langle a_1, a_2\rangle \rtimes MM^{-1}$ for some group V and it follows that S_{P_2} is a maximal order satisfying the ascending chain condition on one-sided ideals and $\mathrm{Spec}^0(S_{P_2}) = \{a_1a_2S_{P_2}\}$. Since $a_1a_2 \notin P_3$ and a_1a_2 is central, we get that $a_1a_2 \in \mathrm{U}(S_{P_3})$. Hence $DD^{-1} \subseteq \mathrm{U}(S_{P_3})$. Thus

$T = (DD^{-1}) \rtimes M \subseteq S_{P_3}$ and T also is a maximal order satisfying the ascending chain condition on one-sided ideals with $Q = (DD^{-1}) \rtimes x^2M$ the unique minimal prime ideal of T. Now, if $g \in S_{P_3}$, then there exists a G-conjugacy class C contained in S so that $Cg \subseteq S$ and $C \not\subseteq P_3 = x^2S$. Let $I = TC$, the ideal of T generated by C. Then $I \not\subseteq Q$ and $Ig \subseteq T$. Hence, $I^*g \subseteq T$, where I^* is the divisorial closure of I in the maximal order T. Because I is not contained in the unique minimal prime ideal of T, it follows that $I^* = T$, and thus $g \in T$. Hence we have shown that $S_{P_3} = DD^{-1} \rtimes M$. Therefore $\mathrm{Spec}^0(S_{P_3}) = \{x^2 S_{P_3}\}$. So all the localized monoids S_{P_i} have a unique minimal prime (and this is in each case generated by a normal element of S). We thus get from Theorem 7.2.7 that $K[S]$ is a Noetherian maximal order. □

We note a difference with the ring situation. Namely the monoid S_{P_1} is not a localization with respect to an Ore set in S. Indeed, suppose the contrary. That is, suppose $S_{P_1} = SC^{-1}$, with C an Ore set in S. Since $b_2a_2^{-1} \in \mathrm{U}(S_{P_1})$, we thus get that $b_2a_2^{-1}s_1s_2 \in C \subseteq S$ for some $s_1 \in D$, $s_2 \in M$. In particular, $b_2a_2^{-1}s_1 \in D$. Then it is easy to see that $s_1 \notin \langle a_1, b_2\rangle$. Now, if $s_1 \in b_1D$ then $s_1 = b_1s_1'$ for some $s_1' \in D$. Hence, because $a_1a_2 = b_1b_2$,

$$a_1s_1's_2 = b_2a_2^{-1}b_1s_1's_2 = b_2a_2^{-1}s_1s_2 \in C.$$

Consequently $a_1 \in \mathrm{U}(S_{P_1})$, which contradicts (10.21). So $s_1 \notin b_1D$ and hence $s_1 \in a_2D$, because we know that $s_1 \notin \langle a_1, b_2\rangle$. It then follows that $b_2s \in C$ for some $s \in S$. Therefore, $b_2 \in \mathrm{U}(S_{P_1})$, again leading to a contradiction.

As remarked at the end of Section 7.1, if T is a Krull monoid with only one minimal prime ideal then $\mathrm{N}(T) = \mathrm{U}(T)\langle n\rangle$. In the next example we show that the converse is false. For this we consider another action of M on D. The elements x and y act as follows:

$$a_1^x = a_2,\ a_2^x = a_1,\ b_1^x = b_2,\ b_2^x = b_1,$$

$$a_1^y = a_2,\ a_2^y = a_1,\ b_1^y = b_2,\ b_2^y = b_1.$$

Let us denote this action of M on D by γ.

Example 10.5.4. Let ρ be the congruence relation on $D \rtimes_\gamma M$ generated by the relations:

$$x^4 = y^4 = a_1a_2 = b_1b_2,\ b_1 = xy\,,\ b_2 = yx.$$

Let $S = (D \rtimes_\gamma M)/\rho$. Then $K[S]$ is a Noetherian maximal order which is a PI domain, with $\mathrm{N}(S) = \langle x^2 \rangle$ and

$$\mathrm{Spec}^0(S) = \{P_1 = (\{x^2, a_2 x, a_1 y\}), P_2 = (\{x^2, a_1 x, a_2 y\})\},$$

where (X) denotes the ideal of S generated by a subset X. Furthermore, $\mathrm{Spec}^0(S_{P_1}) = \{x^2 S_{P_1}\}$ and $\mathrm{Spec}^0(S_{P_2}) = \{x^2 S_{P_2}\}$.

Proof. Let $H = DD^{-1}$ and let $G = \mathrm{gr}(H \cup \{x\})$, where $x^4 = a_1 a_2$, $a_1^x = a_2$, $a_2^x = a_1$, $b_1^x = b_2$, $b_2^x = b_1$. So H is a normal subgroup of G and G/H is a cyclic group of order four. We claim that G is a torsion free group. Since $H \cong \mathbb{Z}^3$, every periodic element of G has order dividing 4. Say, $(ag)^4 = 1$ for some $a \in H$ and $g \in \{1, x, x^2, x^3\}$. Then $a a^g a^{g^2} a^{g^3} g^4 = 1$. As $g^2 \in \mathrm{Z}(G)$, we get $(aa^g)^2 g^4 = 1$. Let $a = a_1^{\alpha_1} a_2^{\alpha_2} b_1^{\beta_1}$ for some integers $\alpha_1, \alpha_2, \beta_1$. If $g = 1$ then it easily follows that $\alpha_1 = \alpha_2 = \beta_1 = 0$, so $a = 1$. Further, $a^g = a$ if $g = x^2$ and $a^g = a_1^{\alpha_2} a_2^{\alpha_1} b_2^{\beta_1}$ if $g = x, x^3$. So $(a_1^{\alpha_1} a_2^{\alpha_2} b_1^{\beta_1})^4 a_1^2 a_2^2 = 1$ if $g = x^2$, $(a_1^{\alpha_1+\alpha_2} a_2^{\alpha_1+\alpha_2} b_1^{\beta_1} b_2^{\beta_1})^2 a_1 a_2$ if $g = x$, and $(a_1^{\alpha_1+\alpha_2} a_2^{\alpha_1+\alpha_2} b_1^{\beta_1} b_2^{\beta_1})^2 a_1^3 a_2^3$ if $g = x^3$. As $a_1 a_2 = b_1 b_2$, we get $a_1^{4\alpha_1+2} a_2^{4\alpha_2+2} b_1^{4\beta_1} = 1$ if $g = x^2$, $(a_1 a_2)^{2(\alpha_1+\alpha_2+\beta_1)+1} = 1$ if $g = x$ and $(a_1 a_2)^{2(\alpha_1+\alpha_2+\beta_1)+3} = 1$ if $g = x^3$. Since a_1, a_2, b_1 are free generators of H, the respective exponents must be 0. This contradiction shows that indeed G is torsion free.

In G, define $y = x^{-1} b_1$. Then a_1, a_2, b_1, b_2, x, y satisfy all defining relations for S. Therefore we have a natural homomorphism $\phi : S \longrightarrow G$. We verify that this is a monomorphism, and hence we may consider S as a submonoid of G. For simplicity, we use the same notation for the generators of S and of G. From the defining relations it follows that $S = \langle a_1, a_2, b_1, b_2 \rangle \{1, x, x^2, y\}$. Let $a, a' \in \langle a_1, a_2, b_1, b_2 \rangle$ and assume that

$$a g_1 = a' g_2 \tag{10.22}$$

in G for some $g_1, g_2 \in \{1, x, x^2, y\}$. Then $g_1 = g_2$, because $1, x, x^2, y$ is a transversal for H in G. Hence $a = a'$ in G and, as ϕ is one-to-one on $\langle a_1, a_2, b_1, b_2 \rangle$, we then get $a = a'$ in S. So (10.22) holds in S. Thus, indeed ϕ is an embedding and hence G can be viewed as a group of quotients of S. Moreover, $S = D \cup Dx \cup Dx^2 \cup Dy$ is a disjoint union.

Applying Lemma 10.5.2 to $D \rtimes_\gamma M$, we obtain that S satisfies the ascending chain condition on one-sided ideals. Let A denote the abelian group $\mathrm{gr}(H \cup \{x^2\})$. Clearly A is of finite index in G and $S \cap A = \langle a_1, a_2, b_1, b_2, x^2 \rangle$ is G-invariant. Every ideal of S contains a power of the central element x^2. Hence x^2 belongs to every prime ideal of S. Moreover, the defining relations easily imply that

$$a_1 x S a_2 x \subseteq x^2 S.$$

It follows that every prime ideal of S either contains $\{x^2, a_1x\}$ or $\{x^2, a_2x\}$. Also

$$a_1xSa_1y \subseteq x^2S.$$

So, if a_1x does not belong to a minimal prime ideal P of S, then a_2x and a_1y belong to P. Let $P_1 = (\{x^2, a_2x, a_1y\})$. Now $S \setminus P_1$ can be written in the generalized matrix form, see Section 4.5,

$$S \setminus P_1 = \begin{bmatrix} \prec a_1, b_1 \succ & \langle a_1, b_1 \rangle x \\ y\langle a_1, b_1 \rangle & \prec a_2, b_2 \succ \end{bmatrix} \cup \{1\}. \tag{10.23}$$

Using standard matrix arguments it follows that P_1 is a prime ideal of S. Since $P_1 \subseteq P$ and P is minimal, we get $P = P_1$.

On the other hand, if a_1x belongs to a minimal prime P, then $\{x^2, a_1x\} \subseteq P$. Since also

$$a_2ySa_2x \subseteq x^2S \text{ and } a_2ySa_1y \subseteq x^2S,$$

we get that either $a_2y \in P$ or $\{a_2x, a_1y\} \subseteq P$. The latter case gives $P_1 = P$. So, assume that $\{x^2, a_1x, a_2y\} \subseteq P$. Let $P_2 = (\{x^2, a_1x, a_2y\})$. As above, it follows from the following generalized matrix form of $S \setminus P_2$

$$S \setminus P_2 = \begin{bmatrix} \prec a_1, b_2 \succ & \langle a_1, b_2 \rangle y \\ x\langle a_1, b_2 \rangle & \prec a_2, b_1 \succ \end{bmatrix} \cup \{1\} \tag{10.24}$$

that P_2 is a prime ideal of S. So we have shown that $\mathrm{Spec}^0(S) = \{P_1, P_2\}$.

In order to get the required information on S_{P_1} note that $C_{a_1b_1} = \{a_1b_1, a_2b_2\} \not\subseteq P_1$ and

$$C_{a_1b_1}\{b_2^{-1}a_2,\ a_2^{-1}b_2,\ a_1b_1^{-1},\ b_1a_1^{-1},\ x^{-1}a_1,\ y^{-1}a_2\} \subseteq S.$$

Thus $\{b_2^{-1}a_2 = b_1a_1^{-1},\ a_2^{-1}b_2 = a_1b_1^{-1},\ x^{-1}a_1,\ y^{-1}a_2\} \subseteq S_{P_1}$ and consequently $a_2x = a_2x^{-1}x^2 = x^{-1}a_1x^2 \in S_{P_1}$ and $a_1y = a_1y^{-1}y^2 = y^{-1}a_2y^2 \in S_{P_1}$. It follows that $S_{P_1}P_1 = P_1S_{P_1} = x^2S_{P_1}$.

We now show that $I(P_1) = x^2S_{P_1}$ (here $I(P_1)$ is defined as in Section 7.1) and $\mathrm{Spec}^0(S_{P_1}) = \{x^2S_{P_1}\}$. Indeed, we know that $I(P_1) = \{g \in S_{P_1} \mid Cg \subseteq P_1$ for some G-conjugacy class $C \subseteq S$, and $C \not\subseteq P_1\}$ and by Lemma 7.1.2 $I(P_1)$ is a prime ideal of S_{P_1}. Now let $g \in I(P_1)$ and let $C \subseteq S$ be a conjugacy class so that $Cg \subseteq P_1$ and $C \not\subseteq P_1$. Then $Cg \subseteq P_1 \subseteq x^2S_{P_1}$. So $Cx^{-2}g \subseteq S_{P_1}$. As $C \subseteq S$, it follows from the definition of S_{P_1} that $x^{-2}g \in S_{P_1}$. So $g \in x^2S_{P_1}$ and therefore $I(P_1) \subseteq x^2S_{P_1}$. As x^2 lies in every prime ideal of S_{P_1} (as it is central and belongs to every ideal of S_{P_1}), we thus get $I(P_1) = x^2S_{P_1}$ and $I(P_1)$ is the only minimal prime ideal of S_{P_1}.

Since $S \cap A$ is G-invariant, we know from Lemma 7.1.3 that $S_{P_1} = S_{(P_1)}$. By (10.23), the only conjugacy classes in $S \cap A$ that are not contained in P_1 are of the type $\{a_1^\alpha b_1^\beta,\ a_2^\alpha b_2^\beta\}$ and, by (10.24), the only conjugacy classes in $S \cap A$ that are not contained in P_2 are of the type $\{a_1^\alpha b_2^\beta,\ a_2^\alpha b_1^\beta\}$, for some $\alpha, \beta \geq 1$. Suppose $a \in S_{P_1} \cap H$. Write $a = a_1^{\alpha_1} a_2^{\alpha_2} b_1^{\beta_1}$ for some integers $\alpha_1, \alpha_2, \beta_1$. Then $\{a_1^{\alpha_1+\delta} a_2^{\alpha_2} b_1^{\beta_1+\delta},\ a_1^{\alpha_1} a_2^{\alpha_2+\delta} b_1^{\beta_1} b_2^\delta\} = \{a_1^\delta b_1^\delta,\ a_2^\delta b_2^\delta\} a \subseteq S$ for some $\delta \geq 1$. Therefore, from the description of D in Example 4.5.13 as a set of words in a free group, it easily follows that $\alpha_2 \geq 0$ and $\alpha_1 + \beta_1 \geq 0$. This implies that $a \in D \operatorname{gr}(b_1 a_1^{-1})$. Hence $S_{P_1} \cap H = D \operatorname{gr}(b_1 a_1^{-1})$. In particular, it is finitely generated and hence $K[S_{P_1}]$ is Noetherian by Lemma 4.1.3. Since also $G = S_{P_1} \operatorname{gr}(x^2)$, we get from Lemma 8.5.5 that S_{P_1} is a maximal order. Similarly, one shows that S_{P_2} is a maximal order, $S_{P_2} \cap H = D \operatorname{gr}(b a_1 b_2^{-1})$, $K[S_{P_2}]$ is Noetherian and $\operatorname{Spec}^0(S_{P_2}) = \{x^2 S_{P_2}\}$.

Let $a \in S_{P_1} \cap S_{P_2} \cap H$. Then, by the previous paragraph, we know that $a = a_1^{\alpha_1} a_2^{\alpha_2} b_1^{\beta_1}$, where $\alpha_2 \geq 0$ and $\alpha_1 + \beta_1 \geq 0$, and similarly $a = a_1^{\gamma_1} a_2^{\gamma_2} b_2^{\delta_2}$, where $\gamma_2 \geq 0$ and $\gamma_1 + \delta_2 \geq 0$. Then $a = a_1^{\gamma_1+\delta_2} a_2^{\gamma_2+\delta_2} b_1^{-\delta_2}$. Hence $\alpha_1 = \gamma_1 + \delta_2, \alpha_2 = \gamma_2 + \delta_2$ and $\beta_1 = -\delta_2$. Therefore $\alpha_i, \gamma_i \geq 0$ and either $\delta_2 \geq 0$ or $\beta_1 \geq 0$. Consequently, $a \in S$. It follows that $S_{P_1} \cap S_{P_2} \cap H = S \cap H = D$.

Next we prove that $S = S_{P_1} \cap S_{P_2}$. Let $g \in S_{P_1} \cap S_{P_2}$. Write $g = ch$, where $c \in H$ and $h \in \{1, x, x^2, y\}$. We know that there exists $\delta \geq 1$ so that $\{a_1^\delta b_1^\delta,\ a_2^\delta b_2^\delta,\ a_1^\delta b_2^\delta,\ a_2^\delta b_1^\delta\}\ ch \subseteq S$. In view of the disjoint union presentation $S = D \cup Dx \cup Dx^2 \cup Dy$, this implies that $\{a_1^\delta b_1^\delta,\ a_2^\delta b_2^\delta,\ a_1^\delta b_2^\delta,\ a_2^\delta b_1^\delta\}\ c \subseteq S$. Therefore $c \in S_{P_1} \cap S_{P_2} \cap H = D$. This yields $g = ch \in S$, so the equality $S = S_{P_1} \cap S_{P_2}$ follows.

Hence we have shown that indeed $S = S_{P_1} \cap S_{P_2}$. Since $K[S]$ is a Noetherian PI domain by Theorem 7.0.1, from Theorem 7.2.5 it now follows that $K[S]$ is a maximal order.

Finally, we show that $\mathrm{N}(S) = \langle x^2 \rangle$. Consider the $\mathbb{Z}$-gradation on S in which x and y are of degree one, and each a_i, b_j is of degree two. So assume

$$p = a_1^{\alpha_1} a_2^{\alpha_2} b_1^{\beta_1} b_2^{\beta_2} g \in \mathrm{N}(S),$$

with $g \in \{1, x, x^2, y,\}$ and nonnegative $\alpha_1, \alpha_2, \beta_1, \beta_2$. To prove $p \in \langle x^2 \rangle$ we may suppose (cancelling a power of x^2, if necessary) that $p \notin x^2 S$. So $g = 1, x$ or y. Now, there exists $h \in S$ so that

$$a_1^{\alpha_1} a_2^{\alpha_2} b_1^{\beta_1} b_2^{\beta_2} gx \quad = \quad h a_1^{\alpha_1} a_2^{\alpha_2} b_1^{\beta_1} b_2^{\beta_2} g. \tag{10.25}$$

Then h has degree one (so $h = y$ or x) and it also must be in the coset of x (with respect to the normal subgroup H). Hence $h = x$. So (10.25) yields

$$a_1^{\alpha_1} a_2^{\alpha_2} b_1^{\beta_1} b_2^{\beta_2} gx = a_1^{\alpha_2} a_2^{\alpha_1} b_1^{\beta_2} b_2^{\beta_1} xg$$

and similarly

$$a_1^{\alpha_1} a_2^{\alpha_2} b_1^{\beta_1} b_2^{\beta_2} gy = a_1^{\alpha_2} a_2^{\alpha_1} b_1^{\beta_2} b_2^{\beta_1} yg.$$

If $g = x$, then we obtain

$$a_1^{\alpha_1} a_2^{\alpha_2} b_1^{\beta_1} b_2^{\beta_2} = a_1^{\alpha_2} a_2^{\alpha_1} b_1^{\beta_2} b_2^{\beta_1}$$

and, as $xy = b_1$ and $yx = b_2$,

$$a_1^{\alpha_1} a_2^{\alpha_2} b_1^{\beta_1+1} b_2^{\beta_2} = a_1^{\alpha_2} a_2^{\alpha_1} b_1^{\beta_2} b_2^{\beta_1+1}$$

It follows that $b_1 = b_2$, a contradiction. Similarly, $g = y$ leads to a contradiction. So $g = 1$. Since we also have

$$a_1^{\alpha_1} a_2^{\alpha_2} b_1^{\beta_1} b_2^{\beta_2} = a_1^{\alpha_2} a_2^{\alpha_1} b_1^{\beta_2} b_2^{\beta_1},$$

it follows that

$$a_1^{\alpha_2-\alpha_1} b_1^{\beta_2-\beta_1} = a_2^{\alpha_2-\alpha_1} b_2^{\beta_2-\beta_1}.$$

Since $a_1 a_2 = b_1 b_2$ and $\mathrm{gr}(a_1, a_2, b_1)$ is free abelian of rank 3, it easily follows that $\alpha_1 = \alpha_2$ and $\beta_1 = \beta_2$ and thus $p \in \{(a_1 a_2)^\delta \mid \delta = 0, 1, 2, \ldots\}$. Hence $\mathrm{N}(S) = \langle x^2 \rangle$. □

10.6 Comments

The results of Section 10.1 come from [17]. The results of Section 10.2 are taken from [16, 85]. Gateva-Ivanova's list of all four generated binomial monoids in [46, Proposition 4.20] is much longer than the one given in Proposition 10.2.1, this because we only list the non-isomorphic.

In [37] a computer generated list is given of all groups of I-type on n generators with $n \leq 8$. It turns out that for $n = 3$ there are 5 such groups of which 4 are decomposable as products of groups on less generators, for $n = 4$ there are 23 and 18 are decomposable, for $n = 5$ there are 88 and 87 are decomposable, for $n = 6$ there are 595 and 585 are decomposable, for $n = 7$ there are 3456 and 3455 are decomposable, for $n = 8$ there are 34528 and 34430 are decomposable. For $n = 2$ there are 2 groups, one of which is indecomposable. The latter is the group $\mathrm{gr}(x, y \mid x^2 = y^2)$ handled in Example 8.1.3. Among all these groups there are only 49 which are indecomposable and not retractable (in the sense of Definition 8.2.11). Only two such groups are four generated and one of them has been handled in Example 8.2.14 and the others are eight generated.

The results in Section 10.3 come from [17], those of Section 10.4 are taken from [16, 85] and the examples given in Section 10.5 are studied in [84].

Bibliography

[1] Amitsur S.A. and Small L.W., GK – dimensions of corners and ideals, Israel J. Math. 69 (1990), 152–160.

[2] Anan'in A.Z., An intriguing story about representable algebras, in: Ring Theory, pp. 31–38, Weizmann Science Press, Jerusalem, 1989.

[3] Anderson D.D. and Anderson D.F., Divisibility properties of graded domains, Canad. J. Math. 34 (1982), 196–215.

[4] Anderson D.F., Graded Krull domains, Comm. Algebra 7 (1979), 79–106.

[5] Anderson D.F., The divisor class group of a semigroup ring, Comm. Algebra 8 (1980), 467–476.

[6] Anh P.N., Gould V. and Marki L., Completely 0-simple semigroups of left quotients of a semigroup, Int. J. Algebra Comput. 6 (1996), 143–153.

[7] Artin M. and Schelter W., Graded algebras of global dimension 3, Adv. Math. 66 (1987), 171–216.

[8] Bass H., The degree of polynomial growth of finitely generated nilpotent groups, Proc. London Math. Soc. 25 (1972), 603–614.

[9] Bergman G.M., The diamond lemma for ring theory, Adv. Math. 29 (1978), 178–218.

[10] Borel A., Linear Algebraic Groups, Springer-Verlag, New York, 1991.

[11] Bourbaki N., Algèbre, ch. 3 Algèbre Multilinéaire, Hermann, Paris, 1958.

[12] Brown K.A., Height one primes of polycyclic group rings, J. London Math. Soc. 32 (1985), 426–438.

[13] Brown K.A., Class groups and automorphism groups of group rings, Glasgow Math. J. 28 (1986), 79–86.

[14] Brown K.A., Corrigendum and addendum to 'Height one primes of polycyclic group rings', J. London Math. Soc. 38 (1988), 421–422.

[15] Brown K.A., Marubayashi H. and Smith P.F., Group rings which are v-HC orders and Krull orders, Proc. Edinburgh Math. Soc. 34 (1991), 217–228.

[16] Cedó F., Jespers E. and Okniński J., Semiprime quadratic algebras of Gelfand-Kirillov dimenension one, J. Algebra Appl. 3 (2004), 283–300.

[17] Cedó F., Jespers E. and Okniński J., The Gelfand-Kirillov dimension of quadratic algebras satisfying the cyclic condition, Proc. Amer. Math. Soc. 134 (2006), 653-663.

[18] Chamarie M., Anneaux de Krull non commutatifs, Thèse, Université Claude-Bernard, Lyon I, 1981.

[19] Chamarie M., Anneaux de Krull non commutatifs, J. Algebra 72 (1981), 210–222.

[20] Charlap S.C., Bieberbach groups and flat manifolds, Springer-Verlag, New York, 1986.

[21] Chatters A.W., Non-commutative unique factorisation domains, Math. Proc. Cambridge Philos. Soc. 95 (1984), 49–54.

[22] Chatters A.W., Unique factorisation in P.I. group-rings, J. Austral. Math. Soc. Ser. A 59 (1995), 232–243.

[23] Chatters A.W. and Clark J., Group rings which are unique factorisation rings, Comm. Algebra 19 (1991), 585–598.

[24] Chatters A.W. and Jordan D.A., Non-commutative unique factorisation rings, J. London Math. Soc. 33 (1986), 22–32.

[25] Chin W. and Quinn D., Rings graded by polycyclic-by-finite groups, Proc. Amer. Math. Soc. 102 (1988), 235–241.

[26] Chouinard II L.G., Krull semigroups and divisor class groups, Canad. J. Math. 23 (1981), 1459–1468.

[27] Clifford A.H. and Preston G.B., Algebraic Theory of Semigroups, American Mathematical Society, Providence, vol.1, 1961; vol.2, 1967.

[28] Cohen M. and Rowen L., Group graded rings, Comm. Algebra 11 (1983), 1253–1270.

[29] Cohn P.M., Skew Field Constructions, London Math. Soc. Lect. Note Ser 27, Cambridge, 1977.

[30] Decruyenaere F., Jespers E. and Wauters P., On commutative principal ideal semigroup rings, Semigroup Forum 43 (1991), 367–377.

[31] Decruyenaere F., Jespers E and Wauters P., Semigroup algebras and direct sums of domains, J. Algebra 138 (1991), 505–514.

[32] Dehornoy P., Complete positive group presentations, J. Algebra 268 (2003), 156–197.

[33] Drinfeld V.G., On some unsolved problems in quantum group theory, in: Quantum Groups, Lect. Notes Math. 1510, pp. 1–8, Springer-Verlag, 1992.

[34] Drozd Ju.A. and Kirichenko V.V., Finite Dimensional Algebras, Springer-Verlag, Berlin, 1994.

[35] Eisenbud D., Commutative Algebra. With a view toward algebraic geometry, Graduate Texts in Mathematics 150, Springer-Verlag, New York, 1995.

[36] Etingof P., Guralnick R. and Soloviev A., Indecomposable set-theoretical solutions to the quantum Yang-Baxter equation on a set with a prime number of elements, J. Algebra 249 (2001), 709–719.

[37] Etingof P., Schedler T. and Soloviev A., Set-theoretical solutions of the quantum Yang-Baxter equation, Duke Math. J. 100 (1999), 169–209.

[38] Faith C., Algebra II, Ring Theory, Springer-Verlag, New York, 1976.

[39] Fisher J.L. and Sehgal S.K., Principal ideal group rings, Comm. Algebra 4 (1976), 319–325.

[40] Fossum R., The Divisor Class Group of a Krull Domain, Springer-Verlag, New York, 1973.

[41] Fountain J. and Gould V.A.R., Completely 0-simple semigroups of quotients II, in: Contributions to General Algebra 3, pp. 115–124, Verlag Hölder-Pichler-Tempsky, 1985.

[42] Fountain J. and Petrich M., Completely 0-simple semigroups of quotients, J. Algebra 101 (1986), 365–402.

[43] Fountain J. and Petrich M., Completely 0-simple semigroups of quotients III, Math. Proc. Cambridge Philos. Soc. 105 (1989), 263–275.

[44] Gateva-Ivanova T., Noetherian properties and growth of some associative algebras, in: Effective Methods in Algebraic Geometry, 143–157, Progr. Math. 94, Birkhäuser, Boston, 1991.

[45] Gateva-Ivanova T., On the Noetherianity of some associative finitely presented algebras, J. Algebra 138 (1991), 13–35.

[46] Gateva-Ivanova T., Noetherian properties of skew polynomial rings with binomial relations, Trans. Amer. Math. Soc. 343 (1994), 203–219.

[47] Gateva-Ivanova T., Skew polynomial rings with binomial relations, J. Algebra 185 (1996), 710–753.

[48] Gateva-Ivanova T., Conjectures on the set-theoretic solutions of the Yang-Baxter equation, Abstract, Nato Advanced Study Institute, and Eurconference "Rings, Modules and Representations", Constanta 2000, Romania.

[49] Gateva-Ivanova T., Set-theoretic solutions of the Yang-Baxter equation, Mathematics and education in Mathematics, Proceedings of the 29-th Spring Conference of the Union of Bulgarian Mathematicians, Lovetch, 2000, pp. 107–117.

[50] Gateva-Ivanova T., A combinatorial approach to the set-theoretic solutions of the Yang-Baxter equation, J. Math. Phys. 45 (2004), 3828–3858.

[51] Gateva-Ivanova T., Jespers E. and Okniński J., Quadratic algebras of skew type and the underlying monoids, J. Algebra 270 (2003), 635–659.

[52] Gateva-Ivanova T. and Van den Bergh M., Regularity of skew polynomial rings with binomial relations, Talk at the International Algebra Conference, Miskolc, Hungary, 1996.

[53] Gateva–Ivanova T. and Van den Bergh M., Semigroups of I-type, J. Algebra 206 (1998), 97–112.

[54] Gilmer R., Commutative Semigroup Rings, Univ. Chicago Press, Chicago, 1984.

[55] Glastad B. and Hopkins G., Commutative semigroup rings which are principal ideal rings, Comment. Math. Univ. Carolinae 21 (1980), 371–377.

[56] Goodearl K.R. and Warfield R.B., An Introduction to Noncommutative Noetherian Rings, Cambridge Univ. Press, New York, 1989.

[57] Gould V., Left orders in regular $\mathcal{H}$-semigroups II, Glasgow Math. J. 32 (1990), 95–108.

[58] Gould V., Semigroups of left quotients: existence, straightness and locality, J. Algebra 267 (2003), 514–541.

[59] Grigorchuk R.I., Cancellative semigroups of power growth, Mat. Zametki 43 (1988), 305–319 (in Russian).

[60] Gromov M., Groups of polynomial growth and expanding maps, Publ. Math. IHES 53(1) (1981), 53–73.

[61] Hall M., Combinatorial Theory, Wiley, New York, 1986.

[62] Hardy B.R. and Shores T.S., Arithmetical semigroup rings, Canad. J. Math. 32 (1980), 1361–1371.

[63] Hietarinta J., Permutation-type solutions to the Yang-Baxter and other n-simplex equations, J. Phys. A 30 (1997), 4757–4771.

[64] Hotzel E., On semigroups with maximal conditions, Semigroup Forum 11 (1975/76), 337–362.

[65] Hotzel E., On finiteness conditions in semigroups, J. Algebra 60 (1979), 352–370.

[66] Howie J.M., Fundamentals of Semigroup Theory, Oxford University Press, 1995.

[67] Humphreys J.E., Linear Algebraic Groups, Springer-Verlag, New York, 1975.

[68] Huppert B., Endliche Gruppen I, Springer-Verlag, New York, 1967.

[69] Jacobson N., Structure of Rings, Amer. Math. Soc., Providence, 1968.

[70] Jategaonkar A.V., Left Principal Ideal Rings, Lect. Notes Math. 123, Springer-Verlag, New York, 1970.

[71] Jespers E., On geometrical Ω-Krull rings, Comm. Algebra 11 (1983), 771–792.

[72] Jespers E., Jacobson rings and rings strongly graded by an abelian group, Israel J. Math. 63 (1988), 67–78.

[73] Jespers E., Semigroup graded rings and Jacobson rings, Lattices, semigroups, and universal algebra (Lisbon, 1988), pp. 105–114, Plenum, New York, 1990.

[74] Jespers E., Krempa J. and Puczylowski E., On radicals of graded rings, Comm. Algebra 10 (1982), 1849–1854.

[75] Jespers E., Le Bruyn L. and Wauters P., A characterization of central Ω-Krull rings, J. Algebra 81 (1983), 165–179.

[76] Jespers E. and Okniński J., Nilpotent semigroups and semigroup algebras, J. Algebra 169 (1994), 984–1011.

[77] Jespers E. and Okniński J., Semigroup algebras that are principal ideal rings, J. Algebra 183 (1996), 837–863.

[78] Jespers E. and Okniński J., Binomial semigroups, J. Algebra 202 (1998), 250–275.

[79] Jespers E. and Okniński J., Noetherian semigroup algebras, J. Algebra 218 (1999), 543–562.

[80] Jespers E. and Okniński J., On a class of Noetherian algebras, Proc. Roy. Soc. Edinburgh, 129 (A) (1999), 1185–1196.

[81] Jespers E. and Okniński J., Semigroup algebras and maximal orders, Canad. Math. Bull. 42 (1999), 298–306.

[82] Jespers E. and Okniński J., Noetherian semigroup algebras: a survey, Interactions between ring theory and representations of algebras (Murcia), pp. 253–276, Lecture Notes in Pure and Appl. Math., 210, Dekker, New York, 2000.

[83] Jespers E. and Okniński J., Submonoids of polycyclic-by-finite groups and their algebras, Algebras Repres. Theory 4 (2001), 133–153.

[84] Jespers E. and Okniński J., Semigroup algebras and Noetherian maximal orders, J. Algebra 238 (2001), 590–622.

[85] Jespers E. and Okniński J., Quadratic algebras of skew type satisfying the cyclic condition, Int. J. Algebra Comput. 14 (2004), 479–498.

[86] Jespers E. and Okniński J., Quadratic algebras of skew type, in: Algebras, Rings and Their Representations, Proceedings of the International Conference, Lisbon, Portugal, 14–18 July 2003, pp. 72–92, eds. A. Facchini, K.R. Fuller, C.M. Ringel and C. Santa Clara, World Sci., 2005.

[87] Jespers E. and Okniński J., Monoids and groups of I-type, Algebras Repres. Theory 8 (2005), 709–729.

[88] Jespers E. and Okniński J., Noetherian semigroup algebras of submonoids of polycyclic-by-finite groups, Bull. London Math. Soc. 38 (2006), 421-428.

[89] Jespers E. and Smith P.F., Group rings and maximal orders, Methods in Ring Theory (Antwerp, 1983), pp. 185–195, Nato Adv. Sci. Inst. Ser. C: Math. Phys. Sci. 129, Reidel, Dordrecht-Boston, 1984.

[90] Jespers E. and Smith P.F., Integral group rings of torsion-free polycyclic-by-finite groups are maximal orders, Comm. Algebra 13 (1985), 669–680.

[91] Jespers E. and Wang Q., Noetherian unique factorization semigroup algebras, Comm. Algebra 29 (2001), 5701–5715.

[92] Jespers E. and Wang Q., Height one prime ideals in semigroup algebras satisfying a polynomial identity, J. Algebra 248 (2002), 118–131.

[93] Jespers E. and Wauters P., On central Ω-Krull rings and their class groups, Canad. J. Math. 36 (1984), 206–239.

[94] Jespers E. and Wauters P., Principal ideal semigroup rings, Comm. Algebra 23 (1995), 5057–5076.

[95] Justin J., Généralisation du théorème de van der Waerden sur les semi-groupes répétitifs, J. Combin. Theory Ser. A 12 (1972), 357–367.

[96] Kargapolov M.I. and Merzljakov Ju.I., Fundamentals of the Theory of Groups, Springer-Verlag, New York, 1979.

[97] Kassel C., Quantum Groups, Graduate Texts in Mathematics, Springer-Verlag, 1995.

[98] Kozhukhov, I.B. On semigroups with minimal or maximal conditions on left congruences, Semigroup Forum 21 (1980), 337–350.

[99] Kramer X.H., The Noetherian property in some quadratic algebras, Trans. Amer. Math. Soc. 351 (2000), 4295–4323.

[100] Krause G.R. and Lenagan T.H., Growth of Algebras and Gelfand-Kirillov Dimension, Revised edition. Graduate Studies in Mathematics, 22. American Mathematical Society, Providence, RI, 2000.

[101] Lam T.Y., Lectures on Modules and Rings, Springer-Verlag, New York, 1999.

[102] Le Bruyn L., Van den Bergh M. and Van Oystaeyen F., Graded Orders, Birkhäuser Boston, Inc., Boston, MA, 1988.

[103] Letzter E.S. and Lorenz M., Polycyclic-by-finite group rings are catenary, Math. Res. Letters 6 (1999), 183–194.

[104] Lothaire M., Combinatorics on Words, Addison-Wesley, Reading, Mass., 1983.

[105] Lu Jiang-Hua, Yan Min and Zhu Yong-Chang, On the set-theoretical Yang-Baxter equation, Duke Math. J. 104 (2000), 1–18.

[106] Malcev A.I., Nilpotent semigroups, Uchen. Zap. Ivanovsk. Ped. Inst. 4 (1953), 107–111 (in Russian).

[107] Marubayashi H., Noncommutative Krull rings, Osaka J. Math. 12 (1975), 703–714.

[108] Marubayashi H., On bounded Krull prime rings, Osaka J. Math. 13 (1976), 491–501.

[109] Marubayashi H., A characterization of bounded Krull prime rings, Osaka J. Math. 15 (1978), 13–20.

[110] Martin R., Skew group rings and maximal orders, Glasgow Math. J. 37 (1995), 249–263.

[111] Maury C. and Raynaud J., Ordres Maximaux au Sens de K. Asano, Lect. Notes in Math. 808, Springer-Verlag, Berlin, 1980.

[112] McConnell J.C. and Robson J.C., Noncommutative Noetherian Rings, Wiley, New York, 1987.

[113] Merzljakov Ju.I., Rational Groups, Nauka, Moscow, 1986 (in Russian).

[114] Mora T., Gröbner bases for noncommutative polynomial rings, Proc. AAECC-3, Lect. Notes Comp. Sci. 229 (1986), 353–362.

[115] Mora T., An introduction to commutative and noncommutative Gröbner bases, Theor. Comp. Sci. 134 (1994), 131–173.

[116] Nastasescu C., Nauwelaerts E. and Van Oystaeyen F., Arithmetically graded rings revisited, Comm. Algebra 14 (1986), 1991–2017.

[117] Neumann B.H. and Taylor T., Subsemigroups of nilpotent groups, Proc. Roy. Soc., Ser. A 274 (1963), 1–4.

[118] Odesskii A.V., Elliptic algebras, Russ. Math. Surv. 57 (2002), 1127–1162. translation from Usp. Mat. Nauk 57, No. 6 (2002), 87–122.

[119] Okniński J., Semigroup Algebras, Marcel Dekker, New York, 1991.

[120] Okniński J., Linear representations of semigroups, in: Monoids and Semigroups with Applications, pp. 257–277, World Sci., 1991.

[121] Okniński J., Gelfand-Kirillov dimension of Noetherian semigroup algebras, J. Algebra 162 (1993), 302–316.

[122] Okniński J., Nilpotent semigroups of matrices, Math. Proc. Cambridge Philos. Soc. 120 (1996), 617–630.

[123] Okniński J., Semigroups of Matrices, World Sci., Singapore, 1998.

[124] Okniński J., In search for Noetherian algebras, in: Algebra – Representation Theory, NATO ASI, pp. 235–247, Kluwer, 2001.

[125] Okniński J., Prime ideals of cancellative semigroups, Comm. Algebra 32 (2004), 2733–2742.

[126] Okniński J. and Salwa A., Generalised Tits alternative for linear semigroups, J. Pure Appl. Algebra 103 (1995), 211–220.

[127] Passman D.S., Permutation Groups, Benjamin, 1968.

[128] Passman D.S., Observations on group rings, Comm. Algebra 5 (1977), 1119–1162.

[129] Passman D.S., The Algebraic Structure of Group Rings, Wiley, New York, 1977.

[130] Passman D.S., Infinite Crossed Products, Academic Press, 1989.

[131] Polishchuk A. and Positselski L., Quadratic Algebras, University Lecture Series, 37. American Mathematical Society, Providence, RI, 2005.

[132] Promislow S.D., A simple example of a torsion-free, nonunique product group, Bull. London Math. Soc. 20 (1988), 302–304.

[133] Putcha M.S., Linear Algebraic Monoids, London Math. Soc. Lecture Note Ser. 133, Cambridge, 1988.

[134] Renner L.E., Linear Algebraic Monoids, Springer-Verlag, Berlin, 2005.

[135] Rips E. and Segev Y., Torsion-free group without unique product property, J. Algebra 108 (1987), 116–126.

[136] Roseblade J.E., Prime ideals in group rings of polycyclic-by-finite groups, Proc. London Math. Soc. 36 (1978), 385–447.

[137] Rowen L.H., Polynomial Identities in Ring Theory, Academic Press, New York, 1980.

[138] Rump W., A decomposition theorem for square-free unitary solutions of the quantum Yang-Baxter equation, Adv. Math. 193 (2005), 40–55.

[139] Salwa A., Structure of skew linear semigroups, Int. J. Algebra Comput. 3 (1993), 101–113.

[140] Segal D., Polycyclic Groups, Cambridge Univ. Press, Cambridge, 1983.

[141] Shirvani M. and Wehrfritz B.A.F., Skew Linear Groups. London Math. Soc. Lect. Note Ser. 118, Cambridge, 1986.

[142] Shoji, K. On semigroup rings, Proc. 9th Symposium on Semigroups and Related Topics, pp. 28–34, 1986, Naruto Univ, Naruto.

[143] Small L.W., Stafford J.T. and Warfield R.B., Affine algebras of Gelfand-Kirillov one are PI, Math. Proc. Cambridge Philos. Soc. 97 (1995), 407–414.

[144] Smith P.F., Some examples of maximal orders, Math. Proc. Cambridge Philos. Soc. 98 (1985), 19–32.

[145] Smith P.F., New examples of maximal orders, Comm. Algebra 17 (1989), 331–339.

[146] Soloviev A., Non-unitary set-theoretical solutions to the quantum Yang-Baxter equation, Math. Res. Letters 7 (2000), 577–596.

[147] Stafford J.T. and Zhang J.J., Homological properties of graded Noetherian PI rings, J. Algebra 168 (1994), 988–1026.

[148] Stephenson D.R. and Zhang,J.J., Growth of graded Noetherian rings, Proc. Amer. Math. Soc. 125 (1997), 1593–1605.

[149] Strojnowski A., A note on u.p. groups, Comm. Algebra 8 (1980), 231–234.

[150] Tate J. and Van den Bergh M., Homological properties of Sklyanin algebras, Invent. Math. 124 (1996), 619–647.

[151] Tits J., Free subgroups of linear groups, J. Algebra 20 (1972), 250–270.

[152] Wauters P., On some subsemigroups of noncommutative Krull rings, Comm. Algebra 12 (1984), 1751–1765.

[153] Wehrfritz B.A.F., Infinite Linear Groups, Springer-Verlag, 1973.

[154] Weinstein A. and Xu P., Classical solutions of the quantum Yang-Baxter equation, Comm. Math. Phys. 148 (1992), 309–343.

[155] Yang C.N., Some exact results for the many-body problem in one dimension with repulsive delta-function interaction, Phys. Rev. Lett. 19 (1967), 1312–1315.

Notation

What follows is a list of symbols and other notation used in this book. Page references guide the reader to the place where the symbol is defined or first mentioned.

Symbol	*Meaning*
$\mathbb{N}$	nonnegative integers, $0, 1, 2, 3, \ldots$
$\mathbb{Z}$	integers
$\mathbb{Q}$	rational numbers
$\mathbb{R}$	real numbers
$\mathbb{C}$	complex numbers
$\lvert X \rvert$	cardinality of a set X
id_X	identity operator on a set X
$x \rho y$	x is related to y under a relation ρ
$(x, y) \in \rho$	x is related to y under a relation ρ
$\mathrm{char}(K)$	characteristic of a field K
$\mathrm{lin}_{\mathrm{K}}(\mathrm{X})$	K-vector space spanned by a set X
$K\{X\}$	subalgebra of a given algebra A generated by a subset X of A
$\dim_K(V)$	dimension of a K-vector space V
$\mathrm{M}_n(R)$	n-by-n matrices over a ring R
M_j	ideal of matrices of rank at most j in a full linear monoid $\mathrm{M}_n(D)$ with D a division ring
$\mathrm{GL}_n(D)$	full linear group over a division ring D
$\mathrm{im}(f)$	image of a function f
$\ker(f)$	kernel of a homomorphism f
$\mathrm{rank}(a)$	rank of a matrix $a \in \mathrm{M}_n(D)$
$\widehat{I}$	completely 0-simple closure of a uniform semigroup I
$\widehat{S}$	p. 29

$\mathrm{cl}(S)$	strongly π-regular closure of a linear semigroup S
$\overline{S}$	Zariski closure of a semigroup S that is linear over a field
S^1	smallest monoid containing a semigroup S
S^0	a semigroup S with a zero adjoined
S/I	Rees factor semigroup of a semigroup S with respect to an ideal I
$\mathrm{Z}(S)$	center of a semigroup S
$\mathrm{C}_S(X)$	centralizer of a subset X of a semigroup S
$\mathrm{N}(S)$	subsemigroup of normalizing elements in a semigroup S
$\mathrm{U}(S)$	subgroup of invertible elements in a monoid S
$\mathrm{E}(S)$	set of idempotents in a semigroup S
ρ_{can}	least cancellative congruence on a semigroup
$S \rtimes T$, $S \rtimes_\sigma T$	semidirect product of monoids S and T
S^{opp}	opposite semigroup of a semigroup S
$\mathrm{Spec}(S)$	prime spectrum of a semigroup S
$\mathrm{Spec}^0(S)$	set of minimal prime ideals of a semigroup S
$\dim(S)$	prime dimension of a semigroup S
T_P, $T_{(P)}$, $T_{[P]}$	localizations of a monoid T with respect to a prime ideal P, p. 178
$I(P)$	p. 179
$\langle A \rangle$	submonoid generated by a set A in a monoid
$\langle x_1, \ldots, x_n; R\rangle$, $\langle X; R\rangle$	monoid presentation defined by a set R of relations
$\langle x_1, \ldots, x_n \mid \text{relations}\rangle$	monoid presentation defined by the explicitly given generators and relations
$\prec A \succ$	subsemigroup generated by a subset A in a semigroup
$\mathrm{gr}(x_1, \ldots, x_n \mid \text{relations})$	group presentation defined by the explicitly given generators and relations
$\mathrm{gr}(A)$	subgroup generated by a subset A in a group
$\mathrm{rk}(S)$	supremum of the ranks of free abelian subsemigroups of a semigroup S
$S = \mathcal{M}(G, *, *; P)$	semigroup of matrix type over a group G
$S_{(x)}$, $S^{(y)}$	p. 9
$S_{(A)}$, $S^{(B)}$, $S_{(A)}^{(B)}$	p. 9

$\mathcal{M}_n(G)$	monomial matrices of degree n over a group G
$\mathcal{L}$, $\mathcal{R}$, $\mathcal{J}$, $\mathcal{H}$	Green's relations on a semigroup
F_n	free group of rank n
Fa_n	free abelian group of rank n
FM_n	free monoid of rank n
FaM_n	free abelian monoid of rank n
FM_X	free monoid on the set X
$(\mathrm{Fa}_n)_+$	positive cone of Fa_n
Sym_n	symmetric group of degree n
D_∞	infinite dihedral group
$\approx$	retract relation on a monoid (respectively group) of I-type, p. 220
$\mathrm{Ret}^m(G)$	m-th retract of a group G of I-type, p. 223
$\Delta(G)$	finite conjugacy center of a group G
$\Delta_H(G)$	p. 47
$\mathrm{C}_H(a)$	p. 47
$\mathrm{N}_G(H)$	normalizer of a subgroup H in a group G
C_g	conjugacy class of an element g in a group G
$\Delta^+(G)$	torsion subgroup of $\Delta(G)$
G^+	set of periodic elements of a group G
$[G:H]$	index of a subgroup H in a group G
$[x,y]$	commutator $x^{-1}y^{-1}xy$ of elements x and y in a group
$[H,H]$	commutator subgroup of a group H
$D_n(K[G])$	n-th dimension subgroup
$\mathrm{h}(G)$	Hirsch rank of a polycyclic-by-finite group G
$\mathrm{pl}(G)$	plinth length of a polycyclic-by-finite group G
$\mathrm{M}_X(R)$	$X \times X$ matrices over R, for a set X
$\mathrm{M}_X^{fin}(R)$	$X \times X$ matrices over R with finitely many nonzero entries
$\mathcal{M}(R,*,*;P)$	Munn ring over a ring R
$\mathcal{J}(R)$	Jacobson radical of a ring R
$\mathcal{B}(R)$	prime radical of a ring R
$\mathrm{Z}(R)$	center of a ring R
$\mathrm{U}(R)$	unit group of a ring R
$\mathrm{N}(R)$	set of normalizing elements of a ring R
$\mathrm{Spec}(R)$	prime spectrum of a ring R
$\mathrm{Spec}^0(R)$	set of minimal prime ideals of a ring R

$\mathrm{Spec}^1(R)$	set of height one prime ideals of R
$\mathrm{Spec}^1_h(K[S])$	set of height one prime ideals of $K[S]$ that intersect S
$\mathrm{height}(P)$	height of a prime ideal P
$\mathrm{ra}_X(A)$, $\mathrm{la}_X(A)$	right, left annihilators in $X \subseteq R$ of a subset A of a ring R (or a semigroup R with zero)
$\mathrm{ann}_X(A)$	two-sided annihilator in $X \subseteq R$ of a subset A of a ring R (or a semigroup R with zero)
RX^{-1}	ring of right quotients of a ring R with respect to a right Ore set X of regular elements
$\mathrm{Q}_{cl}(R)$, $\mathrm{Q}(R)$	classical ring of (right) quotients of R
$\mathrm{clKdim}(R)$	classical Krull dimension of a ring R
$\mathrm{GK}(R)$	Gelfand-Kirillov dimension of a K-algebra R
$(A :_r B)$, $(A :_l B)$	p. 55, p. 58
$(R : I)$, I^{-1}	p. 55, p. 58
I^*	divisorial closure of an ideal I, p. 55, p. 58
$I * J$	divisorial product of two divisorial ideals
$\mathrm{D}(R)$	group of divisorial ideals of a ring or a cancellative semigroup R
$R[S]$	semigroup ring of a semigroup S over a ring R
$R_0[S]$	contracted semigroup ring of S over R
I_G	graded part of an ideal in a G-graded ring
$\mathrm{supp}(\alpha)$	support of an element α of a semigroup ring
$I(\rho)$	ideal of $K[S]$ determined by a congruence ρ on a semigroup S
$\sim_Q$	congruence relation on a semigroup S defined by an ideal Q in $K[S]$, p. 154
$\omega(K[S])$	augmentation ideal of $K[S]$
S_Y	p. 268
D_Y	p. 268
S_i	p. 268
S'_i	p. 270

Index

Algebra and Applications

1. C.P. Millies and S.K. Sehgal: *An Introduction to Group Rings.* 2002
ISBN HB 1-4020-0238-6; PB 1-4020-0239-4
2. P.A. Krylov, A.V. Mikhalev and A.A. Tuganbaev: *Endomorphism Rings of Abelian Groups.* 2003
ISBN 1-4020-1438-4
3. J.F. Carlson, L. Townsley, L. Valero-Elizondo and M. Zhang: *Cohomology Rings of Finite Groups.* Calculations of Cohomology Rings of Groups of Order Dividing 64. 2003
ISBN 1-4020-1525-9
4. K. Kiyek and J.L. Vicente: *Resolution of Curve and Surface Singularities.* In Characteristic Zero. 2004
ISBN 1-4020-2028-7
5. U. Ray: *Automorphic Forms and Lie Superalgebras.* 2006
ISBN 1-4020-5009-7
6. A. Garcia and H. Stichtenoth: *Topics in Geometry, Coding Theory and Cryptography*. 2006
ISBN 1-4020-5333-9
7. E. Jespers and J. Okniński: *Noetherian Semigroup Algebras*. 2007
ISBN 1-4020-5809-8